Lecture Notes in Computer Science 9646

Commenced Publication in 1973
Founding and Former Series Editors:
Gerhard Goos, Juris Hartmanis, and Jan van Leeuwen

More information about this series at http://www.springer.com/series/7410

David Pointcheval · Abderrahmane Nitaj
Tajjeeddine Rachidi (Eds.)

Progress in Cryptology – AFRICACRYPT 2016

8th International Conference on Cryptology in Africa
Fes, Morocco, April 13–15, 2016
Proceedings

 Springer

Editors
David Pointcheval
Ecole Normale Supérieure
Paris
France

Abderrahmane Nitaj
University of Caen
Caen
France

Tajjeeddine Rachidi
Al Akhawayn University in Ifrane
Ifrane
Morocco

ISSN 0302-9743   ISSN 1611-3349 (electronic)
Lecture Notes in Computer Science
ISBN 978-3-319-31516-4   ISBN 978-3-319-31517-1 (eBook)
DOI 10.1007/978-3-319-31517-1

Library of Congress Control Number: 2016933535

LNCS Sublibrary: SL4 – Security and Cryptology

This Springer imprint is published by Springer Nature
The registered company is Springer International Publishing AG Switzerland

Preface

The 8th International Conference on the Theory and Application of Cryptographic Techniques in Africa, Africacrypt 2016, took place April 13–15, 2016, in Fès, Morocco. The conference was organized by Al Akhawayn University in Ifrane, in cooperation with the International Association for Cryptologic Research (IACR).

The conference received 65 submissions, all of which were reviewed by the Program Committee. Each paper was assigned at least three reviewers, while submissions co-authored by Program Committee members were reviewed by at least four reviewers.

The Program Committee was helped by reports from 48 external reviewers. After highly interactive discussions and a careful deliberation, the Program Committee selected 18 papers for presentation (less than 28 % acceptance rate). The program was completed with invited talks: "Computing on Encrypted Data" by Vinod Vaikuntanathan from MIT and "A New Methodology of Constructing Functional Encryption" by Tatsuaki Okamoto from NTT. We are very grateful to them for accepting our invitation.

We would like to thank everyone who contributed to the success of Africacrypt 2016. We are deeply grateful to the Program Committee for their hard work, enthusiasm, and conscientious efforts to ensure that each paper received a thorough and fair review. These thanks arc of course extended to the external reviewers, listed on the following pages, who took the time to help during the evaluation process. We would also like to thank Thomas Baignères and Matthieu Finiasz for writing the iChair software and Springer for agreeing to an accelerated schedule for printing the proceedings.

Our thanks also go to the local Organizing Committee for their commitment and hard work, in order to make the conference an enjoyable experience. They also go to Driss Ouaouicha, President of Al Akhawayn University, and Dean Kevin Smith for their unconditional support. We are deeply grateful to the sponsors Microsoft, Al Akhawayn University, HPS Morocco, the Région Fès-Meknès, ENS, Paris, France, and the European Research Council under the European Union's Seventh Framework Programme (FP7/2007-2013 Grant Agreement no. 339563 – CryptoCloud), for financially supporting the conference.

Last but not least, we wish to thank the participants, submitters, authors, presenters, and invited speakers, and Program Committees who over the past seven editions have made Africacrypt a highly recognized forum in which researchers can interact and share their work and knowledge with others, for the overall growth and development of cryptology research in the world, and Africa in particular.

April 2016

David Pointcheval
Abderrahmane Nitaj
Tajjeedine Rachidi

AFRICACRYPT 2016

8th International Conference on Cryptology in Africa
Fès, Morocco, April 13–15, 2016

General Chair

Tajjeeddine Rachidi   Al Akhawayn University in Ifrane, Morocco

Program Chairs

David Pointcheval   ENS, Paris, France
Abderrahmane Nitaj   University of Caen, France

Program Committee

Muhammad Rezal Kamel Ariffin
Abdelhak Azhari
Hussain Benazza
Colin Boyd
Dario Catalano
Jie Chen
Sherman S.M. Chow
Jean-Sébastien Coron
Itai Dinur
Léo Ducas
Orr Dunkelman
Dario Fiore
Pierre-Alain Fouque
Georg Fuchsbauer
Essam Ghadafi
Tetsu Iwata
Seny Kamara
Benoît Libert
David M'Raihi

Mark Manulis
Jesper Buus Nielsen
Ayoub Otmani
Duong Hieu Phan
Tajjeeddine Rachidi
Magdy Saeb
Palash Sarkar
Peter Schwabe
Francesco Sica
Djiby Sow
Ron Steinfeld
François-Xavier Standaert
Christine Swart
Isamu Teranishi
Mehdi Tibouchi
Susan Thomson
Xiaoyun Wang
Amr M. Youssef

External Reviewers

Ahmed Abdelkhalek
Khalid Abdelmoumen
Hamza Abusalah
Riham Altawy
Toshinori Araki
Sanjay Bhattacherjee
Olivier Blazy
Raphaël Bost
Andrea Cerulli
Donghoon Chang
Jingxian Chen
Joan Daemen
Gareth Davies
Mario Di Raimondo
Ali El Kaafarani
Nadia El Mrabet
Reza Rezaeian Farashahi
Emmanuel Fouotsa
Mohona Ghosh
Junqing Gong
Aurore Guillevic
Haruna Higo
Takanori Isobe
Bootle Jonatthan
Saqib Kakvi

Dmitry Khovratovich
Paul Kirchner
Russell W.F. Lai
Alko Meijer
Marine Minier
Pradeep Kumar Mishra
Fabrice Mouhartem
Sekan Mulayim
Kazuma Ohara
Ludovic Perret
Romain Poussier
Sebastian Ramacher
Michal Rybar
Minoru Saeki
Amin Sakzad
Le Su
Miaomiao Tian
Mohamed Tolba
Frederik Vercauteren
Damien Vergnaud
Kai Zhang
Tao Zhang
Yongjun Zhao

Contents

Lattices

Efficient (Ideal) Lattice Sieving
Using Cross-Polytope LSH

Anja Becker[1](✉) and Thijs Laarhoven[2]

[1] EPFL, Lausanne, Switzerland
anja.becker@epfl.ch
[2] TU/e, Eindhoven, The Netherlands
mail@thijs.com

Abstract. Combining the efficient cross-polytope locality-sensitive hash family of Terasawa and Tanaka with the heuristic lattice sieve algorithm of Micciancio and Voulgaris, we show how to obtain heuristic and practical speedups for solving the shortest vector problem (SVP) on both arbitrary and ideal lattices. In both cases, the asymptotic time complexity for solving SVP in dimension n is $2^{0.298n+o(n)}$.

For any lattice, hashes can be computed in polynomial time, which makes our CPSieve algorithm much more practical than the SphereSieve of Laarhoven and de Weger, while the better asymptotic complexities imply that this algorithm will outperform the GaussSieve of Micciancio and Voulgaris and the HashSieve of Laarhoven in moderate dimensions as well. We performed tests to show this improvement in practice.

For ideal lattices, by observing that the hash of a shifted vector is a shift of the hash value of the original vector and constructing rerandomization matrices which preserve this property, we obtain not only a linear decrease in the space complexity, but also a linear speedup of the overall algorithm. We demonstrate the practicability of our cross-polytope ideal lattice sieve ICPSieve by applying the algorithm to cyclotomic ideal lattices from the ideal SVP challenge and to lattices which appear in the cryptanalysis of NTRU.

Keywords: (Ideal) lattices · Shortest vector problem · Sieving algorithms · Locality-sensitive hashing

1 Introduction

Lattice-Based Cryptography. Lattices are discrete additive subgroups of $\mathbb{R}^n$. More concretely, given a basis $B = \{\boldsymbol{b}_1, \ldots, \boldsymbol{b}_n\} \subset \mathbb{R}^n$, the lattice generated by B, denoted by $\mathcal{L} = \mathcal{L}(B)$, is defined as the set of all integer linear combinations

---

The full version of the paper, including appendices, is presented on the ePrint Archive in [12].

A. Becker—This work was supported by the Swiss National Science Foundation under grant numbers 200021-126368 and 200020-153113.

T. Laarhoven—Part of this work was done while the second author was visiting EPFL.

D. Pointcheval et al. (Eds.): AFRICACRYPT 2016, LNCS 9646, pp. 3–23, 2016.
DOI: 10.1007/978-3-319-31517-1_1

of the basis vectors: $\mathcal{L} = \{\sum_{i=1}^{n} \mu_i \boldsymbol{b}_i : \mu_i \in \mathbb{Z}\}$. The security of lattice-based cryptography relies on the hardness of certain hard lattice problems, such as the shortest vector problem (SVP): given a basis B of a lattice, find a shortest non-zero vector $\boldsymbol{v} \in \mathcal{L}$, where shortest is defined in terms of the Euclidean norm. The length of a shortest non-zero vector is denoted by $\lambda_1(\mathcal{L})$. A common relaxation of SVP is the approximate shortest vector problem (SVP$_\delta$): given a basis B of $\mathcal{L}$ and an approximation factor $\delta > 1$, find a non-zero vector $\boldsymbol{v} \in \mathcal{L}$ whose norm does not exceed $\delta \cdot \lambda_1(\mathcal{L})$.

Although SVP and SVP$_\delta$ with constant approximation factor δ are well-known to be NP-hard under randomized reductions [4,31], choosing parameters in lattice cryptography remains a challenge [20,38,53] as e.g. (i) the actual computational complexity of SVP and SVP$_\delta$ is still not very well understood; and (ii) for efficiency, lattice-based cryptographic primitives such as NTRU [26] commonly use special, structured lattices, for which solving SVP and SVP$_\delta$ may potentially be much easier than for arbitrary lattices.

SVP Algorithms. To improve our understanding of these hard lattice problems, which may ultimately help us strengthen (or lose) our faith in lattice cryptography, the only solution seems to be to analyze algorithms that solve these problems. Studies of algorithms for solving SVP already started in the 1980s [18,30,52] when it was shown that a technique called enumeration can solve SVP in superexponential time ($2^{\Omega(n \log n)}$) and polynomial space. In 2001 Ajtai et al. showed that SVP can actually be solved in single exponential time ($2^{\Theta(n)}$) with a technique called sieving [5], which requires a single exponential space complexity as well. Even more recently, two new methods were invented for solving SVP based on using Voronoi cells [45] and on using discrete Gaussian sampling [2]. These methods also require a single exponential time and space complexity.

Sieving Algorithms. Out of the latter three methods with a single exponential time complexity, sieving still seems to be the most practical to date. We give a summary about sieving in [12, Sect. 2.2]. The provable time exponent for sieving may be as high as $2^{2.465n+o(n)}$ [25,49,54] (compared to $2^{2n+o(n)}$ for the Voronoi cell algorithm, and $2^{n+o(n)}$ for the discrete Gaussian combiner), but various heuristic improvements to sieving since 2001 [10,46,49,64,65] have shown that in practice sieving may be able to solve SVP in time and space as little as $2^{0.378n+o(n)}$. Other works on sieving have further shown how to parallelize and speed up sieving in practice with various polynomial speedups [14,19,29,41–43,48,56,57], and how sieving can be made even faster on certain structured, ideal lattices used in lattice cryptography [14,29,57]. Ultimately both Ishiguro et al. [29] and Bos et al. [14] managed to solve an 128-dimensional ideal SVP challenge [51] using a modified version of the GaussSieve [46], which is currently still the highest dimension in which a challenge from the ideal lattice challenge was successfully solved.

Sieving and Locality-Sensitive Hashing. Even more recently, a new line of research was initiated which combines the ideas of sieving with a technique from the literature of nearest neighbor searching, called locality-sensitive hashing (LSH) as described in [12, Sect. 2.3] and [28]. This led to a practical algorithm with heuristic time and space complexities of only $2^{0.337n+o(n)}$ (the Hash-Sieve [34,43]), and an algorithm with even better asymptotic complexities of only $2^{0.298n+o(n)}$ (the SphereSieve [35]). However, for both methods the polynomial speedups that apply to the GaussSieve for ideal lattices [14,29,57] do not seem to apply, and the latter algorithm may be of limited practical interest due to large hidden order terms in the LSH technique and the fact that this technique seems incompatible with the GaussSieve [46] and only works with the less practical NV-sieve [49]. Understanding the possibilities and limitations of sieving with LSH, as well as finding new ways to efficiently apply similar techniques to ideal lattices remains an open problem.

Our Contributions. In this work we show how to obtain practical, exponential speedups for sieving (in particular for the GaussSieve algorithm [14,29,46]) using the cross-polytope LSH technique first introduced by Terasawa and Tanaka in 2007 [63] and very recently further analyzed by Andoni et al. [9]. Our results are two-fold:

Arbitrary lattices. For arbitrary lattices, using polytope LSH leads to a practical sieve with heuristic time and space complexities of $2^{0.298n+o(n)}$. The exact trade-off between the time and memory is shown in Fig. 1. The low polynomial cost of computing hashes and the fact that this algorithm is based on the GaussSieve (rather than the NV-sieve [49]) indicate that this algorithm is more practical than the SphereSieve [35], while in moderate dimensions this method will be faster than both the GaussSieve and the HashSieve due to its better asymptotic time complexity.

Ideal lattices. For ideal lattices commonly used in cryptography, we show how to obtain similar polynomial speedups and decreases in the space complexity as in the GaussSieve [14,29,57]. In particular, both the time and space for solving SVP decrease by a factor $\Theta(n)$, and the cost of computing hashes decreases by a quasi-linear factor $\Theta(n/\log n)$ using Fast Fourier Transforms.

These results emphasize the potential of sieving for solving high-dimensional instances of SVP, which in turn can be used inside lattice basis reduction algorithms like BKZ [59,60] to find short (rather than shortest) vectors in even higher dimensions. As a consequence, these results will be an important guide for estimating the long-term security of lattice-based cryptography, and in particular for selecting parameters in lattice-based cryptographic primitives.

Outline. The paper is organized as follows. In Sect. 2 we recall some background on lattices, sieving, locality-sensitive hashing, and the polytope LSH family of Terasawa and Tanaka [63]. Section 3 describes how to combine these techniques to solve SVP on arbitrary lattices, and how this leads to an asymptotic time (and space) complexity of $2^{0.298n+o(n)}$. Section 4 describes how to make the resulting

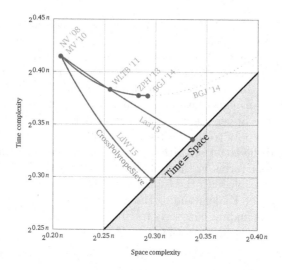

Fig. 1. The heuristic space-time trade-off of various previous heuristic sieving algorithms from the literature (the red points and curves), and the heuristic trade-off between the space and time complexities obtained with our algorithm (the blue curve). The referenced papers are: NV'08 [49] (the NV-sieve), MV'10 [46] (the GaussSieve), WLTB'11 [64] (two-level sieving), ZPH'13 [65] (three-level sieving), BGJ'14 [10] (the decomposition approach), Laa'15 [34] (the HashSieve), LdW'15 [35] (the SphereSieve). Note that the trade-off curve for the CPSieve (the blue curve) overlaps with the asymptotic trade-off of the SphereSieve of [35](Color figure online).

algorithm even faster for lattices with a specific ideal structure, such as some of the lattices of the ideal lattice challenge [51] and lattices appearing in the cryptanalysis of NTRU [26].

2   Preliminaries

2.1   Lattices

Let us first recall some basics on lattices. As mentioned in the introduction, we let $\mathcal{L} = \mathcal{L}(B)$ denote the lattice generated by the basis $B = \{b_1, \ldots, b_n\} \subset \mathbb{R}^n$, and the shortest vector problem asks to find a vector of length $\lambda_1(\mathcal{L})$, i.e. a shortest non-zero vector in the lattice. Lattices are additive groups, and so if $v, w \in \mathcal{L}$, then also $\lambda_v v + \lambda_w w \in \mathcal{L}$ for $\lambda_v, \lambda_w \in \mathbb{Z}$.

Within the set of all lattices there is a subset of ideal lattices, which are defined in terms of ideals of polynomial rings. Given a ring $R = \mathbb{Z}[X]/(g)$ where $g \in \mathbb{Z}[X]$ is a degree-n monic polynomial, we can represent a polynomial $v(X) = \sum_{i=1}^{n} v_i X^{i-1}$ in this ring by a vector $v = (v_1, \ldots, v_n)$. Then, given a set of generators $b_1, \ldots, b_k \in R$, we define the ideal $I = \langle b_1, \ldots, b_k \rangle$ by the properties (i) if $a, b \in I$ then also $\lambda a + \mu b \in I$ for scalars $\lambda, \mu \in \mathbb{Z}$; and (ii) if $a \in R$ and $b \in I$ then $a \cdot b \in I$. Note that when these polynomials are translated to vectors, the first property corresponds exactly to the property of a lattice, while the second property makes this an ideal lattice. In terms of lattices, the second property can equivalently be written as:

$$(v_1, \ldots, v_n) \in \mathcal{L} \;\Leftrightarrow\; (w_1, \ldots, w_n) \in \mathcal{L}, \text{ where } w \equiv X \cdot v \bmod g \text{ in } R. \qquad (1)$$

In this paper we will restrict our attention to a few specific choices of g as follows:

Cyclic lattices: If $g(X) = X^n - 1$ and $v = (v_1, \ldots, v_n)$, then $w \equiv X \cdot v$ implies that $w = (v_n, v_1, \ldots, v_{n-1})$, i.e. multiplying a polynomial in the ring by X corresponds to a right-shift (with carry) of the corresponding vector, and so any cyclic shift of a lattice vector is also in the lattice.

Negacyclic lattices: For the case $g(X) = X^n + 1$ we similarly have that multiplying a polynomial by X in the ring corresponds to a right-shift with carry, but in this case an extra minus sign appears with the carry: $w \equiv X \cdot v$ implies that $w = (-v_n, v_1, \ldots, v_{n-1})$.

Whereas the above descriptions of cyclic and negacyclic lattices are quite general, below we list two instances of these lattices that appear in practice which have certain additional properties.

NTRU lattices: Cyclic lattices most notably appear in the cryptanalysis of NTRU [26], where the polynomial ring is $R = \mathbb{Z}_q[x]/(X^p - 1)$ where p, q are prime. Due to the modular ring, the corresponding lattice is not quite cyclic but rather "block-cyclic". The NTRU lattice is formed by the $n = 2p$ basis vectors $b_i = (q \cdot e_i \| 0)$ for $i = 1, \ldots, p$ and $b_{p+i} = (h_i \| e_i)$ for $i = 1, \ldots, p$, where e_i corresponds to the ith unit vector, and h_i corresponds to the ith cyclic shift of the public key h generated from the private key f, g (see [26] for details). In this case, if $v = (v_1 \| v_2) \in \mathcal{L}$ is a lattice vector, then also shifting both v_1 and v_2 to the right or left leads to a lattice vector. Finding a shortest non-zero vector in this lattice corresponds to finding the secret key $(f \| g)$ and breaking the underlying cryptosystem.

Power-of-two cyclotomic lattices: Negacyclic lattices commonly appear in lattice cryptography, where $n = 2^k$ is a power of 2 so that, among others, g is irreducible. The 128-dimensional ideal lattice attacked by Ishiguro et al. [29] and Bos et al. [14] from the ideal lattice challenge [51] also belongs to this class of lattices. Lattices of this form previously appeared in the context of lattice cryptography in e.g. [22, 40, 62].

2.2 Finding Nearest Neighbors with LSH

The near(est) neighbor problem is the following [28]: Given a long list L of n-dimensional vectors, i.e., $L = \{w_1, w_2, \ldots, w_N\} \subset \mathbb{R}^n$, preprocess L in such a way that, when later given a target vector $v \notin L$, one can efficiently find an element $w \in L$ which is close(st) to v. While in low (fixed) dimensions n there are ways to trivially answer these queries in time sub-linear or even logarithmic in the list size N, in high dimensions it seems hard to do better than with a naive brute-force list search of time $O(N)$. This inability to efficiently store and query lists of high-dimensional objects is sometimes referred to as the "curse of dimensionality" [28]. Fortunately, if we know that e.g. there is a significant gap between what is meant by "nearby" and "far away," then there are ways to preprocess L such that queries can be answered in time sub-linear in N, using locality-sensitive hash families.

To use these LSH families to find nearest neighbors, we can use the following method first described in [28] and outlined in the full version of this paper [12, Sect. 2.3]. First, we choose $t \cdot k$ random hash functions $h_{i,j} \in \mathcal{H}$, and we use the AND-composition to combine k of them at a time to build t different hash functions $h_1, \ldots, h_t$. Then, given the list L, we build t different hash tables $T_1, \ldots, T_t$, where for each hash table T_i we insert $\boldsymbol{w}$ into the bucket labeled $h_i(\boldsymbol{w})$. Finally, given the vector $\boldsymbol{v}$, we compute its t images $h_i(\boldsymbol{v})$, gather all the candidate vectors that collide with $\boldsymbol{v}$ in at least one of these hash tables (an OR-composition) in a list of candidates, and search this set of candidates for a nearest neighbor.

Clearly, the quality of this algorithm for finding nearest neighbors depends on the quality of the underlying hash family and on the parameters k and t. Larger values of k and t amplify the gap between the probabilities of finding 'good' (nearby) and 'bad' (faraway) vectors, which makes the list of candidates shorter, but larger parameters come at the cost of having to compute many hashes (during the preprocessing and querying phases) and having to store many hash tables in memory. The following lemma shows how to balance k and t such that the overall time complexity is minimized.

Lemma 1. *[28] Let $\mathcal{H}$ be an (r_1, r_2, p_1, p_2)-sensitive hash family. Then, for a list L of size N, taking*

$$\rho = \frac{\log(1/p_1)}{\log(1/p_2)}, \qquad k = \frac{\log(N)}{\log(1/p_2)}, \qquad t = O(N^\rho), \qquad (2)$$

with high probability we can either (a) find an element $\boldsymbol{w}^* \in L$ with $D(\boldsymbol{v}, \boldsymbol{w}^*) \leq r_2$, or (b) conclude that with high probability, no elements $\boldsymbol{w} \in L$ with $D(\boldsymbol{v}, \boldsymbol{w}) > r_1$ exist, with the following costs:

1. *Time for preprocessing the list: $O(N^{1+\rho} \log_{1/p_2} N)$.*
2. *Space complexity of the preprocessed data: $O(N^{1+\rho})$.*
3. *Time for answering a query $\boldsymbol{v}$: $O(N^\rho)$.*
    - *Hash evaluations of the query vector $\boldsymbol{v}$: $O(N^\rho)$.*
    - *List vectors to compare to the query vector $\boldsymbol{v}$: $O(N^\rho)$.*

Although Lemma 1 only shows how to choose k and t to minimize the time complexity, we can also tune k and t so that we use more time and less space. In a way this algorithm can be seen as a generalization of the naive brute-force search method, as $k = 0$ and $t = 1$ corresponds to checking the whole list for nearby vectors in linear time and linear space.

2.3 Cross-Polytope Locality-Sensitive Hashing

Whereas the previous subsections covered techniques previously used in [34] and [35], we deviate from these papers by the choice of hash function. The hash function we will use is the one originally described by Terasawa and Tanaka [63] using simplices and orthoplices (cross polytopes), later analyzed by Andoni et al. [9]. The n-dimensional cross-polytope is defined by the vertices $\{\pm e_i\}$, and the corresponding hash function based on using the n-dimensional cross-polytope is defined by finding the vector $\boldsymbol{h} \in \{\pm e_i\}$ which is closest to the target vector $\boldsymbol{v}$. Alternatively, the hash function is defined as:

$$h(\boldsymbol{x}) = \pm \arg\max_i |x_i| \in \{\pm 1, \pm 2, \ldots, \pm n\}, \tag{3}$$

where the sign is equal to the sign of the absolute largest coordinate; if $\boldsymbol{v} = (3, -5)$ then $h(\boldsymbol{v}) = -2$ and $h(-\boldsymbol{v}) = 2$. Two vectors then have the same hash value if (i) the position of the absolute largest coordinate is the same, and (ii) the sign of this coordinate is the same for both vectors.

As this only defines one hash function rather than an entire hash family, we need to somehow rerandomize the hash function, which is done as follows. We denote by $\mathcal{A}$ the distribution on the space of $n \times n$ real matrices where each entry is drawn from a standard normal distribution $\mathcal{N}(0, 1)$. In other words, the distribution $\mathcal{A}$ outputs matrices $A = (a_{i,j}) \in \mathbb{R}^{n \times n}$ where $a_{i,j} \sim \mathcal{N}(0, 1)$ for all i, j. Then, by first multiplying a vector $\boldsymbol{v}$ with a random matrix $A \sim \mathcal{A}$ and then applying the base hash function h, we obtain a hash family $\mathcal{H}$ as

$$\mathcal{H} = \left\{ h_A : h_A(\boldsymbol{x}) \triangleq h(A\boldsymbol{x}), A \sim \mathcal{A} \right\}. \tag{4}$$

Using this hash family, we define probabilities by varying the matrix A, e.g.,

$$\mathbb{P}\left[h(\boldsymbol{v}) = h(\boldsymbol{w})\right] \triangleq \mathbb{P}_{h_A \sim \mathcal{H}}\left[h_A(\boldsymbol{v}) = h_A(\boldsymbol{w})\right] = \mathbb{P}_{A \sim \mathcal{A}}\left[h_A(\boldsymbol{v}) = h_A(\boldsymbol{w})\right]. \tag{5}$$

As suggested by experiments in [63], the above hash function family performs very well in practice for distinguishing between vectors with small and large angles (note that $\mathcal{H}$ is scale-invariant; $h(\lambda \boldsymbol{v}) = h(\boldsymbol{v})$ for arbitrary $\lambda > 0$). Terasawa and Tanaka already indicated that it seems to perform better than Charikar's angular or hyperplane hash family [15]. A recent study of Andoni et al. [9] shows that indeed it provably performs very well, leading to the following result on collision probabilities.

Lemma 2 (Cross-polytope Locality-sensitive Hashing). *[9, Theorem 1] Let $\theta = \theta(\boldsymbol{v}, \boldsymbol{w})$ denote the angle between two vectors $\boldsymbol{v}$ and $\boldsymbol{w}$. Then, for large n,*

$$\mathbb{P}_{h \sim \mathcal{H}}\left[h(\boldsymbol{v}) = h(\boldsymbol{w})\right] = \exp\left[(-\ln n)\tan^2\left(\frac{\theta}{2}\right) + O(\log\log n)\right]. \tag{6}$$

For comparison later, we finally recall that for the spherical LSH family $\mathcal{S}$ described in [7] and used in the SphereSieve [35], we have the following result regarding collision probabilities.

Lemma 3 (Spherical locality-sensitive hashing). *[7, Lemma 3.3] Let $\theta = \theta(\boldsymbol{v}, \boldsymbol{w})$ denote the angle between two vectors $\boldsymbol{v}$ and $\boldsymbol{w}$. Then, for large n,*

$$\mathbb{P}_{h \sim \mathcal{S}}[h(\boldsymbol{v}) = h(\boldsymbol{w})] = \exp\left[-\frac{\sqrt{n}}{2} \tan^2\left(\frac{\theta}{2}\right)(1 + o(1))\right]. \tag{7}$$

Note that the leading-term dependence on θ in both spherical LSH and cross-polytope LSH is the same while the term in n is decreased from a former $\sqrt{n}/2$ to $\ln n$.

3   CPSieve: Sieving in Arbitrary Lattices

To combine sieving (the GaussSieve of Micciancio and Voulgaris) with locality-sensitive hashing (the cross-polytope LSH family of Terasawa and Tanaka) we will make the following changes to the GaussSieve, similar to [34,35]:

- Instead of building a list of pairwise-reduced lattice vectors, we store each vector in t hash tables $T_1, \ldots, T_t$.
- For each hash table T_i, we combine k hash functions $h_{i,1}, \ldots, h_{i,k}$ into one function h_i with an AND-composition.
- To reduce a new vector with the vectors which are already in the hash tables, we only compare it to those vectors that have the same hash value in one or more of these t hash tables (OR-composition).
- When a vector is removed from the list and added to the stack, it is removed from all t hash tables before it is modified and added to S.
- When a vector is added to the list, it is inserted in the t hash tables in the buckets corresponding to its t hash values.

The main difference with previous work [34,35] lies in the choice of the hash function family, which in this paper is the efficient and asymptotically superior cross-polytope LSH, rather than the asymptotically worse angular or hyperplane LSH [15,34] or the less practical spherical LSH [8,35]. This leads to the CPSieve algorithm for which we provide a pseudocode in [12, Sect. 3].

3.1   Solving SVP in Time and Space $2^{0.298n + O(n)}$

To analyze the resulting algorithm and to choose suitable parameters k and t, what matters most is the performance of the underlying locality-sensitive hash functions; the better these functions are at separating reducible from unreducible pairs of vectors, the fewer hash functions and hash tables we will need and the faster the algorithm will be. In particular, as described in various literature on locality-sensitive hashing, to estimate the performance of the LSH family one should consider the parameter $\rho = \frac{\log 1/p_1}{\log 1/p_2}$.

Note that the LSH family $\mathcal{H}$ described in Sect. 2.3 has 'performance parameter' ρ as follows, where the collision probabilities $p_{1,2}$ correspond to certain angles $\theta_{1,2}$ between pairs of vectors:

$$\rho_{\mathcal{H}} = \log(1/p_1)/\log(1/p_2) = \tan^2(\theta_1/2)/\tan^2(\theta_2/2)(1 + o(1)). \tag{8}$$

Comparing this result to Andoni et al.'s spherical hash functions $h \in \mathcal{S}$ [7,8] used in the SphereSieve [35], which have a collision probability of

$$\mathbb{P}_{h \sim \mathcal{S}}[h(\boldsymbol{v}) = h(\boldsymbol{w})] = \exp\left[-\frac{\sqrt{n}}{2}\tan^2\left(\frac{\theta}{2}\right)(1 + o(1))\right], \tag{9}$$

it is clear that also this spherical LSH family $\mathcal{S}$ achieves a ρ of

$$\rho_{\mathcal{S}} = \log(1/p_1)/\log(1/p_2) = \tan^2(\theta_1/2)/\tan^2(\theta_2/2)(1 + o(1)). \tag{10}$$

In terms of analyzing the effects of the use of either of these hash families on sieving, this implies that both families achieve asymptotically equivalent exponents; the analysis from [35] to derive the optimal time and space complexities of $2^{0.298n + o(n)}$ also applies here, thus leading to the following result.

Theorem 1. *The here presented CPSieve heuristically solves SVP in time and space $2^{0.2972n + o(n)}$ using the following parameters:*

$$k = \Theta(n/\log n), \qquad t = 2^{0.0896n + o(n)}. \tag{11}$$

By varying k and t, we further obtain the trade-off between the time and space complexities indicated by the solid blue curve in Fig. 1.

Proof. As the dependence on θ in the collision probabilities for $\mathcal{H}$ and $\mathcal{S}$ is the same, the analysis from [35, Appendix A] also applies to $\mathcal{H}$. The only impact of the different factor in the exponent of the collision probability (in terms of n) is the value of k, which after a similar analysis (where it should hold that the number of buckets roughly equals the eventual list size, i.e., $\Theta(n^k) \sim 2^{\Theta(n)}$) turns out to lead to the given expression for k.

Note that a major difference between the two hash families $\mathcal{H}$ and $\mathcal{S}$ is that computing a single hash value (for one hash function, before amplification) costs $2^{\Theta(\sqrt{n})}$ time for $\mathcal{S}$ and only at most $O(n^2)$ time for $\mathcal{H}$ (due to the matrix-vector multiplication by a random Gaussian matrix A). So by replacing $\mathcal{S}$ by $\mathcal{H}$, the cost of computing hashes goes down from subexponential (but superpolynomial) to only at most quadratic in n. Especially for large n, this means cross-polytope hashing will be orders of magnitude faster than spherical hashing, and may be competitive with the angular hashing of Charikar [15] used in the HashSieve [34,43].

3.2   Practical Aspects of the CPSieve

Although this theoretical result already offers a substantial (albeit subexponential) improvement over the SphereSieve, and an exponential improvement over other sieve algorithms, to make the resulting algorithm truly practical we would like to further reduce the worst-case quadratic cost of computing hashes.

Theoretically, to compute hashes we first multiply a target vector $\boldsymbol{v}$ by a fully random Gaussian matrix A where each entry $a_{i,j}$ is drawn from the same

Gaussian distribution, and then look for the largest coordinate of $v' = Av$; the index of the largest coordinate of v' will be the hash value. Note that finding this largest coordinate, given v', can be done in worst-case linear time, and so the main bottleneck in computing hashes lies in computing the product Av. As also described in [1,34,37], in practice it may be possible to reduce the amount of entropy in the hash functions (the "randomness") without significantly affecting the performance of the scheme. As long as the amount of entropy is high enough that we can build sufficiently many random, independent hash functions, the algorithm will generally still work fine. Some possibilities to reduce the complexity of computing hashes in practice are:

– Use low-precision floating-point matrices A.
– Use sparse random projection matrices.
– Use structured matrices that allow for fast matrix-vector multiplication.

Using structured matrices that allow for e.g. the use of Fast Fourier Transforms for computing matrix-vector multiplications may significantly reduce the cost of computing a hash value from $O(n^2)$ to $O(n \log n)$.

Probing. The idea of probing, where various hash buckets in each hash table are traversed and checked for reductions with v (rather than only the bucket labeled $h(v)$), can also be applied to the CPSieve. For a given vector v, the highest-quality bucket (the bucket most likely to contain vectors for reductions) is the one labeled $h(v)$, containing other vectors which also have the same index of the largest coordinate. It is not hard to see that the second-best bucket for reductions with v is exactly the bucket corresponding to the second-largest absolute coordinate of v. For instance, if $v = (3, -1, 8, -5, 11)$ then the vectors whose largest coordinate is the fifth coordinate are most likely to be useful for reductions, and the next best option to check is those vectors whose largest coordinate is the third coordinate. By checking multiple buckets in each hash table (rather than just one bucket), we may reduce the number of hash tables and the overall space complexity by a polynomial factor at almost no cost. As an example, we performed tests in dimensions 65, 69 and 72, respectively. We set $k = 2$ and test buckets according to four hash values obtained from the two largest coefficients of the two levels. The result is a reduction of the memory to store the hash tables in comparison to a single hash value: We can decrease the number of hash tables such that the resulting memory requirement is at 84 %, 75 % and 64 %, respectively, for the three dimensions in test. For further details on clever (multi-)probing techniques for the cross-polytope LSH family $\mathcal{H}$, as well as ways to use structured matrices to reduce the quadratic cost of hashing, see [9].

3.3 Relation with Angular Hashing and a Practical Trade-Off

To put the hash family $\mathcal{H}$ into context, recall that the angular hash family of Charikar [15] used in the HashSieve [34] is defined as follows: one samples a random vector $r \in \mathbb{R}^n$ (its length is irrelevant), and assigns a hash value to a vector v based on whether the inner product $v \cdot r$ is positive ($h(v) = 1$) or not ($h(v) = 0$). Equivalently, we apply a suitable random projection to v, and check whether v_1 is positive ($h(v) = 1$) or not ($h(v) = 0$).

In this way it is easy to see some similarities with cross-polytope hashing, where all (instead of only one) entries of v are compared and the index of the maximum of these entries (and the sign of the maximum entry) is used as the hash value. This suggests a natural generalization of both angular and cross-polytope hashing as follows:

$$\tilde{h}_m(x) = \pm \arg\max_{i \in \{1,\dots,m\}} |x_i|. \qquad (1 \le m \le n) \qquad (12)$$

Using random Gaussian projection matrices A and setting $m = 1$ then exactly corresponds to the angular hashing technique of Charikar, while with rerandomizations and $m = n$ we obtain the cross-polytope LSH family. This generalization with arbitrary m is also equivalent to first applying a random projection onto a low-dimensional subspace and then using the standard full-dimensional cross-polytope hash function in this low-dimensional space.

Note that although the CPSieve is asymptotically faster than the HashSieve, for the HashSieve the practical cost of computing hash values is much lower. To formalize this potential trade-off, note that for arbitrary m the hash function $\tilde{h}_m$ has $2m$ possible outcomes, and we eventually choose the parameter k to (asymptotically) satisfy that the total number of hash buckets in each hash table is roughly the same as the number of vectors in the system, i.e., $(2m)^k \approx 2^{0.21n}$. For given m, this translates to a condition on k as $k \approx \frac{0.21n}{\log_2 m + 1}$. For actually computing hash values (for the moment ignoring the cost of the rerandomizations) we need to go through m of the vector coordinates to find the largest one in absolute value, incurring a cost of about m comparisons. In total, this means that for one hash table (which uses k hash functions) the cost of computing a vector's hash bucket is

$$(\text{Cost of computing the right bucket}) \approx k \cdot m \approx 0.21n \cdot \left\lceil \frac{m}{\log_2 m + 1} \right\rceil. \qquad (13)$$

This suggests that to bring down the polynomial factors of computing hashes, we should choose m as small as possible, i.e. $m = 1$; this also explains why in low dimensions the HashSieve may outperform the CPSieve due to smaller polynomial terms. On the other hand, as m increases the asymptotic exponent of the algorithm's time complexity decreases from $0.337n + o(n)$ (the HashSieve) to $0.298n + o(n)$ (the CPSieve), so for high dimensions it is clear that setting $m = n$ is best. For moderate dimensions one might find the best option to be somewhere in between these two extremes. Experimentally we verified this to be the case for $n = 50$, where we heuristically found the best choice of m to lie significantly closer to $m = n$ than to $m = 1$; for fixed t, it seems we can slightly reduce the time complexity by less than 20% by choosing m slightly less than n, e.g. $m \approx 2n/3$.

3.4   Experimental Results

We first show that already in mid-size dimensions ($n > 50$), we observe that the costs are similar to the asymptotic estimate for small choices of k. For a given dimension, we can vary the parameters t and k and observe varying numbers of vector comparisons, changes of the list size and number of hash computations. For example, let us fix the number t of hash tables, $t \in [80; 120]$. We can now choose different values for k in practice that influence the probability that a candidate is a valid vector for reduction. A smaller k leads to a less restrictive hash value such that more vectors need to be checked for reduction. Increasing k produces a more restrictive hash value and we might need to increase the number t of hash tables to find good collisions; otherwise the list size may increase drastically, leading to a higher time complexity as well. Varying the parameters means trading time against memory as illustrated in Figs. 2 and 3.[1]

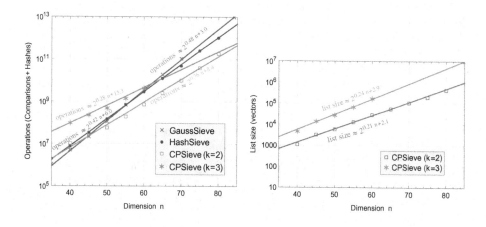

Fig. 2. Number of operations (comparisons + hashes).

Fig. 3. Number of vectors in the list of the CPSieve for optimal $t \in [80; 120]$.

Setting first $k = 2$, we performed experiments on random lattices in dimensions $n = 40$ to 80 with varying $t \in [80; 120]$ and observed an interpolated time complexity of around $0.36n + o(n)$ in logarithmic scale as illustrated by the lower (green) line in Fig. 2. The advantage of this choice is a reduced list size which lies close to $0.21n + o(n)$ as depicted in Fig. 3. If we wish to reduce the number of computations and to approach the minimal asymptotic time, we need to increase k (and t) with n which leads to larger list sizes of around $0.24n + o(n)$ in our experiments (cf. Fig. 3). For $k = 3$ we observe a better approximation of the

---

[1] The figures represent the collected data at the time of submission. More fine grained tests w.r.t. the dimension and the various parameter choices are in progress an will be included in the final version.

heuristic running time of $0.298n + o(n)$ as shown in Fig. 2 by the upper (orange) line. The observed cost lies slightly below the asymptotic estimate.

Figure 2 also shows how various algorithms from the literature compare, including (i) the GaussSieve, which performs an exhaustive search over the list L; (ii) the HashSieve, which uses hash tables based on angular LSH; and (iii) our new CPSieve algorithm, with parameters $k = 2, 3$. As indicated by the theoretical cost, the new CPSieve performs clearly better in terms of the asymptotic exponent, and this also appears from the experiments: the linear interpolation for the data based on the CPSieve in Fig. 2 has a significantly smaller slope than both the GaussSieve and the HashSieve. In dimensions below 60 the polynomial factors for sieving still play an important role in practice, and therefore the absolute number of operations for CPSieve lies partially above the GaussSieve and/or the angular HashSieve.

Overall we see that the new algorithm has a distinguished lower increase in the complexity in practice compared to the traditional GaussSieve and the angular HashSieve, and the crossover points are already in low dimensions. As the gap between the CPSieve and other algorithms will only increase as n increases, this clearly highlights the potential of the CPSieve on arbitrary lattices.

Remark for Sieving Based on Spherical Locality Sensitive Filtering. Recently, the work [11] presented a new technique for speeding up sieving algorithms, which presents slightly better asymptotics for solving SVP in high dimensions compared to the CPSieve presented here ($2^{0.293n+o(n)}$ vs. $2^{0.298n+o(n)}$). Note that if for simplicity we assume the $o(n)$ terms in the exponents of both algorithms are the same, then the improvement over the CPSieve would be a factor 2 every 200 dimensions. Experiments with this new algorithm in [44] suggest that this algorithm actually might not perform as well in practice as originally thought (it only overtakes the HashSieve in dimension around 90), and clearly this factor 2 speedup every 200 dimensions compared to the CPSieve is not only negligible in general, but also is more than compensated for on ideal lattices by the speedup of our algorithm described in the next section.

4   ICPSieve: Sieving in Ideal Lattices

While the CPSieve is very capable of solving the shortest vector problem on arbitrary lattices, it was already shown in various papers [14, 29, 57] that for certain ideal lattices it is possible to obtain substantial polynomial speed-ups to sieving in practice, which may make sieving even more competitive with e.g. enumeration-based SVP solvers. As ideal lattices are commonly used in lattice cryptography, and our main goal is to estimate the complexity of SVP on lattices that are actually used in lattice cryptography, it is important to know if our proposed CPSieve can be sped up on ideal lattices as well. We will show that this is indeed the case, using similar techniques as in [14, 29, 57] but where we need to do some extra work to make sure these speed-ups apply here as well.

4.1   Ideal GaussSieve

For the ideal lattices mentioned in the preliminaries, cyclic shifts of a vector are also in the lattice (modulo minus signs) and have the same Euclidean norm. As first described by Schneider [57], this property can be used in the GaussSieve as follows. First, note that any vector v can be viewed as representing n vectors, namely its n shifted versions $v, v_{(1)}, v_{(2)}, \ldots, v_{(n-1)}$, where we write $x_{(s)} = (x_{n-s+1}, \ldots, x_n, x_1, \ldots, x_{n-s})$ for the sth cyclic right-shift of $x = (x_1, \ldots, x_n)$. Similarly, another vector w represents n different lattice vectors $w, w_{(1)}, w_{(2)}, \ldots, w_{(n-1)}$.

Non-ideal GaussSieve: In the standard GaussSieve, we would treat these $2n$ shifts of v and w as different vectors, and we would store all of them in the system, leading to a storage cost of $2n$ vectors. Furthermore, to make sure that the list remains pairwise reduced, all $\binom{2n}{2} \approx 2n^2$ pairs of vectors are compared for reductions, leading to a time cost of approximately $2n^2$ vector comparisons.

Ideal GaussSieve: To make use of the cyclic structure of certain ideal lattices, the main idea of the ideal GaussSieve is that comparing $v_{(s)}$ to $w_{(s')}$ is the same as comparing $v_{(s-s')}$ to w for any s, s': there exist shifts of v and w that can (cannot) reduce each other if and only if there exists a shift of v that can reduce (be reduced by) w. So instead of storing all $2n$ shifts, we only store the two representative vectors v and w in the system (storage cost of 2 vectors), and more importantly, to see if any of the shifts of v and w can reduce each other we only compare all n shifts of v to the single vector w stored in memory (n comparisons). To make sure that also v (w) and its own cyclic shifts are pairwise reduced, we further need $n/2$ ($n/2$) comparisons to compare v to $v_{(s)}$ (w to $w_{(s)}$) for $s = 1, \ldots, n/2$. In total, we therefore need $n + n/2 + n/2 = 2n$ comparisons to reduce v, w and all their cyclic shifts.

Overall, this shows that in cyclic and negacyclic lattices, the memory cost of the GaussSieve goes down by a factor n, and the number of inner products that we compute to make sure the list is pairwise reduced also goes down by a factor approximately n. Although only polynomial, a factor 100 speedup and using 100 times less memory in dimension 100 can be very useful.

4.2   Hashing Shifted Vectors is Shifting Hashes of Vectors

To see how we can obtain similar improvements for the CPSieve, let us first look at the basic hash function $h(x) = \pm \arg\max_i x_i$. Suppose we have a cyclic lattice, and for some lattice vector v we have $h(v) = i$ for some $i \in \{1, \ldots, n\}$. Due to the choice of the hash function, we know that if we shift the entries of v to the right by s positions to get $v_{(s)}$, then the hash of this vector will increase by s as well, modulo n:

$$h(v_{(s)}) = [h(v) + s] \bmod n, \tag{14}$$

where the result of the modular addition is assumed to lie in $\{1, \ldots, n\}$. As a result, we know that $h(v) = h(w)$ if and only if $h(v_{(s)}) = h(w_{(s)})$ for any s.

For the basic hash function h, this property allows us to use a similar trick as in the ideal GaussSieve: we only store one representative of w in the hash tables, and for reducing v we compare all n shifts $v_{(s)}$ to the lattice vectors in their corresponding buckets $h(v_{(s)})$. We are then guaranteed that if any pair of vectors $v_{(s)}$ and $w_{(s')}$ can be reduced and have the same hash value, we will encounter this reduction when we compare $v_{(s-s')}$ and w as they will also have the same hash values and can reduce each other.

4.3   Ideal Rerandomizations Through Circulant Matrices

While this shows that the basic hash function h has this nice property that allows us to obtain the linear decreases in the time and space complexity similar to the ideal GaussSieve, to make this algorithm work we will need many different hash functions from $\mathcal{H}$ for each of the hash tables for the AND- and OR-compositions; in particular, the number of hash tables t (and therefore also the number of hash functions) increases exponentially with n. And once we apply a random rotation to a vector, we may lose the property described in (14):

$$h_A(v_{(s)}) = h(Av_{(s)}) \stackrel{?}{=} [h(Av) + s] \bmod n = [h_A(v) + s] \bmod n, \qquad (15)$$

The second equality is crucial here, as without preserving the property that the hash of a shift of a vector equals the shift of the hash of a vector, it might be that there exists a pair of vectors $v_{(s)}$ and $w_{(s')}$ that can be reduced *and has the same hash value*, while we will not reduce $v_{(s-s')}$ and w because they have different hash values. If that happens, then not all $2n$ shifts of both vectors are pairwise reduced, which implies that the 'quality' of the list goes down, so the list size goes up, and we lose the factor n speedup again.

To guarantee that the second equality in (15) is always an equality, we would like to make sure that $Av_{(s)} = (Av)_{(s)}$, i.e., multiplying a shifted vector by A is the same as shifting the vector which has already been multiplied by A. After all, in that case we would have

$$h_A(v_{(s)}) = h(Av_{(s)}) = h((Av)_{(s)}) = [h(Av) + s] \bmod n = [h_A(v) + s] \bmod n, \qquad (16)$$

where the second equality follows from the condition $Av_{(s)} = (Av)_{(s)}$ and the third equality follows from the property (14) of the base hash function h. So if we can guarantee that $Av_{(s)} = (Av)_{(s)}$ for all v and s, then also these rerandomized hash functions satisfy the property we need to obtain a linear speedup. Now, it is not hard to see that $Av_{(s)} = (Av)_{(s)}$ for all v and s is equivalent to the fact that A is circulant; substituting $v = e_1$ and varying $s = 1, \ldots, n$ tells us that $a_{i,j} = a_{1,[j-i+1] \bmod n}$ for all i and j. In other words, we are free to choose the first row of A, and the ith row of the matrix is then defined as the $(i-1)$th cyclic shift of A.

So finally, the question becomes: can we simply impose the condition that A is circulant? While proving that the answer is yes or no seems hard, experimentally the answer seems to be yes: by only generating the first rows of each rerandomization matrix A at random from a standard Gaussian distribution, and then deriving the remaining entries of A from the first row, we obtain circulant matrices which appear to be as suitable for random rotations as fully random Gaussian matrices. The resulting circulant matrices on average appear to be as orthogonal as non-circulant ones, thus preserving relative norms and distances between vectors, and do not seem to perform worse in our experiments than non-circulant matrices.

Remark 1. The angular/hyperplane hash function of the HashSieve [15,34], as well as the spherical hash functions in the SphereSieve [7,35] do not have the properties mentioned above, and so while it may be possible to obtain the trivial decrease in the space complexity of a factor n, it seems impossible to obtain the factor n time speedup described above that applies to the GaussSieve and to the CPSieve.

Remark 2. By using circulant matrices, computing hashes of shifted vectors (to compare all shifts of a target vector v against the vectors in the hash tables) can be done by shifting the hash of the original vector. Also, one can compute the product of a circulant matrix with an arbitrary vector in $O(n \log n)$ time using Fast Fourier Transforms [24] instead of $O(n^2)$ time, which for large n may further reduce the overall time complexity of the algorithm. However, the even faster random rotations described in [9] which may be useful for the non-ideal case do not apply here, as we need A to be circulant to obtain the factor n speedup.

4.4   Power-of-2 Cyclotomic Ideal Lattices ($X^n + 1$)

For our experiments we will consider two specific classes of ideal lattices, the first of which is the class of ideal lattices over the ring $\mathbb{Z}[X]/(X^n + 1)$ where n is a power of 2. These are negacyclic lattices, and so for any lattice vector v all its $2n$ shifts are in the lattice as well, and $v_{(n)} = -v$. As for comparisons in the GaussSieve/CPSieve we usually compare both $\pm v$ to candidate vectors w, in this case this corresponds to going through all $2n$ shifts of a target vector v (which all have different hash values) and searching the hash buckets for vectors that may reduce these vectors. In short, for each new target vector taken from the stack, the algorithm will proceed as described in [12, Sect. 4.4]. For convenience, we will assume that negative partial hash values $h_{i,j}(v) < 0$ are replaced by $h'_{i,j}(v) = n - h_{i,j}(v)$, so that the partial hash values always lie in the range $1, \ldots, 2n$ and are consecutive hash values of consecutive shifted vectors.

4.5   NTRU Lattices ($X^n - 1$)

The lattice basis of an NTRU encryption scheme [26,27] can be described by a prime power p, the ring $R = \mathbb{Z}_q[X]/(X^p - 1)$, a small power q of two and two polynomials $f, g \in R$ with small coefficient, for example in $\{-1, 0, 1\}$. We require that f is invertible in R and set $h = g/f \mod q$. The public basis is then given by p, q and h as the $n \times n$ matrix M (where $n = 2p$) as depicted in Fig. 4.

Note that not only (f, g) but also all block-wise rotations (fX^k, gX^k) are short vectors in the lattice. More generally, we observe that each block of $p = n/2$ entries of a lattice vector can be shifted (without minus sign) to obtain another valid lattice vector.

For these lattices we can apply similar techniques as in the previous subsection, but in this case we only have $n/2$ shifts of a vector in n dimensions; the speedups and memory gains are not equal to the dimension, but only to half the dimension of the lattice we are trying to tackle. The improvement we expect with respect to the non-ideal case will therefore be less than for the power-of-2 lattices described above.

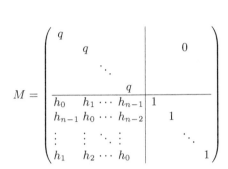

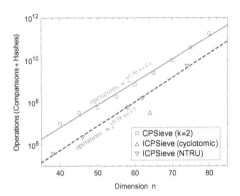

Fig. 4. Public NTRU lattice basis.

Fig. 5. Operations for arbitrary vs. ideal lattices.

4.6 Experiments for Ideal Lattices

For testing the performance of SVP algorithms on ideal lattices, we focused on NTRU lattices where $n = 2p$ and p is prime, and on negacyclic lattices where $n = 2^s$ is a power of 2, which can be generated with the ideal lattice challenge generator [51]. For the NTRU lattices we considered values $n = 38, 46, 58, 62, 74$, while for the cyclotomic lattices we restricted our experiments to only $n = 64$; for $n = 32$ the data will be unreliable as the algorithm terminates very quickly and the basis reduction sometimes already finds a shortest vector, while $n = 128$ is out of reach for our single-core proof-of-concept implementation; investigating the costs of solving the 128-dimensional ideal lattice challenge with the ICP-Sieve, as done in [14,29], is left for future work. The limited set of experiments performed as expected, and the results are shown in Fig. 5 in comparison to the random, non-ideal complexities of the CPSieve. The costs in the ideal case are decreased by a factor linear in n as we make use of the (block) cyclic structure of the respective ideal lattices as outlined in the previous subsections. We expect an analogue observation for different choices of the parameters. Note that for cyclotomic lattices we get a better exponent as the speedup and memory improvement are equal to n, rather than $n/2$ for NTRU lattices.

Acknowledgments. The authors thank Léo Ducas, Nicolas Gama and Benne de Weger for various useful discussions on this and related topics during a visit at EPFL. The authors also thank the authors of [9] for providing an early draft for use here. The second author acknowledges Memphis Depay, Meilof Veeningen and Niels de Vreede for their inspiration.

References

1. Achlioptas, D.: Database-friendly random projections. In: PODS, pp. 274–281 (2001)
2. Aggarwal, D., Dadush, D., Regev, O., Stephens-Davidowitz, N.: Solving the shortest vector problem in 2^n time via discrete Gaussian sampling. In: STOC (2015)
3. Ajtai, M.: Generating hard instances of lattice problems (extended abstract). In: STOC, pp. 99–108 (1996)
4. Ajtai, M.: The shortest vector problem in L_2 is NP-hard for randomized reductions (extended abstract). In: STOC, pp. 10–19 (1998)
5. Ajtai, M., Kumar, R., Sivakumar, D.: A sieve algorithm for the shortest lattice vector problem. In: STOC, pp. 601–610 (2001)
6. Andoni, A., Indyk, P.: Near-optimal hashing algorithms for approximate nearest neighbor in high dimensions. In: FOCS, pp. 459–468 (2006)
7. Andoni, A., Indyk, P., Nguyen, H.L., Razenshteyn, I.: Beyond locality-sensitive hashing. In: SODA, pp. 1018–1028 (2014)
8. Andoni, A., Razenshteyn, I.: Optimal data-dependent hashing for approximate near neighbors. In: STOC (2015)
9. Andoni, A., Indyk, P., Laarhoven, T., Razenshteyn, I., Schmidt, L.: Practical and optimal LSH for angular distance. In: NIPS (2015)
10. Becker, A., Gama, N., Joux, A.: A sieve algorithm based on overlattices. In: ANTS, pp. 49–70 (2014)
11. Becker, A., Ducas, L., Gama, N., Laarhoven, T.: New directions in nearest neighbor searching with applications to lattice sieving. In: SODA (2016)
12. Becker, A., Laarhoven, T.: Efficient (ideal) lattice sieving using cross-polytope LSH. In: Cryptology ePrint Archive, Report 2015/823. This is the full version of this paper (2015)
13. Bernstein, D.J., Buchmann, J., Dahmen, E.: Post-Quantum Cryptography. Springer, Berlin (2009)
14. Bos, J.W., Naehrig, M., van de Pol, J.: Sieving for shortest vectors in ideal lattices: a practical perspective. Cryptology ePrint Archive, Report 2014/880 (2014)
15. Charikar, M.S.: Similarity estimation techniques from rounding algorithms. In: STOC, pp. 380–388 (2002)
16. Conway, J.H., Sloane, N.J.A.: Sphere Packings, Lattices and Groups. Springer, New York (1999)
17. Datar, M., Immorlica, N., Indyk, P., Mirrokni, V.S.: Locality-sensitive hashing scheme based on p-stable distributions. In: SOCG, pp. 253–262 (2004)
18. Fincke, U., Pohst, M.: Improved methods for calculating vectors of short length in a lattice. Math. Comput. **44**(170), 463–471 (1985)

19. Fitzpatrick, R., Bischof, C., Buchmann, J., Dagdelen, Ö., Göpfert, F., Mariano, A., Yang, B.-Y.: Tuning gausssieve for speed. In: Aranha, D.F., Menezes, A. (eds.) LATINCRYPT 2014. LNCS, vol. 8895, pp. 288–305. Springer, Heidelberg (2015)
20. Gama, N., Nguyen, P.Q.: Predicting lattice reduction. In: Smart, N.P. (ed.) EURO-CRYPT 2008. LNCS, vol. 4965, pp. 31–51. Springer, Heidelberg (2008)
21. Gama, N., Nguyen, P.Q., Regev, O.: Lattice enumeration using extreme pruning. In: Gilbert, H. (ed.) EUROCRYPT 2010. LNCS, vol. 6110, pp. 257–278. Springer, Heidelberg (2010)
22. Garg, S., Gentry, C., Halevi, S.: Candidate multilinear maps from ideal lattices. In: Johansson, T., Nguyen, P.Q. (eds.) EUROCRYPT 2013. LNCS, vol. 7881, pp. 1–17. Springer, Heidelberg (2013)
23. Gentry, C.: Fully homomorphic encryption using ideal lattices. In: STOC, pp. 169–178 (2009)
24. Golub, G.H., Van Loan, C.F.: Matrix Computations. John Hopkins University Press, Baltimore (2012)
25. Hanrot, G., Pujol, X., Stehlé, D.: Algorithms for the shortest and closest lattice vector problems. In: Chee, Y.M., Guo, Z., Ling, S., Shao, F., Tang, Y., Wang, H., Xing, C. (eds.) IWCC 2011. LNCS, vol. 6639, pp. 159–190. Springer, Heidelberg (2011)
26. Hoffstein, J., Pipher, J., Silverman, J.H.: NTRU: a ring-based public key cryptosystem. In: Buhler, J.P. (ed.) ANTS 1998. LNCS, vol. 1423, pp. 267–288. Springer, Heidelberg (1998)
27. Hoffstein, J., Pipher, J., Silverman, J.H.: NSS: an NTRU lattice-based signature scheme. In: Pfitzmann, B. (ed.) EUROCRYPT 2001. LNCS, vol. 2045, pp. 211–228. Springer, Heidelberg (2001)
28. Indyk, P., Motwani, R.: Approximate nearest neighbors: towards removing the curse of dimensionality. In: STOC, pp. 604–613 (1998)
29. Ishiguro, T., Kiyomoto, S., Miyake, Y., Takagi, T.: Parallel Gauss Sieve algorithm: solving the SVP challenge over a 128-dimensional ideal lattice. In: PKC, pp. 411–428 (2014)
30. Kannan, R.: Improved algorithms for integer programming and related lattice problems. In: STOC, pp. 193–206 (1983)
31. Khot, S.: Hardness of approximating the shortest vector problem in lattices. In: FOCS, pp. 126–135 (2004)
32. Klein, P.: Finding the closest lattice vector when it's unusually close. In: SODA, pp. 937–941 (2000)
33. Kleinjung, T.: Private communication (2014)
34. Laarhoven, T.: Sieving for shortest vectors in lattices using angular locality-sensitive hashing. In: CRYPTO, pp. 3–22 (2015)
35. Laarhoven, T., de Weger, B.: Faster sieving for shortest lattice vectors using spherical locality-sensitive hashing. In: Lauter, K., Rodríguez-Henríquez, F. (eds.) LATINCRYPT 2015. LNCS, vol. 9230, pp. 101–118. Springer, Heidelberg (2015)
36. Lenstra, A.K., Lenstra, H.W., Lovász, L.: Factoring polynomials with rational coefficients. Math. Ann. **261**(4), 515–534 (1982)
37. Li, P., Hastie, T.J., Church, K.W.: Very sparse random projections. In: KDD, pp. 287–296 (2006)
38. Lindner, R., Peikert, C.: Better key sizes (and attacks) for LWE-based encryption. In: Kiayias, A. (ed.) CT-RSA 2011. LNCS, vol. 6558, pp. 319–339. Springer, Heidelberg (2011)
39. Lv, Q., Josephson, W., Wang, Z., Charikar, M., Li, K.: Multi-probe LSH: efficient indexing for high-dimensional similarity search. In: VLDB, pp. 950–961 (2007)

40. Lyubashevsky, V., Peikert, C., Regev, O.: On ideal lattices and learning with errors over rings. In: Gilbert, H. (ed.) EUROCRYPT 2010. LNCS, vol. 6110, pp. 1–23. Springer, Heidelberg (2010)
41. Mariano, A., Timnat, S., Bischof, C.: Lock-free GaussSieve for linear speedups in parallel high performance SVP calculation. In: SBAC-PAD (2014)
42. Mariano, A., Dagdelen, Ö., Bischof, C.: A comprehensive empirical comparison of parallel ListSieve and GaussSieve. In: APCI&E (2014)
43. Mariano, A., Laarhoven, T., Bischof, C.: Parallel (probable) lock-free HashSieve: a practical sieving algorithm for the SVP. In: ICPP (2015)
44. Mariano, A., Laarhoven, T.: A parallel variant of LDSieve and the tractability of sieving algorithms for the SVP. In preparation (2016)
45. Micciancio, D., Voulgaris, P.: A deterministic single exponential time algorithm for most lattice problems based on Voronoi cell computations. In: STOC, pp. 351–358 (2010)
46. Micciancio, D., Voulgaris, P.: Faster exponential time algorithms for the shortest vector problem. In: SODA, pp. 1468–1480 (2010)
47. Micciancio, D., Walter, M.: Fast lattice point enumeration with minimal overhead. In: SODA, pp. 276–294 (2015)
48. Milde, B., Schneider, M.: A parallel implementation of GaussSieve for the shortest vector problem in lattices. In: Malyshkin, V. (ed.) PaCT 2011. LNCS, vol. 6873, pp. 452–458. Springer, Heidelberg (2011)
49. Nguyen, P.Q., Vidick, T.: Sieve algorithms for the shortest vector problem are practical. J. Math. Crypt. **2**(2), 181–207 (2008)
50. Panigraphy, R.: Entropy based nearest neighbor search in high dimensions. In: SODA, pp. 1186–1195 (2006)
51. Plantard, T., Schneider, M.: Ideal lattice challenge (2014). http://latticechallenge. org/ideallattice-challenge/
52. Pohst, M.E.: On the computation of lattice vectors of minimal length, successive minima and reduced bases with applications. ACM SIGSAM Bull. **15**(1), 37–44 (1981)
53. van de Pol, J., Smart, N.P.: Estimating key sizes for high dimensional lattice-based systems. In: Stam, M. (ed.) IMACC 2013. LNCS, vol. 8308, pp. 290–303. Springer, Heidelberg (2013)
54. Pujol, X., Stehlé, D.: Solving the shortest lattice vector problem in time $2^{2.465n}$. Cryptology ePrint Archive, Report 2009/605 (2009)
55. Regev, O.: On lattices, learning with errors, random linear codes, and cryptography. In: STOC, pp. 84–93 (2005)
56. Schneider, M.: Analysis of Gauss-Sieve for solving the shortest vector problem in lattices. In: Katoh, N., Kumar, A. (eds.) WALCOM 2011. LNCS, vol. 6552, pp. 89–97. Springer, Heidelberg (2011)
57. Schneider, M.: Sieving for shortest vectors in ideal lattices. In: Youssef, A., Nitaj, A., Hassanien, A.E. (eds.) AFRICACRYPT 2013. LNCS, vol. 7918, pp. 375–391. Springer, Heidelberg (2013)
58. Schneider, M., Gama, N., Baumann, P., Nobach, L.: SVP challenge (2014). http://latticechallenge.org/svp-challenge
59. Schnorr, C.-P.: A hierarchy of polynomial time lattice basis reduction algorithms. Theor. Comput. Sci. **53**(2), 201–224 (1987)
60. Schnorr, C.-P., Euchner, M.: Lattice basis reduction: improved practical algorithms and solving subset sum problems. Math. Program. **66**(2), 181–199 (1994)
61. Shoup, V.: Number Theory Library (NTL), v6.2 (2014). http://www.shoup.net/ ntl/

62. Stehlé, D., Steinfeld, R.: Making NTRU as secure as worst-case problems overideal lattices. In: EUROCRYPT, pp. 27–47 (2011)
63. Terasawa, K., Tanaka, Y.: Spherical LSH for approximate nearest neighbor search on unit hypersphere. In: Dehne, F., Sack, J.-R., Zeh, N. (eds.) WADS 2007. LNCS, vol. 4619, pp. 27–38. Springer, Heidelberg (2007)
64. Wang, X., Liu, M., Tian, C., Bi, J.: Improved Nguyen-Vidick heuristic sieve algorithm for shortest vector problem. In: ASIACCS, pp. 1–9 (2011)
65. Zhang, F., Pan, Y., Hu, G.: A three-level sieve algorithm for the shortest vector problem. In: Lange, T., Lauter, K., Lisoněk, P. (eds.) SAC 2013. LNCS, vol. 8282, pp. 29–47. Springer, Heidelberg (2014)

On the Hardness of LWE with Binary Error: Revisiting the Hybrid Lattice-Reduction and Meet-in-the-Middle Attack

Johannes Buchmann[1]([⊠]), Florian Göpfert[1], Rachel Player[2], and Thomas Wunderer[1]

[1] Technische Universität Darmstadt, Darmstadt, Germany
{buchmann,fgoepfert,twunderer}@cdc.informatik.tu-darmstadt.de
[2] Information Security Group, Royal Holloway, University of London, Egham, UK
Rachel.Player.2013@live.rhul.ac.uk

Abstract. The security of many cryptographic schemes has been based on special instances of the Learning with Errors (LWE) problem, e.g., Ring-LWE, LWE with binary secret, or LWE with ternary error. However, recent results show that some subclasses are weaker than expected. In this work we show that *LWE with binary error*, introduced by Micciancio and Peikert, is one such subclass. We achieve this by applying the Howgrave-Graham attack on NTRU, which is a combination of lattice techniques and a Meet-in-the-Middle approach, to this setting. We show that the attack outperforms all other currently existing algorithms for several natural parameter sets. For instance, for the parameter set $n = 256$, $m = 512$, $q = 256$, this attack on LWE with binary error only requires 2^{85} operations, while the previously best attack requires 2^{117} operations. We additionally present a complete and improved analysis of the attack, using analytic techniques. Finally, based on the attack, we give concrete hardness estimations that can be used to select secure parameters for schemes based on LWE with binary error.

Keywords: Learning with errors · Lattice-based cryptography · Cryptanalysis · NTRU · Hybrid attack

1 Introduction

The Learning with Errors problem (LWE) is one of the most important problems in lattice-based cryptography. A huge variety of schemes, ranging from basic primitives like signature [18] and encryption schemes [32] to highly advanced schemes like group signatures [30] and fully homomorphic encryption [12], base their security on the LWE assumption. Understanding the concrete hardness of LWE is therefore important for selecting parameters.

Many cryptographic schemes are based on the hardness of special LWE instances like Ring-LWE [34], or LWE with ternary error [22]. Understanding the hardness of subclasses of the LWE problem and identifying those that are easy to

© Springer International Publishing Switzerland 2016
D. Pointcheval et al. (Eds.): AFRICACRYPT 2016, LNCS 9646, pp. 24–43, 2016.
DOI: 10.1007/978-3-319-31517-1_2

solve is therefore an important task. In fact, several recent results [15,19,20,29] show that some subclasses are easier than expected.

We show that the subclass LWE with binary error, which has been considered before in several papers [1,35], fits into this category. To show that LWE with binary error is considerably easier than expected, we modify the hybrid lattice-reduction and meet-in-the-middle attack by Howgrave-Graham [25] (refered to as hybrid attack in the following), apply it to this setting, and analyze its complexity. In order to compare our approach to existing ones, we apply known attacks on LWE to the binary error setting and analyze their complexities in this case. Our comparison shows that the hybrid attack is much faster than existing methods such as the enumeration attack [32,33], or the embedding approach [4] for several natural parameter sets. Figure 1 illustrates our improvement, by comparing the runtime of the best previously known attack with the hybrid attack, where $m = 2n$ samples from an LWE distribution with binary error are given and n is the dimension of the secret vector. For example, in the case of $n = 256$ and $q = 256$, the hardness of the problem drops from 117 to 85 bits, which is a significant improvement. A detailed comparison between the hybrid attack and previous approaches is given in Table 1 in Sect. 4.

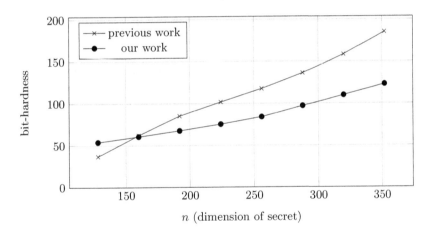

Fig. 1. Hardness of LWE instances with number of samples $m = 2n$ and modulus $q = 256$ before and after this work

The hybrid attack can also be seen as an improvement of an idea sketched by Bai and Galbraith [9]. However, Bai and Galbraith did not provide an analysis of their suggestion, and the analysis of Howgrave-Graham is partly based on experiments. A theoretical analysis of the hybrid attack that is not based on experimental results has been presented by Hirschhorn et al. in [24]. However, their analysis requires an additional assumption.

In this work we present a complete and improved analysis based on the same assumptions used in [25] without the additional assumption of [24], that does not require experimental support. For this reason, we introduce new analytic techniques. Our new analysis can also be applied to the Howgrave-Graham attack, as well as to the attack mentioned by Bai and Galbraith (see [9]). In addition, we show how to use our techniques to analyze the decoding attack on LWE with binary error.

Related Work. A number of recent works have highlighted the importance of considering the hardness of variants of LWE. For example, certain choices of rings lead to weak instances of the Ring-LWE problem [15,19,20]. Additionally, Laine and Lauter [29] provide a polynomial time attack for LWE instances with an exponentially large modulus q and a sufficiently narrow Gaussian error. The existence of such weak instances shows the necessity of studying the hardness of special instances of the LWE problem separately.

The hardness of LWE with binary error has been considered in some detail. So far, there are known attacks that require access to superlinearly many samples (i.e., $m > \mathcal{O}(n)$), and hardness results when the crypanalyst is given a sublinear number of additional samples (i.e., $m = n + \mathcal{O}(n/\log(n))$), where n is the dimension of the secret vector. More precisely, the problem can be solved in polynomial time using the algorithm of Arora and Ge [6], when the number of samples is $m = \mathcal{O}(n^2)$ (see, e.g., [1]). Furthermore, Albrecht et al. [1] showed that LWE with binary error can be solved in subexponential time using an improved version of the Arora-Ge attack, if the attacker has access to a quasi-linear number of samples, e.g., $m = \mathcal{O}(n \log \log n)$. On the other hand, Micciancio and Peikert [35] proved that LWE with binary error reduces to worst-case lattice problems when the number of samples is restricted to $n + \mathcal{O}(n/\log(n))$. We close the margin between these hardness results on the one side and the weakness results on the other side by presenting an attack that runs with only n additional samples.

The idea of Bai and Galbraith which we build upon is to guess the first r components of the secret vector and apply a lattice attack on the remaining problem [9]. As noted in [5], this strategy enables the transformation of any algorithm for solving LWE into another one whose complexity is bounded by the cost of exhaustive search. Howgrave-Graham's algorithm [25], which we apply here, involves a Meet-in-the-Middle component to speed up this guessing: this was not considered in either of [5,9]. The existence of a Meet-in-the-Middle approach for solving LWE (without combining with any another algorithm) was mentioned in [9] and such an algorithm was presented in [5]. In Sect. 4 we show that it is much more efficient to combine a Meet-in-the-Middle approach with a decoding attack than to solve LWE with binary error entirely by a Meet-in-the-Middle approach.

Structure. In Sect. 2 we give some notation and required preliminaries. In Sect. 3 we describe how to apply the hybrid attack to LWE with binary error and analyze its complexity. In Sect. 4 we apply other possible attacks on LWE to the binary error case, analyze their complexities, and compare the results to the hybrid attack.

2   Notation and Preliminaries

Notation. In this work vectors are denoted in bold lowercase letters, e.g., $\mathbf{a}$, and matrices in bold uppercase letters, e.g., $\mathbf{A}$. For a vector $\mathbf{v} \in \mathbb{R}^n$ we write $\mathbf{v} \bmod q$ for its unique representative modulo q in $[-\lfloor \frac{q}{2} \rfloor, \frac{q}{2})^n$. Logarithms are base two unless stated otherwise, and $\ln(x)$ denotes the natural logarithm of x.

Learning with Errors. The Learning with Errors (LWE) problem, introduced by Regev [41], is a computational problem, whose presumed hardness is the basis for several cryptographic constructions, e.g., [39–41]. In this work, we consider the variant *LWE with binary error.*

Problem Statement 1 (LWE with binary error). *Let n, q be positive integers, $\mathcal{U}$ be the uniform distribution on $\{0, 1\}$ and $\mathbf{s} \xleftarrow{\$} \mathcal{U}^n$ be a secret vector in $\{0, 1\}^n$. We denote by $L_{\mathbf{s}, \mathcal{U}}$ the probability distribution on $\mathbb{Z}_q^n \times \mathbb{Z}_q$ obtained by choosing $\mathbf{a} \in \mathbb{Z}_q^n$ uniformly at random, choosing $e \xleftarrow{\$} \mathcal{U}$ and returning $(\mathbf{a}, \langle \mathbf{a}, \mathbf{s} \rangle + e) \in \mathbb{Z}_q^n \times \mathbb{Z}_q$.*
LWE with binary error is the problem of recovering $\mathbf{s}$ from m samples $(\mathbf{a}_i, \langle \mathbf{a}_i, \mathbf{s}_i \rangle + e_i) \in \mathbb{Z}_q^n \times \mathbb{Z}_q$ sampled according to $L_{\mathbf{s}, \mathcal{U}}$, with $i \in \{1, \ldots, m\}$.

Note that Regev defined LWE with a secret vector $\mathbf{s}$ chosen uniformly at random from the whole of $\mathbb{Z}_q^n$. However, it is well-known that LWE with arbitrarily distributed secret can be transformed to LWE with secret distributed according to the error distribution. Consequently, most cryptographic constructions are based on LWE where secret and error are identically distributed, and we focus on this case in this work.

Lattices and Bases. A lattice is a discrete additive subgroup of $\mathbb{R}^m$. A set of linearly independent vectors $\mathbf{B} = \{\mathbf{b}_1, \ldots, \mathbf{b}_n\} \subset \mathbb{R}^m$ is called a basis of a lattice Λ, if $\Lambda = \Lambda(\mathbf{B})$, where

$$\Lambda(\mathbf{B}) = \{\mathbf{x} \in \mathbb{R}^m \mid \mathbf{x} = \sum_{i=1}^{n} \alpha_i \mathbf{b}_i \text{ for } \alpha_i \in \mathbb{Z}\}.$$

The dimension of a lattice Λ is defined as the cardinality of some (equivalently any) basis of Λ. For the rest of this work we restrict our studies to lattices in $\mathbb{R}^m$ whose dimension is maximal, e.g., m, which are called full-ranked lattices. The fundamental parallelepiped of a lattice basis $\mathbf{B} = \{\mathbf{b}_1, \ldots, \mathbf{b}_m\} \subset \mathbb{R}^m$ is given by

$$\mathcal{P}(\mathbf{B}) = \{\mathbf{x} \in \mathbb{R}^m \mid \mathbf{x} = \sum_{i=1}^{m} \alpha_i \mathbf{b}_i \text{ for } -1/2 \leq \alpha_i < 1/2\}.$$

The determinant of a lattice $\Lambda(\mathbf{B})$ for a basis $\mathbf{B}$ is defined as the m dimensional volume of its fundamental parallelepiped. Note that the determinant of the lattice is independent of the choice of the basis.

Every lattice of dimension $m \geq 2$ has infinitely many different bases. A measure for the quality of a basis is provided by the Hermite delta. A lattice basis $\mathbf{B} = \{\mathbf{b}_1, \ldots, \mathbf{b}_m\}$ has Hermite delta δ if $\|\mathbf{b}_1\| = \delta^m \det(\Lambda)^{1/m}$.

Differing estimates exist in the literature for the number of operations of a basis reduction necessary to achieve a certain Hermite delta δ (see for example [5,16,32,33,37]). Throughout this work we will use the estimate given by Lindner and Peikert [32]. This is that the number of operations needed to achieve a certain Hermite delta δ is around

$$\mathrm{ops}_{\mathrm{BKZ}}(\delta) = 2^{1.8/\log_2(\delta)-110} \cdot 2.3 \cdot 10^9. \tag{1}$$

A lattice Λ satisfying $q \cdot \mathbb{Z}^m \subset \Lambda \subset \mathbb{R}^m$ is a q-ary lattice. For a matrix $\mathbf{A} \in \mathbb{Z}_q^{m \times n}$, we define the q-ary lattice

$$\Lambda_q(\mathbf{A}) := \{\mathbf{v} \in \mathbb{Z}^m \mid \exists \mathbf{w} \in \mathbb{Z}^n : \mathbf{A}\mathbf{w} = \mathbf{v} \mod q\}.$$

If $m \geq n$ and all column vectors $\mathbf{A} \in \mathbb{Z}_q^{m \times n}$ are linearly independent over $\mathbb{Z}_q$, we have $\det(\Lambda_q(\mathbf{A})) = q^{m-n}$.

The closest vector problem is the problem of recovering the lattice vector closest to a given target vector, given also a basis of the lattice. One can consider a relaxation, namely a close vector problem, where the inputs are the same (a basis and a target vector), and the task is to recover a lattice vector which is sufficiently close to the target.

Babai's Nearest Plane. The hybrid attack uses Babai's nearest plane algorithm [7] (denoted by NP in the following) as subroutine. It gets a lattice basis $\mathbf{B} \subset \mathbb{Z}^m$ and a target vector $\mathbf{t} \in \mathbb{R}^m$ as input and outputs a vector $\mathbf{e} \in \mathbb{R}^m$ such that $\mathbf{t} - \mathbf{e} \in \Lambda(\mathbf{B})$, which we denote by $\mathrm{NP}_{\mathbf{B}}(\mathbf{t}) = \mathbf{e}$. If the used lattice basis is clear from the context, we omit it in the notation and simply write $\mathrm{NP}(\mathbf{t})$. A detailed explanation of nearest plane can be found in Babai's original work [7] and Lindner and Peikert's follow up work [32]. The output of nearest plane plays an important role in the analysis of the hybrid attack and can be understood without knowing details about the algorithm itself. It depends on the Gram-Schmidt basis of the input basis $\mathbf{B}$, which is defined as $\overline{\mathbf{B}} = \{\overline{\mathbf{b}_1}, \ldots, \overline{\mathbf{b}_n}\}$ with

$$\overline{\mathbf{b}_i} = \mathbf{b}_i - \sum_{j=1}^{i-1} \frac{\langle \overline{\mathbf{b}_j}, \mathbf{b}_i \rangle}{\langle \overline{\mathbf{b}_j}, \overline{\mathbf{b}_j} \rangle} \overline{\mathbf{b}_j},$$

where $\overline{\mathbf{b}_1} = \mathbf{b}_1$. We will use the following result from [8].

Lemma 1. *For a lattice basis $\mathbf{B}$ with Gram-Schmidt basis $\overline{\mathbf{B}}$ and a target vector $\mathbf{t}$ as input, the nearest plane algorithm returns the unique vector $\mathbf{e} \in \mathcal{P}(\overline{\mathbf{B}})$ that satisfies $\mathbf{t} - \mathbf{e} \in \Lambda(\mathbf{B})$.*

Lemma 1 shows that analyzing the output of the nearest plane algorithm requires to estimate the lengths of the basis vectors of the corresponding Gram-Schmidt basis. The established way to do this is via the the following heuristic (see Lindner and Peikert [32] for more details).

Heuristic 1 (Geometric Series Assumption). *Let* $\{\mathbf{b}_1 \ldots \mathbf{b}_m\} \subset \mathbb{Z}^m$ *be a reduced basis with Hermite delta* δ *of an* m-*dimensional lattice with determinant* D. *Also let* $\overline{\mathbf{b}}_i$ *denote the basis vectors of the corresponding Gram-Schmidt basis. Then the length of* $\overline{\mathbf{b}}_i$ *is approximated by*

$$\|\overline{\mathbf{b}}_i\| \approx \delta^{-2(i-1)+m} D^{\frac{1}{m}}.$$

3 The Attack

In this section we present and analyze the hybrid attack on LWE with binary error. The attack is described in Algorithm 1 of Sect. 3.1. In Theorem 1 of Sect. 3.2 we analyze the expected runtime of the hybrid attack. Section 3.3 shows how to optimize the attack parameters and perform a trade-off between precomputation and the actual attack in order to minimize the runtime of the attack.

3.1 The Hybrid Attack

In the following we describe the hybrid attack on LWE with binary error. The attack is presented in Algorithm 1.

Let $m, n, q \in \mathbb{N}$ and let

$$(\mathbf{A}, \mathbf{b} = \mathbf{A}\tilde{\mathbf{s}} + \mathbf{e} \mod q) \tag{2}$$

with $\mathbf{A} \in \mathbb{Z}_q^{m \times n}, \mathbf{b} \in \mathbb{Z}_q^m, \tilde{\mathbf{s}} \in \{0,1\}^n$ and $\mathbf{e} \in \{0,1\}^m$ be an LWE instance with binary error $\mathbf{e}$ and binary secret $\tilde{\mathbf{s}}$. In order to obtain a smaller error vector we can subtract the vector $(1/2) \cdot \mathbf{1}$ consisting of all $1/2$ entries from Eq. (2). This yields a new LWE instance $(\mathbf{A}, \mathbf{b}' = \mathbf{A}\tilde{\mathbf{s}} + \mathbf{e}' \mod q)$, where $\mathbf{b}' = \mathbf{b} - (1/2) \cdot \mathbf{1}$ and $\mathbf{e}' = \mathbf{e} - (1/2) \cdot \mathbf{1}$. The new error vector $\mathbf{e}'$ now has norm $\sqrt{m/4}$ instead of the expected norm $\sqrt{m/2}$ of the original error vector $\mathbf{e}$. For $r \in \{1, \ldots, n-1\}$, we can split the secret $\tilde{\mathbf{s}} = \begin{pmatrix} \mathbf{v} \\ \mathbf{s} \end{pmatrix}$ and the matrix $\mathbf{A} = (\mathbf{A}_1 | \mathbf{A}_2)$ into two parts and rewrite this LWE instance as

$$\mathbf{b}' = (\mathbf{A}_1 | \mathbf{A}_2) \begin{pmatrix} \mathbf{v} \\ \mathbf{s} \end{pmatrix} + \mathbf{e}' = \mathbf{A}_1 \mathbf{v} + \mathbf{A}_2 \mathbf{s} + \mathbf{e}' \mod q, \tag{3}$$

where $\mathbf{v} \in \{0,1\}^r, \mathbf{s} \in \{0,1\}^{n-r}, \mathbf{A}_1 \in \mathbb{Z}_q^{m \times r}, \mathbf{A}_2 \in \mathbb{Z}_q^{m \times (n-r)}, \mathbf{b}' = \mathbf{b} - (1/2) \cdot \mathbf{1} \in \mathbb{Q}^m$, and $\mathbf{e}' = \mathbf{e} - (1/2) \cdot \mathbf{1} \in \{-1/2, 1/2\}^m$.

The main idea of the attack is to guess $\mathbf{v}$ and solve the remaining LWE instance $(\mathbf{A}_2, \tilde{\mathbf{b}} = \mathbf{b}' - \mathbf{A}_1\mathbf{v} = \mathbf{A}_2\mathbf{s} + \mathbf{e}' \mod q)$, which has binary secret $\mathbf{s}$ and error $\mathbf{e}' \in \{-1/2, 1/2\}^m$. The new LWE instance obtained in this way turns out to be considerably easier to solve, since the determinant $\det(\Lambda_q(\mathbf{A}_2)) = q^{m-n+r}$ of the new lattice is significantly bigger than the determinant $\det(\Lambda_q(\mathbf{A})) = q^{m-n}$ of the original lattice (see Sect. 6.1 of [9]). The newly obtained LWE instance is solved by solving a close vector problem in the lattice $\Lambda_q(\mathbf{A}_2)$. In more detail, $\tilde{\mathbf{b}} = \mathbf{A}_2\mathbf{s} + q\mathbf{w} + \mathbf{e}'$ for some vector $\mathbf{w} \in \mathbb{Z}^m$ is close to the lattice

---

Algorithm 1. The Hybrid Attack

Input : $q, r \in \mathbb{Z}$

$\quad\quad \mathbf{A} = (\mathbf{A}_1 | \mathbf{A}_2)$, where $\mathbf{A}_1 \in \mathbb{Z}_q^{m \times r}, \mathbf{A}_2 \in \mathbb{Z}_q^{m \times (n-r)}$

$\quad\quad \mathbf{b} \in \mathbb{Z}_q^m$

$\quad\quad \mathbf{B}$, a lattice basis of $\Lambda_q(\mathbf{A}_2)$

1   calculate $c = \lfloor r/4 \rfloor$;

2   calculate $\mathbf{b}' = \mathbf{b} - (1/2) \cdot \mathbf{1}$;

3   **while** *true* **do**

4   guess a binary vector $\mathbf{v}_1 \in \{0,1\}^r$ with c ones ;

5   calculate $\mathbf{x}_1 = -\operatorname{NP}_\mathbf{B}(-\mathbf{A}_1 \mathbf{v}_1) \in \mathbb{R}^m$;

6   calculate $\mathbf{x}_2 = \operatorname{NP}_\mathbf{B}(\mathbf{b}' - \mathbf{A}_1 \mathbf{v}_1) \in \mathbb{R}^m$;

7   store $\mathbf{v}_1$ in all the boxes addressed by $\mathcal{A}_{\mathbf{x}_1}^{(r)} \cup \mathcal{A}_{\mathbf{x}_2}^{(r)}$;

8   **for** *all* $\mathbf{v}_2 \neq \mathbf{v}_1$ *in all the boxes addressed by* $\mathcal{A}_{\mathbf{x}_1}^{(r)} \cup \mathcal{A}_{\mathbf{x}_2}^{(r)}$ **do**

9   Set $\mathbf{v} = \mathbf{v}_1 + \mathbf{v}_2$ and calculate $\mathbf{x} = (1/2) \cdot \mathbf{1} + \operatorname{NP}_\mathbf{B}(\mathbf{b}' - \mathbf{A}_1 \mathbf{v}) \in \mathbb{R}^m$;

10   **if** $\mathbf{x} \in \{0,1\}^m$ *and* $\exists \tilde{\mathbf{s}} \in \{0,1\}^n : \mathbf{b} = \mathbf{A}\tilde{\mathbf{s}} + \mathbf{x} \mod q$ **then**

11   **return** $\mathbf{x}$;

---

vector $\mathbf{A}_2 \mathbf{s} + q\mathbf{w} \in \Lambda_q(\mathbf{A}_2)$ since $\mathbf{e}'$ is small. Hence $\mathbf{e}'$ can be found by running the nearest plane algorithm in combination with a sufficient basis reduction as a precomputation (see [32]).

The guessing of $\mathbf{v}$ is sped up by a Meet-in-the-Middle approach, i.e., guessing binary vectors $\mathbf{v}_1 \in \{0,1\}^r$ and $\mathbf{v}_2 \in \{0,1\}^r$ such that $\mathbf{v} = \mathbf{v}_1 + \mathbf{v}_2$. In order to recognize matching guesses $\mathbf{v}_1$ and $\mathbf{v}_2$ that sum up to $\mathbf{v}$, one searches for collisions in (hash) boxes. The addresses of these boxes are determined in the following way.

Definition 1. *Let* $m \in \mathbb{N}$. *For a vector* $\mathbf{x} \in \mathbb{R}^m$ *the set* $\mathcal{A}_\mathbf{x}^{(m)} \subset \{0,1\}^m$ *is defined as*

$$\mathcal{A}_\mathbf{x}^{(m)} = \left\{ \mathbf{z} \in \{0,1\}^m \,\middle|\, \begin{matrix} (\mathbf{z})_i = 1 \text{ for all } i \in \{1, \ldots, m\} \text{ with } (\mathbf{x})_i > -1/2, \text{ and} \\ (\mathbf{z})_i = 0 \text{ for all } i \in \{1, \ldots, m\} \text{ with } (\mathbf{x})_i < -1/2 \end{matrix} \right\}.$$

Intuitively, for $\mathbf{x}_2$ obtained during Algorithm 1, the set $\mathcal{A}_{\mathbf{x}_2}^{(m)}$ captures all the possible sign vectors of $\mathbf{x}_2$ added up with a vector in $\{-1/2, 1/2\}^m$ (where 1 represents a non-negative and 0 a negative sign). For $\mathbf{x}_1$ obtained during Algorithm 1, the set $\mathcal{A}_{\mathbf{x}_1}^{(m)}$ consists only of the sign vector of $\mathbf{x}_1$. This is due to the fact that $\mathbf{x}_2 \in \mathbb{Z}^m + \{1/2\}^m$, whereas $\mathbf{x}_1 \in \mathbb{Z}^m$. This leads to the desired collisions, as can be seen in the upcoming Lemma 3.

3.2 Runtime Analysis

In this section we analyze the runtime and success probability of the attack presented in Algorithm 1. We start by presenting our main result.

Theorem 1. *Let* $n, m, q, c \in \mathbb{N}$, *and* $1 \leq \delta \in \mathbb{R}$ *be fixed. Consider the following input distribution of* $(q, r, \mathbf{A}, \mathbf{b}, \mathbf{B})$ *for Algorithm 1. The modulus* q *and the attack*

parameter $r = 4c$ are fixed, $\mathbf{A} = (\mathbf{A}_1|\mathbf{A}_2)$, *where* $\mathbf{A}_1 \xleftarrow{\$} \mathbb{Z}_q^{m \times r}$, $\mathbf{A}_2 \xleftarrow{\$} \mathbb{Z}_q^{m \times (n-r)}$,
$\mathbf{b} = \mathbf{A}\begin{pmatrix} \mathbf{v} \\ \mathbf{s} \end{pmatrix} + \mathbf{e} \mod q$, *where* $\mathbf{v} \xleftarrow{\$} \{0,1\}^r$, $\mathbf{s} \xleftarrow{\$} \{0,1\}^{n-r}$, $\mathbf{e} \xleftarrow{\$} \{0,1\}^m$, *and*
$\mathbf{B}$ *is some lattice basis of* $\Lambda_q(\mathbf{A}_2)$ *with Hermite delta* δ. *Let all notations be as in the above description of the input distribution. Assume that the approximations given in Heuristics 2 and 4 are in fact equations and that* $\mathrm{NP}_{\mathbf{B}}(\mathbf{b} - (1/2) \cdot \mathbf{1} - \mathbf{A}_1 \mathbf{v}) = \mathbf{e} - (1/2) \cdot \mathbf{1}$. *Then, if Algorithm 1 terminates, it finds a valid binary error vector of the LWE with binary error instance* $(\mathbf{A}, \mathbf{b})$. *The probability that Algorithm 1 terminates is at least*

$$p_0 = 2^{-r} \binom{r}{2c}.$$

In case that Algorithm 1 terminates, the expected number of operations is

$$2^{16} \binom{r}{c} \left(p \binom{2c}{c} \right)^{-1/2},$$

with

$$p = \prod_{i=1}^{m} \left(1 - \frac{1}{r_i B(\frac{m-1}{2}, \frac{1}{2})} J(r_i, m) \right),$$

where $B(\cdot, \cdot)$ *denotes the Euler beta function (see [38]),*

$$J(r_i, m) = \begin{cases} \int_{-r_i-1}^{r_i-1} \int_{-1}^{z+r_i} (1-y^2)^{\frac{m-3}{2}} dy dz \\ \quad + \int_{r_i-1}^{-r_i} \int_{z-r_i}^{z+r_i} (1-y^2)^{\frac{m-3}{2}} dy dz & \text{for } r_i < \frac{1}{2} \\ \int_{-r_i-1}^{-r_i} \int_{-1}^{z+r_i} (1-y^2)^{\frac{m-3}{2}} dy dz & \text{for } r_i \geq \frac{1}{2}, \end{cases}$$

and

$$r_i = \frac{\delta^{-2(i-1)+m} q^{\frac{m-n+r}{m}}}{2\sqrt{m/4}}.$$

Remark 1. *Algorithm 1 gets some basis* $\mathbf{B}$ *as input. This basis has a certain quality, given by the Hermite delta* δ. *In practice, we can improve the attack by providing a basis with better, i.e., smaller, Hermite delta. We achieve this by running a basis reduction (e.g., BKZ) on* $\mathbf{B}$ *in a precomputation step (see Sect. 3.3).*

We postpone the proof of Theorem 1 to the end of this subsection, since we first need to develop some necessary tools. We start by giving a definition of a notion which is crucial to our analysis. We then give a useful lemma.

Definition 2. *Let* $m \in \mathbb{N}$. *A vector* $\mathbf{x} \in \mathbb{Z}^m$ *is called* $\mathbf{y}$-*admissible for some vector* $\mathbf{y} \in \mathbb{Z}^m$ *if* $\mathrm{NP}(\mathbf{x}) = \mathrm{NP}(\mathbf{x} - \mathbf{y}) + \mathbf{y}$.

Intuitively, $\mathbf{x}$ being $\mathbf{y}$-admissible means that running the nearest plane algorithm on $\mathbf{x}$ and running it on $\mathbf{x} - \mathbf{y}$ yields the same lattice vector, since then we have $\mathbf{x} - \mathrm{NP}(\mathbf{x}) = (\mathbf{x} - \mathbf{y}) - \mathrm{NP}(\mathbf{x} - \mathbf{y})$.

Lemma 2. *Let $\mathbf{t}_1 \in \mathbb{R}^m, \mathbf{t}_2 \in \mathbb{R}^m$ be two arbitrary target vectors. Then the following are equivalent.*

1. $\mathrm{NP}(\mathbf{t}_1) + \mathrm{NP}(\mathbf{t}_2) = \mathrm{NP}(\mathbf{t}_1 + \mathbf{t}_2)$.
2. $\mathbf{t}_1$ is $\mathrm{NP}(\mathbf{t}_1 + \mathbf{t}_2)$-admissible.
3. $\mathbf{t}_2$ is $\mathrm{NP}(\mathbf{t}_1 + \mathbf{t}_2)$-admissible.

A proof of this lemma can be found in the full version [13].

As we will see in our analysis, the expected runtime heavily depends on the following probability. Let all notations be as in Theorem 1 and $\mathbf{e}' = \mathbf{e} - (1/2) \cdot \mathbf{1}$. For

$$W = \{\mathbf{w} \in \{0,1\}^r : \text{ exactly } c \text{ entries of } \mathbf{w} \text{ are } 1\} \tag{4}$$

we define

$$p := \begin{cases} \Pr_{\mathbf{v}_1 \leftarrow W}[-\mathbf{A}_1\mathbf{v}_1 \text{ is } \mathbf{e}'\text{-admissible}|\mathbf{v} - \mathbf{v}_1 \in W] & \text{if } \Pr_{\mathbf{v}_1 \leftarrow W}[\mathbf{v} - \mathbf{v}_1 \in W] > 0 \\ 0 & \text{else.} \end{cases}$$

$$\tag{5}$$

Note that the hybrid attack requires that nearest plane called on the target vector $\mathbf{b} - (1/2) \cdot \mathbf{1} - \mathbf{A}_1\mathbf{v}$ returns the correct shifted error vector $\mathbf{e} - (1/2) \cdot \mathbf{1}$. However, this is not a big restriction in practice, since this probability is bigger than the probability that the same vector is $\mathbf{e}'$-admissible. To see why, recall that nearest plane returns the correct error vector if and only if it lies in the fundamental parallelepiped $\Lambda(\mathbf{B})$. On the other hand, Heuristic 3 states that the probability that $\mathbf{b} - (1/2) \cdot \mathbf{1} - \mathbf{A}_1\mathbf{v}$ is $\mathbf{e}'$-admissible is approximately the probability that the sum of a random point in $\Lambda(\mathbf{B})$ and the error vector is still in $\Lambda(\mathbf{B})$. Consequently, we expect that $\mathrm{NP}_{\mathbf{B}}(\mathbf{b} - (1/2) \cdot \mathbf{1} - \mathbf{A}_1\mathbf{v}) = \mathbf{e} - (1/2) \cdot \mathbf{1}$ holds with high probability for all realistic attack parameters.

Note that the analysis of the attack on the NTRU encryption proposed by Howgrave-Graham [25] also requires to calculate the probability p. In the original work, this is done experimentally. Replacing this probability estimation with the analytic methodology presented in the following removes the dependency on experimental support in the analysis of the hybrid attack. A first mathematical calculation of the probability p has already been presented by Hirschhorn et al. in [24]. However, their analysis requires an additional assumption that we no longer need.

Success Probability. In this subsection we determine the probability that Algorithm 1 terminates. We start by giving a sufficient condition for this event.

Lemma 3. *Let all notations be as in Theorem 1 and let $\mathbf{b}' = \mathbf{b} - (1/2) \cdot \mathbf{1}$ and $\mathbf{e}' = \mathbf{e} - (1/2) \cdot \mathbf{1}$. Assume that $\mathbf{v}_1$ and $\mathbf{v}_2$ are guessed in separate loops of Algorithm 1 and satisfy $\mathbf{v}_1 + \mathbf{v}_2 = \mathbf{v}$. Also let $\mathbf{t}_1 = -\mathbf{A}_1\mathbf{v}_1$ and $\mathbf{t}_2 = \mathbf{b}' - \mathbf{A}_1\mathbf{v}_2$ and assume $\mathrm{NP}(\mathbf{t}_1) + \mathrm{NP}(\mathbf{t}_2) = \mathrm{NP}(\mathbf{t}_1 + \mathbf{t}_2) = \mathbf{e}'$ holds. Then $\mathbf{v}_1$ and $\mathbf{v}_2$ collide in at least one box chosen during Algorithm 1 and the algorithm outputs the error vector $\mathbf{e}$ of the given LWE instance.*

Proof: According to the notation used in Algorithm 1, let $x_1 = -NP(t_1)$ correspond to v_1 and $x_2 = NP(t_2)$ correspond to v_2. By assumption we have $x_1 = x_2 - e'$. Using the definition it is easy to verify that x_1 and x_2 share at least one common address, since $e' \in \{-1/2, 1/2\}^m$. Therefore v_1 and v_2 collide in at least one box. Again by assumption, we obtain $x = NP(b' - A_1 v) = NP(t_1 + t_2) = e'$. Hence the algorithm outputs the error vector e.   ∎

In the following lemma we give a lower bound on the probability that Algorithm 1 terminates.

Lemma 4. *Let all notations be as in Theorem 1 and let $b' = b - (1/2) \cdot 1$ and $e' = e - (1/2) \cdot 1$. Assume that if v has exactly $2c$ one-entries, then $p > 0$, where p is as defined in Eq. (5). If $NP(b' - A_1 v) = e'$, then Algorithm 1 terminates with probability at least*

$$p_0 = 2^{-r} \binom{r}{2c}.$$

Proof: We show that Algorithm 1 terminates if v consists of exactly $2c$ one-entries. The probability of this happening is exactly p_0, since there are 2^r binary vectors of length r, and $\binom{r}{2c}$ of them have exactly $2c$ one-entries. Assume that v consists of exactly $2c$ one-entries. The claim follows directly from Lemmas 2 and 3. Since $p > 0$ there exist binary vectors $v_1, v_2 \in \{0, 1\}^r$, each containing exactly c one-entries, such that $v_1 + v_2 = v$ and $-A_1 v_1$ is e'-admissible. These vectors will eventually be guessed during Algorithm 1 if it does not terminate before. By Lemma 2 they satisfy

$$NP(-A_1 v_1) + NP(b' - A_1 v_2) = NP(b' - A_1 v) = e'.$$

Lemma 3 now guarantees that Algorithm 1 then outputs the error vector e.   ∎

Estimating the Number of Loops. The next step is to estimate the number of loops until the attack terminates.

Heuristic 2. *Let all notations be as in Theorem 1 and let $b' = b - (1/2) \cdot 1$ and $e' = e - (1/2) \cdot 1$. Assume that $NP(b' - A_1 v) = e'$, and that v consists of exactly $2c$ one-entries. Then the expected number of loops of Algorithm 1 is*

$$L \approx \binom{r}{c} \left(p \binom{2c}{c} \right)^{-1/2},$$

and the probability p, as given in Eq. (5), is

$$p \approx \prod_{i=1}^{m} \left(1 - \frac{1}{r_i B(\frac{m-1}{2}, \frac{1}{2})} J(r_i, m) \right),$$

with $B(\cdot, \cdot)$, $J(\cdot, \cdot)$, and r_i defined as in Theorem 1.

In the following, we justify the heuristic. Assume that $\mathbf{v}$ consists of exactly $2c$ one-entries. In addition to W (see Eq. (4)), define the set

$$V = \{\mathbf{v}_1 \in W : \mathbf{v} - \mathbf{v}_1 \in W \text{ and } -\mathbf{A}_1\mathbf{v}_1 \text{ is } \mathbf{e}'\text{-admissible}\}.$$

Note that W is the set from which Algorithm 1 samples the vectors $\mathbf{v}_1$. Lemma 3 shows that the attack succeeds if two vectors $\mathbf{v}_1, \mathbf{v}_2 \in V$ satisfying $\mathbf{v}_1 + \mathbf{v}_2 = \mathbf{v}$ are sampled in different loops of Algorithm 1. Since otherwise the probability of success is close to zero, for simplicity we assume that the attack is only successful in this case. Therefore we need to estimate the necessary number of loops in Algorithm 1 until some $\mathbf{v}_1, \mathbf{v}_2 \in V$ with $\mathbf{v}_1 + \mathbf{v}_2 = \mathbf{v}$ are found. Note that by Lemma 2 if $\mathbf{v}_1 \in V$, then also $\mathbf{v}_2 = \mathbf{v} - \mathbf{v}_1 \in V$.

We start by calculating the probability that a vector sampled during Algorithm 1 lies in V. By definition of p, this probability is given by

$$\Pr_{\mathbf{v}_1 \xleftarrow{\$} W} [\mathbf{v}_1 \in V] = p_1 p, \text{ where } p_1 := \Pr_{\mathbf{v}_1 \xleftarrow{\$} W} [\mathbf{v} - \mathbf{v}_1 \in W].$$

Therefore we expect to sample a vector $\mathbf{v}_1 \in V$ every $\frac{1}{p_1 p}$ loops in Algorithm 1. The above equation also implies $p_1 p = \frac{|V|}{|W|}$, which gives us

$$|V| = p_1 p |W| = p_1 p \binom{r}{c}.$$

The probability p_1 is given by $p_1 = \binom{2c}{c} / \binom{r}{c}$, see the full version [13]. Therefore by the birthday paradox, the expected number of loops in Algorithm 1 until some $\mathbf{v}_1, \mathbf{v}_2 \in V$ with $\mathbf{v}_1 + \mathbf{v}_2 = \mathbf{v}$ are found can be estimated by

$$L \approx \frac{1}{p_1 p} \sqrt{|V|} = \frac{\sqrt{\binom{r}{c}}}{\sqrt{p_1 p}} = \binom{r}{c} \left(p \binom{2c}{c} \right)^{-1/2}.$$

It remains to approximate the probability p which we do in the following. Let $\mathbf{v}_1 \in \{0,1\}^r$ and $\mathbf{B}$ be some basis of $\Lambda_q(\mathbf{A}_2)$. By Lemma 1 there exist unique $\mathbf{u}_1, \mathbf{u}_2 \in \Lambda_q(\mathbf{A}_2)$ such that $\mathrm{NP}_{\mathbf{B}}(-\mathbf{A}_1\mathbf{v}_1) = -\mathbf{A}_1\mathbf{v}_1 - \mathbf{u}_1 \in \mathcal{P}(\overline{\mathbf{B}})$ and $\mathrm{NP}_{\mathbf{B}}(-\mathbf{A}_1\mathbf{v}_1 - \mathbf{e}') + \mathbf{e}' = -\mathbf{A}_1\mathbf{v}_1 - \mathbf{u}_2 \in \mathbf{e}' + \mathcal{P}(\overline{\mathbf{B}})$. Without loss of generality, in the following we assume $\mathbf{u}_1 = \mathbf{0}$, or equivalently $-\mathbf{A}_1\mathbf{v}_1 \in \mathcal{P}(\overline{\mathbf{B}})$. Now $-\mathbf{A}_1\mathbf{v}_1$ is $\mathbf{e}'$-admissible if and only if $\mathbf{u}_2 = \mathbf{u}_1 = \mathbf{0}$, which is equivalent to $\mathbf{e}' + \mathbf{A}_1\mathbf{v}_1 \in \mathcal{P}(\overline{\mathbf{B}})$. Therefore p is equal to the probability that $\mathbf{e}' + \mathbf{A}_1\mathbf{v}_1 \in \mathcal{P}(\overline{\mathbf{B}})$, which we determine in the following.

There exists some orthonormal transformation that aligns $\mathcal{P}(\overline{\mathbf{B}})$ along the standard axes of $\mathbb{R}^m$. By applying this transformation, we may therefore assume that $\mathcal{P}(\overline{\mathbf{B}})$ is aligned along the standard axes of $\mathbb{R}^m$ and that in consequence $\mathbf{e}'$ is a uniformly random vector of length $\sqrt{m/4}$. Because $\mathbf{A}_1$ is uniformly random in $\mathbb{Z}_q^{m \times r}$ we may further assume that $\mathbf{A}_1\mathbf{v}_1$ is uniformly random in $\mathcal{P}(\overline{\mathbf{B}})$, since without loss of generality we assume $\mathbf{A}_1\mathbf{v}_1 \in \mathcal{P}(\overline{\mathbf{B}})$. This gives rise to the following heuristic.

Heuristic 3. *The probability p as defined in Eq. 5 (with respect to a reduced basis with Hermite delta δ) is*

$$p \approx \Pr_{\mathbf{t} \xleftarrow{\$} R, \mathbf{e}' \xleftarrow{\$} S_m(\sqrt{m/4})} [\mathbf{t} + \mathbf{e}' \in R],$$

where

$$S_m(\sqrt{m/4}) = \{\mathbf{x} \in \mathbb{R}^m \mid \|\mathbf{x}\| = \sqrt{m/4}\}$$

is the surface of a sphere with radius $\sqrt{m/4}$ centered around the origin and

$$R = \{\mathbf{x} \in \mathbb{R}^m \mid \forall i \in \{1, \ldots, m\} : -R_i/2 \le x_i < R_i/2\}$$

is the search rectangle with edge lengths

$$R_i = \delta^{-2(i-1)+m} q^{\frac{m-n+r}{m}}.$$

In the heuristic, the edge lengths are implied by the Geometric Series Assumption.

We continue calculating the approximation of p given in Heuristic 3. Let R and R_i be as defined in Heuristic 3. We can rewrite the approximation given in Heuristic 3 as

$$p \approx \Pr_{t_i \xleftarrow{\$} [-R_i/2, R_i/2], \mathbf{e}' \xleftarrow{\$} S_m(\sqrt{m/4})} [\forall i \in \{1, \ldots, m\} : t_i + e_i' \in [-R_i/2, R_i/2]].$$

Rescaling everything by a factor of $1/\sqrt{m/4}$ leads to

$$p \approx \Pr_{t_i \xleftarrow{\$} [-r_i, r_i], \mathbf{e}' \xleftarrow{\$} S_m(1)} [\forall i \in \{1, \ldots, m\} : t_i + e_i' \in [-r_i, r_i]],$$

where

$$r_i = \frac{R_i}{2\sqrt{m/4}} = \frac{\delta^{-2(i-1)+m} q^{\frac{m-n+r}{m}}}{2\sqrt{m/4}}. \tag{6}$$

Unfortunately, the distributions of the coordinates of $\mathbf{e}$ are not independent, which makes calculating p extremely complicated. In practice, however, the probability that $e_i \in [-R_i/2, R_i/2]$ is big for all but the last few indices i. This is due to the fact that by the Geometric Series Assumption typically only the last values R_i are small. Consequently, we expect the dependence of the remaining entries not to be strong. This assumption was already established by Howgrave-Graham [25] and appears to hold for all values of R_i appearing in practice.

It is therefore reasonable to assume that

$$p \approx \prod_{i=1}^{m} \Pr_{t_i \xleftarrow{\$} [-r_i, r_i], e_i' \xleftarrow{\$} D_m} [t_i + e_i' \in [-r_i, r_i]],$$

were D_m denotes the distribution on the interval $[-1,1]$ obtained by the following experiment: sample a vector $\mathbf{w}$ uniformly at random on the unit sphere and then output the first (equivalently, any arbitrary but fixed) coordinate of $\mathbf{w}$.

Next we explore the density function of D_m. The probability that $e_i' \leq x$ for some $-1 < x < 0$, where $e_i' \overset{\$}{\leftarrow} D_m$, is given by the ratio of the surface area of a hyperspherical cap of the unit sphere in $\mathbb{R}^m$ with height $h = 1 + x$ and the surface area of the unit sphere. This is illustrated in the full version [13] for $m = 2$. The surface area of a hyperspherical cap of the unit sphere in $\mathbb{R}^m$ with height $h < 1$ is given by (see [31])

$$A_m(h) = \frac{1}{2} A_m I_{2h-h^2}\left(\frac{m-1}{2}, \frac{1}{2}\right),$$

where $A_m = 2\pi^{m/2}/\Gamma(m/2)$ is the surface area of the unit sphere and

$$I_x(a,b) = \frac{\int_0^x t^{a-1}(1-t)^{b-1}dt}{B(a,b)}$$

is the regularized incomplete beta function (see [38]) and $B(a,b)$ is the Euler beta function.

Consequently, for $-1 < x < 0$, we have

$$\Pr_{e_i' \overset{\$}{\leftarrow} D_m} [e_i' \leq x] = \frac{A_m(1+x)}{A_m}$$

$$= \frac{1}{2} I_{1-x^2}\left(\frac{m-1}{2}, \frac{1}{2}\right)$$

$$= \frac{1}{2B(\frac{m-1}{2}, \frac{1}{2})} \int_0^{1-x^2} t^{\frac{m-3}{2}}(1-t)^{-1/2}dt$$

$$= \frac{1}{B(\frac{m-1}{2}, \frac{1}{2})} \int_{-1}^{x} (1-t^2)^{\frac{m-3}{2}} dt. \tag{7}$$

Together with

$$\Pr_{t_i \overset{\$}{\leftarrow} [-r_i, r_i]} [t_i \leq x] = \int_{-r_i}^{x} \frac{1}{2r_i} dy,$$

we can use a convolution to obtain

$$\Pr_{t_i \overset{\$}{\leftarrow} [-r_i, r_i], e_i' \overset{\$}{\leftarrow} D_m} [t_i + e_i' \leq x] = \frac{1}{2r_i B(\frac{k-1}{2}, \frac{1}{2})} \int_{-r-1}^{x} \int_{\max(-1, z-r_i)}^{\min(1, z+r_i)} (1-y^2)^{\frac{m-3}{2}} dy dz.$$

Since

$$\Pr_{t_i \overset{\$}{\leftarrow} [-r_i, r_i], e_i' \overset{\$}{\leftarrow} D_m} [t_i + e_i' \in [-r_i, r_i]] = 1 - 2\left(\Pr_{t_i \overset{\$}{\leftarrow} [-r_i, r_i], e_i' \overset{\$}{\leftarrow} D_m} [t_i + e_i' < -r_i] \right),$$

it suffices to calculate the integral

$$J(r_i, m) = \int_{-r_i-1}^{-r_i} \int_{\max(-1, z-r_i)}^{z+r_i} (1-y^2)^{\frac{m-3}{2}} \, dy \, dz \tag{8}$$

in order to calculate p. We calculated the integral symbolically using sage [42], which allows an efficient calculation of p.

Time Spend per Loop Cycle. With the estimation of the number of loops given, the remaining task is to estimate the time spend per loop cycle. Each cycle consists of four steps:

1. Guessing a binary vector.
2. Running the nearest plane algorithm (twice).
3. Calculating $\mathcal{A}_{\mathbf{x}_1}^{(r)} \cup \mathcal{A}_{\mathbf{x}_1'}^{(r)}$.
4. Dealing with collisions in the boxes.

We assume that the runtime of one inner loop of Algorithm 1 is dominated by the runtime of the nearest plane algorithm, as argued in the following. It is well known that sampling a binary vector is extremely fast. Furthermore, note that only very few of the 2^n addresses contain a vector, since filling a significant proportional would take exponential time. Consequently, collisions are extremely rare, and lines 8–11 of Algorithm 1 do not contribute much to the overall runtime.

An estimation by Howgrave-Graham [25] shows that for typical instances, the runtime of the nearest plane algorithm exceeds the time spent for storing the collision. We therefore omit the latter from our considerations.

Lindner and Peikert [32] estimated the time necessary to run the nearest plane algorithm to be about 2^{-16} seconds, which amounts to about 2^{15} bit operations on their machine. This leads to the following heuristic for the runtime of the attack.

Heuristic 4. *The average number of operations per inner loop in Algorithm 1 is $N \approx 2^{16}$.*

Total Runtime. We are now able to prove our main theorem.

Proof (Theorem 1): By definition, every output of Algorithm 1 is a valid binary error vector of the given LWE with binary error instance. The rest follows directly from Lemma 4, Heuristics 2 and 4. ∎

3.3   Minimizing the Expected Runtime

As mentioned in Remark 1, we can perform a basis reduction to obtain a lattice basis with smaller Hermite delta δ before running the actual attack in order to speed up the attack. We perform a binary search for the δ such that the estimated runtimes of both the basis reduction and the actual attack are about equal. We also need to optimise r, the Meet-in-the-Middle dimension, which we do numerically, as there are only finitely many r to check. We refer the reader to the full version [13] for further details on the choice of δ and r.

4 Comparison

In this section we consider other approaches to solve LWE with binary error and compare these algorithms to Algorithm 1. In particular we give upper bounds for the runtimes of the algorithms. A comparison of the most practical attacks, including the hybrid attack, is given in Table 1.

Much of the analyses below are in a similar spirit to that given in the survey [5] for methods of solving standard LWE. However we are often able to specifically adapt the analysis for the binary error case. Note that to solve LWE with binary error, in addition to algorithms for standard LWE, one may also be able to apply algorithms for the related Inhomogeneous Short Integer Solution problem. A discussion of these algorithms is given in [10].

4.1 Number of Samples

Recall that for reducing LWE with binary error to worst-case problems on lattices, one must restrict the number of samples to be $m = n\left(1 + \Omega(1/\log n)\right)$ [35, Theorem 1.2]. On the other hand, with slightly more than linear samples, such as $m = \mathcal{O}(n \log \log n)$, the algorithm given in [1] is subexponential. Therefore if a scheme bases its security on the hardness of LWE with binary error, it is reasonable to expect that one has only access to at most linearly many samples. We assume this is the case in our analysis below. For concreteness, we fix $m = 2n$.

4.2 Algorithms for Solving LWE

There are several approaches one could use to solve LWE or its variants (see the survey [5]). One may employ combinatorial algorithms such as the BKW [2,11]

Table 1. Comparison of attacks on LWE with binary error using at most $m = 2n$ samples. $\log_2(T_{\text{attack}})$ denotes the bit operations required to perform the algorithm described in 'attack'. For algorithms requiring lattice reduction, we choose whichever is the fewer of $m = 2n$ or the 'optimal subdimension' $m = \sqrt{n \log(q)/\log(\delta)}$ [36].

Instance	n	q	$\log_2(T_{\text{Hybrid attack}})$	$\log_2(T_{\text{Decoding}})$	$\log_2(T_{\text{uSVP}})$	$\log_2$ $(T_{\text{Distinguishing}})$
I	128	256	**55**	67	82	**37**
II	160	256	**61**	77	122	**62**
III	192	256	**68**	88	162	**85**
IV	224	256	**76**	**102**	165	109
V	256	256	**85**	**117**	203	132
VI	288	256	**98**	**136**	254	154
VII	320	256	110	**158**	327	176
VIII	352	256	**123**	**185**	443	198

algorithm and its variants [3,17,23,27]. However, all these algorithms require far more samples than are available in the binary error case, and are therefore ruled out. We also omit a Meet-in-the-Middle attack [5] or attacks based on the algorithm of Arora and Ge [1,6], as they will be slower than other methods. We consider them in the full version [13] for completeness.

Distinguishing Attack. One can solve LWE via a distinguishing attack as described in [32,36]. The idea is to find a short vector $\|\mathbf{v}\|$ in the scaled dual lattice of $\mathbf{A}$, i.e. the lattice $\Lambda = \{\mathbf{w} \in \mathbb{Z}_q^m \mid \mathbf{w}\mathbf{A} \equiv 0 \mod q\}$. Then, if the problem is to distinguish $(\mathbf{A}, \mathbf{b})$ where $\mathbf{b}$ is either formed as an LWE instance $\mathbf{b} = \mathbf{A}\mathbf{s} + \mathbf{e}$ or is uniformly random, one can use this short vector $\mathbf{v}$ as follows. Consider $\langle \mathbf{v}, \mathbf{b} \rangle = \langle \mathbf{v}, \mathbf{e} \rangle$ if $\mathbf{b}$ is from an LWE instance, which as the inner product of two short vectors, is small mod q. On the other hand, if $\mathbf{b}$ is uniform then $\langle \mathbf{v}, \mathbf{b} \rangle$ is uniform on $\mathbb{Z}_q$ so these cases can be distinguished if $\mathbf{v}$ is suitably small.

We determine how small a $\mathbf{v}$ which must be found as follows. Recall that our errors are chosen uniformly at random from $\{0, 1\}$. So they follow a Bernoulli distribution with parameter $1/2$, and have expectation $1/2$ and variance $1/4$. Consider the distribution of $\langle \mathbf{v}, \mathbf{e} \rangle$. Since the errors e_i are chosen independently, its expectation is $\frac{1}{2}\sum_{i=1}^{m} v_i$ and its variance is $\frac{1}{4}\sum_{i=1}^{m} v_i^2$. Since $\langle \mathbf{v}, \mathbf{e} \rangle$ is the sum of many independent random variables, asymptotically it follows a normal distribution with those parameters. Since the distinguishing attack success is determined by the variance and not the mean, and we can account for the mean, we assume it is zero. Then we can use the result of [32] to say that we can distinguish a Gaussian from uniform with advantage close to $\exp(-\pi(\|\mathbf{v}\| \cdot s/q)^2)$, where s is the width parameter of the Gaussian. In our case $s^2 = 2\pi \cdot \frac{1}{4}$ so we can distinguish with advantage close to $\epsilon = \exp(-\pi^2 \|\mathbf{v}\|^2 / 2q^2)$. Therefore to distinguish with advantage ϵ we require a vector $\mathbf{v}$ of length $\|\mathbf{v}\| = q \cdot \frac{\sqrt{2\ln(1/\epsilon)}}{\pi}$.

We calculate a basis of the scaled dual lattice Λ and find a short vector $\mathbf{v} \in \Lambda$ by lattice basis reduction. With high probability the lattice Λ has rank m and volume q^n [5,36]. By definition of the Hermite delta we therefore have $\|\mathbf{v}\| = \delta^m q^{n/m}$. So the Hermite delta we require to achieve for the attack to succeed with advantage ϵ is given by $\delta^m q^{n/m} = q \cdot \frac{\sqrt{2\ln(1/\epsilon)}}{\pi}$. Assuming that the number of samples m is large enough to use the 'optimal subdimension' $m = \sqrt{n\log(q)/\log(\delta)}$ [36], we rearrange to obtain

$$\log \delta = \frac{\left(\log(q) + \log\left(\frac{\sqrt{2\ln(1/\epsilon)}}{\pi} \right) \right)^2}{4n\log(q)}.$$

To establish the estimates for the runtime of this attack given in Table 1, we assume one has to run the algorithm about $1/\epsilon$ times to succeed, and consider δ as a function of ϵ. The overall running time is then given by $1/\epsilon$ multiplied the estimated time, according to Eq. (1), to achieve $\delta(\epsilon)$. We pick the optimal ϵ such that this overall running time is minimized.

It is possible that we do not have enough samples to use the 'optimal subdimension', in which case we use $m = 2n$. For details, see the full version [13].

Reducing to uSVP. One may solve LWE via Kannan's embedding technique [26], thus seeing an LWE instance as a unique shortest vector problem instance. This technique is used in [4,9]. We follow analogously the analysis in [4,5] for the LWE with binary error case and obtain that we require a Hermite delta of $\log(\delta) = \frac{[\log(q) - \log(2\tau\sqrt{\pi e})]^2}{4n\log(q)}$ for this attack to succeed. The number of operation necessary to achieve this Hermite delta is estimated using Eq. (1). A comprehensive analysis can be found in the full version [13].

Decoding. The decoding approach for solving LWE was first described in [32] and is based on Babai's nearest plane algorithm [7]. The aim is to recover the error vector (so seeing LWE as a Bounded Distance Decoding instance). Recall (Lemma 1) that the error vector can be recovered using Babai's algorithm if it lies within the fundamental parallelepiped of the Gram-Schmidt basis. The idea of Lindner and Peikert in [32] is to widen the search parallelepiped to

$$\mathcal{P}_{\text{decoding}} = \{\mathbf{x} \in \mathbb{Z}^m \mid \mathbf{x} = \sum_{i=1}^{n} \alpha_i d_i \overline{\mathbf{b}}_i \text{ for } -1/2 \leq \alpha_i < 1/2\},$$

where $d_1, \ldots, d_m$ are integers chosen by the attacker.

Following the analysis of Lindner and Peikert, we estimate that an attack on a reduced basis with Hermite delta δ requires about $2^{15} \cdot \prod_{i=1}^{m} d_i$ operations. However, the analysis of the success probability is more complicated. By definition of search parallelepiped, the attack succeeds if (and only if) the error $\mathbf{e}$ lies in the search rectangle $\mathcal{P}_{\text{decoding}}$. Under the same assumption as in Sect. 3.2 (and using the same error transformation), this probability can be estimated via

$$p_{\text{decoding}} \approx \prod_{i=1}^{m} \left(\Pr_{\mathbf{e}_i \xleftarrow{\$} D_m} [\mathbf{e}_i \in [-r_i, r_i]] \right)$$

where

$$r_i = d_i \frac{\delta^{-2(i-1)+m} q^{\frac{m-n}{m}}}{2\sqrt{m/4}}.$$

Together with Eq. (7), this leads to

$$p_{\text{decoding}} \approx \prod_{i=1}^{m} \left(1 - \frac{2}{B(\frac{m-1}{2}, \frac{1}{2})} \int_{-1}^{-r_i} (1 - t^2)^{\frac{m-3}{2}} dt \right)$$

A standard way to increase the runtime of the attack is to use basis reduction (like BKZ2.0) as precomputation. Predicting the runtime of BKZ2.0 according to Eq. (1) leads to the runtime estimation

$$T_{\text{decoding}} \approx \frac{2^{1.8/\log_2(\delta) - 110} \cdot 2.3 \cdot 10^9 + 2^{15} \prod_{i=1}^{m} d_i}{p_{\text{decoding}}}.$$

Using the same numeric optimization techniques as presented above to minimize the expected runtime leads to the complexity estimates given in Table 1.

Acknowledgements. Player was supported by an ACE-CSR PhD grant. This work has been co-funded by the DFG as part of project P1 within the CRC 1119 CROSSING. We thank Sean Murphy for useful discussions and comments.

References

1. Albrecht, M.R., Cid, C., Faugère, J., Fitzpatrick, R., Perret, L.: Algebraic algorithms for LWE problems. In: IACR Cryptology ePrint Archive 2014, p. 1018 (2014)
2. Albrecht, M.R., Cid, C., Faugère, J., Fitzpatrick, R., Perret, L.: On the complexity of the BKW algorithm on LWE. Des. Codes Crypt. **74**(2), 325–354 (2015)
3. Albrecht, M.R., Faugère, J., Fitzpatrick, R., Perret, L.: Lazy modulus switching for the BKW algorithm on LWE. In: Krawczyk [28], pp. 429–445
4. Albrecht, M.R., Fitzpatrick, R., Göpfert, F.: On the efficacy of solving LWE by reduction to unique-SVP. In: Lee, H.-S., Han, D.-G. (eds.) ICISC 2013. LNCS, vol. 8565, pp. 293–310. Springer, Heidelberg (2014)
5. Albrecht, M.R., Player, R., Scott, S.: On the concrete hardness of learning with errors. J. Math. Cryptology **9**(3), 169–203 (2015)
6. Arora, S., Ge, R.: New algorithms for learning in presence of errors. In: Aceto, L., Henzinger, M., Sgall, J. (eds.) ICALP 2011, Part I. LNCS, vol. 6755, pp. 403–415. Springer, Heidelberg (2011)
7. Babai, L.: On Lovász' lattice reduction and the nearest lattice pointproblem. In: Mehlhorn, K. (ed.) STACS 85. LNCS, vol. 182, pp. 13–20. Springer, Berlin (1985)
8. Babai, L.: On Lovász' lattice reduction and the nearest lattice point problem. Combinatorica **6**(1), 1–13 (1986)
9. Bai, S., Galbraith, S.D.: Lattice decoding attacks on binary LWE. In: Susilo, W., Mu, Y. (eds.) ACISP 2014. LNCS, vol. 8544, pp. 322–337. Springer, Heidelberg (2014)
10. Bai, S., Galbraith, S.D., Li, L., Sheffield, D.: Improved exponential-time algorithms for inhomogeneous-sis. In: IACR Cryptology ePrint Archive 2014, p. 593 (2014)
11. Blum, A., Kalai, A., Wasserman, H.: Noise-tolerant learning, the parity problem, and the statistical query model. J. ACM **50**(4), 506–519 (2003)
12. Brakerski, Z., Vaikuntanathan, V.: Efficient fully homomorphic encryption from(standard) LWE. In: Ostrovsky, R. (eds.) FOCS 2011, pp. 97–106, Palm Springs, CA, USA. IEEE Computer Society , 22–25 October 2011
13. Buchmann, J., Göpfert, F., Player, R., Wunderer, T.: On the hardness of LWE with binary error: revisiting the hybridlattice-reduction and meet-in-the-middle attack. Cryptology ePrint Archive, Report 2016/089 (2016). http://eprint.iacr.org/
14. Canetti, R., Garay, J.A. (eds.): CRYPTO 2013, Part I. LNCS, vol. 8042. Springer, Heidelberg (2013)
15. Chen, H., Lauter, K.E., Stange, K.E.: Attacks on search RLWE. IACR Cryptology ePrint Archive 2015, p. 971 (2015)
16. Chen, Y., Nguyen, P.Q.: BKZ 2.0: better lattice security estimates. In: Lee, D.H., Wang, X. (eds.) ASIACRYPT 2011. LNCS, vol. 7073, pp. 1–20. Springer, Heidelberg (2011)

17. Duc, A., Tramèr, F., Vaudenay, S.: Better algorithms for LWE and LWR. In: Oswald, E., Fischlin, M. (eds.) EUROCRYPT 2015. LNCS, vol. 9056, pp. 173–202. Springer, Heidelberg (2015)

18. Ducas, L., Durmus, A., Lepoint, T., Lyubashevsky, V.: Lattice signatures and bimodal gaussians. In: Canetti, R., Garay, J.A. (eds.) [14], pp. 40–56

19. Eisenträger, K., Hallgren, S., Lauter, K.: Weak Instances of PLWE. In: Joux, A., Youssef, A. (eds.) SAC 2014. LNCS, vol. 8781, pp. 183–194. Springer, Heidelberg (2014)

20. Elias, Y., Lauter, K.E., Ozman, E., Stange, K.E.: Provably weak instances of ring-LWE. In: Gennaro, R., Robshaw, M. (eds.) [21], pp. 63–92

21. Gennaro, R., Robshaw, M. (eds.): CRYPTO 2015. LNCS, vol. 9215. Springer, Berlin (2015)

22. Güneysu, T., Lyubashevsky, V., Pöppelmann, T.: Practical lattice-based cryptography: a signature scheme for embedded systems. In: Prouff, E., Schaumont, P. (eds.) CHES 2012. LNCS, vol. 7428, pp. 530–547. Springer, Heidelberg (2012)

23. Guo, Q., Johansson, T., Stankovski, P.: Coded-BKW: solving LWE using lattice codes. In: Gennaro, R., Robshaw, M. (eds.) [21], pp. 23–42

24. Hirschhorn, P.S., Hoffstein, J., Howgrave-Graham, N., Whyte, W.: Choosing NTRUEncrypt parameters in light of combined lattice reduction and MITM approaches. In: Abdalla, M., Pointcheval, D., Fouque, P.-A., Vergnaud, D. (eds.) ACNS 2009. LNCS, vol. 5536, pp. 437–455. Springer, Heidelberg (2009)

25. Howgrave-Graham, N.: A hybrid lattice-reduction and meet-in-the-middle attack against NTRU. In: Menezes, A. (ed.) CRYPTO 2007. LNCS, vol. 4622, pp. 150–169. Springer, Heidelberg (2007)

26. Kannan, R.: Minkowski's convex body theorem and integer programming. Math. Oper. Res. **12**(3), 415–440 (1987)

27. Kirchner, P., Fouque, P.: An improved BKW algorithm for LWE with applications to cryptography and lattices. In: Gennaro, R., Robshaw, M. (eds.) [21], pp. 43–62

28. Krawczyk, H. (ed.): PKC 2014. LNCS, vol. 8383. Springer, Heidelberg (2014)

29. Laine, K., Lauter, K.E.: Key recovery for LWE in polynomial time. IACR Cryptology ePrint Archive 2015, p. 176 (2015)

30. Langlois, A., Ling, S., Nguyen, K., Wang, H.: Lattice-based group signature scheme with verifier-local revocation. In: Krawczyk, H. (ed.) [28], pp. 345–361

31. Li, S.: Concise formulas for the area and volume of a hyperspherical cap. Asian J. Math. Stat. **4**(1), 66–70 (2011)

32. Lindner, R., Peikert, C.: Better key sizes (and attacks) for LWE-based encryption. In: Kiayias, A. (ed.) CT-RSA 2011. LNCS, vol. 6558, pp. 319–339. Springer, Heidelberg (2011)

33. Liu, M., Nguyen, P.Q.: Solving BDD by enumeration: an update. In: Dawson, E. (ed.) CT-RSA 2013. LNCS, vol. 7779, pp. 293–309. Springer, Heidelberg (2013)

34. Lyubashevsky, V., Peikert, C., Regev, O.: On ideal lattices and learning with errors over rings. J. ACM **60**(6), 43 (2013)

35. Micciancio, D., Peikert, C.: Hardness of SIS and LWE with small parameters. In: Canetti, R., Garay, J.A. (eds.) [14], pp. 21–39

36. Micciancio, D., Regev, O.: Lattice-based cryptography. Springer, Berlin (2009)

37. Micciancio, D., Walter, M.: Practical, predictable lattice basis reduction. IACR Cryptology ePrint Archive 2015, p. 1123 (2015)

38. Olver, F.W.: NIST Handbook of Mathematical Functions. Cambridge University Press, Cambridge (2010)

39. Peikert, C.: Public-key cryptosystems from the worst-case shortest vector problem: extended abstract. In: Mitzenmacher, M. (eds.) STOC 2009, pp. 333–342, Bethesda, MD, USA. ACM, May 31–June 2, 2009

40. Peikert, C., Waters, B.: Lossy trapdoor functions and their applications. SIAM J. Comput. **40**(6), 1803–1844 (2011)

41. Regev, O.: On lattices, learning with errors, random linear codes, and cryptography. In: Gabow, H.N., Fagin, R. (eds.) Proceedings of the 37th Annual ACM Symposium on Theory of Computing, pp. 84–93, Baltimore, MD, USA. ACM, 22–24 May 2005

42. Stein, W., et al.: Sage Mathematics Software (Version 6.3). The Sage Development Team (2014). http://www.sagemath.org

An Efficient Lattice-Based Signature Scheme with Provably Secure Instantiation

Sedat Akleylek[1(✉)], Nina Bindel[2(✉)], Johannes Buchmann[2], Juliane Krämer[2], and Giorgia Azzurra Marson[2]

[1] Ondokuz Mayis University, Samsun, Turkey
sedat.akleylek@bil.omu.edu.tr
[2] Technische Universität Darmstadt, Darmstadt, Germany
{nbindel,buchmann,jkraemer}@cdc.informatik.tu-darmstadt.de,
giorgia.marson@cased.de

Abstract. In view of the expected progress in cryptanalysis it is important to find alternatives for currently used signature schemes such as RSA and ECDSA. The most promising lattice-based signature schemes to replace these schemes are (CRYPTO 2013) and GLP (CHES 2012). Both come with a security reduction from a lattice problem and have high performance. However, their parameters are not chosen according to their provided security reduction, i.e., the *instantiation* is not provably secure. In this paper, we present the first lattice-based signature scheme with good performance when provably secure instantiated. To this end, we provide a tight security reduction for the new scheme from the ring learning with errors problem which allows for provably secure and efficient instantiations. We present experimental results obtained from a software implementation of our scheme. They show that our scheme, when provably secure instantiated, performs comparably with BLISS and the GLP scheme.

Keywords: Lattice-based cryptography · Tightness · Ideal lattices · Signatures · Ring learning with errors

1 Introduction

Electronic signatures are essential for cybersecurity. For example, they provide authenticity proofs for billions of software downloads daily on the Internet. In recent years, lattice-based signatures such as BLISS [23] or the GLP [28] signature scheme have become an interesting alternative to the schemes that are currently being used in practice, like RSA and ECDSA. Providing such alternatives is very important in view of the expected progress in cryptanalysis of RSA and ECDSA, in particular by quantum computers.

The lattice-based signature schemes BLISS and GLP have two important properties. They have *good performance*, i.e., they can compete with RSA and ECDSA. Also, they are *provably secure*: they allow for security reductions from

© Springer International Publishing Switzerland 2016
D. Pointcheval et al. (Eds.): AFRICACRYPT 2016, LNCS 9646, pp. 44–60, 2016.
DOI: 10.1007/978-3-319-31517-1_3

lattice problems that are expected to be hard even in the presence of quantum computers.

Provable security is a very strong security argument. In this paper, we go one step further and present an R-LWE-based signature scheme which has a security property which we consider to be even stronger: *good performance with provably secure instantiation*. By this property we mean that the parameters are chosen according to the security reduction and at the same time allow for good performance. This implies the following: suppose that parameters are constructed for a certain security level. By virtue of the security reduction these parameters correspond to an instance of the ring learning with errors problem (R-LWE). Since the parameters were chosen according to the security reduction, this reduction provably guarantees that our scheme has the selected security level as long as the corresponding R-LWE instance is intractable. In other words, hardness statements for R-LWE instances have a provable consequence for the security levels of our scheme. Currently, both BLISS and GLP do not allow for good performance and provably secure instantiation at the same time. Choosing parameters according to the security reductions for these schemes reduces their performance significantly (see for example [11,17]).

We note that our scheme has another potential advantage over BLISS. BLISS uses Gaussian sampling, which is generally assumed to be vulnerable to timing attacks [14,20], while GLP and our scheme use uniform sampling during signature generation which appears to not have this vulnerability.

Our signature scheme is based on the design of Bai and Galbraith [10] and its optimizations by Dagdelen *et al.* [20]. The reason why our scheme allows for good performance with provably secure instantiation is that we are able to give a tight security reduction from the R-LWE problem to our scheme. The proof of this result is an optimized adaption of the tightness proof in [6] to the R-LWE setting which allows for better tightness bounds. To demonstrate that our scheme has good performance, we present experimental results which are based on a software implementation. These results show that our scheme, when provably secure instantiated, performs comparably with BLISS and the GLP scheme without provably secure instantiation.

Related Work. The first lattice-based signature scheme with tight security reduction is the GPV signature scheme [27]. Its instantiations are provably secure, but not efficient. Most of the recent lattice-based signature schemes [10,20,23, 28,38] come neither with a tight reduction nor with provably secure instantiation. The security of all those schemes was proven by applying the powerful Forking Lemma [41], which inherently results in a non-tight security reduction.

Abdalla *et al.* [1] circumvent the Forking Lemma and use an approach inspired by the proof idea introduced by Katz and Wang [32]. However, their tight reduction demands an impractically large choice of the modulus. Recently, Alkim *et al.* [6] also used the approach by Katz and Wang [32] to provide a tight security reduction from the learning with errors problem over standard lattices (LWE) to an improved variant of the Bai-Galbraith signature scheme [10,20].

Instantiations of their scheme are provably secure, but they yield larger key sizes and worse run times than the BLISS and GLP signature scheme.

Organization. After stating notations and definitions in Sect. 2, we describe the signature scheme in Sect. 3. In Sect. 4, we analyze the hardness of R-LWE and we explain the derivation of the parameter sets. Our implementation is described in Sect. 5. We give our experimental results and compare them with BLISS and GLP in Sect. 6.

2 Preliminaries

2.1 Notation

Let $k \in \mathbb{N}$. Throughout this paper we define $n = 2^k \in \mathbb{N}$. Let $q \in \mathbb{N}$ be a prime with $q = 1 \pmod{2n}$. We denote by $\mathbb{Z}_q$ the finite field $\mathbb{Z}/q\mathbb{Z}$ and identify an element in $\mathbb{Z}_q$ with its representative in $(-\lfloor q/2 \rfloor, \lceil q/2 \rceil]$, and we write $\pmod{q}$ to denote the unique representative in $\mathbb{Z}_q$. We define the ring $\mathcal{R} = \mathbb{Z}[x]/(x^n + 1)$ and denote the set of its units by $\mathcal{R}^\times$. Further, we define $\mathcal{R}_q = \mathbb{Z}_q[x]/(x^n + 1)$, $\mathcal{R}_{q,[B]} = \{\sum_{i=0}^{n-1} a_i x^i \in \mathcal{R}_q \mid i \in [0, n-1], a_i \in [-B, B]\}$ for $B \in [0, q/2]$, and $\mathbb{B}_{n,\omega} = \{\mathbf{v} \in \{0,1\}^n \mid ||\mathbf{v}||^2 = \omega\}$ for $\omega \in [0, n]$. We denote polynomials by lower case letters (e.g., p) and (column) vectors by bold lower case letters (e.g., $\mathbf{v}$). Without further mentioning, we use the symbol $\mathbf{p}$ to denote the coefficient vector of a polynomial p. We denote matrices by bold upper case letters (e.g., $\mathbf{M}$) and the transpose of a matrix $\mathbf{M}$ by $\mathbf{M}^T$. We indicate the Euclidean norm of a vector $\mathbf{v} \in \mathbb{R}^n$ by $||\mathbf{v}||$. All logarithms are in base 2.

Rounding Operators. Let $d \in \mathbb{N}$ and $c \in \mathbb{Z}$. We denote by $[c]_{2^d}$ the unique representative of c modulo 2^d in the set $(-2^{d-1}, 2^{d-1}] \subset \mathbb{Z}$. Let $\lfloor \cdot \rceil_d$ be the rounding operator defined as $\lfloor \cdot \rceil_d : \mathbb{Z} \to \mathbb{Z}, c \mapsto (c - [c]_{2^d})/2^d$. We naturally extend these definitions to vectors and polynomials by applying $\lfloor \cdot \rceil_d$ and $[\cdot]_{2^d}$ to each component of the vector and to each coefficient of the polynomial, respectively. We abbreviate $\lfloor v \pmod{q} \rceil_d$ by $\lfloor v \rceil_{d,q}$.

Algorithms and Distributions. If $\mathcal{A}$ is a randomized algorithm we denote by $y \leftarrow \mathcal{A}(x)$ the output of $\mathcal{A}$ on input x and randomly chosen (internal) coins. For an oracle $\mathcal{O}$ we write $\mathcal{A}^{\mathcal{O}}$ to indicate that $\mathcal{A}$ has access to that oracle. Let $\sigma \in \mathbb{R}_{>0}$. The centered discrete Gaussian distribution $\mathcal{D}_\sigma$ on $\mathbb{Z}$ with standard deviation σ is defined as follows: for every $z \in \mathbb{Z}$ the probability of z is given by $\rho_\sigma(z)/\rho_\sigma(\mathbb{Z})$, where $\rho_\sigma(z) = \exp(\frac{-z^2}{2\sigma^2})$ and $\rho_\sigma(\mathbb{Z}) = 1 + 2\sum_{z=1}^{\infty} \rho_\sigma(z)$. We denote by $d \leftarrow \mathcal{D}_\sigma$ the operation of sampling an element d with Gaussian distribution $\mathcal{D}_\sigma$. When writing $\mathbf{v} \leftarrow \mathcal{D}_\sigma^n$ we mean sampling each component of the vector $\mathbf{v}$ with Gaussian distribution. To simplify the notation we indicate sampling all coefficients of a polynomial $a \in \mathcal{R}$ with Gaussian distribution by $a \leftarrow \mathcal{D}_\sigma^n$ as well. Similarly, for a finite set S we write $s \leftarrow \mathcal{U}(S)$, or simply $s \leftarrow_\$ S$, to indicate that an element s is sampled uniformly at random from S.

Lattices and Gaussian Heuristic. Let $n \geq k > 0$. A k-dimensional lattice Λ is a discrete additive subgroup of $\mathbb{R}^n$ containing all integer linear combinations of k linearly independent vectors $\{\mathbf{b}_1, \ldots, \mathbf{b}_k\} = \mathbf{B}$, i.e., $\Lambda = \Lambda(\mathbf{B}) = \{\mathbf{Bx} \mid \mathbf{x} \in \mathbb{Z}^k\}$. The determinant of a lattice is defined by $\det(\Lambda(\mathbf{B})) = \sqrt{\det(\mathbf{B}^\top \mathbf{B})}$.

Throughout this paper we are mostly concerned with q-ary lattices. $\Lambda \in \mathbb{Z}^n$ is called a *q-ary* lattice if $q\mathbb{Z} \subset \Lambda$ for some $q \in \mathbb{Z}$. Let $\mathbf{A} \leftarrow_\$ \mathbb{Z}_q^{m \times n}$. We define the q-ary lattices $\Lambda_q^\perp(\mathbf{A}) = \{\mathbf{x} \in \mathbb{Z}^n \mid \mathbf{Ax} = \mathbf{0} \pmod{q}\}$ and $\Lambda_q(\mathbf{A}) = \{\mathbf{x} \in \mathbb{Z}^n \mid \exists \mathbf{s} \in \mathbb{Z}^m \text{ s.t. } \mathbf{x} = \mathbf{A}^\top \mathbf{s} \pmod{q}\}$. Furthermore, for $\mathbf{u} \in \mathbb{Z}_q^m$ we define cosets $\Lambda_{\mathbf{u},q}^\perp(\mathbf{A}) = \{x \in \mathbb{Z}^n \mid \mathbf{Ax} = \mathbf{u} \pmod{q}\}$, i.e., $\Lambda_{\mathbf{u}}^\perp(\mathbf{A}) = \Lambda_{\mathbf{0},q}^\perp(\mathbf{A})$. One can consider $\Lambda_{\mathbf{u},q}^\perp(\mathbf{A})$ as a *shifted lattice* by a vector $\mathbf{u}$, i.e., $\Lambda_{\mathbf{u},q}^\perp(\mathbf{A}) = \Lambda_q^\perp(\mathbf{A}) + \mathbf{y}$ where $\mathbf{y} \in \mathbb{Z}^m$ is an integer solution of $\mathbf{Ax} = \mathbf{u} \pmod{q}$.

Let S be a measurable set and let $\Lambda \subset \mathbb{Z}^n$ be a lattice. The *Gaussian heuristic* approximates the number of lattice points in the set S by $|S \cap \Lambda| = \frac{vol(S)}{det(\Lambda)}$.

2.2   The Learning with Errors Problem Over Rings

Given the isomorphism $\Phi_q : \mathbb{Z}_n \rightarrow \mathcal{R}_q$ with $(a_0, \ldots, a_{n-1}) \mapsto a_0 + a_1 x + \ldots + a_{n-1}x^{n-1}$, $\mathcal{R}_q$ is isomorphic to $\mathbb{Z}_q^n$ as a $\mathbb{Z}$-module. Therefore, we can identify a polynomial $a = a_0 + a_1 x + \ldots + a_{n-1}x^{n-1} \in \mathcal{R}_q$ with its coefficient vector $\mathbf{a} = (a_0, \ldots, a_{n-1})^T$. We define the rotation of a vector $\mathbf{a} = (a_0, \ldots, a_{n-1})^T$ to be the coefficient vector of $ax \in \mathcal{R}_q$, i.e., $\mathrm{rot}(\mathbf{a}) = (-a_{n-1}, a_0, \ldots, a_{n-2})^T$. Furthermore, we define the rotation matrix of a polynomial a as $\mathrm{Rot}(a) = (\mathbf{a}, \mathrm{rot}(\mathbf{a}), \mathrm{rot}^2(\mathbf{a}), \ldots, \mathrm{rot}^{n-1}(\mathbf{a})) \in \mathbb{Z}_q^{n \times n}$. Polynomial multiplication of $a, b \in \mathcal{R}_q$ is equivalent to the matrix-vector multiplication $\mathrm{Rot}(a)\mathbf{b}$ in $\mathbb{Z}_q$. It can be easily shown that $a \in \mathcal{R}_q$ is invertible, i.e., $a \in \mathcal{R}_q^\times$, if and only if $\mathrm{rank}(\mathrm{Rot}(a)) = n$.

We define the learning with errors distribution and the ring learning with errors problem (R-LWE) in the following.

Definition 1 (Learning with Errors Distribution). *Let $n, q > 0$ be integers, $s \in \mathcal{R}_q$, and χ be a distribution over $\mathcal{R}$. We define by $\mathcal{D}_{s,\chi}$ the R-LWE distribution which outputs $(a, \langle a, s \rangle + e) \in \mathcal{R}_q \times \mathcal{R}_q$, where $a \leftarrow_\$ \mathcal{R}_q$ and $e \leftarrow \chi$.*

Since our signature scheme is based on the decisional R-LWE problem, we omit the definition of the search version and state only the decisional learning with errors problem.

Definition 2 (Ring Learning with Errors Problem). *Let $n, q > 0$ be integers and $q = 2^k$ for some $k \in \mathbb{N}_{>0}$ and χ be a distribution over $\mathcal{R}$. Moreover, define $\mathcal{O}_\chi$ to be an oracle, which upon input polynomial $s \in \mathcal{R}_q$ returns samples from the learning with errors distribution $\mathcal{D}_{s,\chi}$. The ring learning with errors problem R-LWE$_{n,m,q,\chi}$ is (t, ε)-hard if for any probabilistic polynomial time (PPT) algorithm $\mathcal{A}$, running in time t and making at most m queries to its oracle, it holds that*

$$Adv_{n,q,\chi}^{\text{R-LWE}}(\mathcal{A}) = \left| \Pr\left[\mathcal{A}^{\mathcal{O}_\chi(s)(\cdot)} = 1 \right] - \Pr\left[\mathcal{A}^{\mathcal{U}(\mathbb{Z}_q^n \times \mathbb{Z}_q)(\cdot)} = 1 \right] \right| \leq \varepsilon,$$

where the probabilities are taken over the random choices of $s \leftarrow \mathcal{U}(\mathcal{R}_q)$, the random choice of the distribution $\mathcal{D}_{s,\chi}$, as well as the random coins of $\mathcal{A}$.

The R-LWE assumption comes with a worst-case to average-case reduction to problems over ideal lattices [39]. Furthermore, it was shown in [7] that the learning with errors problem remains hard if one chooses the secret distribution to be the same as the error distribution. We write R-LWE$_{n,m,q,\sigma}$ if χ is the discrete Gaussian distribution with standard deviation σ.

3   Description and Security of the Signature Scheme

In this section, we present our signature scheme and we prove it to be unforgeable against a chosen-message attack—shortly ufcma-secure (cf. Appendix A, Fig. 3)—as long as R-LWE is computationally hard. We recall basic definitions and notations about signatures schemes in Appendix A. We name our scheme ring-TESLA since it is based on the signature scheme TESLA by Alkim et $al.$ [6].

Our signature scheme is parametrized by the integers $n \in \mathbb{N}_{>0}$, ω, d, B, q, U, L, κ, and the security parameter λ with $n > \kappa > \lambda$, by the Gaussian distribution $\mathcal{D}_\sigma$ with standard deviation σ, by the hash function $H : \{0,1\}^* \rightarrow \{0,1\}^\kappa$, and by the encoding function $F : \{0,1\}^\kappa \rightarrow \mathbb{B}_{n,\omega}$. The encoding function F takes the (binary) output of the hash function H and maps it to a vector of length n and weight ω. For more information about the encoding function see [28]. Furthermore, let $a_1, a_2 \in \mathcal{R}_q^\times$ be two uniformly sampled polynomials which are publicly known as global constants. They can be shared among arbitrary many signers.

The secret key sk consists of three small polynomials s, e_1, and e_2; the public key pk is given by two polynomials $t_1 = a_1 s + e_1$ and $t_2 = a_2 s + e_2$. To ensure that signatures are short and verified correctly, we use a procedure checkE similar to the one introduced by Dagdelen et $al.$ [20]. Let $max_k(\cdot)$ be a function that takes as input a vector and returns its k-th largest entry. The key polynomials e_1, e_2 are rejected during checkE if $\sum_{k=1}^\omega max_k(\mathbf{e_i})$ is greater then L for at least one of e_1 or e_2. Otherwise e_1, e_2 are accepted. To sign a message μ, first a random polynomial $y \in \mathcal{R}_{q,[B]}$ is chosen. Afterwards, the most significant bits of $a_1 y$ and $a_2 y$ and the message are hashed to a value c. The signature of μ consists of the hash value c and the polynomial $z = sc + y$. To hide the secret, rejection sampling is applied. For verification of the signature (c, z), the size of z and the equality of c and $H(\lfloor a_1 z - t_1 c \rceil_d, \lfloor a_2 z - t_2 c \rceil_d, \mu)$ is checked. The signature scheme ring-TESLA is depicted in detail in Fig. 1. We present parameter sets in Table 1 and their derivation in Sect. 4.

In our security reduction we follow an idea introduced by Katz and Wang [32] that can be summarized at follows: assume there exists an algorithm $\mathcal{A}$ that forges a signature given a valid public key, i.e., an LWE tuple. In contrast, given a random key $\mathcal{A}$ forges a signature only with very small probability. Hence, the security reduction distinguishes whether its own challenge tuple is an LWE tuple or not by the different behavior of the algorithm $\mathcal{A}$.

Theorem 1. *Let* $n, \omega, d, B, q, U, L,$ *and* σ *be arbitrary parameters satisfying the constraints described in Sect. 4. Assume that the Gaussian heuristic holds for*

KeyGen($1^\lambda; a_1, a_2$) :

1 $s, e_1, e_2 \leftarrow D_\sigma^n$
2 If checkE(e_1) = 0 $\vee$ checkE(e_2) = 0
3 $\quad$ Restart
4 $t_1 \leftarrow a_1 s + e_1 \pmod{q}$
5 $t_2 \leftarrow a_2 s + e_2 \pmod{q}$
6 sk $\leftarrow (s, e_1, e_2)$
7 pk $\leftarrow (t_1, t_2)$
8 Return (sk, pk)

Sign($\mu; a_1, a_2, s, e_1, e_2$) :

9 $y \leftarrow_\$ \mathcal{R}_{q,[B]}$
10 $v_1 \leftarrow a_1 y \pmod{q}$
11 $v_2 \leftarrow a_2 y \pmod{q}$
12 $c' \leftarrow H\left(\lfloor v_1 \rceil_{d,q}, \lfloor v_2 \rceil_{d,q}, \mu\right)$
13 $c \leftarrow F(c')$
14 $z \leftarrow y + sc$
15 $w_1 \leftarrow v_1 - e_1 c \pmod{q}$
16 $w_2 \leftarrow v_2 - e_2 c \pmod{q}$
17 If $|[w_1]_{2^d}|, |[w_2]_{2^d}| \notin \mathcal{R}_{2^d - L} \vee z \notin \mathcal{R}_{B-U}$:
18 $\quad$ Restart
19 Return (z, c')

Verify($\mu; z, c'; a_1, a_2, t_1, t_2$)

19 $c \leftarrow F(c')$
20 $w_1' \leftarrow a_1 z - t_1 c \pmod{q}$
21 $w_2' \leftarrow a_2 z - t_2 c \pmod{q}$
22 $c'' \leftarrow H\left(\lfloor w_1' \rceil_{d,q}, \lfloor w_2' \rceil_{d,q}, \mu\right)$
23 If $c' = c'' \wedge z \in \mathcal{R}_{B-U}$:
24 $\quad$ Return 1
25 Else: Return 0

Fig. 1. Specification of the signature scheme ring-TESLA

lattice instances defined by the parameters above. For every ufcma-*adversary* $\mathcal{A}$ *that runs in time* $t_\mathcal{A}$, *asks at most* q_s *and* q_h *queries to the signing oracle and the hash oracle, respectively, and forges a valid signature of the signature scheme* ring-TESLA *with probability* $\varepsilon_\mathcal{A}$, *there exists a distinguisher* $\mathcal{D}$ *that runs in time* $t_\mathcal{D} = t_\mathcal{A} + \mathcal{O}(q_s \kappa^2 + q_h)$ *and breaks the* R-LWE$_{n,2,q,\sigma}$ *problem (in the random oracle model) with success probability*

$$\varepsilon_\mathcal{D} \geq \varepsilon_\mathcal{A}\left(1 - \frac{q_s q_h 2^{(d+1)2n}}{(2B+1)^n q^n}\right) - \frac{q_h 2^{dn}(2B - 2U + 1)^n + (28\sigma + 1)^{3n}}{q^{2n}}.$$

Proof sketch. We show how to turn any successful forger $\mathcal{A}$ against the signature scheme ring-TESLA into a distinguisher $\mathcal{D}$ for the R-LWE problem. The distinguisher obtains two R-LWE samples from its sampling oracle $\mathcal{O}_\chi(s)$ (cf. Definition 2) and embeds them into a public key pk. Thus, $\mathcal{D}$ simulates the ufcma game (cf. Fig. 3, Appendix A). When $\mathcal{A}$ returns a forgery (μ, σ), $\mathcal{D}$ checks whether σ is a valid signature for message μ under key pk: if so, it outputs 1 as a guess that $\mathcal{O}_\chi(s)$ presented two R-LWE tuples, otherwise it outputs 0.

To derive the explicit relation between $\mathcal{D}$ and $\mathcal{A}$'s success probabilities $\varepsilon_{\mathcal{D}}$ and $\varepsilon_{\mathcal{A}}$ as indicated in the theorem statement, we show that (i) $\mathcal{D}$ provides a good simulation of the ufcma game for $\mathcal{A}$. In particular, we show that the simulated signatures look like genuine ones. And we prove, (ii) $\mathcal{D}$'s simulation does not abort too often. Formal proofs of both facts, (i) and (ii), require several technical lemmas that we state and prove in the full version of this paper [2]. For proving fact (i), we observe that $\mathcal{D}$ simulates signatures $\sigma = (z, c')$ by choosing z and c' uniformly at random from appropriate spaces. By applying rejection sampling and the fact that c' is the output of a random oracle, we show that simulated signatures are statistically indistinguishable from genuine ones. Concerning fact (ii), we first note that $\mathcal{D}$'s signing simulation needs to program the random oracle H, which may lead to inconsistencies in case one of $\mathcal{A}$'s signature requests results in programming a hash value $H(x)$ for which x was already queried. Such an occurrence causes a premature termination of the simulation. In [2], we prove that the latter happens only with small probability.   □

As described in [24, Sect. 3.3], the probability that a polynomial chosen uniformly random in $\mathcal{R}_q$ is in the subset of multiplicative invertible elements of $\mathcal{R}_q$ is given by $\Pr\left[a \in \mathcal{R}_q^{\times}\right] = (1 - 1/q)^n$, where the probability is taken over random choices of $a \leftarrow_{\$} \mathcal{R}_q$. This probability is overwhelming for our choices of q and n in the signature scheme presented in this paper. Thus, it is justified to sample the polynomials a_1 and a_2 uniformly random in $\mathcal{R}_q^{\times}$ instead of $\mathcal{R}_q$ as defined in the R-LWE problem.

Relation to Former Security Reductions. The scheme ring-TESLA is based on the signature scheme by Bai and Galbraith [10] with a tight security reduction by Alkim *et al.* [6]. Essentially, we convert the scheme by Bai and Galbraith to a scheme over ideal lattices. Our security reduction follows the proof strategy of [6]. We emphasize that lifting the security statements for the original (lattice-based) scheme to our (ideal lattice-based) scheme is not trivial. For example, it is unclear whether distributions remain the same when lemmata are applied on rotation matrices instead of matrices chosen uniformly random; in some cases even improvements can be made. Indeed, we could sharpen the bound given in [6, Lemma 2]. Our corresponding result is stated in the full version of this paper [2]. Moreover, we formulate and prove a similar lemma to [10, Lemma 3] for ideal lattices and we state explicitly which property related to the Gaussian heuristic is necessary to prove the statement. Likewise, Bai and Galbraith make use of the Gaussian heuristic in their corresponding proof. The methods used in our security reduction resemble those formalized by Abdalla *et al.* [1]. Abdalla *et al.* define four properties of identification schemes for which they give a black-box-transformation to signature schemes with tight security reduction. Applying their black-box-transformation to a lattice-based signature scheme led to inefficiently large parameters as stated by the authors [1]. Hence, we prove unforgeability of ring-TESLA more directly— without passing through an intermediate identification scheme—by following the proof technique introduced by Katz and Wang [32]. This yields practical instantiation as we show in Sect. 4.

4 Selecting Parameters

The reductionist approach to prove security of a given cryptosystem essentially consists in building an efficient reduction that turns any successful adversary against the cryptosystem into one that solves some computationally hard problem. The hardness of breaking the cryptosystem and of solving the underlying problem are often expressed asymptotically. When a scheme is to be deployed in the real world, however, for a security analysis to be realistic it is essential that run times and success probabilities are estimated in a more explicit way. Moreover, given a (concrete and) tight security reduction, the security of the scheme is about the same as the hardness of the underlying computational assumption when the scheme is instantiated according to the reduction. In contrast, if only a non-tight reduction is available, larger security parameters shall be used in order to achieve a specific level of security. As a consequence, it is often hard to tell whether a provably secure scheme with a non-tight reduction effectively provides the claimed level of security and performance.

In this section, we propose our choice of provably secure parameters for different levels of bit-security for the signature scheme presented in this paper and we explain how we estimate the hardness of the ring learning with errors problem.

4.1 Derivation of Parameters for Different Security Levels

The security reduction given in Sect. 3 provides a *tight* reduction to the hardness of R-LWE and bounds *explicitly* the forging probability with the success probability of the reduction. More formally, let $\varepsilon_\mathcal{A}$ and $t_\mathcal{A}$ denote the success probability and the runtime of a forger $\mathcal{A}$ against our signature scheme and let $\varepsilon_\mathcal{D}$ and $t_\mathcal{D}$ denote analogous quantities for the reduction $\mathcal{D}$ presented in the proof of Theorem 1. We can write the explicit relations $\varepsilon_\mathcal{D} \geq c_1 \varepsilon_\mathcal{A} + c_2$ and $t_\mathcal{D} \leq c_3 t_\mathcal{A} + c_4$, where c_1, c_2, c_3, c_4 are constants which are fixed for a concrete instantiation of the signature scheme. We say that R-LWE is *n-bit hard* if $t_\mathcal{D}/\varepsilon_\mathcal{D} \geq 2^n$; similarly, we say that the signature scheme is *m-bit secure* if $t_\mathcal{A}/\varepsilon_\mathcal{A} \geq 2^m$.

Given an explicit security reduction and the assumed bit-hardness of R-LWE, we can compute the bit-security of the signature scheme. In our case, we instantiate the signature scheme such that the constants c_1, c_2, and c_3 are less than $2^{-\lambda}$. Thus, the bit-hardness of the R-LWE instance is the same as the bit-security of our signature instantiated as described below. To ensure both correctness and security of our signature, the following dependencies must hold.

Let λ be the security parameter. We choose a hash function $H : \{0,1\}^* \to \{0,1\}^\kappa$ with $\kappa > \lambda$ to ensure that the hash function gives at least a bit-hardness of λ. We instantiate the hash function for our parameter sets with SHA-256. Furthermore, security relies on the encoding function $F : \{0,1\}^\kappa \to \mathbb{B}_{n,\omega}$. Following Bai and Galbraith [10], we require F to be close to an injective function. That means that the probability of mapping two different values to the same output is smaller than or equal to $2^{-\lambda}$. We choose ω such that $2^\kappa \geq |\mathbb{B}_{n,\omega}| = 2^\omega \binom{n}{\omega}$. To use efficient polynomial multiplication, i.e., the number theoretic transform (NTT) in the ring $\mathcal{R}_q$, we restrict ourselves to a polynomial degree of a power of 2, i.e., $n = 2^k$ for $k \in \mathbb{N}$. Choosing the Gaussian parameter σ, we can compute the system parameters to give a concrete instantiation of ring-TESLA with λ-bit security.

To apply rejection sampling we choose $U = 14\sqrt{\omega}\sigma$ and $B \geq 14(n-1)\sqrt{\omega}\sigma$. The rejection probability is given by $M = \left(\frac{2(B-U)+1}{2B+1}\right)^n$. We select the rounding value d to be larger than $\log(B)$ and such that the acceptance probability in the first part of Step 17 in Fig. 1 is greater than or equal to 0.4, i.e., $(1 - 2L/2^d)^m \geq 0.4$. The bound L is important during the key generation as well as during the sign procedure. We choose L such that we reject only very few of the possible key pairs in checkE. For example, we achieve an acceptance probability of almost 100 % in KeyGen and an acceptance probability of 0.34 in Sign for parameter ring-TESLA-II. At last, the modulus q has to be greater than or equal to $\left(\frac{2^{(d+1)2n+\kappa}}{(2B)^n}\right)^{1/n}$ and greater than or equal to $4B$. The theoretical size of the secret key is given by $3n\lceil\log(14\sigma)\rceil$ bits. The public key is represented by $2n\lceil\log(q)\rceil$ bits and the length of the signature is $n\lceil\log(2B-2U)\rceil + \kappa$ bits. Given the concrete instantiations in Table 1, we get a signature size of 1,488 byte, a public key size of 3,328 byte, and a secret key size of 1,920 byte for parameters chosen such that the signature scheme is 128-bit secure. In Table 1 we also propose instantiations for 80 bit of security. For comparison, we depict our signature and key sizes together with the corresponding values of BLISS [23] and the GLP [28] signature scheme in Table 2.

4.2   Hardness Estimation of the R-LWE Problem

Since the introduction of the learning with errors problem over rings [39], it is an open question whether the R-LWE is as hard as the LWE problem. Recently,

Table 1. Parameter sets for our signature scheme in comparison; the hardness of the LWE instance is defined by the dimension n, the modulus q, and the Gaussian parameter σ; derivation of L, ω, B, U, d is explained in Sect. 4.1; pk and sk denote the public and private key, resp.

		Parameter selection							
Parameter Set	Security (bit)	n	σ	L	ω	B	U	d	q
ring-TESLA-I	80	512	30	814	11	$2^{21}-1$	993	21	8399873
ring-TESLA-II	128	512	52	2766	19	$2^{22}-1$	3173	23	39960577

		Acceptance prob.		pk Size	sk Size	Signature Size
		KeyGen	Sign	(byte)	(byte)	(byte)
ring-TESLA-I	80	0.5	0.23	3,072	1,728	1,418
ring-TESLA-II	128	0.99	0.34	3,328	1,920	1,488

the cyclic structure of ideal lattices has been exploited by Garg et al. [26], by Campbell et al. [16], by Cramer et al. [19], and by Elias et al. [25]. However, up to now, these novel results are not known to be directly applicable to most of the proposed ideal-lattice-based signature schemes. Hence, as the R-LWE problem can be seen as an instantiation of the LWE problem, we estimate the hardness of R-LWE via state-of-the-art attacks against LWE. We explain four basic attacks on LWE: the embedding approach, the decoding attack, the algorithm by Blum, Kalai, and Wassermann [13], and the Arora-Ge-Algorithm [8]. We briefly describe the algorithms next. The most efficient practical approaches to solve LWE are the embedding approach and the decoding attack.

During the *decoding attack*, an LWE instance $(\mathbf{A}, \mathbf{As}+\mathbf{e})$ is seen as an instance of the bounded distance decoding problem (BDD). The idea of the attack is to reduce the lattice by algorithms such as the BKZ algorithm [18] first, and to find the closest lattice vector to a target vector via the nearest plane algorithm by Babai [9] (or improved variants such as by Linder and Peikert [36] or Liu and Nguyen [37]) afterwards. The closest vector corresponds to $\mathbf{As}$ of the LWE instance, such that the secret can be easily discovered.

The *embedding approach* is to solve an LWE instance by reducing it to an instance of the (unique) shortest vector problem. There are different ways to define a lattice that contains the error term of an LWE instance (e.g., [4,10,12]). In the end, the short error term is found as a shortest vector of the constructed lattice via basis reductions such as BKZ [18] and LLL [18,35], or directly via sieving algorithms [34,40] or enumeration [5]. Recent results [15,31,42] exploit the cyclic structure of ideal lattices to improve sieving algorithms. However, the improved sieving algorithms are still slower than the enumeration approach on instances currently used for signatures.

Further, there are two non-lattice approaches to solve LWE, namely the attack based on the algorithm by *Blum, Kalai,* and *Wassermann* (BKW) [13] and the algorithm by *Arora and Ge* [8]. Both algorithms require a (very) large number of LWE samples to be applied efficiently. Although the number of required samples was crucially reduced, for both BKW [21,30,33] and the Arora-Ge algorithm [3], our proposed instances give far less LWE samples than required for the attacks. Hence, we only take the decoding attack and the embedding approach into account when estimating the bit-security of our instances.

We estimate the hardness of our chosen LWE instances based on [4,10,36]. We propose parameters for two different levels of security: 80-bit security (ring-TESLA-I) and 128-bit security (ring-TESLA-II). The embedding attack yields 166 bit of security and the result of the decoding attack is a bit security of 139 on the instances in ring-TESLA-II.

5   Software Implementation

The implementation of the proposed scheme targets the Intel Haswell micro architecture. We perform benchmarks on a machine with an Intel Core i7-5820K (Haswell) CPU at 3300 MHz and 16 GB of RAM. In our software we use the benefits of AVX2 instructions, where multiplication, addition, and subtraction instructions have one cycle throughput for eight doubles. The software is compiled with gcc-4.7 with optimization code. The experimental results are obtained by using gcc-4.7 with "-Ofast" optimization since it enables all "-O3" optimizations together with turning on "-ffast-math". This optimization helps us to reduce the timing results significantly. The performance of our implementation mainly depends on the number of rejections during Sign and KeyGen and on the time a single polynomial multiplication takes. The derivation of the number of rejections is explained in Sect. 4.1. We optimized the time for multiplication by choosing the most suitable multiplication algorithm for different cases as it is explained below.

Polynomial Multiplication. In the presented scheme two types of polynomial multiplication occur: standard and sparse polynomial multiplication. For standard polynomial multiplication we use the number theoretic transform (NTT) since NTT performs polynomial multiplication with a quasilinear complexity, i.e., $O(n \log n)$. Thus, the parameter sets are selected in such a way that NTT is applicable, i.e., $q = 1 \pmod{2n}$, where n is a power of 2. In our implementation, we store the integer in double format in a word. Then, after arithmetic operations in NTT, it is expected to fit in a double, i.e., $\log(\log(n)q) + \log(q) < 54$. To avoid an overflow one needs to make extra reduction operations when using ring-TESLA-II because $\log(q)$ is represented by 26 bits. This, of course, results in a drawback of the performance. NTT with extra modulo q reduction would need almost 28383 cycles for $n = 512$ and $\omega = 19$ as chosen in ring-TESLA-II. Without extra reductions, the average cycle count of NTT developed for ring-TESLA-I is 10625. Barrett reduction is preferred over reducing the coefficients because of the modular structure. The hybrid approach of using NTT and sparse polynomial multiplication requires more inverse NTT operations since sparse polynomial multiplication is applicable only in the integer domain.

Recall that the weight of c, i.e., the number of 1's, is ω. Then, the multiplication operations in the signature generation phase (Step 14, 15, and 16: sc, e_1c, and e_2c) and in the signature verification phase (Step 20 and 21: t_1c and t_2c) can be considered as sparse polynomial multiplications because of the number of nonzero elements in c. In order to speed up, we use the sparse polynomial multiplication given in Fig. 2. The complexity of the algorithm in Fig. 2 depends on the nonzero coefficients of $b(x)$. Note that polynomial multiplication is performed by using only additions if one of the multiplicands is sparse. The required number of additions and subtractions is $(\omega n + n)$. The last for-loop is designed for polynomial reduction modulo $x^n + 1$. There is only one reduction modulo q of the coefficients. This improves the runtime and complexity. Sparse multiplication requires almost 3650 cycles.

We place our implementation of ring-TESLA in public domain. It can be found under https://www.cdc.informatik.tu-darmstadt.de/cdc/personen/nina-bindel.

Input: array $d = [i_1, ..., i_\omega]$, poly $a(x) = \sum_{i=0}^{n-1} a_i x^i$, poly $b(x) = \sum_{i=0}^{n-1} b_i x^i$; with $a_i \in \mathbb{Z}_q$, $b_i \in \{0, 1\}$, $d[k] = i_k$ such that $b_{i_k} = 1$
Output: poly $c(x) = a(x)b(x)$

1 Set all coefficients of $c(x)$ to 0
2 for $i = 0, ..., \omega - 1$:
3   for $j = 0, ..., n - 1$:
4      $c_{j+d[i]} \leftarrow c_{j+d[i]} + a_j$
5 for $i = 0, ..., n - 1$
6    $c_i \leftarrow c_i - c_{i+n} \pmod{p}$
7 Return $c(x)$

Fig. 2. Sparse polynomial multiplication

6 Performance Analysis

We performed our benchmarks on a machine with an Intel Core i7-5820K (Haswell) CPU at 3300 MHz and 16 GB of RAM, while disabling Turbo Boost and hyper threading. In our measurement we considered two parameter sets: ring-TESLA-I and ring-TESLA-II with 80 and 128 bits of security, respectively. Our benchmarks are averaged[1] over 10,000 runs of Sign and Verify. We summarize benchmarks for our proposed parameter sets and state-of-the-art ideal-lattice-based signature schemes in Table 2. We emphasize once more that our parameter sets are the only ones in Table 2 which are chosen according to the given security reduction, cf. Sect. 4. Nevertheless, we achieve good performance with respect to time and space. In the following, we compare sizes and run times for 80 and 128 bits of security.

For *low security of 80-bit*, key and signature sizes of GLP-I are smaller than those of our proposed parameters. Our run time of Sign is a factor of 1.19 faster than GLP. As Table 2 indicates, the software implementations of ring-TESLA and of the GLP signature scheme are optimized for micro architectures. For *medium security of 128-bit* the instantiation of our scheme gives smallest key sizes. Signature sizes are comparably good. We emphasize that we report the signature size used in the publicly available software implementation of BLISS-I and BLISS-II[2]. Those sizes differ from the theoretical signature sizes presented in [23], which are 700 and 625 bytes for BLISS-I and BLISS-II, respectively, because signatures are not compressed in the BLISS software. To our knowledge, there is no implementation of BLISS available that compresses the signature sizes. The signature size of ring-TESLA are also obtained from our implementation.

---

[1] Sometimes benchmarks are given as the median instead of the average value. Due to the rejection sampling, taking the median value of our experiments would be overly optimistic for Sign.

[2] bliss.di.ens.fr.

Table 2. Comparison of our results with the software implementations of the signature schemes BLISS [22,23] and GLP [20,28,29]. To indicate the considered platforms Intel Core i5-3210M (Ivy Bridge), Intel Core i7-5820K (Haswell), and Intel Core 3.4 GHz we use shortcuts A, B, and C, respectively. Sizes of signatures, signing and verification keys are indicated in Bytes. We abbreviate 'Decisional Compact Knapsack problem' by DCK. In the benchmarks of GLP we include the improvements by Dagdelen et al. presented in [20]. In the benchmarks of BLISS we include the improvements by Ducas presented in [22].

80-bit security	GLP [20, 28, 29]	ring-TESLA-I (this paper)
Assumption	DCK	R-LWE
CPU	A	B
Signing key size	256	1,728
Verification key size	1,536	3,072
Signature size	1,186	1,568
Sign cycle counts	452,223	370,880
Verify cycle counts	34,004	94,124

128-bit security	BLISS [22, 23] BLISS-I	BLISS-II	ring-TESLA-II (this paper)
Assumption	R-SIS, NTRU		R-LWE
CPU	C		B
Signing key size	2,048	2,048	1,920
Verification key size	7,168	7,168	3,328
Signature size	1,559	1,514	1,568
Sign cycle counts	351,333	582,857	510,981
Verify cycle counts	102,000	102,000	167,791

The time-optimized implementation of BLISS-I by Ducas [22] is only a factor of 1.45 faster than our implementation. We note that our signature scheme uses uniform sampling during Sign. In contrast, BLISS uses Gaussian sampling, which might be vulnerable to timing attacks [14,20]. Up to now, available implementations of BLISS do not protect against timing-attacks. It would be very interesting to compare our implementation with an optimized and timing-attack-protected implementation of BLISS.

In summary, our signature scheme has good performance compared to state-of-the-art ideal-lattice-based signature schemes, while it is instantiated provably secure. Hence, when real world security matters our presented scheme is a very interesting choice.

Acknowledgment. This work has been cofunded by the DFG as part of project P1 and P2 within the CRC 1119 CROSSING.

A   Extended Definitions and Security Notions

A.1   Syntax, Functionality, and Security of Signature Schemes

A signature scheme with key space $\mathcal{K}$, message space $\mathcal{M}$, and signature space $\mathcal{S}$, is a tuple $\Sigma = (\mathsf{KeyGen}, \mathsf{Sign}, \mathsf{Verify})$ of algorithms defined as follows.

- The (probabilistic) key generation algorithm on input the security parameter 1^λ returns a key pair $(\mathsf{sk}, \mathsf{pk}) \in \mathcal{K}$. We write $(\mathsf{sk}, \mathsf{pk}) \leftarrow \mathsf{KeyGen}(1^\lambda)$ and call sk the secret or signing key and pk the public or verification key.
- The (probabilistic) signing algorithm takes as input a signing key sk, a message $\mu \in \mathcal{M}$, and outputs a signature $\sigma \in \mathcal{S}$. We write $\sigma \leftarrow \mathsf{Sign}(\mathsf{sk}, \mu)$.
- The verification algorithm, on input a verification key pk, a message $\mu \in \mathcal{M}$, and a signature $\sigma \in \mathcal{S}$, returns a bit b: if $b = 1$ we say that the algorithm accepts, otherwise we say that it rejects. We write $b \leftarrow \mathsf{Verify}(\mathsf{pk}, \mu, \sigma)$.

We require (perfect) correctness of the signature scheme: for every security parameter λ, every choice of the randomness of the probabilistic algorithms, every key pair $(\mathsf{sk}, \mathsf{pk}) \leftarrow \mathsf{KeyGen}(1^\lambda)$, every message $\mu \in \mathcal{M}$, and every signature $\sigma \leftarrow \mathsf{Sign}(\mathsf{sk}, \mu)$, $\mathsf{Verify}(\mathsf{pk}, \mu, \sigma) = 1$ holds.

We target the standard security requirement for signature schemes, namely *unforgeability under chosen-message attack* (ufcma). The corresponding experiment involving an adversary $\mathcal{A}$ against a signature scheme Σ is depicted in Fig. 3. Since we prove security of the scheme presented in Sect. 3 in the random oracle model, we reproduce a corresponding ufcma experiment which grants $\mathcal{A}$ access to a random oracle H. Given the experiment, we say that a signature scheme Σ is (t, q_s, q_h, ϵ)-*unforgeable under chosen-message attack* if every adversary $\mathcal{A}$ which runs in time t and poses at most q_s queries to the signing oracle and q_h queries to the random oracle has advantage

$$\mathrm{Adv}_{\Sigma}^{\mathsf{ufcma}}(\mathcal{A}) = \Pr\left[\mathrm{Expt}_{\Sigma,\mathcal{A}}^{\mathsf{ufcma}} = 1\right] \leq \epsilon .$$

$\mathrm{Expt}_{\Sigma,\mathcal{A}}^{\mathsf{ufcma}}(1^\lambda)$:

1    $(\mathsf{sk}, \mathsf{pk}) \leftarrow \mathsf{KeyGen}(1^\lambda)$
2    $(\mu^*, \sigma^*) \leftarrow \mathcal{A}(1^\lambda, \mathsf{pk})^{\mathcal{O}_{\mathsf{Sign}}(\cdot), H(\cdot)}$
3   If $\mathsf{Verify}(\mathsf{pk}, \mu^*, \sigma^*) = 1 \wedge \mu^* \notin \mathcal{Q}_S$:
4   Return 1
5   Else: Return 0

If $\mathcal{A}$ queries $\mathcal{O}_{\mathsf{Sign}}(\mu)$:

6    $\mathcal{Q}_S \leftarrow \mathcal{Q}_S \cup \{\mu\}$
7    $\sigma \leftarrow \mathsf{Sign}(\mathsf{sk}, \mu)$
8   Return σ to $\mathcal{A}$

Fig. 3. Security experiment of unforgeability under chosen-message attack for an adversary $\mathcal{A}$ against a signature scheme $\Sigma = (\mathsf{KeyGen}, \mathsf{Sign}, \mathsf{Verify})$ in the random oracle model (i.e., all parties including $\mathcal{A}$ have access to a public function H with uniformly distributed output).

References

1. Abdalla, M., Fouque, P.-A., Lyubashevsky, V., Tibouchi, M.: Tightly-secure signatures from lossy identification schemes. In: Pointcheval, D., Johansson, T. (eds.) EUROCRYPT 2012. LNCS, vol. 7237, pp. 572–590. Springer, Heidelberg (2012)
2. Akleylek, S., Bindel, N., Buchmann, J., Krämer, J., Marson, G.A.: An efficient lattice-based signature scheme with provably secure instantiation. Cryptology ePrint Archive, Report 2016/030 (2016). http://eprint.iacr.org/
3. Albrecht, M., Cid, C., Faugère, J.-C., Fitzpatrick, R., Perret, L.: Algebraic algorithms for LWE problems. Cryptology ePrint Archive, Report 2014/1018 (2014). http://eprint.iacr.org/2014/1018/
4. Albrecht, M.R., Fitzpatrick, R., Göpfert, F.: On the efficacy of solving LWE by reduction to unique-SVP. In: Lee, H.-S., Han, D.-G. (eds.) ICISC 2013. LNCS, vol. 8565, pp. 293–310. Springer, Heidelberg (2014)
5. Albrecht, M.R., Player, R., Scott, S.: On the concrete hardness of learning with errors. Cryptology ePrint Archive, Report 2015/046 (2015). http://eprint.iacr.org/
6. Alkim, E., Bindel, N., Buchmann, J., Dagdelen, O.: Tesla: tightly-secure efficient signatures from standard lattices. Cryptology ePrint Archive, Report 2015/755 (2015). http://eprint.iacr.org/
7. Applebaum, B., Cash, D., Peikert, C., Sahai, A.: Fast cryptographic primitives and circular-secure encryption based on hard learning problems. In: Halevi, S. (ed.) CRYPTO 2009. LNCS, vol. 5677, pp. 595–618. Springer, Heidelberg (2009)
8. Arora, S., Ge, R.: New algorithms for learning in presence of errors. In: Aceto, L., Henzinger, M., Sgall, J. (eds.) ICALP 2011, Part I. LNCS, vol. 6755, pp. 403–415. Springer, Heidelberg (2011)
9. Babai, L.: A Las Vegas-NC algorithm for isomorphism of graphs with bounded multiplicity of eigenvalues. In: 27th FOCS, pp. 303–312. IEEE Computer Society Press, Toronto, 27–29 October 1986
10. Bai, S., Galbraith, S.D.: An improved compression technique for signatures based on learning with errors. In: Benaloh, J. (ed.) CT-RSA 2014. LNCS, vol. 8366, pp. 28–47. Springer, Heidelberg (2014)
11. Bellare, M., Rogaway, P.: The exact security of digital signatures - how to sign with RSA and Rabin. In: Maurer, U.M. (ed.) EUROCRYPT 1996. LNCS, vol. 1070, pp. 399–416. Springer, Heidelberg (1996)
12. Bernstein, D.J., Buchmann, J., Dahmen, E. (eds.): Post-Quantum Cryptography. Mathematics and Statistics Springer-11649; ZDB-2-SMA. Springer, Heidelberg (2009)
13. Blum, A., Kalai, A., Wasserman, H.: Noise-tolerant learning, the parity problem, and the statistical query model. In: 32nd ACM STOC, pp. 435–440. ACM Press, Portland, 21–23 May 2000
14. Bos, J.W., Costello, C., Naehrig, M., Stebila, D.: Post-quantum key exchange for the TLS protocol from the ring learning with errors problem. In: 2015 IEEE Symposium on Security and Privacy, pp. 553–570. IEEE Computer Society Press, San Jose, 17–21 May 2015
15. Bos, J.W., Naehrig, M., van de Pol, J.: Sieving for shortest vectors in ideal lattices: a practical perspective. Cryptology ePrint Archive, Report 2014/880 (2014). http://eprint.iacr.org/2014/880
16. Campbell, P., Groves, M., Shepherd, D., SOLILOQUY: a cautionary tale. In: ETSI 2nd Quantum-Safe Crypto Workshop (2014). http://docbox.etsi.org/Workshop/2014/201410_CRYPTO/S07_Systems_and_Attacks/S07_Groves_Annex.pdf

17. Chatterjee, S., Menezes, A., Sarkar, P.: Another look at tightness. In: Miri, A., Vaudenay, S. (eds.) SAC 2011. LNCS, vol. 7118, pp. 293–319. Springer, Heidelberg (2012)
18. Chen, Y., Nguyen, P.Q.: BKZ 2.0: better lattice security estimates. In: Lee, D.H., Wang, X. (eds.) ASIACRYPT 2011. LNCS, vol. 7073, pp. 1–20. Springer, Heidelberg (2011)
19. Cramer, R., Ducas, L., Peikert, C., Regev, O.: Recovering short generators of principal ideals in cyclotomic rings. Cryptology ePrint Archive, Report 2015/313 (2015). http://eprint.iacr.org/2015/313
20. Dagdelen, Ö., El Bansarkhani, R., Göpfert, F., Güneysu, T., Oder, T., Pöppelmann, T., Sánchez, A.H., Schwabe, P.: High-speed signatures from standard lattices. In: Aranha, D.F., Menezes, A. (eds.) LATINCRYPT 2014. LNCS, vol. 8895, pp. 84–102. Springer, Heidelberg (2015)
21. Duc, A., Tramèr, F., Vaudenay, S.: Better algorithms for LWE and LWR. In: Oswald, E., Fischlin, M. (eds.) EUROCRYPT 2015. LNCS, vol. 9056, pp. 173–202. Springer, Heidelberg (2015)
22. Ducas, L.: Accelerating Bliss: the geometry of ternary polynomials. Cryptology ePrint Archive, Report 2014/874 (2014). http://eprint.iacr.org/2014/874
23. Ducas, L., Durmus, A., Lepoint, T., Lyubashevsky, V.: Lattice signatures and bimodal gaussians. In: Canetti, R., Garay, J.A. (eds.) CRYPTO 2013, Part I. LNCS, vol. 8042, pp. 40–56. Springer, Heidelberg (2013)
24. El Bansarkhani, R., Buchmann, J.: High performance lattice-based CCA-secure encryption. Cryptology ePrint Archive, Report 2015/042 (2015). http://eprint.iacr.org/2015/042
25. Elias, Y., Lauter, K.E., Ozman, E., Stange, K.E.: Ring-LWE cryptography for the number theorist. Cryptology ePrint Archive, Report 2015/758 (2015). http://eprint.iacr.org/2015/758
26. Garg, S., Gentry, C., Halevi, S.: Candidate multilinear maps from ideal lattices. In: Johansson, T., Nguyen, P.Q. (eds.) EUROCRYPT 2013. LNCS, vol. 7881, pp. 1–17. Springer, Heidelberg (2013)
27. Gentry, C., Peikert, C., Vaikuntanathan, V.: Trapdoors for hard lattices and new cryptographic constructions. In: Ladner, R.E., Dwork, C. (eds.) 40th ACM STOC, pp. 197–206. ACM Press, Victoria, 17–20 May 2008
28. Güneysu, T., Lyubashevsky, V., Pöppelmann, T.: Practical lattice-based cryptography: a signature scheme for embedded systems. In: Prouff, E., Schaumont, P. (eds.) CHES 2012. LNCS, vol. 7428, pp. 530–547. Springer, Heidelberg (2012)
29. Güneysu, T., Oder, T., Pöppelmann, T., Schwabe, P.: Software speed records for lattice-based signatures. In: Gaborit, P. (ed.) PQCrypto 2013. LNCS, vol. 7932, pp. 67–82. Springer, Heidelberg (2013)
30. Guo, Q., Johansson, T., Stankovski, P.: Coded-BKW: solving LWE using lattice codes. In: Gennaro, R., Robshaw, M. (eds.) CRYPTO 2015. LNCS, vol. 9215, pp. 23–42. Springer, Heidelberg (2015)
31. Ishiguro, T., Kiyomoto, S., Miyake, Y., Takagi, T.: Parallel Gauss Sieve algorithm: solving the SVP challenge over a 128-dimensional ideal lattice. In: Krawczyk, H. (ed.) PKC 2014. LNCS, vol. 8383, pp. 411–428. Springer, Heidelberg (2014)
32. Katz, J., Wang, N.: Efficiency improvements for signature schemes with tight security reductions. In: Jajodia, S., Atluri, V., Jaeger, T. (eds.) ACM CCS 2003, pp. 155–164. ACM Press, Washington D.C., 27–30 October 2003
33. Kirchner, P., Fouque, P.-A.: An improved BKW algorithm for LWE with applications to cryptography and lattices. In: Gennaro, R., Robshaw, M. (eds.) CRYPTO 2015. LNCS, vol. 9215, pp. 43–62. Springer, Heidelberg (2015)

34. Laarhoven, T.: Sieving for shortest vectors in lattices using angular locality-sensitive hashing. In: Gennaro, R., Robshaw, M. (eds.) CRYPTO 2015. LNCS, vol. 9215, pp. 3–22. Springer, Heidelberg (2015)

35. Lenstra, A., Lenstra, H., Lovász, L.: Factoring polynomials with rational coefficients. Math. Ann. **261**, 515–534 (1982)

36. Lindner, R., Peikert, C.: Better key sizes (and attacks) for LWE-based encryption. In: Kiayias, A. (ed.) CT-RSA 2011. LNCS, vol. 6558, pp. 319–339. Springer, Heidelberg (2011)

37. Liu, M., Nguyen, P.Q.: Solving BDD by enumeration: an update. In: Dawson, E. (ed.) CT-RSA 2013. LNCS, vol. 7779, pp. 293–309. Springer, Heidelberg (2013)

38. Lyubashevsky, V.: Lattice signatures without trapdoors. In: Pointcheval, D., Johansson, T. (eds.) EUROCRYPT 2012. LNCS, vol. 7237, pp. 738–755. Springer, Heidelberg (2012)

39. Lyubashevsky, V., Peikert, C., Regev, O.: On ideal lattices and learning with errors over rings. In: Gilbert, H. (ed.) EUROCRYPT 2010. LNCS, vol. 6110, pp. 1–23. Springer, Heidelberg (2010)

40. Micciancio, D., Voulgaris, P.: Faster exponential time algorithms for the shortest vector problem. In: Charika, M. (ed) 21st SODA, pp. 1468–1480. ACM-SIAM, Austin, 17–19 January 2010

41. Pointcheval, D., Stern, J.: Security arguments for digital signatures and blind signatures. J. Cryptology **13**(3), 361–396 (2000)

42. Schneider, M.: Sieving for shortest vectors in ideal lattices. In: Nitaj, A., Hassanien, A.E., Youssef, A. (eds.) AFRICACRYPT 2013. LNCS, vol. 7918, pp. 375–391. Springer, Heidelberg (2013)

Elliptic Curves

A Fast and Compact FPGA Implementation of Elliptic Curve Cryptography Using Lambda Coordinates

Burak Gövem[1], Kimmo Järvinen[1], Kris Aerts[2], Ingrid Verbauwhede[1], and Nele Mentens[1,2(✉)]

[1] KU Leuven ESAT/COSIC and iMinds,
Kasteelpark Arenberg 10 bus 2452, 3001 Leuven-Heverlee, Belgium
{burak.govem,kimmo.jarvinen,ingrid.verbauwhede,
nele.mentens}@esat.kuleuven.be
[2] KU Leuven Technology Campus Diepenbeek, ES&S, Diepenbeek, Belgium
kris.aerts@kuleuven.be

Abstract. Elliptic curve cryptography (ECC) provides high security with shorter keys than other public-key cryptosystems and it has been successfully used in security critical embedded systems. We present an FPGA-based coprocessor that communicates with the host processor via a 32-bit bus. It implements ECC over an elliptic curve that offers roughly 128-bit security. It is the first hardware implementation that uses the recently introduced lambda coordinates and the Galbraith-Lin-Scott (GLS) technique with fast endomorphisms. One scalar multiplication requires 65,000 clock cycles with a maximum clock frequency of 274 MHz on a Xilinx Virtex-5 FPGA, which gives a computation time of 0.24 ms. The area utilization is 1552 slices and 4 BlockRAMs. Our coprocessor compares favorably to other published works both in terms of speed and area, which makes it a good choice for embedded systems that rety public-key cryptography.

1 Introduction

Many embedded systems are used in applications where security and safety are of utmost importance. Such security-critical embedded systems include airplanes, cars, medical devices, home automation systems, military devices, etc. They require that confidentiality, integrity, and authenticity are ensured by using strong cryptography. Cryptography is often computationally demanding and efficient implementation of cryptographic computations is a topic that has been an active research field during the last couple of decades. In particular, public-key cryptosystems are challenging to implement efficiently and securely in embedded systems because they are computationally more demanding than many other forms of encryption. Public-key cryptosystems are essential parts of a variety of cryptosystems because they are required, for example, for computing digital signatures. Therefore, techniques for their fast computation with small amounts of resources are needed in embedded systems in practice.

© Springer International Publishing Switzerland 2016
D. Pointcheval et al. (Eds.): AFRICACRYPT 2016, LNCS 9646, pp. 63–83, 2016.
DOI: 10.1007/978-3-319-31517-1_4

Elliptic curve cryptography (ECC) is a type of public-key cryptography that was introduced in the mid-1980s. ECC has many benefits compared to other forms of public-key cryptography. The main benefit is that high security levels can be achieved with shorter keys than for other public-key cryptosystems (such as RSA, Diffie-Hellman, ElGamal, etc.). ECC uses $2n$-bit keys for achieving roughly n bits of security. For example, RSA requires significantly longer keys: 1,024 and 3,072 bits for 80-bit and 128-bit security levels, respectively [34]. ECC implementations have also proven to be faster than RSA implementations [15]. ECC is also widely included in multiple standards, which is a significant advantage in commercial applications.

Field-programmable gate arrays (FPGA) have been popular implementation platforms for cryptography and a plethora of ECC implementations for FPGAs are available in the literature. While many of them target primarily to hardware acceleration of ECC by optimizing speed with very loose area constraints (see, e.g., [1,4,11,20,24]), there also exist many implementations that are suitable for security-critical embedded systems including, e.g., [26,28,41,42]. In them, the primary optimization target is typically either area or speed-area ratio. A vast majority of such designs has been introduced for the 80-bit security level, which should have been used only up to 2010 as recommended by the National Institute of Standards and Technology (NIST) of the United States [33]. Most of the publications target the NIST curves specified in [35]. These curves date back to the 1990s and they cannot utilize certain state-of-the-art optimizations that have been introduced in the recent years. Several new curves have been introduced allowing more efficient computations (see, e.g., [7–10,13,14,17,37]). Although many studies about software performance of these curves are available, hardware implementations are still largely missing from the literature (for some exceptions, see, e.g., [2,3,5,6,16,40]). In particular, to the best of our knowledge, there are no hardware implementations of the new λ-coordinates [37] combined with the Galbraith-Lin-Scott (GLS) technique for binary curves [17] which was shown to be very efficient in software [37]. The implementation in [2] uses the same curve without λ-coordinates, focusing on speed maximization with very loose area constraints. It is widely known that there is often a difference between efficiency in software and hardware and some curves that are fast in software may not be as efficient in hardware, and vice versa. Good examples are prime and binary curves, of which the former are better in software and the latter in hardware. Also NIST has shown interest in the new curves and techniques and there appears to be considerations to standardize new elliptic curves (see, e.g., the call for papers of the NIST workshop on elliptic curves [36]). Hence, it is important to shed light on hardware performance of the new curves and techniques in order to complete the picture about their efficiency.

In this paper, we present an FPGA coprocessor for ECC which is designed primarily for security-critical embedded systems. Our coprocessor communicates with a host processor via a 32-bit interface which allows easy integration to various systems. The coprocessor implements ECC for the 128-bit security level which matches, e.g., the security offered by the 128-bit version of the Advanced

Encryption Standard (AES) [32]. Most other publications consider the significantly less secure 80-bit security level as discussed above. The coprocessor is designed so that it is both fast and compact and, thus, meets the requirements of various security-critical embedded systems. Our implementation uses the elliptic curve and parameters as well as many of the state-of-the-art optimizations introduced by Oliveira et al. [37] in 2013. To the best of our knowledge, our implementation is the first FPGA-based ECC implementation that uses λ-coordinates from [37] and the GLV/GLS technique from [13,14,17]. We compiled our architecture for Xilinx Virtex-4, Virtex-5, and Spartan-6 FPGAs and the results show that our coprocessor compares favorably with the related work available in the literature although it offers a significantly higher security level. Our results also show that λ-coordinates and the GLS technique provide good results not only in software but also in hardware.

The paper is structured as follows. Section 2 presents the preliminaries of ECC and the algorithms we implement in our coprocessor. Section 3 describes the architecture of our coprocessor. Section 4 presents the results on Xilinx FPGAs and compares them to other relevant ECC implementations available in the literature. We end with conclusions and discussion on certain topics for future research in Sect. 5.

2   Preliminaries

This section provides background on ECC in general and on GLS curves and λ-coordinates that we implement in this paper in particular.

2.1   Elliptic Curve Cryptography

The use of elliptic curves for public-key cryptography was independently proposed by Miller [31] and Koblitz [21] in the mid-1980s. ECC achieves high security levels with significantly shorter key lengths than other public-key cryptosystems such as RSA or ElGamal. Hence, ECC has become a popular choice for public-key cryptography especially in embedded systems.

Elliptic curves defined over a finite field $\mathbb{F}_q$ are used in cryptography. The points (x, y) that satisfy the equation of an elliptic curve combined with a special point called the point-at-infinity $\mathcal{O}$ form an additive Abelian group E, where $\mathcal{O}$ is the zero element. Let $P_1, P_2 \in E$. The group operation $P_1 + P_2$ is called point addition if $P_1 \neq \pm P_2$ and point doubling if $P_1 = P_2$. The most important operation of every elliptic curve cryptosystem is the scalar multiplication:

$$Q = kP = \underbrace{P + P + \ldots + P}_{k \text{ times}} \tag{1}$$

where k is an integer in the interval $[1, r-1]$ where r is the order of P (the smallest positive integer for which $rP = \mathcal{O}$). The security of ECC is based on the computational difficulty of the elliptic curve discrete logarithm problem (ECDLP), which is the problem of finding k when given Q and P. The ECDLP is believed to be infeasible to solve if the parameters of the system are chosen properly (similarly as integer factorization in the case of RSA). A secure elliptic curve over a $2m$-bit finite field is believed to offer roughly m bits of security. Hence, ECC offers roughly 128-bit security with 256-bit keys whereas, for example, RSA would require approximately 3,072-bit keys [34].

Computing (1) consists of several hierarchical levels, which are depicted in the ECC pyramid of Fig. 1. The scalar multiplication algorithm is on the top and it computes (1) with a series of point arithmetic operations (typically, point additions and point doublings). The algorithms that implement the point operations with series of finite field operations are in the middle. The algorithms for computing finite field operations including multiplication, addition (subtraction), and inversion (division) are in the bottom. In the following, we discuss the hierarchical levels and provide descriptions of our choices for implementing them from the bottom to the top.

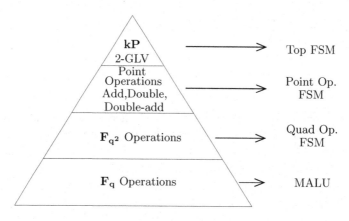

Fig. 1. ECC hierarchical pyramid. The components in our architecture (see Sect. 3) for implementing the hierarchical levels of the pyramid are shown on the right.

2.2 Finite Field Arithmetic

Either prime fields, where q is a prime p, or binary extension fields, where $q = 2^m$, are typically used for ECC. Prime fields are more commonly used in software implementations. Binary fields allow significantly more efficient implementations in hardware because they employ carry-free arithmetic operations. The inclusion of carry-free instructions in modern processors has enabled extremely fast implementations using binary fields also in software [43]. Elliptic curves defined over prime and binary fields are called prime and binary curves, respectively.

In this paper, we follow the approach of [37] and use the binary field $\mathbb{F}_{2^{254}}$ which can be constructed as a quadratic extension of the binary extension field $\mathbb{F}_{2^{127}}$ by using the irreducible polynomial $g(u) = u^2 + u + 1$. That is, we set $\mathbb{F}_{2^{127}}[u] \, / \, u^2 + u + 1$. An element $a \in \mathbb{F}_{2^{254}}$ is represented as $a = a_0 + a_1 u$ where

$a_0, a_1 \in \mathbb{F}_{2^{127}}$. Arithmetic operations in $\mathbb{F}_{2^{254}}$ can be decomposed into operations in $\mathbb{F}_{2^{127}}$ and they are computed as follows [37]:

$$a + b = a_0 + b_0 + (a_1 + b_1)u \tag{2}$$
$$a \times b = (a_0 b_0 + a_1 b_1) + (a_0 b_1 + a_1 b_0 + a_1 b_1)u \tag{3}$$
$$a^2 = a_0^2 + a_1^2 + a_1^2 u \tag{4}$$
$$a^{-1} = (a_0 + a_1)t^{-1} + a_1 t^{-1} u \tag{5}$$

where $t = a_0 a_1 + a_0^2 + a_1^2$. Let $\mathbf{a}$, $\mathbf{m}$, $\mathbf{s}$, and $\mathbf{i}$ denote the costs of addition, multiplication, squaring, and inversion in $\mathbb{F}_{2^{127}}$ and let $\mathbf{A}$, $\mathbf{M}$, $\mathbf{S}$, and $\mathbf{I}$ denote the costs of the respective operations in $\mathbb{F}_{2^{254}}$. The above equations give that $\mathbf{A} = 2\mathbf{a}$, $\mathbf{M} = 4\mathbf{m} + 3\mathbf{a}$, $\mathbf{S} = 2\mathbf{s} + \mathbf{a}$, and $\mathbf{I} = \mathbf{i} + 3\mathbf{m} + 2\mathbf{s} + 3\mathbf{a}$.

We construct $\mathbb{F}_{2^{127}}$ by setting $\mathbb{F}_2[x] / p(x)$ where $p(x)$ is the irreducible trinomial $p(x) = x^{127} + x^{63} + 1$. Addition in $\mathbb{F}_{2^{127}}$ is a bitwise exclusive-or (xor) of the bit vectors representing the elements. Multiplication is carried out by computing a multiplication of polynomials in $\mathbb{F}_2[x]$ followed by a reduction modulo $p(x)$. Because squaring in $\mathbb{F}_2[x]$ can be performed by adding zeros between each bit of the bit vector, squaring in $\mathbb{F}_{2^{127}}$ contains only rewiring followed by the reduction modulo $p(x)$ when implemented in hardware. Inversion can be computed as an exponentiation consisting of multiplications and squarings in $\mathbb{F}_{2^{127}}$ by using the Itoh-Tsujii algorithm [19]. We use a variant of the Itoh-Tsujii inversion from [37] that utilizes the optimal addition chain $(1, 2, 3, 6, 12, 24, 48, 96, 120, 126)$ and requires $126\mathbf{s} + 9\mathbf{m}$ in $\mathbb{F}_{2^{127}}$.

2.3 Point Representation with λ-Coordinates

Point addition and point doubling are the basic point operations. If the points are represented in affine coordinates by using two coordinates (x, y), then both point addition and point doubling require an inversion, which is a very expensive operation as shown in Sect. 2.2. Hence, projective coordinates, where points are represented with three coordinates as (X, Y, Z), are commonly used in practical implementations of ECC. They allow computing point additions and point doublings without inversions (but with an increased number of other operations). A single inversion is required in the end of computing (1) in order to obtain the affine coordinates of the result point Q. In the case of binary curves, popular choices have been standard projective coordinates, where $x = X/Z$ and $y = Y/Z$, and López-Dahab (LD) coordinates [27], where $x = X/Z$ and $y = Y/Z^2$.

In 2013, Oliveira et al. [37] proposed a new coordinate system called λ-coordinates, where points are represented as (x, λ) so that $\lambda = x + \frac{y}{x}$. We refer to this coordinate system as affine λ-coordinates. The projective version of λ-coordinates represents a point with three coordinates (X, L, Z) so that $x = X/Z$ and $\lambda = L/Z$. They result in the fastest formulae that are currently available for computing point arithmetic on binary Weierstrass curves. A point addition and point doubling require (excluding additions) $8\mathbf{M} + 2\mathbf{S}$ and $5\mathbf{M} + 4\mathbf{S}$ (including one multiplication by a constant), respectively. One of the main benefits of λ-coordinates compared to LD coordinates is that they allow efficient combination

---

Algorithm 1. Double-and-add (left-to-right)

Require: Scalar $k = \sum_{i=0}^{n-1} k_i 2^i \in [0, r-1]$ with $k_{n-1} = 1$, base point P
Ensure: Result point $Q = kP$
$\quad Q \leftarrow P$
$\quad$ **for** $i = n-2$ **down to** 0 **do**
$\quad\quad Q \leftarrow 2Q$
$\quad\quad$ **if** $k_i = 1$ **then**
$\quad\quad\quad Q \leftarrow Q + P$
$\quad\quad$ **end if**
$\quad$ **end for**
$\quad$ **return** Q

---

of point doubling and point addition operations. Computing $2P_1 + P_2$ requires only $11\mathbf{M} + 6\mathbf{S}$ (including one multiplication with a constant).

2.4 Scalar Multiplication on the GLS Curves

In this paper, we use the same curve that was used in [37]. It is a Weierstrass curve over $\mathbb{F}_{2^{254}}$ defined by the following equation:

$$y^2 + xy = x^3 + ax^2 + b \tag{6}$$

where $a = a_0 + a_1 u, b = b_0 + b_1 u \in \mathbb{F}_{2^{254}}$ such that $a_0 = b_1 = 0$, $a_1 = 1$, and b_0 is a specific element in $\mathbb{F}_{2^{127}}$.

Scalar multiplication defined by (1) can be computed with a series of point additions and point doublings. The simplest option is to use the double-and-add algorithm given in Algorithm 1. It scans through the bits of k and performs a point doubling for every bit and an additional point addition if the bit is one. Let k be an n-bit integer and let $h(k)$ be its Hamming weight (the number of ones in the binary expansion). Then, one scalar multiplication requires $n-1$ point doublings and $h(k) - 1$ point additions, where $h(k) \approx n/2$.

One way of improving the speed of scalar multiplications is to utilize efficiently computable endomorphisms. Menezes and Vanstone [30] showed how point doublings can be replaced with the Frobenius endomorphisms $(x, y) \mapsto (x^2, y^2)$ on certain supersingular elliptic curves over $\mathbb{F}_{2^m}$, but these curves were found to be cryptographically weak [29]. In 1991, Koblitz [22] introduced a secure class of nonsupersingular elliptic curves over $\mathbb{F}_{2^m}$ which have the advantage of the Frobenius endomorphisms after certain conversions are computed for the scalar k. These curves are commonly known as Koblitz curves and they are nowadays included in many standards (e.g., in [35]). In 2001, Gallant, Lambert, and Vanstone (GLV) [14] introduced a specific class of elliptic curves over $\mathbb{F}_p$ which allows utilizing efficiently computable endomorphisms also for prime curves. Galbraith, Lin, and Scott (GLS) [13] generalized the GLV technique to a broader class of elliptic curves defined over $\mathbb{F}_{p^2}$. The GLS curves were generalized for binary curves over $\mathbb{F}_{2^{2m}}$ by Hankerson et al. [17]. In this paper, we focus on these variants of the GLS curves and, in particular, the curve considered by Oliveira et al. in [37].

The GLS technique allows splitting the computation of (1) with an n-bit k into $k_1 P + k_2 \psi(P)$ where k_1 and k_2 are approximately $n/2$-bit integers. That is, instead of the single scalar multiplication, one computes a sum of two smaller scalar multiplications, which can be computed efficiently with the so called Shamir's trick (see below). We skip the mathematical subtleties and merely state that the efficiently computable endomorphism ψ of the GLS technique is based on a composition of the Frobenius endomorphism and endomorphisms between E and its quadratic twist. Interested readers can find details, e.g., from [13,14,17,37]. We use $\psi(P)$ as defined in [37] as follows:

$$\psi : (x, \lambda) \mapsto ((x_0 + x_1) + x_1 u, (\lambda_0 + \lambda_1) + (\lambda_1 + 1)u) \qquad (7)$$

where $x = x_0 + x_1 u$ and $\lambda = \lambda_0 + \lambda_1 u$ with $x_0, x_1, \lambda_0, \lambda_1 \in \mathbb{F}_{2^{127}}$. Hence, $\psi(P)$ requires only three additions (3a) in $\mathbb{F}_{2^{127}}$.

The integer k needs to be decomposed into $n/2$-bit k_1 and k_2 such that $k \equiv k_1 + k_2\delta \pmod{r}$ where δ is an integer such that $\psi(P) = \delta P$ for all $P \in E$. Such a decomposition can be found by using techniques for finding the GLV decomposition given in [14]. The decomposition algorithm can be simplified for specific curve parameters and we use the decomposition algorithm used by Oliveira et al. [37] in the C code that is publicly available[1]. It finds k_1 and k_2 by computing:

$$k_1 = k_l + k_h - \beta \lfloor (k\beta)/2^{254} \rfloor \qquad (8)$$
$$k_2 = k_h\beta - (2^{127} - 1)\lfloor (k\beta)/2^{254} \rfloor \qquad (9)$$

where β is a 64-bit constant specific for the curve and k_l and k_h are the lowest and highest 127-bit words of the 254-bit k.

Shamir's trick (see, e.g., [18]) is a technique that allows evaluating a sum of two scalar multiplications $k_1 P + k_2 \psi(P)$ simultaneously. We call this operation double scalar multiplication. Let k_1 and k_2 be $n/2$-bit integers. If the double scalar multiplication is computed with two separate scalar multiplications, then it requires $n - 2$ point doublings and $h(k_1) + h(k_2) - 1 \approx n/2 - 1$ point additions. Shamir's trick arranges k_1 and k_2 into a $2 \times n/2$-bit matrix and precomputes the point $P + \psi(P)$. The double scalar multiplication is computed by scanning through the columns of the matrix. A point doubling is computed for every column and a point addition is computed for all nonzero columns. If the column is $\frac{1}{0}$, $\frac{0}{1}$ or $\frac{1}{1}$, then one adds either P, $\psi(P)$ or $P + \psi(P)$, respectively. Hence, Shamir's trick requires $n/2 - 1$ point doublings and, on average, $\frac{3}{8}n$ point additions for the double scalar multiplication.

In [37], Oliveira et al. used a parallelization technique that splits the scalar multiplication in two parallel computations: one based on the double-and-add and the other on halve-and-add. Halve-and-add computes point halvings $Q \leftarrow \frac{1}{2}Q$ instead of point doublings. They also utilized the window nonadjacent form

---

[1] https://github.com/floodyberry/supercop/tree/master/crypto_dh/gls254.

(w-NAF) for representing k_1 and k_2 in the GLV encoding which leads to a smaller number of point additions but requires precomputations and extra storage for the precomputed points. We decided not to use these optimizations in our implementation because implementing them would lead to a significant growth of the control logic and the parallelization technique would also require another unit for field arithmetic (see MALU in Sect. 3.4).

To summarize, we implement the scalar multiplication by using the GLS technique which splits an n-bit k into $n/2$-bit k_1 and k_2 so that both are given in standard binary representation. We precompute $P+\psi(P)$. We then use Shamir's trick for evaluating the double scalar multiplication by computing point doublings and combined point doublings and additions. The point arithmetic is performed in projective λ-coordinates by using the formulae from [37]. Finite field arithmetic is computed in $\mathbb{F}_{2^{254}}$ by decomposing the operations into operations in $\mathbb{F}_{2^{127}}$ as shown in (2)–(5).

3   Architecture

We present a coprocessor architecture for FPGAs that implements (1) using λ-coordinates and the fast GLS endomorphism. In this architecture, the finite field processing unit (FFPU) performs operations in $\mathbb{F}_{2^{127}}$ and three control units (i.e., finite state machines (FSM)) drive the FFPU in a hierarchical manner according to the ECC hierarchy shown in Fig. 1. This provides a natural way to decompose the complex control logic required for computing (1) into a set of smaller FSM, which can be implemented efficiently in FPGAs. Additionally, smaller processing and control units for integer arithmetic perform the scalar decomposition. A register file and block RAMs (BRAM) are used for storing temporary results. The register file stores frequently used temporary variables. The base point, the result of a scalar multiplication and the curve constant a are also stored in the BRAM. In the following, we describe the architecture of the coprocessor in a hierachical manner according to Fig. 1.

3.1   Top Level FSM

Our implementation is designed to be used as an ECC coprocessor. The architecture of the coprocessor is shown in Fig. 2. The host processor initiates the coprocessor by sending the base point P and scalar k over a 32-bit bus. The base point is stored in the BRAM and the scalar is stored in the registers. Next, the coprocessor starts the precomputation. The scalar multiplication FSM (the Top FSM in Fig. 2) manages the point operation FSM, shifts the scalar registers k_1 and k_2, and determines the next point operation. The scalar multiplication FSM also organizes precomputations, which are the scalar decomposition, computation of the endomorphism $\psi(P)$, the point addition for computing $P+\psi(P)$, finding the first nonzero column of the scalar matrix, and performing the coordinate transformations. First, $\psi(P)$ is calculated and stored in the BRAM and, then, it is added to P. Since the combined point doubling and point addition requires the other input point to be given in affine λ coordinates, the point $P + \psi(P)$ is converted to affine λ-coordinates and stored in the BRAM. After that, the scalar decomposition that computes (8) and (9) starts and when it is ready, the precomputation ends with finding the first nonzero column of the scalar matrix. Because all points and the scalars are ready, the scalar multiplication starts after this. As discussed in Sect. 2.4, the double scalar multiplication $k_1 P + k_2 \psi(P)$ is implemented using Shamir's trick. The accumulator point Q is initialized with P if the first nonzero column is $\begin{smallmatrix}1\\0\end{smallmatrix}$, with $\psi(P)$ if $\begin{smallmatrix}0\\1\end{smallmatrix}$, and with $P + \psi(P)$ if $\begin{smallmatrix}1\\1\end{smallmatrix}$. After this, either point doublings (for zero columns) or the combined point doublings and additions (for nonzero columns) are performed depending on the bits of k_1 and k_2 and the points to be added are determined as above. After the scalar multiplication is finished, the result point Q is first converted to affine λ-coordinates and then to affine coordinates and it is stored into a specific location in the BRAM, where it is available to the host processor. The scalar multiplication algorithm implemented by the coprocessor is shown in Algorithm 2.

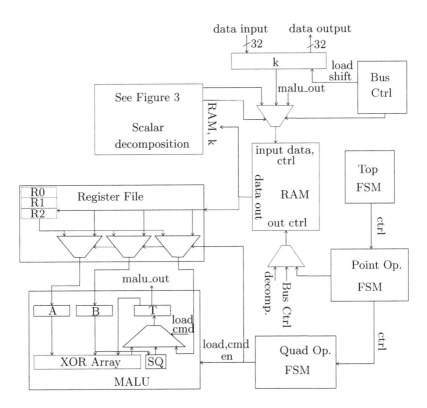

Fig. 2. Architecture

Algorithm 2. Scalar multiplication on a GLS curve, $Q = kP$

Require: Scalars $k_1 = (k_{1,n_1-1} \ldots k_{1,0})$, $k_2 = (k_{2,n_2-1} \ldots k_{2,0})$, base point P
Ensure: Result point $Q = kP = k_1 P + k_2 \psi(P)$

$\quad P_1 \leftarrow \psi(P)$ // 3a in $\mathbb{F}_{2^{127}}$
$\quad P_2 \leftarrow P + P_1$ // Algorithm 3
$\quad n \leftarrow \max(n_1, n_2)$
$\quad$**if** $k_{1,n-1} = 1$ **and** $k_{2,n-1} = 1$ **then**
$\quad\quad Q \leftarrow P_2$
$\quad$**else if** $k_{1,n-1} = 0$ **and** $k_{2,n-1} = 1$ **then**
$\quad\quad Q \leftarrow P_1$
$\quad$**else if** $k_{1,n-1} = 1$ **and** $k_{2,n-1} = 0$ **then**
$\quad\quad Q \leftarrow P$
$\quad$**end if**
$\quad$**for** $i = n - 2$ **down to** 0 **do**
$\quad\quad$**if** $k_{1,i} = 1$ **and** $k_{2,i} = 1$ **then**
$\quad\quad\quad Q \leftarrow 2Q + P_2$ // Algorithm 5
$\quad\quad$**else if** $k_{1,i} = 0$ **and** $k_{2,i} = 1$ **then**
$\quad\quad\quad Q \leftarrow 2Q + P_1$ // Algorithm 5
$\quad\quad$**else if** $k_{1,i} = 1$ **and** $k_{2,i} = 0$ **then**
$\quad\quad\quad Q \leftarrow 2Q + P$ // Algorithm 5
$\quad\quad$**else**
$\quad\quad\quad Q \leftarrow 2Q$ // Algorithm 4
$\quad\quad$**end if**
$\quad$**end for**
$\quad Q = (x, y) \leftarrow \mathrm{affine}(Q)$ // $\mathbf{I} + 3\mathbf{M} + \mathbf{A}$ in $\mathbb{F}_{2^{254}}$
$\quad$**return** Q

3.2 Point Operations FSM

Point operations (point doubling, point addition and combined point doubling and addition) and coordinate conversion are implemented in the point operation FSM. Point addition, point doubling and combined point doubling and addition require $5\mathbf{M} + 2\mathbf{S} + 5\mathbf{A}$, $5\mathbf{M} + 4\mathbf{S} + 5\mathbf{A}$, and $11\mathbf{M} + 6\mathbf{S} + 9\mathbf{A}$ operations in $\mathbb{F}_{2^{254}}$, respectively. Affine coordinates to affine λ-coordinates, affine λ-coordinates to affine coordinates, and projective λ-coordinates to affine λ-coordinates conversions require $\mathbf{I} + \mathbf{M} + \mathbf{A}$, $\mathbf{M} + \mathbf{A}$ and $\mathbf{I} + 2\mathbf{M}$, respectively. One of the multiplications in both point doubling and combined point doubling and addition is a multiplication by a constant (the curve parameter a).

The point operations FSM fetches input operands for an operation in $\mathbb{F}_{2^{254}}$ and writes them into registers of the register file. After the operation in $\mathbb{F}_{2^{254}}$ is finished, the point operation FSM writes the result of the operation to the BRAM unless the result is required only for the next operation. In that case, the writing is skipped and the point operation FSM proceeds to the next operation that will operate directly on the result in the register file. Results are written to the register file by default. For some cases results are both needed in the next operation and in later operations. Therefore, the result is stored in the BRAM to be used in the later operations and the next operation uses the result in the register file in order to save clock cycles. Temporary variable $R3$ that is shown in Algorithms 3, 4 and 5 is a register in the register file and all other temporary variables are in the BRAM.

Point addition is used only once when $P + \psi(P)$ is computed during the precomputation. Two points in affine λ-coordinates are added in this step. The point addition formula for adding $P = (x_P, \lambda_P)$ and $Q = (x_Q, \lambda_Q)$ is shown in (10). It returns the point $P+Q = (X_{P+Q}, L_{P+Q}, Z_{P+Q})$. An operation sequence for computing (10) is shown in Algorithm 3 in the Appendix.

$$
\begin{aligned}
A &= \lambda_P + \lambda_Q, \\
B &= (x_P + x_Q)^2, \\
X_{P+Q} &= A \cdot x_P \cdot x_Q \cdot A, \\
Z_{P+Q} &= A \cdot B, \\
L_{P+Q} &= (A \cdot x_Q + B)^2 + A \cdot B \cdot (\lambda_P + 1)
\end{aligned}
\tag{10}
$$

The point doubling formula for a point in projective λ-coordinates $Q = (X_Q, L_Q, Z_Q)$ is shown in (11). It returns the point $2Q = (X_{2Q}, L_{2Q}, Z_{2Q})$. An operation sequence for computing (11) is shown in Algorithm 4 in the Appendix.

$$
\begin{aligned}
T &= L_Q{}^2 + L_Q \cdot Z_Q + a \cdot Z_Q{}^2, \\
X_{2Q} &= T^2, \\
Z_{2Q} &= T \cdot Z_Q{}^2, \\
L_{2Q} &= (X_Q \cdot Z_Q)^2 + X_{2Q} + T \cdot (L_Q \cdot Z_Q) + Z_{2Q}
\end{aligned}
\tag{11}
$$

The efficiency of combined point doubling and addition is one of the reasons why λ-coordinates give faster results [37]. Computing them separately takes $13\mathbf{M} + 6\mathbf{S}$ in total, whereas the combined point doubling and addition requires only $11\mathbf{M} + 6\mathbf{S}$. Therefore, combined point doubling and addition saves 2 multiplications in $\mathbb{F}_{2^{254}}$. Since the double scalar multiplication with Shamir's trick requires significantly more combined point doublings and additions (for 75 % of the columns, on average) than point doublings (for 25 % of the columns), this trick significantly reduces the overall execution time. A formula for combined point doubling and addition with the inputs $Q = (X_Q, L_Q, Z_Q)$ and $P = (x_P, \lambda_P)$ is shown in (12). It returns the point $2Q+P = (X_{2Q+P}, L_{2Q+P}, Z_{2Q+P})$. An operation sequence for computing (12) is shown in Algorithm 5 in the Appendix.

$$
\begin{aligned}
A &= X_Q{}^2 \cdot Z_Q{}^2 + T \cdot (L_Q{}^2 + (a + 1 + \lambda_P) \cdot Z_Q{}^2), \\
B &= (x_P \cdot Z_Q{}^2 + T)^2, \\
X_{2Q+P} &= (x_P \cdot Z_Q{}^2) \cdot A^2, \\
Z_{2Q+P} &= A \cdot B \cdot Z_Q{}^2, \\
L_{2Q+P} &= T \cdot (A + B)^2 + (\lambda_P + 1) \cdot Z_{2Q+P}
\end{aligned}
\tag{12}
$$

3.3   Quadratic Extension FSM

Operations in the quadratic extension field (multiplication, constant multiplication, squaring, addition and inversion in $\mathbb{F}_{2^{254}}$) and the endomorphism $\psi(P)$ are implemented by the quadratic operations FSM. The quadratic operations FSM drives the MALU and generates the addresses of input operands of the MALU and generates the output write address for the BRAM. Operation costs of multiplication, squaring, addition and inversion are given in Sect. 2.2. In order to save time, constant multiplication with the curve parameter a is optimized. As discussed in Sect. 2.4, $a = u$ for this curve (i.e., $a_0 = 0$ and $a_1 = 1$). Therefore, multiplication $c = a \cdot b$ simplifies to $c = u \cdot (b_0 + b_1 \cdot u) = b_0 \cdot u + b_1 \cdot u^2 \equiv b_1 + (b_0 + b_1) \cdot u$ (mod $g(u)$). Therefore, multiplication by a only requires the swapping of the two halves followed by an addition in $\mathbb{F}_{2^{127}}$. I.e., the cost is only **a** and we save **4m+2a** compared to a general multiplication. The constant multiplication is used once in every iteration of the double scalar multiplication because it is needed once both in point doubling and combined point doubling and addition. The endomorphism ψ is performed as shown in (7) and it takes three modular additions in $\mathbb{F}_{2^{127}}$. The Itoh-Tsujii inversion algorithm is implemented for coordinate conversions. Itoh-Tsujii inversion uses an addition chain of length 9 as given in [37].

3.4   Modular Arithmetic Logic Unit (MALU)

Operations in $\mathbb{F}_{2^{127}}$ (addition, multiplication and squaring) are performed by the MALU which is imported from [39]. The MALU in [39] supports multiplication and addition in $\mathbb{F}_{2^m}$. The MALU is basically a most-significant digit first digit-serial modular multiplier over $\mathbb{F}_{2^m}$ with digit size d. It is adjustable for different irreducible polynomials and digit sizes, which are fixed at the time of implementation. The support for additions is added to the multiplier architecture with a very low cost by utilizing resource sharing. The latency of multiplication is $\lceil \frac{m}{d} \rceil + 2$ clock cycles, where m is the bit size of the operands which are elements in $\mathbb{F}_{2^m}$. The digit size of the MALU is chosen to be $d = 16$ for this implementation as it provides a good tradeoff between area and latency. Therefore, one multiplication in $\mathbb{F}_{2^{127}}$ takes $\lceil \frac{127}{16} \rceil + 2 = 10$ clock cycles. Addition takes 3 clock cycles. The MALU in [39] does not include a dedicated squaring circuitry and, therefore, squaring takes the same time as multiplication. Squaring in $\mathbb{F}_{2^m}$ is a very simple operation and support for it can be added with a very small overhead. Because the point operations involve several squarings, adding the support for squaring provides a significant speedup. Hence, we extended the MALU from [39] with a dedicated squarer. One squaring takes only 3 clock cycles which makes it as fast as an addition. For all field operations, one clock cycle is consumed by loading the inputs and another one goes to a one-stage pipeline for the inputs of the MALU in order to shorten the critical path. The remaining cycles are the actual operation time of the MALU.

3.5 Scalar Decomposition

The scalar decomposition is computed in the scalar-decomposition module which contains an 8-bit ALU that performs integer addition, subtraction and multiplication operations. An FSM splits operations with large operands into 8-bit operations and another FSM controls the execution of the decomposition equations given in (8) and (9). Scalar splitting is performed as described in [13]. After rearranging the equations, maximum operand sizes reduce to 192 bits for addition and subtraction and 128 bits for multiplication. These large operands are processed in 8-bit pieces by the ALU. The integer multiplication is implemented using Algorithm 5.1 given in [38]. Since the decomposition is executed in the beginning, it uses free space in the BRAM. The integer multiplication uses one of the dedicated multipliers in the FPGA (one DSP block). The architecture of the scalar-decomposition module is shown in Fig. 3.

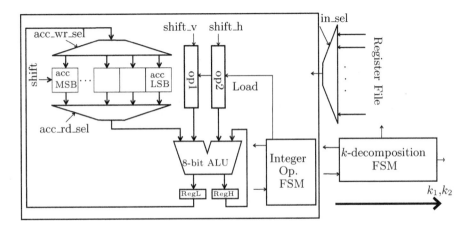

Fig. 3. The architecture for computing the decomposition of k

3.6 Optimizations

Point operations are optimized for minimum execution time and the temporary variable count is optimized based on this minimum timing constraint. Since the most complex equations stand for the combined point doubling and addition, it has the largest temporary variable count, which is 7 in our case. Temporary variables are stored in the BRAM excluding one of them ($R3$) which is in the register file. The base point P, $\psi(P)$, $P + \psi(P)$, and curve constants are also stored in the BRAM. When the MALU is processing, one of the operands is stored in the register file. Also the output from the MALU is stored into the registers (two 127-bit registers) so that it can be used for the next operation, which is usually the case. Another 127-bit register is used for storing temporary results of multiplication and inversion in $\mathbb{F}_{2^{254}}$. Therefore, there are in total three 127-bit registers in the register file. The latencies of point operations are given

Table 1. Point operation latencies (clock cycles)

Operation	M $3m + 4a$	S $2s + a$	A $2a$	Estimated cycles	Actual cycles
Point doubling and addition	11	6	9	$33m + 12s + 68a = 531$	565
Point doubling	5	4	5	$15m + 8s + 34a = 237$	258
Point addition	5	2	5	$15m + 4s + 32a = 258$	278

in Table 1. The differences between estimated and actual cycle counts arise from access times to the BRAM between operations.

The control logic is pipelined in order to shorten the critical path and to ensure that the critical path is in the processing units (i.e., in the MALU or the 8-bit ALU). Output multiplexers of the register file are pipelined before feeding the data into the MALU. Although this pipeline increases the execution time of every field operation by one clock cycle, the maximum frequency increases significantly. The slice count does not increase significantly because the slice based structure of FPGAs allows to use the flip-flops of a slice essentially for free if the LUT of the slice is already used. Other pipelined paths are address and control signals of the quadratic operations FSM that are controlling the register file and the BRAM. The 8-bit ALU also includes a one-stage pipeline. We experimented with different word sizes for the ALU and selected 8 bits because it was the optimal choice considering the critical path and area.

4   Results and Comparison

The architecture of the coprocessor was described in VHDL. This code was compiled with the Xilinx ISE 12.2 tool for Virtex-5 XC5VLX85-3FF676, Virtex-4 XC4VLX200-11FF1513 and Spartan-6 XC6SLX45T-3FGG484 FPGAs. Simulations for verifying the functionality of the implementations were performed with ModelSim software. Detailed area results after placement and routing are given in Table 2 for Virtex-5 XC5VLX85-3FF676. We compiled the design also for Virtex-4 and Spartan-6 FPGAs. Virtex-4 was selected in order to provide a fair comparison with previous works and Spartan-6 shows the performance of our architecture on a low-cost FPGA, which is commonly used in security-critical embedded systems.

The maximum clock frequency after synthesis is 274.982 MHz for the Virtex-5 implementation. This frequency was added as a design constraint for the place &route and it was able to meet this timing goal. One scalar multiplication takes on average around 61,300 clock cycles, but the exact latency depends on the scalar (the length and the number of nonzero columns in the scalar matrix). Therefore, the total time for an entire scalar multiplication is around 0.223 ms. Precomputation, scalar decomposition and coordinate transformations are also included in the given execution time.

Table 2. Hardware area cost on a Virtex-5 XC5VLX85-3FF676

Module	FF	LUT	BRAM	DSP48
MALU	399	1263	-	-
Scalar-decomposition	667	1025	-	1
Register File	381	1018	-	-
Top FSM	696	911	-	-
Point Op. FSM	44	171	-	-
Quad Op. FSM	42	142	-	-
Bus Controller	10	16	-	-
RAM	-	-	6	-
Total	2239	4546	6	1

There are many FPGA implementations of elliptic curve scalar multiplications over binary fields. However, most of them do not utilize curves with fast endomorphisms. The most popular curve has been the NIST B-163 curve from [35]. It provides a lower 80-bit security level, but because finite field arithmetic in $\mathbb{F}_{2^{163}}$ cannot be decomposed into a smaller field, the complexity of FFPU is actually larger. However, the latency of computing (1) on curves with lower security levels is expected to be significantly shorter because fewer operations are computed. These differences should be considered when the designs are compared.

A summary of related work and our implementations is shown in Table 3. The design proposed by Sinha Roy et al. in [41] uses LD coordinates for a binary field EC implementation. They implemented a Karatsuba hybrid multiplier for $\mathbb{F}_{2^{163}}$. The design by Ansari and Hasan in [1] uses a Montgomery multiplier and represent points in projective coordinates. The work by Liu et al. in [24] is a very fast implementation for NIST B-163. It computes a scalar multiplication in only $9\,\mu s$. However, it occupies a huge area due to the extensive use of parallelism in the processor. The implementation presented by Lutz and Hasan in [28] computes a scalar multiplication on NIST K-163, which is a Koblitz curve over $\mathbb{F}_{2^{163}}$, and represents points in LD coordinates. A scalable design that implements all NIST Koblitz curves was proposed by Loi and Ko in [26]. In [42], various NIST curve implementations were proposed by Sutter et al. Their architecture uses three multipliers and, therefore, it gives fast results at the expense of area. To the best of our knowledge, the only other implementation that uses the GLS technique is an implementation described by Azarderakhsh and Karabina in [2], but they do not use λ-coordinates. The implementation provides results on the same security level but it targets high speed at the expense of area which makes comparisons difficult. However, we expect that also their implementation would benefit from the use of λ-coordinates through the reduction of the number of field operations. When our results are compared to the other works in Table 3, it must be borne in mind that our implementation offers about 128 bits of secu-

Table 3. Comparison with related works

Design	Platform	Time (μs)	Slice	LUT	FF	Curve	Coordinate system
Sinha Roy et al. [41]	Virtex 5	9.5	3513	10195	-	B-163	LD
Ansari and Hasan [1]	Virtex 2	41	-	8300	1100	B-163	Projective
Liu et al. [24]	Virtex 4	9	10413	-	-	B-163	Projective
Lutz and Hasan [28]	Virtex 2	446	-	7362	1930	K-163	LD
Loi [26]	Virtex 4	603	2431	3815	1219	K-233	Mixed
Sutter et al. [42]	Virtex 5	19.89	6487	22340	-	B-233	Projective
Chelton and Benaissa [11]	Virtex 4	19.55	16209	-	-	B-163	Projective
Azarderakhsh and Karabina [2]	Virtex 4	16.85	12043	-	-	GLS-254	Projective
This work	Virtex 4	317	3985	7112	2247	GLS-254	λ
This work	Virtex 5	223	1552	4546	2239	GLS-254	λ
This work	Spartan-6	370	1546	4983	2315	GLS-254	λ

rity whereas the ones for B/K-163 and B/K-233 curves offer only about 80 or 112 bits of security, respectively. The security level has significant effects on both the hardware complexity and the total execution time.

As can be seen from Table 3, our design is faster than some related works, although it offers more security. Also, the area consumption of our design is very low compared to other FPGA implementations available in the literature. Hence, our design is suitable for embedded systems that require fast scalar multiplications with a small footprint. In such applications, low-cost FPGAs are more feasible implementation platforms than high-end FPGAs such as the Virtex family FPGAs. Hence, we also give implementation result for a low-cost FPGA from the Spartan-6 family in Table 3. Our results show that the new technique of using λ-coordinates with the GLS endomorphism introduced in [37] results in a fast and compact ECC coprocessor that can be used in various security-critical embedded systems. Note that a slice consists of several LUTs and FFs. In Table 3, we give both the number of slices and the number of LUTs and FFs in order to compare to other implementations that only give one of both.

5   Conclusion

We presented a fast and compact FPGA implementation of ECC with a 128-bit security level. Our implementation uses many of the optimizations used in the software implementation presented in [37] by Oliveira et al. To the best of our knowledge, it is the first FPGA implementation that utilizes λ-coordinates and the GLS decomposition.

We demonstrated that these techniques do not only offer significant advantages in software implementations but that improvements can be obtained also for hardware implementations. An especially significant advantage in the case of hardware is that finite field arithmetic is decomposed so that it can be performed in a small field, where m is of the size of the security level. This results in small areas and high maximum clock frequencies. In the case of many other curves (e.g., the NIST curves), field arithmetic is performed in fields with sizes of twice the security level. We showed also that the GLS decompositions can be computed efficiently in hardware, although extra hardware is required for the integer operations that are not natively supported by field arithmetic units (such as the MALU that we used in this paper). This extra cost is small compared to the benefits that can be obtained from the faster scalar multiplication.

The target applications for our implementation are various security-critical embedded systems that require high-security public-key cryptography. Hence, primary concerns are small area and good speed-area ratio and not necessarily the lowest computation time or highest throughput (scalar multiplications per second). In fact, the highly optimized software implementation reported in [37] requires only 47,900 clock cycles per scalar multiplication on Intel Sandy Bridge processors which run at significantly higher clock frequencies (e.g., 3.4 GHz) compared to our implementation (274 MHz) and, therefore, it outperforms our FPGA implementation. However, this kind of performance is completely out of the reach of embedded software implementations. For example, a single scalar multiplication requires some tens of milliseconds on a 32-bit processor [12] and more than 0.5 s on an 8-bit microcontroller [23,25]. Thus, our implementation provides major performance advantages compared to software implementations typically used in security-critical embedded systems.

We foresee several directions for future research on this topic. A careful parameter space exploration for our architecture would allow finding the optimum parameters (e.g., the digit-size of the MALU) which would allow finding the optimum speed-area ratio. Countermeasures against side-channel attacks need to be implemented in order to use our implementation in applications where side-channel attacks are a threat. In particular, timing and operation patterns need to be constant in order to thwart timing attacks and simple power analysis or electromagnetic attacks. Further speedups can be obtained by using precomputations (e.g., w-NAF) and by using the two-core parallelization technique based on double-and-add and halve-and-add from [37]. Further increase in throughput can be achieved by instantiating multiple cores in a single FPGA. For instance, we extrapolated that about 25 parallel cores would fit into the largest Virtex-5 FPGA (filled up to 80 %). Hence, our implementation can be valuable also for accelerating cryptographic computations.

Acknowledgments. K. Järvinen is an FWO Pegasus Marie Curie Fellow funded by Fonds Wetenschappelijk Onderzoek—the Research Foundation of Flanders. Further, this work was supported in part by the Research Council KU Leuven (C16/15/058) and IOF project EDA-DSE (HB/13/020).

A   Algorithms for Point Operations

Algorithm 3. Point addition $P + Q$

1 $R3 \leftarrow x_P + x_Q$	7 $R3 \leftarrow R1 + B$
2 $RL_Q \leftarrow B = R3^2$	8 $R1 \leftarrow R3^2$
3 $R3, R0 \leftarrow A = \lambda_P \cdot \lambda_Q$	9 $RZ_Q \leftarrow Z_{P+Q} = R0 \cdot RL_Q$
4 $R3, R1 \leftarrow R3 \cdot x_Q$	10 $R3 \leftarrow \lambda_P + 1$
5 $R3 \leftarrow R3 \cdot x_P$	11 $R3 \leftarrow RZ_Q \cdot R3$
6 $RX_Q \leftarrow R3 \cdot R0$	12 $RL_Q \leftarrow L_{P+Q} = R1 + R3$

Algorithm 4. Point doubling $2Q$

1 $R0 \leftarrow L_Q^2$	8 $R2 \leftarrow R3^2$
2 $R3, RL_Q \leftarrow L_Q \cdot Z_Q$	9 $RX_Q \leftarrow X_{2Q} = R0^2$
3 $R0 \leftarrow R0 + R3$	10 $RZ_Q \leftarrow Z_{2Q} = R0 \cdot R1$
4 $R1 \leftarrow Z_Q^2$	11 $R3 \leftarrow R0 \cdot RL_Q$
5 $R3 \leftarrow a \cdot R1$	12 $R3 \leftarrow R3 + R2$
6 $R0 \leftarrow T = R0 + R3$	13 $R3 \leftarrow R3 + RX_Q$
7 $R3 \leftarrow X_Q \cdot Z_Q$	14 $RL_Q \leftarrow L_{2Q} = R3 + RZ_Q$

Algorithm 5. Combined point doubling and addition $Q = 2Q + P$

1 $R0 \leftarrow L_Q^2$	14 $R0 \leftarrow A = R0 + R3$
2 $R3 \leftarrow L_Q \cdot Z_Q$	15 $R3, RX_Q \leftarrow x_P \cdot RZ_Q$
3 $RL_Q \leftarrow R0 + R3$	16 $R3 \leftarrow R1 + R3$
4 $R3, RZ_Q \leftarrow Z_Q^2$	17 $R2 \leftarrow B = R3^2$
5 $R3 \leftarrow a \cdot R3$	18 $R3 \leftarrow R0^2$
6 $R1 \leftarrow T = RL_Q + R3$	19 $RX_Q \leftarrow X_{2Q+P} = R3 \cdot RX_Q$
7 $R3, RL_Q \leftarrow 1 + \lambda_P$	20 $R3 \leftarrow R0 \cdot R2$
8 $R3 \leftarrow a + R3$	21 $R3, RZ_Q \leftarrow Z_{2Q+P} = R3 \cdot RZ_Q$
9 $R3 \leftarrow R3 \cdot RZ_Q$	22 $RL_Q \leftarrow RL_Q \cdot R3$
10 $R3, R0 \leftarrow R0 + R3$	23 $R3 \leftarrow R0 + R2$
11 $R0 \leftarrow R1 \cdot R3$	24 $R3 \leftarrow R3^2$
12 $R3 \leftarrow X_Q^2$	25 $R3 \leftarrow R1 \cdot R3$
13 $R3 \leftarrow R3 \cdot RZ_Q$	26 $RL_Q \leftarrow L_{2Q+P} = R3 \cdot RL_Q$

The formulas used for deriving the algorithms are available in Explicit-Formulas Database (https://hyperelliptic.org/EFD/g12o/auto-shortw-lambda.html). It also includes verification scripts (Sage).

References

1. Ansari, B., Hasan, M.A.: High-performance architecture of elliptic curve scalar multiplication. IEEE Trans. Comput. **57**(11), 1443–1453 (2008)
2. Azarderakhsh, R., Karabina, K.: A new double point multiplication method and its implementation on binary elliptic curves with endomorphisms. Technical report CACR 2012–24, University of Waterloo, Centre for Applied Cryptographic Research (2012)

3. Azarderakhsh, R., Reyhani-Masoleh, A.: Efficient FPGA implementations of point multiplication on binary Edwards and generalized Hessian curves using Gaussian normal basis. IEEE Trans. VLSI Syst. **20**(8), 1453–1466 (2012)
4. Azarderakhsh, R., Reyhani-Masoleh, A.: High-performance implementation of point multiplication on Koblitz curves. IEEE Trans. Circuits Syst. II Express Briefs **60**(1), 41–45 (2013)
5. Baldwin, B., Moloney, R., Byrne, A., McGuire, G., Marnane, W.P.: A hardware analysis of twisted Edwards curves for an elliptic curve cryptosystem. In: Becker, J., Woods, R., Athanas, P., Morgan, F. (eds.) ARC 2009. LNCS, vol. 5453, pp. 355–361. Springer, Heidelberg (2009)
6. Batina, L., Hogenboom, J., Mentens, N., Moelans, J., Vliegen, J.: Side-channel evaluation of FPGA implementations of binary Edwards curves. In: Proceedings of the 17th IEEE International Conference on Electronics, Circuits, and Systems (ICECS 2010), pp. 1248–1251. IEEE (2010)
7. Bernstein, D.J.: Curve25519: new Diffie-Hellman speed records. In: Yung, M., Dodis, Y., Kiayias, A., Malkin, T. (eds.) PKC 2006. LNCS, vol. 3958, pp. 207–228. Springer, Heidelberg (2006)
8. Bernstein, D.J., Birkner, P., Joye, M., Lange, T., Peters, C.: Twisted Edwards curves. In: Vaudenay, S. (ed.) AFRICACRYPT 2008. LNCS, vol. 5023, pp. 389–405. Springer, Heidelberg (2008)
9. Bernstein, D.J., Lange, T., Rezaeian Farashahi, R.: Binary Edwards curves. In: Oswald, E., Rohatgi, P. (eds.) CHES 2008. LNCS, vol. 5154, pp. 244–265. Springer, Heidelberg (2008)
10. Bos, J.W., Costello, C., Longa, P., Naehrig, M.: Selecting elliptic curves for cryptography: an efficiency and security analysis. Cryptology ePrint Archive, Report 2014/130 (2014). http://eprint.iacr.org/
11. Chelton, W.N., Benaissa, M.: Fast elliptic curve cryptography on FPGA. IEEE Trans. VLSI Syst. **16**(2), 198–205 (2008)
12. De Clercq, R., Uhsadel, L., Van Herrewege, A., Verbauwhede, I.: Ultra low-power implementation of ECC on the ARM Cortex-M0+. In: Proceedings of the The 51st Annual Design Automation Conference on Design Automation Conference (DAC 2014), pp. 1–6. ACM (2014)
13. Galbraith, S.D., Lin, X., Scott, M.: Endomorphisms for faster elliptic curve cryptography on a large class of curves. J. Cryptology **24**(3), 446–469 (2011)
14. Gallant, R.P., Lambert, R.J., Vanstone, S.A.: Faster point multiplication on elliptic curves with efficient endomorphisms. In: Kilian, J. (ed.) CRYPTO 2001. LNCS, vol. 2139, pp. 190–200. Springer, Heidelberg (2001)
15. Gura, N., Patel, A., Wander, A., Eberle, H., Shantz, S.C.: Comparing elliptic curve cryptography and RSA on 8-bit CPUs. In: Joye, M., Quisquater, J.-J. (eds.) CHES 2004. LNCS, vol. 3156, pp. 119–132. Springer, Heidelberg (2004)
16. Hamilton, M., Marnane, W.P.: FPGA implementation of an elliptic curve processor using the GLV method. In: Proceedings of the International Conference on Reconfigurable Computing and FPGAs (ReConFig 2009), pp. 249–254. IEEE (2009)
17. Hankerson, D., Karabina, K., Menezes, A.: Analyzing the Galbraith-Lin-Scott point multiplication method for elliptic curves over binary fields. IEEE Trans. Comput. **58**(10), 1411–1420 (2009)
18. Hankerson, D., Menezes, A., Vanstone, S.: Guide to Elliptic Curve Cryptography. Springer, New York (2004)
19. Itoh, T., Tsujii, S.: A fast algorithm for computing multiplicative inverses in $GF(2^m)$ using normal bases. Inf. Comput. **78**(3), 171–177 (1988)

20. Järvinen, K.: Optimized FPGA-based elliptic curve cryptography processor for high-speed applications. Integr. VLSI J. **44**(4), 270–279 (2011)
21. Koblitz, N.: Elliptic curve cryptosystems. Math. Comput. **48**(177), 203–209 (1987)
22. Koblitz, N.: CM-curves with good cryptographic properties. In: Feigenbaum, J. (ed.) CRYPTO 1991. LNCS, vol. 576, pp. 279–287. Springer, Heidelberg (1992)
23. Liu, A., Ning, P.: TinyECC: a configurable library for elliptic curve cryptography in wireless sensor networks. In: Proceedings of the International Conference on Information Processing in Sensor Networks (IPSN 2008), pp. 245–256. IEEE (2008)
24. Liu, S., Ju, L., Cai, X., Jia, Z., Zhang, Z.: High performance FPGA implementation of elliptic curve cryptography over binary fields. In: 2014 IEEE 13th International Conference on Trust, Security and Privacy in Computing and Communications (TrustCom), pp. 148–155. IEEE (2014)
25. Liu, Z., Seo, H., Großschädl, J., Kim, H.: Efficient implementation of NIST-compliant elliptic curve cryptography for sensor nodes. In: Qing, S., Zhou, J., Liu, D. (eds.) ICICS 2013. LNCS, vol. 8233, pp. 302–317. Springer, Heidelberg (2013)
26. Loi, K.C., Ko, S.B.: High performance scalable elliptic curve cryptosystem processor for Koblitz curves. Microprocess. Microsyst. **37**(4), 394–406 (2013)
27. López, J., Dahab, R.: Improved algorithms for elliptic curve arithmetic in $GF(2^n)$. In: Tavares, S., Meijer, H. (eds.) SAC 1998. LNCS, vol. 1556, pp. 201–212. Springer, Heidelberg (1999)
28. Lutz, J., Hasan, A.: High performance FPGA based elliptic curve cryptographic co-processor. In: International Conference on Information Technology: Coding and Computing, 2004 (ITCC 2004), Proceedings, vol. 2, pp. 486–492. IEEE (2004)
29. Menezes, A.J., Okamoto, T., Vanstone, S.A.: Reducing elliptic curve logarithms to logarithms in a finite field. IEEE Trans. Inf. Theory **39**(5), 1639–1646 (1993)
30. Menezes, A.J., Vanstone, S.A.: Elliptic curve cryptosystems and their implementation. J. Cryptology **6**(4), 209–224 (1993)
31. Miller, V.S.: Use of elliptic curves in cryptography. In: Williams, H.C. (ed.) CRYPTO 1985. LNCS, vol. 218, pp. 417–426. Springer, Heidelberg (1986)
32. National Institute of Standards and Technology (NIST): Advanced encryption standard (AES). Federal Information Processing Standard, FIPS PUB 197 (2001)
33. National Institute of Standards and Technology (NIST): Recommendation for transitioning the use of cryptographic algorithms and key lengths. NIST Special Publication 800–131A Transitions (2011)
34. National Institute of Standards and Technology: (NIST): Recommendation for key management – Part 1: General. NIST Special Publication 800–57 (2012)
35. National Institute of Standards and Technology (NIST): Digital signature standard (DSS). Federal Information Processing Standard, FIPS PUB 186–4 (2013)
36. National Institute of Standards and Technology (NIST): NIST workshop onelliptic curve cryptography standards. Call for papers (2015). http://www.nist.gov/itl/csd/ct/upload/CFP-Elliptic-Curve-Crypto-June2015.pdf
37. Oliveira, T., López, J., Aranha, D.F., Rodríguez-Henríquez, F.: Lambda coordinates for binary elliptic curves. In: Bertoni, G., Coron, J.-S. (eds.) CHES 2013. LNCS, vol. 8086, pp. 311–330. Springer, Heidelberg (2013)
38. Rodríguez-Henríquez, F., Saqib, N.A., Perez, A.D., Koç, C.K.: Cryptographic Algorithms on Reconfigurable Hardware. Springer, New York (2007)
39. Sakiyama, K., Batina, L., Mentens, N., Preneel, B., Verbauwhede, I.: Small-footprint ALU for public-key processors for pervasive security. In: Workshop on RFID Security – RFIDSec 2006 (2006)

40. Sasdrich, P., Güneysu, T.: Efficient elliptic-curve cryptography using Curve25519 on reconfigurable devices. In: Goehringer, D., Santambrogio, M.D., Cardoso, J.M.P., Bertels, K. (eds.) ARC 2014. LNCS, vol. 8405, pp. 25–36. Springer, Heidelberg (2014)
41. Sinha Roy, S., Rebeiro, C., Mukhopadhyay, D.: Theoretical modeling of elliptic curve scalar multiplier on LUT-based FPGAs for area and speed. IEEE Trans. VLSI Syst. **21**(5), 901–909 (2013)
42. Sutter, G.D., Deschamps, J., Imaña, J.L.: Efficient elliptic curve point multiplication using digit-serial binary field operations. IEEE Trans. Industr. Electron. **60**(1), 217–225 (2013)
43. Taverne, J., Faz-Hernández, A., Aranha, D.F., Rodríguez-Henríquez, F., Hankerson, D., López, J.: Software implementation of binary elliptic curves: impact of the carry-less multiplier on scalar multiplication. In: Preneel, B., Takagi, T. (eds.) CHES 2011. LNCS, vol. 6917, pp. 108–123. Springer, Heidelberg (2011)

Three Dimensional Montgomery Ladder, Differential Point Tripling on Montgomery Curves and Point Quintupling on Weierstrass' and Edwards Curves

Srinivasa Rao Subramanya Rao$^{(\boxtimes)}$

Mathematical Sciences Institute (MSI),
The Australian National University (ANU), Canberra, Australia
srinivasa.subramanya.anu@gmail.com

Abstract. Elliptic Curve Cryptography is an important alternative to traditional public key schemes such as RSA. This paper presents

(i) a simultaneous triple scalar multiplication algorithm to compute the x-coordinate of $kP + lQ + uR$ on a Montgomery Curve E_m defined over $\mathbb{F}_p$ which is about 15 to 22 % faster than the straight forward method of doing the same. The algorithm, motivated by Bernstein's paper on Differential Addition Chains, where the author proposes various 2-dimensional differential addition chains and asks for 3-dimensional versions to be constructed, can be generalized to other elliptic curve forms with differential addition formula,

(ii) a formula for Differential point tripling on Montgomery Curves which is slightly better than computing $3P$ as $2P + P$ and relevant in the implementation of Montgomery's PRAC and

(iii) an improvement in Mishra and Dimitrov's point Quintupling algorithm for Weierstrass' curves and an efficient Quintupling algorithm for Edwards Curves.

Keywords: Lucas chains · Differential addition chains · Montgomery curves · Edwards curves · Multiexponentiation · Double scalar multiplication · Schoenmakers' algorithm · Akishita's algorithm · Bernstein's algorithm · Triple scalar multiplication · Differential point tripling · Point quintupling

1 Introduction

Security in smart devices and mobile networks require an efficient implementation of cryptographic algorithms owing to the computational, bandwidth, power and memory constraints experienced in these environments. With its smaller key sizes, Elliptic curve cryptography (ECC) is increasingly seen as an alternative to traditional public key algorithms such as RSA, especially in constrained environments such as mobile devices. Thus while ECC is attractive for the success of

© Springer International Publishing Switzerland 2016
D. Pointcheval et al. (Eds.): AFRICACRYPT 2016, LNCS 9646, pp. 84–106, 2016.
DOI: 10.1007/978-3-319-31517-1_5

lightweight applications including mobile/embedded applications, RFID and in the context of "Internet for Things", optimized low-cost ECC implementations are crucial for the success of light weight cryptography.

The computation of exponentiation in a group is at the core of most discrete log based public key cryptosystems. Further, multi-exponentiation (also known as simultaneous exponentiation) in an Abelian group is a commonly used computation in cryptography, for example in signature verification algorithms and identification schemes. (Chaps. 7 and 9 in [1]). In groups with an additive notation such as Elliptic curve groups over a finite field $\mathbb{F}_p$, an exponentiation becomes a multiplication and thus simultaneous exponentiation becomes simultaneous scalar multiplication. A straight forward method to compute the double exponentiation $x_1^{n_1} * x_2^{n_2} \in G$ (where G is an Abelian group and $x_1, x_2 \in G$ and the exponents $n_1, n_2 \in Z$) is to compute $x_1^{n_1}$ and $x_2^{n_2}$ separately and then multiply them. The Strauss-Shamir method ([2,3], Algorithm 9.23 in [4]) scans the bit representations of n_1 and n_2 simultaneously from left to right and makes use of precomputed group elements to compute $x_1^{n_1} * x_2^{n_2}$, thus reducing the number of multiplications required to compute the desired product. The Joint Sparse Form [5] introduced by Solinas in 2001 makes use of signed representations of the exponents to improve the Strauss-Shamir method and are useful in groups where inverses of group elements can be computed efficiently such as Elliptic curve groups. The problem of minimizing the number of multiplications whilst computing an exponentiation, such as $x_1^{n_1}$ or $x_2^{n_2}$, can be reduced to minimizing the number of additions in an abstraction known as an addition chain. A finite sequence of integers $a_0, a_1, \ldots a_r$ is called an addition chain (Sect. 4.63 in [9]) for a_r if for each element a_i, there exists a_j and a_k in the sequence such that

$$a_i = a_j + a_k, \text{ for some } k \leq j < i \tag{1}$$

for all $i = 1, 2, \ldots, r$. Addition chains can be used to efficiently compute either a single exponentiation or multi-exponentiation (by using Strauss' method). Addition chains are applicable both in the context of multiplicative groups and additive groups such as Elliptic curve groups over a finite field.

In 1987, Montgomery proposed a special type of an elliptic curve, now known as Montgomery form of an Elliptic curve or simply Montgomery curve [6]. The arithmetic on a Montgomery curve relies on x-coordinate only arithmetic and also requires the 'difference' of two group elements (points) to be known prior to the computation of addition of these two elements. Thus ordinary addition chains and improvements of these chains cannot be directly utilized for scalar multiplication on Montgomery curves. A special form of an addition chain called Lucas chains is useful in this context. A Lucas chain is a restricted variant of an addition chain where the indices in Eq. (1) above are such that either $j = k$ or the difference $a_k - a_j$ is already part of the chain. A special case of Lucas chains occur when either $j = k$ or $a_k - a_j = a_0 = 1$ and this is called the binary chain. The well known Montgomery ladder presented in Sect. 2 is a binary chain. A Lucas chain is also known as a *Differential Addition Chain(DAC)* in the literature [10]. The Strauss-Shamir method for simultaneous scalar multiplication cannot be immediately used in the context of DACs. However, this technique can be adapted to DACs as shown by Schoenmakers, who constructed the first algorithm to produce two dimensional (double scalar) DACs in 2000. This algorithm was published in [8] by Stam in 2003. Akishitha's algorithm to construct two dimensional DACs was published in 2001 [7]. Shoenmakers' and Akishita's algorithm to construct two dimensional DACs produces binary chains. Bernstein proposed new algorithms to construct two dimensional DACs in 2006 along with a summary of previously known algorithms [10]. These included binary chain as well as Euclidean chain algorithms (algorithms using the Euclidean GCD scheme to construct DACs). In [14], Azarderakhsh et al. propose another DAC.

A natural question to ask is, if one could construct triple scalar multiplication analogues of the two dimensional DACs listed above. A practical motivation to construct such multi scalar multiplication algorithms arises in the implementation of some digital signature and identification schemes and their elliptic curve analogues. Chapter 11 in [18] covers some of these signature schemes. The Okamoto Identification scheme (Refer to Sect. 9.3 in [1]) requires a triple scalar multiplication operation to be performed by the signature verifier. Triple scalar multiplication can also be utilized in the accelerated verification of ElGamal like Signatures [19]. The need for higher order analogues can be seen in the batch verification of multiple signatures [20]. In [21], Karati et al. propose three methods for randomized batch verification of ECDSA signatures, one of which is based on Montgomery ladders. Simultaneous scalar multiplication in the context of DAC could be utilized to achieve improved running times in the Montgomery ladder signature verification method. Interest in higher order DACs also arises from the recent interest in standardizing Montgomery curves [39] such as Curve25519. Further motivation to construct higher order DACs is found in [10], where Bernstein presents new double scalar DAC algorithms and proposes further research on 3-dimensional versions of these ideas. This exploration can be extended to other double scalar multiplication algorithms too, such as Akishita's and Schoenmakers'. In 2006, Brown in [11] extended Bernstein's ideas to dimensions ≥ 2 but this method is patented [12]. One may utilize an extension of Schoenmakers' algorithm for this purpose, but recently, it was shown by SubramanyaRao in [17] that while this is feasible, it is not an efficient option. Thus, there is a need to construct other triple scalar algorithms. In this paper, we extend Akishita's algorithm to triple scalar multiplication. Our results in this paper are independent of Brown's results in [11].

The construction of Euclidean chain DACs were first proposed by Montgomery via his well known algorithms CFRC and PRAC in [13]. In the implementation of PRAC, point tripling may be useful. As far as we know, no formulas have been presented in the literature for differential point tripling. This paper presents differential point tripling formulae for Montgomery Curves.

Double Base Number Systems (DBNS) was proposed by Dimitrov et al. in [23] for efficient Elliptic curve scalar multiplication. In this context, efficient point quintupling formulae can be utilized. In this paper, we provide new quintupling formulae for Edwards curves improving upon those proposed by Bernstein in [22].

We also improve Mishra and Dimitrov's Quintupling algorithm [25] for Weierstrass' curves over a prime field.

In this paper, we denote a field multiplication, field squaring and field inversion by M, S and I respectively. Multiplication is a little more expensive than squaring. Typically S/M can range from 0.67 to about 0.8 depending on implementation and field size (Sect. 1.3.6 in [16]). The rest of the paper is structured as follows: We begin with a brief background on Montgomery Curve Arithmetic along with brief analysis of the straight forward method of computing $kP + lQ + sR$ on a Montgomery Curve in Sect. 2. In Sect. 3, we motivate the construction of a Montgomery ladder for simultaneous triple scalar multiplication(whilst presenting the ladder algorithm in Appendix D), with comparisons to the straight forward method. We then present the differential tripling formula for Montgomery curve arithmetic in Sect. 4. In Sect. 5, we present point quintupling formulae for Edwards curves and then conclude in Sect. 6.

2   Preliminaries

The Montgomery curve defined over a finite field $\mathbb{F}_p$ is given by $E_m : By^2 = x^3 + Ax^2 + x$. Let $P = (x_1, y_1)$ be a point on E_m. In projective coordinates, P can be written as $P = (X_1, Y_1, Z_1)$ and further let $[n]P = (X_n : Y_n : Z_n)$, where the scalar multiplication by n on E_m is denoted by $[n]$ and $[n]P = \underbrace{P + P + \cdots + P}_{n \text{ times}}$.

The set of points on E_m form an Abelian group and the identity element in this group is denoted as $\mathcal{O}$. The sum $[n+m]P = [n]P + [m]P$ can be computed using the formulae in the table below:

Addition: $(n \neq m)$	Doubling: $(n = m)$
$X_{m+n} = Z_{m-n}((X_m - Z_m)(X_n + Z_n)$ $\qquad + (X_m + Z_m)(X_n - Z_n))^2$ $Z_{m+n} = X_{m-n}((X_m - Z_m)(X_n + Z_n)$ $\qquad - (X_m + Z_m)(X_n - Z_n))^2$	$X_{2n} = (X_n + Z_n)^2(X_n - Z_n)^2$ $4X_n Z_n = (X_n + Z_n)^2 - (X_n - Z_n)^2$ $Z_{2n} = 4X_n Z_n((X_n - Z_n)^2 + ((A+2)/4)(4X_n Z_n))$ $\qquad = 4X_n Z_n((X_n + Z_n)^2 + ((A-2)/4)(4X_n Z_n))$

Thus point addition requires $4M + 2S$ operations and a point doubling requires $3M + 2S$ operations. If $Z_{m-n} = 1$, then point addition would require $3M + 2S$ operations. The above formulae for Montgomery curves were the first differential addition formulae published in the literature. The idea of differential addition has since been extended to other forms of elliptic curves such as

- Lopez and Dahab's [28] extension to Weierstrass curves over $\mathbb{F}_{2^m}$.
- Two independently developed extensions to Weierstrass curves over $\mathbb{F}_p$ - one due to Fisher, Giraud, Knudsen and Seifert [29] and the other due to Brier and Joye [30].
- Bernstein, Lange and Farashahi's [31] extension to Binary Edwards Curves.
- Justus &Loebenberger's [32] extension to Generalized Edwards Curves over $\mathbb{F}_q$ (Characteristic ($\mathbb{F}_q \neq 2$)).
- Devigne and Joye's [33] extension to Binary Huff Curves.

- Hutter, Joye and Sierra's [34] extension to Weierstrass curves over $\mathbb{F}_p$ in homogeneous projective coordinates where the points to be added share the same Z-coordinate.
- Wu, Tang and Feng's [35] extension to a new model of Binary Edwards curves.
- Farashahi and Joye's [36] extension to Generalized Binary Hessian Curves.

The well known Montgomery ladder for scalar multiplication that was initially proposed and utilized for Montgomery curves can be adapted to the examples listed above. The Left-to-Right Montgomery ladder is:

Algorithm 1. Left-to-Right Binary algorithm for Montgomery's ladder

---

INPUT: A point P on E_m and a positive integer $n = (n_t \ldots n_0)_2$
OUTPUT: The point $[n]P$

$P_1 \leftarrow P$ and $P_2 \leftarrow [2]P$
for $i = t - 1$ down to 0 do
    if $n_i = 0$ then
        $P_2 \leftarrow P_2 + P_1$ (P); $P_1 \leftarrow 2P_1$
    else
        $P_1 \leftarrow P_2 + P_1$ (P); $P_2 \leftarrow 2P_2$
    end if
end for
return P_1

In all algorithms in this paper, whenever the difference between two points is required to compute the sum of those points, the difference is indicated in brackets immediately after the addition formula. Thus, in Algorithm 1, we have

$$P_1 \leftarrow P_2 + P_1 \ (P). \tag{2}$$

The notation in Eq. (2) means that when P_2 is added to P_1 and the result is stored in P_1 while the difference required between these two points is P i.e., $P_2 - P_1 = P$. From the above algorithm we can see that, to compute $[n]P$, where $(n_t \ldots n_1 n_0)_2$ is the binary representation of n and $(n_t = 1)$, we hold $\{m_i P, (m_i + 1)P\}$ for $m_i = (n_t \ldots n_i)_2$. If $n_i = 0, m_i P = 2m_{i+1}P$ and $(m_i + 1)P = (m_{i+1} + 1)P + m_{i+1}P$ else $m_i P = (m_{i+1} + 1)P + m_{i+1}P$ and $(m_i + 1)P = 2(m_{i+1} + 1)P$. Beginning from $\{P, 2P\}$, Algorithm 1 computes $\{[n]P, [n+1]P\}$.

Closely following the approach in [7] and letting $|n|$ denote the bit length of n, computation of $[n]P$ on a Montgomery curve E_m would require $(6|n| - 3)M + (4|n| - 2)S$ operations. To compute x-coordinate of $kP + lQ + uR$ on E_m, using the straight forward method, we require the following steps:

1. Compute kP using the above Montgomery ladder.
2. Recover Y-coordinate of kP.

3. Compute lQ using the above Montgomery ladder.
4. Recover Y-coordinate of lQ.
5. Compute uR using the above Montgomery ladder.
6. Recover Y-coordinate of uR.
7. Compute $kP + lQ + uR$ in projective coordinates.
8. Compute x-coordinate(affine) of $kP + lQ + uR$.

We will assume the bit length of all the three scalars k, l and u to be the same. The algorithm for recovery of the Y-coordinate is described in [15] and this costs $(12M + S)$ operations. Then, the computational cost of steps 1, 3 and 5 together is $3\big[(6|k| - 3)M + (4|k| - 2)S\big]$. Steps 2, 4 and 6 together costs $3(12M + S)$. Consistent with [7], the cost of projective addition is $10M + 2S$, and thus the total cost of Step 7 is $2(10M + 2S)$ while step 8 costs $M + I$ where I denotes a field inversion. Thus the cost of computing the x-coordinate of $kP + lQ + sR$ is $(18|k| + 48)M + (12|k| + 3)S + I$.

3   Three-Dimensional Scalar Multiplication on a Montgomery Curve

In this section, we extend Akishita's ideas [7] to compute $kP + lQ + uR$. We define a set of 8 points

$$
G_i = \left\{
\begin{array}{l}
m_iP + n_iQ + s_iR \\
m_iP + n_iQ + (s_i + 1)R \\
m_iP + (n_i + 1)Q + s_iR \\
m_iP + (n_i + 1)Q + (s_i + 1)R \\
(m_i + 1)P + n_iQ + s_iR \\
(m_i + 1)P + n_iQ + (s_i + 1)R \\
(m_i + 1)P + (n_i + 1)Q + s_iR \\
(m_i + 1)P + (n_i + 1)Q + (s_i + 1)R
\end{array}
\right\}
$$

for $m_i = (k_t \ldots k_i)_2$, $n_i = (l_t \ldots l_i)_2$ and $s_i = (u_t \ldots u_i)_2$ where $(k_t \ldots k_1 k_0)_2$, $(l_t \ldots l_1 l_0)_2$ and $(u_t \ldots u_1 u_0)_2$ are binary representations of k, l and u respectively; $m_i = 2m_{i+1}$ or $m_i = (2m_{i+1} + 1)$ depending on whether $k_i = 0$ or $k_i = 1$. Similar relationships hold for n_i and s_i i.e., if $l_i = 0$, $n_i = 2n_{i+1}$ else $n_i = (2n_{i+1} + 1)$; if $u_i = 0$, $s_i = 2s_{i+1}$ else $s_i = (2s_{i+1} + 1)$. Each of the 8 elements in G_i can be written in terms of the elements in G_{i+1}. For instance, when $(k_i, l_i, u_i) = (0, 1, 0)$ we can write $m_i = 2m_{i+1}$, $n_i = (2n_{i+1} + 1)$ and $s_i = 2s_{i+1}$. In this case, as examples, we show a couple of elements of G_i written in terms of elements in G_{i+1} as follows:

$$m_iP + n_iQ + s_iR = (m_{i+1}P + n_{i+1}Q + s_{i+1}R) + (m_{i+1}P + (n_{i+1} + 1)Q + s_{i+1}R) \text{ and}$$
$$(m_i + 1)P + (n_i + 1)Q + s_iR = (m_{i+1}P + (n_{i+1} + 1)Q + s_{i+1}R) +$$
$$((m_{i+1} + 1)P + (n_{i+1} + 1)Q + s_{i+1}R).$$

We can write the other six elements of G_i similarly, in terms of elements of G_{i+1}. However, whilst computing the elements in G_i, we do not want to be using all of the eight elements in G_{i+1} towards computing $(kP + lQ + uR)$, because this would be more expensive than the straight forward computation of $kP + lQ + uR$. Straight forward computation using the binary ladder would require two such elements to be processed for each bit in the binary representation of a scalar and thus a total of six elements need to be processed for every bit in the three scalars taken at a time. Hence, to make our method more cost effective than the straight forward method, the number of elements in each G_{i+1} should be less than 6. It turns out that it is enough to have five elements in each of the G_{i+1} to achieve our goal of computing $(kP + lQ + uR)$. For example, if $(k_i, l_i, u_i) = (0, 0, 0)$, it would suffice to have the following five elements in G_{i+1}:

$$
\left.
\begin{cases}
m_{i+1}P + n_{i+1}Q + s_{i+1}R \\
m_{i+1}P + n_{i+1}Q + (s_{i+1} + 1)R \\
m_{i+1}P + (n_{i+1} + 1)Q + s_{i+1}R \\
(m_{i+1} + 1)P + n_{i+1}Q + s_{i+1}R \\
(m_{i+1} + 1)P + (n_{i+1} + 1)Q + s_{i+1}R
\end{cases}
\right\} \tag{3}
$$

If $(k_i, l_i, u_i) = (0, 1, 0)$, the following 5 elements in G_{i+1} would suffice:

$$
\left.
\begin{cases}
m_{i+1}P + n_{i+1}Q + s_{i+1}R \\
m_{i+1}P + (n_{i+1} + 1)Q + s_{i+1}R \\
m_{i+1}P + (n_{i+1} + 1)Q + (s_{i+1} + 1)R \\
(m_{i+1} + 1)P + (n_{i+1} + 1)Q + s_{i+1}R \\
(m_{i+1} + 1)P + (n_{i+1} + 1)Q + (s_{i+1} + 1)R
\end{cases}
\right\} \tag{4}
$$

Next, we need to construct rules for computing elements of G_i from G_{i+1}. For this, we take into consideration the values of k_{i-1}, l_{i-1} and u_{i-1} in addition to k_i, l_i and u_i. We show this with an example. Let $(k_i, l_i, u_i, k_{i-1}, l_{i-1}, u_{i-1}) = (0, 0, 0, 0, 1, 0)$. Then $m_i = 2m_{i+1}$, $n_i = 2n_{i+1}$, $s_i = 2s_{i+1}$. The five elements of G_{i+1} are the same as those depicted in Eq. (3) above. The five elements of G_i are

$$
\left.
\begin{cases}
m_iP + n_iQ + s_iR \\
m_iP + (n_i + 1)Q + s_iR \\
m_iP + (n_i + 1)Q + (s_i + 1)R \\
(m_i + 1)P + (n_i + 1)Q + s_iR \\
(m_i + 1)P + (n_i + 1)Q + (s_i + 1)R
\end{cases}
\right\} \tag{5}
$$

These five elements of G_i can be computed from those of G_{i+1} as follows:

$$m_i P + n_i Q + s_i R = (m_{i+1} P + n_{i+1} Q + s_{i+1} R) + (m_{i+1} P + n_{i+1} Q + s_{i+1} R),$$

$$m_i P + (n_i + 1)Q + s_i R$$
$$= (m_{i+1} P + n_{i+1} Q + s_{i+1} R) + (m_{i+1} P + (n_{i+1} + 1)Q + s_{i+1} R),$$

$$m_i P + (n_i + 1)Q + (s_i + 1)R$$
$$= (m_{i+1} P + n_{i+1} Q + (s_{i+1} + 1)R) + (m_{i+1} P + (n_{i+1} + 1)Q + s_{i+1} R),$$

$$(m_i + 1)P + (n_i + 1)Q + s_i R$$
$$= (m_{i+1} P + n_{i+1} Q + s_{i+1} R) + ((m_{i+1} + 1)P + (n_{i+1} + 1)Q + s_{i+1} R) \text{ and}$$

$$(m_i + 1)P + (n_i + 1)Q + (s_i + 1)R$$
$$= (m_{i+1} P + n_{i+1} Q + (s_{i+1} + 1)R) + ((m_{i+1} + 1)P + (n_{i+1} + 1)Q + s_{i+1} R).$$

If elements of G_{i+1} were to be listed as $T_0 Tmp$, $T_1 Tmp$, $T_2 Tmp$, $T_3 Tmp$ and $T_4 Tmp$ in the same order in Eq. (3) above and the elements of G_i were to be listed as T_0, T_1, T_2, T_3 and T_4 in the same order as in Eq. (5) above, then the following rules would enable us to compute elements of G_i from those of G_{i+1}:

$$T_0 \leftarrow 2T_0 Tmp,$$
$$T_1 \leftarrow T_2 Tmp + T_0 Tmp \ (Q),$$
$$T_2 \leftarrow T_2 Tmp + T_1 Tmp \ (Q - R),$$
$$T_3 \leftarrow T_4 Tmp + T_0 Tmp \ (P + Q) \text{ and}$$
$$T_4 \leftarrow T_4 Tmp + T_1 Tmp \ (P + Q - R).$$

As in the case of the formulae in the Montgomery ladder, the values in the brackets beside the formula above give the difference between points being added as these differences would be required for differential addition point arithmetic. We derived the Montgomery ladder rules when $(k_i, l_i, u_i, k_{i-1}, l_{i-1}, u_{i-1}) = (0, 0, 0, 0, 1, 0)$. Similarly, rules can be derived for the other 63 possible binary combinations of $(k_i, l_i, u_i, k_{i-1}, l_{i-1}, u_{i-1})$. Whilst we do not explicitly derive these rules here, we list in Appendix A, the five element set G_{i+1} for all combinations of (k_i, l_i, u_i) that was used in the construction of the 3 dimensional extension of Akishita's algorithm, which we present in Appendix D:

Referring to the three dimensional Montgomery ladder algorithm presented in Appendix D, we now analyze this algorithm when applied to Montgomery curves. As in the previous section, we will take the bit lengths of all the three scalars to be the same. Computing $P + Q$ and $P - Q$ in affine coordinates costs $4M + 2S + I$. Similarly points $((P + R), (P - R))$, $((Q + R), (Q - R))$ and $((P + Q + R), (P + Q - R))$ need to be precomputed as well in affine coordinates. Thus the total cost of the precomputation steps in Algorithm 2 is $4 * (4M + 2S + I) = 16M + 8S + 4I$. The cost of a point addition in the above ladder would be $3M + 2S$, as the difference of the points added is in affine form (i.e., $Z = 1$). In the For loop of the above algorithm, either point addition formulae are required four times and point doubling once or alternatively, the point addition formula is required five times per bit of the scalar k. Thus, the cost for every bit of k is $5 * (3M + 2S) = 15M + 10S$ and the total cost of the for loop in the above algorithm is $15(|k|-1)M + 10(|k|-1)S$. The finalization step after the for loop costs $3M + 2S$. Computation of the x-coordinate by $x = X/Z$ costs $M + I$. Thus the total cost of the above algorithm is $(15|k|+5)M + 10|k|S + 5I$. If $|k| = 160$, $S/M = 0.8$ and $I/M = 30$, the cost of the above algorithm is $3835M$.

For the same set of parameters, the cost of the straight forward algorithm as calculated in Sect. 2 is $(18|k|+48)M + (12|k|+3)S + I = 4496M$. Thus simultaneous triple scalar multiplication results in approximately 15 % improvement over the straight forward method. When $|k| = 256$, the improvement is about 22 % as the three dimensional Montgomery ladder costs $6043M$ and the straight forward method costs $7761M$.

As in the case of the one dimensional Montgomery ladder (Algorithm 1), the three dimensional Montgomery ladder (Algorithm 2) can be adapted to work with differential addition extensions to various other forms of elliptic curves(examples listed previously in Sect. 2) and not limited to Montgomery curves alone.

The usage of temporary variables can be improved in the above algorithm (Algorithm 2). Some operations towards the end of the computation can be eliminated. In the last iteration of the for loop in the above algorithm computation of T_2 and T_4 can be done away with, thus resulting in a further saving of at least $6M$ and $4S$ operations. Further, the cost of some finite field additions can be done away with by combining some point additions. For example if one has to compute $T_3 \leftarrow T_4\text{Tmp}+T_0\text{Tmp}$ (P+Q) and $T_4 \leftarrow T_4\text{Tmp}+T_1\text{Tmp}$ (P+Q-R) where $T_0\text{Tmp}=(X_0,Y_0,Z_0)$, $T_1\text{Tmp}=(X_1,Y_1,Z_1)$, $P+Q = (X_2,Y_2,Z_2)$, $(P+Q-R) = (X_3,Y_3,Z_3)$, $T_4\text{Tmp}=(X_4,Y_4,Z_4)$, $T_3=(X_5,Y_5,Z_5)$ and $T_4=(X_6,Y_6,Z_6)$ then

$$X_5 = Z_2[(X_0 - Z_0)(X_4 + Z_4) + (X_0 + Z_0)(X_4 - Z_4)]^2 \text{ and}$$
$$X_6 = Z_3[(X_1 - Z_1)(X_4 + Z_4) + (X_1 + Z_1)(X_4 - Z_4)]^2 \text{ while}$$
$$Z_5 = X_2[(X_0 - Z_0)(X_4 + Z_4) - (X_0 + Z_0)(X_4 - Z_4)]^2 \text{ and}$$
$$Z_6 = X_3[(X_1 - Z_1)(X_4 + Z_4) - (X_1 + Z_1)(X_4 - Z_4)]^2.$$

Thus one could club the computations of T_3 and T_4 together, thereby computing $(X_4 + Z_4)$ and $(X_4 - Z_4)$ just once, thus saving 2 field additions. In general the addition of points T_2+T_0, T_1+T_0 can save 2 field additions and can be extended to saving n field additions whilst computing $T_n + T_0$, $T_{n-1} + T_0$, ..., $T_1 + T_0$. Similar benefits can be obtained when one combines the point addition and doubling operations together. These enhancements can be utilized to improve the performance of the algorithm in this paper.

4 Differential Tripling Formulae for Montgomery Curves

Montgomery's PRAC [13] is an Euclidean chain algorithm that produces differential addition chains. PRAC can be used both to compute single dimensional and two dimensional differential addition chains. PRAC permits the exponent to be either prime or composite. Independent of the exponent being prime or composite, Montgomery provides a list of transformations that can be applied to the exponents in the course of constructing the differential chain. This list can be found in Table 4 of [13]. We reproduce the 7th and 8th transformations here for convenience:

Condition	Action(s)
$d \equiv -e \pmod 3$	$d \leftarrow (d - 2e)/3$ and $T_1 \leftarrow f(A,B,C)$ and $(A,B) \leftarrow (X_3(A)), f(T_1, A, B))$
$d \equiv e \pmod 3$	$d \leftarrow (d - e)/3$ and $(T_1, T_2) \leftarrow (f(A,B,C), f(A,C,B))$ and $(A,B,C) \leftarrow (X_3(A), T_1, T_2)$

The above two transformations call for point tripling.

When PRAC is used with a composite exponent, the transformations suggested by Montgomery was employed by Stam in Algorithm 3.33 in [8]. We reproduce the first two steps of this algorithm here:

ALGORITHM 3.33 (Montgomery's PRAC Algorithm)
Given a base v and an exponent n, this algorithm computes v_n

1. [Make d odd] Let f_2 be the highest power of 2 dividing n. Set $d \leftarrow (n/2^{f_2})$ and $A \leftarrow \delta^{f_2}(v)$.
2. [$d \neq 0 \pmod 3$] Let f_3 be the highest power of 3 dividing n. Set $d \leftarrow (d/3^{f_3})$ and $A \leftarrow \delta^{f_3}(A)$.

$\vdots$ $\qquad$ $\vdots$ $\qquad$ $\vdots$

As seen in step 2 of the algorithm above, if the exponent is a multiple of three and if f_3 is the highest power of 3 dividing the exponent n, then the algorithm requires f_3 triplings. For instance, if $n = 108 = 2^2 * 3^3$, doublings are carried out twice and triplings thrice. The triplings can be achieved by using a doubling and addition together. However, dedicated differential point tripling can be useful, if it is more efficient than a differential doubling and addition taken together.

Now we present differential point tripling formulae for Montgomery Curves. Let $P_1 = (X_1, Y_1, Z_1)$, $P_2 = (X_2, Y_2, Z_2)$ and $P_3 = (X_3, Y_3, Z_3)$ be points on a Montgomery curve E_m with $P_2 = 2P_1$ and $P_3 = 3P_1$. Then,

$$X_3 = X_1\left(\left(X_1^2 - Z_1^2\right)^2 - \left(X_1^2 + Z_1^2 + AX_1Z_1\right)(2Z_1)^2\right)^2 \text{ and}$$
$$Z_3 = Z_1\left(\left(X_1^2 - Z_1^2\right)^2 - \left(X_1^2 + Z_1^2 + AX_1Z_1\right)(2X_1)^2\right)^2.$$

Thus tripling needs $6M + 5S$ operations. The Algorithm to compute the above tripling is provided in part-A of the Appendices. If P_3 is computed as $2P_1 + P_1$ (i.e.,using a point doubling followed by a point addition) we need $(4M + 2S) + (3M + 2S) = 7M + 4S$. Thus the above tripling formulae are efficient as $(6M + 5S) < (7M+4S)$. When $Z_1 = 1$ the above tripling formulae only needs $(3M+4S)$

operations whereas, if $3P$ were to be computed as $2P + P$, (that is, a doubling followed by an addition with $Z_1 = 1$), then $(3M+2S)+(3M+2S) = (6M+4S)$ operations would be required thus resulting in a saving of $3M$.

5   Quintupling Formulae Revisited

The double base number system (DBNS) introduced initially by Dimitrov and Cooklev in [24] was utilized later in the context of elliptic curves in [23]. With this system, an integer n is written as

$$n = \sum_{i=1}^{l} s_i 2^{a_i} 3^{b_i} \qquad \text{or} \qquad n = \sum_{i=1}^{l} s_i 2^{a_i} 5^{b_i} \qquad \text{where } s_i = \pm 1.$$

The above idea can be generalized to a triple base number system where an integer n is represented as $n = \sum_{i-1}^{l} s_i 2^{a_i} 3^{b_i} 5^{c_i}$ where $s_i = \pm 1$. Double and Triple base number system representations, though very short, are not suitable for use in scalar multiplication algorithms. But, if by chance, the three exponents are all simultaneously decreasing, i.e., $a_1 \geq a_2 \geq \ldots a_l$ and $b_1 \geq b_2 \geq \ldots b_l$ and $c_1 \geq c_2 \geq \ldots c_l$, then Horner's rule scalar multiplication algorithm can be easily developed to computed $[n]P$. The simultaneously decreasing exponents can be computed using a greedy algorithm such as Algorithm 1 in [25]. From the double and triple base representations of a scalar n with simultaneously decreasing exponents as depicted above, we can see that fast point doubling, tripling and quintupling algorithms are highly desirable, as this would speed up the computation of $[n]P$. Thus, there has been a keen interest in obtaining faster point doubling, tripling and quintupling algorithms amongst researchers. In [25] the authors provide a fast quintupling algorithm for Affine coordinates on binary Weierstrass curves. In the same paper, the authors also propose a fast quintupling algorithm in Affine and Projective Jacobian coordinates for Weierstrass curves over $\mathbb{F}_p$. In Jacobian coordinates, the cost of the quintupling algorithm provided in [25] is $(15M + 10S)$. This was improved to $(7M + 16S)$ by Giorgi et al. in [26]. The authors in [25] take into account the multiplication by the curve parameter a while computing the cost of their algorithm, whereas the authors in [26] do not take into account the multiplication by the curve parameter a while computing the costs of their algorithm. Thus for comparison purposes we can take the cost of the algorithm in [26] to be $(8M + 16S)$. In [27], Longa et al. provide quintupling algorithm (Jacobian coordinates, Weierstrass curve over $\mathbb{F}_p$) with costs equal to $(10M + 14S)$. When the curve parameter $a = -3$, Mishra's algorithm in [25] costs $(15M + 8S)$, Giorgi's algorithm in [26] costs $(7M + 16S)$ and Longa's algorithm in [27] costs $(11M + 11S)$. Thus while Longa's algorithm in [27] performs better than Mishra's algorithm in [25], Giorgi's algorithm in [26] is the best option. However there are circumstances where the most cost-effective option is not always the best option for every situation. For example, in [37], Abarzua and Theriault, while designing side-channel resistant atomic blocks (for an introduction to atomic blocks, refer to Chap. 29 in [4]) for Weierstrass Elliptic

Curves in Jacobian Coordinates over prime fields, use a $(9M + 7S)$ formulae due to Dimitirov, Imbert and Mishra [23], as a more economical $(7M + 7S)$ formulae due to Longa and Miri [38] could not fit nicely into their atomic block pattern. Giorgi's $8M + 16S$ algorithm in [26] was derived using an automation implementing a directed acyclic graph structure looking for common subexpressions in the formulae and executing several arithmetic transformations. Using

$$2XY = (X + Y)^2 - X^2 - Y^2 \tag{6}$$

the complexity of Mishra's algorithm can be reduced, as shown below. Further in this section, we provide an improved algorithm for Quintupling on Edwards' Curves.

5.1 Mishra and Dimitrov's Algorithm for Quintupling on Weierstrass' Curves

A Weierstrass' curve over a prime field K is given by

$$H_w(K) : y^2 = x^3 + ax + b$$

where $a, b \in K$ and $4a^3 + 27b^2 \neq 0$. In projective Jacobian coordinates, the point $P = (X : Y : Z)$ corresponds to the point $(X/Z^2, Y/Z^3)$ on the above Weierstrass' curve $H_w(K)$. Given that P is a point on H_w and if $[5]P = [5](X : Y : Z) = (X_5, Y_5, Z_5)$, using Division Polynomials, Mishra and Dimitrov provide the following formulae in [25] to compute X_5, Y_5 and Z_5.

$$X_5 = XV^2 - 2YUW,$$
$$Y_5 = Y(E^3(12VL^2 - V^2 - 16L^4) - 64TL^5) \text{ and}$$
$$Z_5 = ZV$$

where

$$T = 8Y^4; \ (\text{Cost} = 2S),$$
$$M = 3X^2 + aZ^4; \ (\text{Cost} = 3S + 1M),$$
$$E = 12XY^2 - M^2; \ (\text{Cost} = 1S + 1M),$$
$$2L = 2ME - 2T; \ (\text{Cost} = 1M),$$
$$U = 4YL; \ (\text{Cost} = 1M),$$
$$V = 4TL - E^3; \ (\text{Cost} = 1S + 2M),$$
$$N = V - 4L^2; \ (\text{Cost} = 1S),$$
$$2W = 2EN; \ (\text{Cost} = 1M),$$
$$X_5 = 4(X.V^2 - 2Y.U.W); \ (\text{Cost} = 3M + 1S),$$
$$Y_5 = 8Y.[E^3.(12V.L^2 - V^2 - 16(L^2)^2) - 64(TL.(L^2)^2)]; \ (\text{Cost} = 4M + 1S)$$
$$\text{and } Z_5 = 2Z.V; \ (\text{Cost} = 1M).$$

Thus the total cost of computing the Quintupling formulae is $(15M + 10S)$. Now using Eq. (6), $12XY^2$, $4YL$, $12V.L^2$, ME and ZV can be computed as

$$12X.Y^2 = 6[(X + Y^2)^2 - X^2 - Y^4]; \ (\text{Cost} = 1S \text{ traded for } 1M),$$
$$4YL = 2[(Y + L)^2 - Y^2 - L^2]; \ (\text{Cost} = 1S \text{ traded for } 1M),$$
$$12V.L^2 = 6[(V + L^2) - V^2 - L^4]; \ (\text{Cost} = 1S \text{ traded for } 1M),$$
$$2ME = [(M + E)^2 - M^2 - E^2]; \ (\text{Cost} = 1S \text{ traded for } 1M) \text{ and}$$
$$2ZV = [(Z + V)^2 - Z^2 - V^2]; \ (\text{Cost} = 1S \text{ traded for } 1M).$$

Thus the cost of the Mishra and Dimitrov Quintupling algorithm can be reduced from $15M + 10S$ to $10M + 15S$. It turns out that one multiplication can further be eliminated from the Mishra and Dimitrov Algorithm. Indeed, in the computation of X_5, YUW is computed where $U = 4YL$ and thus $YUW = 4Y^2LW$. Thus we could alter U to be equal to $4Y^2L$ instead of $4YL$. Now, we could write $4Y^2L$ as

$$U = 4Y^2L = 2[(Y^2 + L)^2 - Y^4 - L^2]; \text{ and thus}$$
$$X_5 = 4(XV^2 - 2UW); \ (\text{New Cost} = 2M + 1S).$$

Thus the cost of the modified Mishra and Dimitrov Algorithm can be reduced to $9M + 15S$ which is just slightly better than the $10M + 14S$ cost of the Longa and Miri Quintupling Algorithm. Further we could compute N^2 as

$$N^2 = (V - 4L^2)^2 = V^2 + 16L^4 - 8VL^2.$$

Now N^2 could be computed without using any extra squarings or multiplications, as V^2, L^4 and VL^2 are computed for other steps in the algorithm, as shown above. Thus $2W = 2EN$ could be computed as

$$2W = [(E + N)^2 - E^2 - N^2].$$

Now, E^2 is also computed in another step in the algorithm, thus effectively replacing $1M$ with a $1S$. Thus, we have reduced the cost of the modified Mishra and Dimitrov algorithm to $8M + 16S$ mainly using Eq. (6).

5.2   Quintupling Formulae on Edwards Curves

In [22], Bernstein et al., amongst other things, provide with two fast algorithms for point quintupling on Edwards curves defined over $\mathbb{F}_p$. An Edward's curve E_d defined over a field k is given by the equation

$$x^2 + y^2 = 1 + dx^2y^2, \qquad \text{where } d \in k\backslash\{0, 1\}.$$

The two quintupling algorithms provided in [22], (we call them Algorithm A and Algorithm B for convenience) costs $(17M + 7S)$ and $(14M + 11S)$ respectively. The authors in [22] conclude that Algorithm A performs better if the S/M ratio i.e., $S/M > 0.75$ while Algorithm B performs better if $S/M < 0.75$.

When $S/M = 0.75$, both algorithms share the same complexity. Here we modify Algorithm B slightly to provide an alternate algorithm (Algorithm C). If the affine point $(X_1/Z_1, Y_1/Z_1)$ represents the point (X_1, Y_1, Z_1) on the homogenized equation of E_d, and if $(X_5, Y_5, Z_5) = 5(X_1, Y_1, Z_1)$, the quintupling algorithm is as below. (We justify the working of Algorithm C in Appendix C).

$$A = X_1^2; \quad B = Y_1^2; \quad C = Z_1^2; \quad D = A + B; \quad E = 2C - D; \quad F = A^2; \quad G = B^2;$$
$$H = F + G; \quad I = D^2 - H; \quad J = E^2; \quad K = G - F; \quad L = K^2; \quad M = 2I.J; \quad N = L + M;$$
$$O = L - M; \quad P = N.O; \quad Q = (E + K)^2 - J - L; \quad R = 2(2JH - L); \quad S = Q.R;$$
$$T = 4Q.O.(D - C); \quad U = R.N; \quad V = U + T; \quad W = U - T; \quad X_5 = 2X_1.(P + B.S).W;$$
$$Y_5 = 2Y_1.(P - A.S).V; \quad Z_5 = Z_1.V.W$$

Algorithm C above for quintupling costs $(15M + 9S)$ and is better than both the quintupling algorithms provided in [22] (Table 1).

Table 1. Edwards curve quintupling formulae summary

Algorithm	Quintupling costs
Bernstein et al. [22] (2007)	$17M + 7S$
Bernstein et al. [22] (2007)	$14M + 11S$
This work (2016)	$15M + 9S$

6 Conclusion

We presented a simultaneous triple scalar multiplication algorithm to compute the x-coordinate of $kP + lQ + uR$ on the Montgomery form Elliptic curve over $\mathbb{F}_p$. This method is about 15 to 22 % faster than the straight forward method of computing $kP + lQ + sR$. The new algorithm in this paper is ready to be used by implementers without the overhead of using a patented method such as Brown's method. The matrix execution phase required in Brown's method is not required in the algorithm presented in this paper and the number of precomputed points required is 8, whereas Brown's method, when used for the 3-dimensional case may require 13 precomputed points. Further, we also presented point tripling formulae on Montgomery curves and outlined a couple of scenarios where these formulae may be useful. We also provided an efficient point quintupling algorithm on Edwards Curves and an improvement of Mishra and Dimitrov's quintupling algorithm over Weierstrass' curves.

Acknowledgments. Many thanks to the anonymous reviewers of Africacrypt 2016 for their valuable feedback.

Appendices

A   Five Element Set G_{i+1} for all Combinations of (k_i, l_i, u_i)

Here, we list the five elements in G_{i+1} for all eight combinations of (k_i, l_i, u_i), that was used to construct the three dimensional Montgomery Ladder presented in Sect. 3 of this paper.

$(k_i, l_i, u_i) = (0,0,0)$:	$(k_i, l_i, u_i) = (0,0,1)$:
$m_{i+1}P + n_{i+1}Q + s_{i+1}R$	$m_{i+1}P + n_{i+1}Q + s_{i+1}R$
$m_{i+1}P + n_{i+1}Q + (s_{i+1} + 1)R$	$m_{i+1}P + n_{i+1}Q + (s_{i+1} + 1)R$
$m_{i+1}P + (n_{i+1} + 1)Q + s_{i+1}R$	$m_{i+1}P + (n_{i+1} + 1)Q + (s_{i+1} + 1)R$
$(m_{i+1} + 1)P + n_{i+1}Q + s_{i+1}R$	$(m_{i+1} + 1)P + n_{i+1}Q + (s_{i+1} + 1)R$
$(m_{i+1} + 1)P + (n_{i+1} + 1)Q + s_{i+1}R$	$(m_{i+1}+1)P+(n_{i+1}+1)Q+(s_{i+1}+1)R$
$(k_i, l_i, u_i) = (0,1,0)$:	$(k_i, l_i, u_i) = (0,1,1)$:
$m_{i+1}P + n_{i+1}Q + s_{i+1}R$	$m_{i+1}P + n_{i+1}Q + s_{i+1}R$
$m_{i+1}P + (n_{i+1} + 1)Q + s_{i+1}R$	$m_{i+1}P + n_{i+1}Q + (s_{i+1} + 1)R$
$m_{i+1}P + (n_{i+1} + 1)Q + (s_{i+1} + 1)R$	$m_{i+1}P + (n_{i+1} + 1)Q + s_{i+1}R$
$(m_{i+1} + 1)P + (n_{i+1} + 1)Q + s_{i+1}R$	$m_{i+1}P + (n_{i+1} + 1)Q + (s_{i+1} + 1)R$
$(m_{i+1}+1)P+(n_{i+1}+1)Q+(s_{i+1}+1)R$	$(m_{i+1}+1)P+(n_{i+1}+1)Q+(s_{i+1}+1)R$
$(k_i, l_i, u_i) = (1,0,0)$:	$(k_i, l_i, u_i) = (1,0,1)$:
$m_{i+1}P + n_{i+1}Q + s_{i+1}R$	$m_{i+1}P + n_{i+1}Q + s_{i+1}R$
$(m_{i+1} + 1)P + n_{i+1}Q + s_{i+1}R$	$m_{i+1}P + n_{i+1}Q + (s_{i+1} + 1)R$
$(m_{i+1} + 1)P + n_{i+1}Q + (s_{i+1} + 1)R$	$(m_{i+1} + 1)P + n_{i+1}Q + s_{i+1}R$
$(m_{i+1} + 1)P + (n_{i+1} + 1)Q + s_{i+1}R$	$(m_{i+1} + 1)P + n_{i+1}Q + (s_{i+1} + 1)R$
$(m_{i+1}+1)P+(n_{i+1}+1)Q+(s_{i+1}+1)R$	$(m_{i+1}+1)P+(n_{i+1}+1)Q+(s_{i+1}+1)R$
$(k_i, l_i, u_i) = (1,1,0)$:	$(k_i, l_i, u_i) = (1,1,1)$:
$m_{i+1}P + n_{i+1}Q + s_{i+1}R$	$m_{i+1}P + n_{i+1}Q + (s_{i+1} + 1)R$
$m_{i+1}P + (n_{i+1} + 1)Q + s_{i+1}R$	$m_{i+1}P + (n_{i+1} + 1)Q + (s_{i+1} + 1)R$
$(m_{i+1} + 1)P + n_{i+1}Q + s_{i+1}R$	$(m_{i+1} + 1)P + n_{i+1}Q + (s_{i+1}R + 1)$
$(m_{i+1} + 1)P + (n_{i+1} + 1)Q + s_{i+1}R$	$(m_{i+1} + 1)P + (n_{i+1} + 1)Q + s_{i+1}R$
$(m_{i+1}+1)P+(n_{i+1}+1)Q+(s_{i+1}+1)R$	$(m_{i+1}+1)P+(n_{i+1}+1)Q+(s_{i+1}+1)R$

B   Derivation of Differential Tripling Formula on Montgomery Curves and an Algorithm for Differential Tripling

We derive the differential point tripling formulae for Montgomery Curves. Let $P_1 = (X_1, Y_1, Z_1)$, $P_2 = (X_2, Y_2, Z_2)$ and $P_3 = (X_3, Y_3, Z_3)$ be points on a Montgomery curve E_m with $P_2 = 2P_1$ and $P_3 = 3P_1$. We can write $P_3 = 3P_1 = 2P_1 + P_1 = P_2 + P_1$. Then,

$$X_2 = (X_1 + Z_1)^2 (X_1 - Z_1)^2 \tag{7}$$

$$Z_2 = 4X_1 Z_1 ((X_1 - Z_1)^2 + ((A + 2)/4)(4X_1 Z_1)) \tag{8}$$

$$X_3 = Z_1 [(X_1 - Z_1)(X_2 + Z_2) + (X_1 + Z_1)(X_2 - Z_2)]^2 \tag{9}$$

$$Z_3 = X_1 [(X_1 - Z_1)(X_2 + Z_2) - (X_1 + Z_1)(X_2 - Z_2)]^2 \tag{10}$$

From Eqs. (7), (8), (9) and (10) we can write

$$X_3 = Z_1 \left[(X_1 - Z_1) \left\{ (X_1 + Z_1)^2 (X_1 - Z_1)^2 + 4X_1 Z_1 (X_1 - Z_1)^2 + ((A + 2)/4)(4X_1 Z_1)^2 \right\} + \right.$$
$$\left. (X_1 + Z_1) \{ (X_1 + Z_1)^2 (X_1 - Z_1)^2 - 4X_1 Z_1 (X_1 - Z_1)^2 - ((A + 2)/4)(4X_1 Z_1)^2 \} \right]^2$$
$$= Z_1 \left[((X_1 + Z_1)(X_1 - Z_1))^2 \{ 2X_1 \} - 4X_1 Z_1 (X_1 - Z_1)^2 \{ 2Z_1 \} - \right.$$
$$\left. ((A + 2)/4)(4X_1 Z_1)^2 \{ 2Z_1 \} \right]^2$$
$$= 4X_1^2 Z_1 \left[((X_1 + Z_1)(X_1 - Z_1))^2 - 4Z_1^2 (X_1 - Z_1)^2 - ((A + 2)/4) \left(16X_1 Z_1^3 \right) \right]^2$$
$$= 4X_1^2 Z_1 \left(\left(X_1^2 - Z_1^2 \right)^2 - \left(X_1^2 + Z_1^2 + AX_1 Z_1 \right)(2Z_1)^2 \right)^2$$

Similarly,

$$Z_3 = X_1 \left[(X_1 - Z_1) \left\{ (X_1 + Z_1)^2 (X_1 - Z_1)^2 + 4X_1 Z_1 (X_1 - Z_1)^2 + ((A + 2)/4)(4X_1 Z_1)^2 \right\} - \right.$$
$$\left. (X_1 + Z_1) \{ (X_1 + Z_1)^2 (X_1 - Z_1)^2 - 4X_1 Z_1 (X_1 - Z_1)^2 - ((A + 2)/4)(4X_1 Z_1)^2 \} \right]^2$$
$$= X_1 \left[((X_1 + Z_1)(X_1 - Z_1))^2 \{ -2Z_1 \} - 4X_1 Z_1 (X_1 - Z_1)^2 \{ 2X_1 \} - \right.$$
$$\left. ((A + 2)/4)(4X_1 Z_1)^2 \{ 2X_1 \} \right]^2$$
$$= 4X_1 Z_1^2 \left[-((X_1 + Z_1)(X_1 - Z_1))^2 + 4X_1^2 (X_1 - Z_1)^2 + ((A + 2)/4) \left(16X_1^3 Z_1 \right) \right]^2$$
$$= 4X_1 Z_1^2 \left(-\left(X_1^2 - Z_1^2 \right)^2 + \left(X_1^2 + Z_1^2 + AX_1 Z_1 \right)(2X_1)^2 \right)^2$$
$$= 4X_1 Z_1^2 \left(\left(X_1^2 - Z_1^2 \right)^2 - \left(X_1^2 + Z_1^2 + AX_1 Z_1 \right)(2X_1)^2 \right)^2$$

Dividing both X_3 and Z_3 by $4X_1 Z_1$ we get, when $(X_1, Y_1) \neq (0, 0)$

$$X_3 = X_1 \left(\left(X_1^2 - Z_1^2 \right)^2 - \left(X_1^2 + Z_1^2 + AX_1 Z_1 \right)(2Z_1)^2 \right)^2$$
$$Z_3 = Z_1 \left(\left(X_1^2 - Z_1^2 \right)^2 - \left(X_1^2 + Z_1^2 + AX_1 Z_1 \right)(2X_1)^2 \right)^2$$

The formulae for X_3 and Y_3 derived above can be computed using the following algorithm:

$T_1 \leftarrow X_1; \quad T_2 \leftarrow Z_1$

$T_1 \leftarrow T_1^2$	$(= X_1^2)$
$T_2 \leftarrow T_2^2$	$(= Z_1^2)$
$T_3 \leftarrow (T_1 - T_2)^2$	$(= (X_1^2 - Z_1^2)^2)$
$T_4 \leftarrow X_1 Z_1$	$(= X_1 Z_1)$
$T_4 \leftarrow A.T_4$	$(= AX_1 Z_1)$
$T_5 \leftarrow T_2 + T_2 + T_2 + T_2$	$(= 4Z_1^2)$
$T_6 \leftarrow T_1 + T_1 + T_1 + T_1$	$(= 4X_1^2)$
$T_4 \leftarrow T_1 + T_2 + T_4$	$(= X_1^2 + Z_1^2 + AX_1 Z_1)$
$T_7 \leftarrow T_4.T_5$	$(= (X_1^2 + Z_1^2 + AX_1 Z_1)(4Z_1^2))$
$T_8 \leftarrow T_4.T_6$	$(= (X_1^2 + Z_1^2 + AX_1 Z_1)(4X_1^2))$
$T_1 \leftarrow (T_3 - T_7)^2$	$(= ((X_1^2 - Z_1^2)^2 - (X_1^2 + Z_1^2 + AX_1 Z_1)(2Z_1)^2)^2)$
$T_2 \leftarrow (T_3 - T_8)^2$	$(= ((X_1^2 - Z_1^2)^2 - (X_1^2 + Z_1^2 + AX_1 Z_1)(2X_1)^2)^2)$
$X_3 \leftarrow X_1.T_1$	$(= X_1((X_1^2 - Z_1^2)^2 - (X_1^2 + Z_1^2 + AX_1 Z_1)(2Z_1)^2)^2)$
$Z_3 \leftarrow Z_1.T_2$	$(= Z_1((X_1^2 - Z_1^2)^2 - (X_1^2 + Z_1^2 + AX_1 Z_1)(2X_1)^2)^2)$

C   Edwards Curve Quintupling Formulae

Algorithms A and B were verified by the authors in [22]. The only difference between Algorithm C presented in this paper and Algorithm B in [22] is in the computation of R. It was computed in Algorithm B as

$$R = ((D + E)^2 - J - H - I)^2 - 2N$$

In Algorithm C, we employ $R = 2(2JH - L)$ as we can rewrite R as follows:

$$
\begin{aligned}
R &= \left\{ \left[(X_1^2 + Y_1^2) + (2Z_1^2 - (X_1^2 + Y_1^2)) \right]^2 \right. \\
&\quad \left. - \left[2Z_1^2 - (X_1^2 + Y_1^2) \right]^2 - \left[X_1^4 + Y_1^4 \right] - 2X_1^2 Y_1^2 \right\}^2 - 2N \\
&= \left[2(X_1^2 + Y_1^2)(2Z_1^2 - (X_1^2 + Y_1^2)) \right]^2 \\
&\quad - 2\left[(Y_1^4 - X_1^4)^2 + 4(X_1^2 Y_1^2)\{2Z_1^2 - (X_1^2 + Y_1^2)\}^2 \right] \\
&= 4\left[2Z_1^2 - (X_1^2 + Y_1^2) \right]^2 \{ X_1^4 + Y_1^4 \} - 2\left[(Y_1^4 - X_1^4)^2 \right] = 4JH - 2L = 2(2JH - L)
\end{aligned}
$$

D   Three Dimensional Montgomery Ladder Algorithm

Algorithm 2. L-R 3-Dimensional Montgomery Ladder

```
INPUT: Points P, Q, R on E_m and positive integers k, l, u;
k = (k_t ··· k_1, k_0)_2, l = (l_t ··· l_1, l_0)_2,
      u = (u_t ··· u_1, u_0)_2;    (at least one of k_t or l_t or u_t = 1).
      Precompute A ← (P + Q), B ← (P − Q), C ← (P + R), D ← (P − R), E ← (Q + R),
      F ← (Q − R), G ← (P + Q + R), H ← (P + Q − R);
OUTPUT: x coordinate of W = kP + lQ + uR.
```

[Initialize]

if $k_t, l_t, u_t = (0, 0, 1)$
    $T_0 \leftarrow \mathcal{O}$, $T_1 \leftarrow R$, $T_2 \leftarrow E$, $T_3 \leftarrow C$,
    $T_4 \leftarrow G$;
else if $k_t, l_t, u_t = (0, 1, 0)$
    $T_0 \leftarrow \mathcal{O}$, $T_1 \leftarrow Q$, $T_2 \leftarrow E$, $T_3 \leftarrow A$,
    $T_4 \leftarrow G$;
else if $k_t, l_t, u_t = (0, 1, 1)$
    $T_0 \leftarrow \mathcal{O}$, $T_1 \leftarrow R$, $T_2 \leftarrow Q$, $T_3 \leftarrow E$,
    $T_4 \leftarrow G$;
else if $k_t, l_t, u_t = (1, 0, 0)$
    $T_0 \leftarrow \mathcal{O}$, $T_1 \leftarrow P$, $T_2 \leftarrow C$, $T_3 \leftarrow A$,
    $T_4 \leftarrow G$;

else if $k_t, l_t, u_t = (1, 0, 1)$
    $T_0 \leftarrow \mathcal{O}$, $T_1 \leftarrow R$, $T_2 \leftarrow P$, $T_3 \leftarrow C$,
    $T_4 \leftarrow G$;
else if $k_t, l_t, u_t = (1, 1, 0)$
    $T_0 \leftarrow \mathcal{O}$, $T_1 \leftarrow Q$, $T_2 \leftarrow P$, $T_3 \leftarrow A$,
    $T_4 \leftarrow G$;
else if $k_t, l_t, u_t = (1, 1, 1)$
    $T_0 \leftarrow R$, $T_1 \leftarrow E$, $T_2 \leftarrow C$, $T_3 \leftarrow A$,
    $T_4 \leftarrow G$;

```
[Process the three scalar bits simultaneously]
for i from t down to 1
```

$T_0\mathtt{Tmp} \leftarrow T_0$, $T_1\mathtt{Tmp} \leftarrow T_1$, $T_2\mathtt{Tmp} \leftarrow T_2$, $T_3\mathtt{Tmp} \leftarrow T_3$, $T_4\mathtt{Tmp} \leftarrow T_4$;

if $(k_i, l_i, u_i, k_{i-1}, l_{i-1}, u_{i-1}) = (0, 0, 0, 0, 0, 0)$

$T_0 \leftarrow 2T_0\mathtt{Tmp}$, $T_1 \leftarrow T_1\mathtt{Tmp}+T_0\mathtt{Tmp}(\mathtt{R})$, $T_2 \leftarrow T_2\mathtt{Tmp}+T_0\mathtt{Tmp}(\mathtt{Q})$, $T_3 \leftarrow T_3\mathtt{Tmp}+T_0\mathtt{Tmp}(\mathtt{P})$,
$T_4 \leftarrow T_4\mathtt{Tmp}+T_0\mathtt{Tmp}(\mathtt{A})$;

else if $(k_i, l_i, u_i, k_{i-1}, l_{i-1}, u_{i-1}) = (0, 0, 0, 0, 0, 1)$

$T_0 \leftarrow 2T_0\mathtt{Tmp}$, $T_1 \leftarrow T_1\mathtt{Tmp}+T_0\mathtt{Tmp}(\mathtt{R})$, $T_2 \leftarrow T_2\mathtt{Tmp}+T_1\mathtt{Tmp}(\mathtt{F})$, $T_3 \leftarrow T_3\mathtt{Tmp}+T_1\mathtt{Tmp}(\mathtt{D})$,
$T_4 \leftarrow T_4\mathtt{Tmp}+T_1\mathtt{Tmp}(\mathtt{H})$;

else if $(k_i, l_i, u_i, k_{i-1}, l_{i-1}, u_{i-1}) = (0, 0, 0, 0, 1, 0)$

$T_0 \leftarrow 2T_0\mathtt{Tmp}$, $T_1 \leftarrow T_2\mathtt{Tmp}+T_0\mathtt{Tmp}(\mathtt{Q})$, $T_2 \leftarrow T_2\mathtt{Tmp}+T_1\mathtt{Tmp}(\mathtt{F})$, $T_3 \leftarrow T_4\mathtt{Tmp}+T_0\mathtt{Tmp}(\mathtt{A})$,
$T_4 \leftarrow T_4\mathtt{Tmp}+T_1\mathtt{Tmp}(\mathtt{H})$;

else if $(k_i, l_i, u_i, k_{i-1}, l_{i-1}, u_{i-1}) = (0, 0, 0, 0, 1, 1)$

$T_0 \leftarrow 2T_0\mathtt{Tmp}$, $T_1 \leftarrow T_1\mathtt{Tmp}+T_0\mathtt{Tmp}(\mathtt{R})$, $T_2 \leftarrow T_2\mathtt{Tmp}+T_0\mathtt{Tmp}(\mathtt{Q})$, $T_3 \leftarrow T_2\mathtt{Tmp}+T_1\mathtt{Tmp}(\mathtt{F})$,
$T_4 \leftarrow T_4\mathtt{Tmp}+T_1\mathtt{Tmp}(\mathtt{H})$;

else if $(k_i, l_i, u_i, k_{i-1}, l_{i-1}, u_{i-1}) = (0, 0, 0, 1, 0, 0)$

$T_0 \leftarrow 2T_0\mathtt{Tmp}$, $T_1 \leftarrow T_3\mathtt{Tmp}+T_0\mathtt{Tmp}(\mathtt{P})$, $T_2 \leftarrow T_3\mathtt{Tmp}+T_1\mathtt{Tmp}(\mathtt{D})$, $T_3 \leftarrow T_4\mathtt{Tmp}+T_0\mathtt{Tmp}(\mathtt{A})$,
$T_4 \leftarrow T_4\mathtt{Tmp}+T_1\mathtt{Tmp}(\mathtt{H})$;

else if $(k_i, l_i, u_i, k_{i-1}, l_{i-1}, u_{i-1}) = (0, 0, 0, 1, 0, 1)$

$T_0 \leftarrow 2T_0\mathtt{Tmp}$, $T_1 \leftarrow T_1\mathtt{Tmp}+T_0\mathtt{Tmp}(\mathtt{R})$, $T_2 \leftarrow T_3\mathtt{Tmp}+T_0\mathtt{Tmp}(\mathtt{P})$, $T_3 \leftarrow T_3\mathtt{Tmp}+T_1\mathtt{Tmp}(\mathtt{D})$,
$T_4 \leftarrow T_4\mathtt{Tmp}+T_1\mathtt{Tmp}(\mathtt{H})$;

else if $(k_i, l_i, u_i, k_{i-1}, l_{i-1}, u_{i-1}) = (0, 0, 0, 1, 1, 0)$

$T_0 \leftarrow 2T_0\mathtt{Tmp}$, $T_1 \leftarrow T_2\mathtt{Tmp}+T_0\mathtt{Tmp}(\mathtt{Q})$, $T_2 \leftarrow T_3\mathtt{Tmp}+T_0\mathtt{Tmp}(\mathtt{P})$, $T_3 \leftarrow T_4\mathtt{Tmp}+T_0\mathtt{Tmp}(\mathtt{A})$,
$T_4 \leftarrow T_4\mathtt{Tmp}+T_1\mathtt{Tmp}(\mathtt{H})$;

else if $(k_i, l_i, u_i, k_{i-1}, l_{i-1}, u_{i-1}) = (0, 0, 0, 1, 1, 1)$

$T_0 \leftarrow T_1\mathtt{Tmp}+T_0\mathtt{Tmp}(\mathtt{R})$, $T_1 \leftarrow T_2\mathtt{Tmp}+T_1\mathtt{Tmp}(\mathtt{F})$, $T_2 \leftarrow T_3\mathtt{Tmp}+T_1\mathtt{Tmp}(\mathtt{D})$,
$T_3 \leftarrow T_4\mathtt{Tmp}+T_0\mathtt{Tmp}(\mathtt{A})$, $T_4 \leftarrow T_4\mathtt{Tmp}+T_1\mathtt{Tmp}(\mathtt{H})$;

else if $(k_i, l_i, u_i, k_{i-1}, l_{i-1}, u_{i-1}) = (0, 0, 1, 0, 0, 0)$

$T_0 \leftarrow T_1\mathtt{Tmp}+T_0\mathtt{Tmp}(\mathtt{R})$, $T_1 \leftarrow 2T_1\mathtt{Tmp}$, $T_2 \leftarrow T_2\mathtt{Tmp}+T_0\mathtt{Tmp},(\mathtt{E})$ $T_3 \leftarrow T_3\mathtt{Tmp}+T_0\mathtt{Tmp}(\mathtt{C})$,
$T_4 \leftarrow T_4\mathtt{Tmp}+T_0\mathtt{Tmp}(\mathtt{G})$;

else if $(k_i, l_i, u_i, k_{i-1}, l_{i-1}, u_{i-1}) = (0, 0, 1, 0, 0, 1)$

$T_0 \leftarrow T_1\mathtt{Tmp}+T_0\mathtt{Tmp}(\mathtt{R})$, $T_1 \leftarrow 2T_1\mathtt{Tmp}$, $T_2 \leftarrow T_2\mathtt{Tmp}+T_1\mathtt{Tmp}(\mathtt{Q})$, $T_3 \leftarrow T_3\mathtt{Tmp}+T_1\mathtt{Tmp}(\mathtt{P})$,
$T_4 \leftarrow T_3\mathtt{Tmp}+T_2\mathtt{Tmp}(\mathtt{B})$;

else if $(k_i, l_i, u_i, k_{i-1}, l_{i-1}, u_{i-1}) = (0, 0, 1, 0, 1, 0)$

$T_0 \leftarrow T_1\mathtt{Tmp}+T_0\mathtt{Tmp}(\mathtt{R})$, $T_1 \leftarrow T_2\mathtt{Tmp}+T_0\mathtt{Tmp}(\mathtt{E})$, $T_2 \leftarrow T_2\mathtt{Tmp}+T_1\mathtt{Tmp}(\mathtt{Q})$,
$T_3 \leftarrow T_4\mathtt{Tmp}+T_0\mathtt{Tmp}(\mathtt{G})$, $T_4 \leftarrow T_3\mathtt{Tmp}+T_2\mathtt{Tmp}(\mathtt{B})$;

else if $(k_i, l_i, u_i, k_{i-1}, l_{i-1}, u_{i-1}) = (0, 0, 1, 0, 1, 1)$

$T_0 \leftarrow T_1\mathtt{Tmp}+T_0\mathtt{Tmp}(\mathtt{R})$, $T_1 \leftarrow 2T_1\mathtt{Tmp}$, $T_2 \leftarrow T_2\mathtt{Tmp}+T_0\mathtt{Tmp}(\mathtt{E})$, $T_3 \leftarrow T_2\mathtt{Tmp}+T_1\mathtt{Tmp}(\mathtt{Q})$,
$T_4 \leftarrow T_3\mathtt{Tmp}+T_2\mathtt{Tmp}(\mathtt{B})$;

else if $(k_i, l_i, u_i, k_{i-1}, l_{i-1}, u_{i-1}) = (0, 0, 1, 1, 0, 0)$

$T_0 \leftarrow T_1\mathtt{Tmp}+T_0\mathtt{Tmp}(\mathtt{R})$, $T_1 \leftarrow T_3\mathtt{Tmp}+T_0\mathtt{Tmp}(\mathtt{C})$, $T_2 \leftarrow T_3\mathtt{Tmp}+T_1\mathtt{Tmp}(\mathtt{P})$,
$T_3 \leftarrow T_4\mathtt{Tmp}+T_0\mathtt{Tmp}(\mathtt{G})$, $T_4 \leftarrow T_3\mathtt{Tmp}+T_2\mathtt{Tmp}(\mathtt{B})$;

else if $(k_i, l_i, u_i, k_{i-1}, l_{i-1}, u_{i-1}) = (0, 0, 1, 1, 0, 1)$

$T_0 \leftarrow T_1\mathtt{Tmp}+T_0\mathtt{Tmp}(\mathtt{R})$, $T_1 \leftarrow 2T_1\mathtt{Tmp}$, $T_2 \leftarrow T_3\mathtt{Tmp}+T_0\mathtt{Tmp}(\mathtt{C})$, $T_3 \leftarrow T_3\mathtt{Tmp}+T_1\mathtt{Tmp}(\mathtt{P})$,
$T_4 \leftarrow T_3\mathtt{Tmp}+T_2\mathtt{Tmp}(\mathtt{B})$;

else if $(k_i, l_i, u_i, k_{i-1}, l_{i-1}, u_{i-1}) = (0, 0, 1, 1, 1, 0)$

$T_0 \leftarrow T_1\mathtt{Tmp}+T_0\mathtt{Tmp}(\mathtt{R})$, $T_1 \leftarrow T_2\mathtt{Tmp}+T_0\mathtt{Tmp}(\mathtt{E})$, $T_2 \leftarrow T_3\mathtt{Tmp}+T_0\mathtt{Tmp}(\mathtt{C})$,
$T_3 \leftarrow T_4\mathtt{Tmp}+T_0\mathtt{Tmp}(\mathtt{G})$, $T_4 \leftarrow T_3\mathtt{Tmp}+T_2\mathtt{Tmp}(\mathtt{B})$;

else if $(k_i, l_i, u_i, k_{i-1}, l_{i-1}, u_{i-1}) = (0, 0, 1, 1, 1, 1)$

$T_0 \leftarrow 2T_1\mathtt{Tmp}$, $T_1 \leftarrow T_2\mathtt{Tmp}+T_1\mathtt{Tmp}(\mathtt{Q})$, $T_2 \leftarrow T_3\mathtt{Tmp}+T_1\mathtt{Tmp}(\mathtt{P})$,
$T_3 \leftarrow T_4\mathtt{Tmp}+T_0\mathtt{Tmp}(\mathtt{G})$, $T_4 \leftarrow T_3\mathtt{Tmp}+T_2\mathtt{Tmp}(\mathtt{B})$;

else if $(k_i, l_i, u_i, k_{i-1}, l_{i-1}, u_{i-1}) = (0, 1, 0, 0, 0, 0)$

$T_0 \leftarrow T_1\mathtt{Tmp}+T_0\mathtt{Tmp}(\mathtt{Q})$, $T_1 \leftarrow T_2\mathtt{Tmp}+T_0\mathtt{Tmp}(\mathtt{E})$, $T_2 \leftarrow 2T_1\mathtt{Tmp}$, $T_3 \leftarrow T_3\mathtt{Tmp}+T_0\mathtt{Tmp}(\mathtt{A})$,
$T_4 \leftarrow T_3\mathtt{Tmp}+T_1\mathtt{Tmp}(\mathtt{C})$;

else if $(k_i, l_i, u_i, k_{i-1}, l_{i-1}, u_{i-1}) = (0, 1, 0, 0, 0, 1)$

$T_0 \leftarrow T_1\mathtt{Tmp}+T_0\mathtt{Tmp}(\mathtt{Q})$, $T_1 \leftarrow T_2\mathtt{Tmp}+T_0\mathtt{Tmp}(\mathtt{E})$, $T_2 \leftarrow T_2\mathtt{Tmp}+T_1\mathtt{Tmp}(\mathtt{R})$,
$T_3 \leftarrow T_4\mathtt{Tmp}+T_0\mathtt{Tmp}(\mathtt{G})$, $T_4 \leftarrow T_4\mathtt{Tmp}+T_1\mathtt{Tmp}(\mathtt{C})$;

else if $(k_i, l_i, u_i, k_{i-1}, l_{i-1}, u_{i-1}) = (0, 1, 0, 0, 1, 0)$

$T_0 \leftarrow T_1\mathtt{Tmp}+T_0\mathtt{Tmp}(\mathtt{Q})$, $T_1 \leftarrow 2T_1\mathtt{Tmp}$, $T_2 \leftarrow T_2\mathtt{Tmp}+T_1\mathtt{Tmp}(\mathtt{R})$ $T_3 \leftarrow T_3\mathtt{Tmp}+T_1\mathtt{Tmp}(\mathtt{P})$,
$T_4 \leftarrow T_4\mathtt{Tmp}+T_1\mathtt{Tmp}(\mathtt{C})$;

else if $(k_i, l_i, u_i, k_{i-1}, l_{i-1}, u_{i-1}) = (0, 1, 0, 0, 1, 1)$

$T_0 \leftarrow T_1\mathtt{Tmp}+T_0\mathtt{Tmp}(\mathtt{Q})$, $T_1 \leftarrow T_2\mathtt{Tmp}+T_0\mathtt{Tmp}(\mathtt{E})$, $T_2 \leftarrow 2T_1\mathtt{Tmp}$, $T_3 \leftarrow T_2\mathtt{Tmp}+T_1\mathtt{Tmp}(\mathtt{R})$,
$T_4 \leftarrow T_4\mathtt{Tmp}+T_1\mathtt{Tmp}(\mathtt{C})$;

else if $(k_i, l_i, u_i, k_{i-1}, l_{i-1}, u_{i-1}) = (0, 1, 0, 1, 0, 0)$

$T_0 \leftarrow T_1\mathtt{Tmp}+T_0\mathtt{Tmp}(\mathtt{Q})$, $T_1 \leftarrow T_3\mathtt{Tmp}+T_0\mathtt{Tmp}(\mathtt{A})$, $T_2 \leftarrow T_4\mathtt{Tmp}+T_0\mathtt{Tmp}(\mathtt{G})$,
$T_3 \leftarrow T_3\mathtt{Tmp}+T_1\mathtt{Tmp}(\mathtt{P})$, $T_4 \leftarrow T_4\mathtt{Tmp}+T_1\mathtt{Tmp}(\mathtt{C})$;

else if $(k_i, l_i, u_i, k_{i-1}, l_{i-1}, u_{i-1}) = (0, 1, 0, 1, 0, 1)$

$T_0 \leftarrow T_1\mathtt{Tmp}+T_0\mathtt{Tmp}(\mathtt{Q})$, $T_1 \leftarrow T_2\mathtt{Tmp}+T_0\mathtt{Tmp}(\mathtt{E})$, $T_2 \leftarrow T_3\mathtt{Tmp}+T_0\mathtt{Tmp}(\mathtt{A})$,

$T_3 \leftarrow T_4\text{Tmp}+T_0\text{Tmp(G)}, \ T_4 \leftarrow T_4\text{Tmp}+T_1\text{Tmp(C)};$
else if $(k_i, l_i, u_i, k_{i-1}, l_{i-1}, u_{i-1}) = (0, 1, 0, 1, 1, 0)$
$\quad T_0 \leftarrow T_1\text{Tmp}+T_0\text{Tmp(Q)}, \ T_1 \leftarrow 2T_1\text{Tmp}, \ T_2 \leftarrow T_3\text{Tmp}+T_0\text{Tmp(A)},$
$\quad T_3 \leftarrow T_3\text{Tmp}+T_1\text{Tmp(P)}, \ T_4 \leftarrow T_4\text{Tmp}+T_1\text{Tmp(C)};$
else if $(k_i, l_i, u_i, k_{i-1}, l_{i-1}, u_{i-1}) = (0, 1, 0, 1, 1, 1)$
$\quad T_0 \leftarrow T_2\text{Tmp}+T_0\text{Tmp(E)}, \ T_1 \leftarrow T_2\text{Tmp}+T_1\text{Tmp(R)}, \ T_2 \leftarrow T_4\text{Tmp}+T_0\text{Tmp(G)},$
$\quad T_3 \leftarrow T_3\text{Tmp}+T_1\text{Tmp(P)}, \ T_4 \leftarrow T_4\text{Tmp}+T_1\text{Tmp(C)};$
else if $(k_i, l_i, u_i, k_{i-1}, l_{i-1}, u_{i-1}) = (0, 1, 1, 0, 0, 0)$
$\quad T_0 \leftarrow T_3\text{Tmp}+T_0\text{Tmp(E)}, \ T_1 \leftarrow T_3\text{Tmp}+T_1\text{Tmp(Q)}, \ T_2 \leftarrow T_3\text{Tmp}+T_2\text{Tmp(R)},$
$\quad T_3 \leftarrow T_4\text{Tmp}+T_0\text{Tmp(G)}, \ T_4 \leftarrow T_4\text{Tmp}+T_2\text{Tmp(C)};$
else if $(k_i, l_i, u_i, k_{i-1}, l_{i-1}, u_{i-1}) = (0, 1, 1, 0, 0, 1)$
$\quad T_0 \leftarrow T_3\text{Tmp}+T_0\text{Tmp(E)}, \ T_1 \leftarrow T_3\text{Tmp}+T_1\text{Tmp(Q)}, \ T_2 \leftarrow 2T_3\text{Tmp}, \ T_3 \leftarrow T_4\text{Tmp}+T_1\text{Tmp(A)},$
$\quad T_4 \leftarrow T_4\text{Tmp}+T_3\text{Tmp(P)};$
else if $(k_i, l_i, u_i, k_{i-1}, l_{i-1}, u_{i-1}) = (0, 1, 1, 0, 1, 0)$
$\quad T_0 \leftarrow T_3\text{Tmp}+T_0\text{Tmp(E)}, \ T_1 \leftarrow T_3\text{Tmp}+T_2\text{Tmp(R)}, \ T_2 \leftarrow 2T_3\text{Tmp}, \ T_3 \leftarrow T_4\text{Tmp}+T_2\text{Tmp(C)},$
$\quad T_4 \leftarrow T_4\text{Tmp}+T_3\text{Tmp(P)};$
else if $(k_i, l_i, u_i, k_{i-1}, l_{i-1}, u_{i-1}) = (0, 1, 1, 0, 1, 1)$
$\quad T_0 \leftarrow T_3\text{Tmp}+T_0\text{Tmp(E)}, \ T_1 \leftarrow T_3\text{Tmp}+T_1\text{Tmp(Q)}, \ T_2 \leftarrow T_3\text{Tmp}+T_2\text{Tmp(R)}, \ T_3 \leftarrow 2T_3\text{Tmp},$
$\quad T_4 \leftarrow T_4\text{Tmp}+T_3\text{Tmp(P)};$
else if $(k_i, l_i, u_i, k_{i-1}, l_{i-1}, u_{i-1}) = (0, 1, 1, 1, 0, 0)$
$\quad T_0 \leftarrow T_3\text{Tmp}+T_0\text{Tmp(E)}, \ T_1 \leftarrow T_4\text{Tmp}+T_0\text{Tmp(G)}, \ T_2 \leftarrow T_4\text{Tmp}+T_1\text{Tmp(A)},$
$\quad T_3 \leftarrow T_4\text{Tmp}+T_2\text{Tmp(C)}, \ T_4 \leftarrow T_4\text{Tmp}+T_3\text{Tmp(P)};$
else if $(k_i, l_i, u_i, k_{i-1}, l_{i-1}, u_{i-1}) = (0, 1, 1, 1, 0, 1)$
$\quad T_0 \leftarrow T_3\text{Tmp}+T_0\text{Tmp(E)}, \ T_1 \leftarrow T_3\text{Tmp}+T_1\text{Tmp(Q)}, \ T_2 \leftarrow T_4\text{Tmp}+T_0\text{Tmp(G)},$
$\quad T_3 \leftarrow T_4\text{Tmp}+T_1\text{Tmp(A)}, \ T_4 \leftarrow T_4\text{Tmp}+T_3\text{Tmp(P)};$
else if $(k_i, l_i, u_i, k_{i-1}, l_{i-1}, u_{i-1}) = (0, 1, 1, 1, 1, 0)$
$\quad T_0 \leftarrow T_3\text{Tmp}+T_0\text{Tmp(E)}, \ T_1 \leftarrow T_3\text{Tmp}+T_2\text{Tmp(R)}, \ T_2 \leftarrow T_4\text{Tmp}+T_0\text{Tmp(G)},$
$\quad T_3 \leftarrow T_4\text{Tmp}+T_2\text{Tmp(C)}, \ T_4 \leftarrow T_4\text{Tmp}+T_3\text{Tmp(P)};$
else if $(k_i, l_i, u_i, k_{i-1}, l_{i-1}, u_{i-1}) = (0, 1, 1, 1, 1, 1)$
$\quad T_0 \leftarrow T_3\text{Tmp}+T_1\text{Tmp(Q)}, \ T_1 \leftarrow 2T_3\text{Tmp}, \ T_2 \leftarrow T_4\text{Tmp}+T_1\text{Tmp(A)}, \ T_3 \leftarrow T_4\text{Tmp}+T_2\text{Tmp(C)},$
$\quad T_4 \leftarrow T_4\text{Tmp}+T_3\text{Tmp(P)};$
else if $(k_i, l_i, u_i, k_{i-1}, l_{i-1}, u_{i-1}) = (1, 0, 0, 0, 0, 0)$
$\quad T_0 \leftarrow T_1\text{Tmp}+T_0\text{Tmp(P)}, \ T_1 \leftarrow T_2\text{Tmp}+T_0\text{Tmp(C)}, \ T_2 \leftarrow T_3\text{Tmp}+T_0\text{Tmp(A)}, \ T_3 \leftarrow 2T_1\text{Tmp},$
$\quad T_4 \leftarrow T_3\text{Tmp}+T_1\text{Tmp(Q)};$
else if $(k_i, l_i, u_i, k_{i-1}, l_{i-1}, u_{i-1}) = (1, 0, 0, 0, 0, 1)$
$\quad T_0 \leftarrow T_1\text{Tmp}+T_0\text{Tmp(P)}, \ T_1 \leftarrow T_2\text{Tmp}+T_0\text{Tmp(C)}, \ T_2 \leftarrow T_4\text{Tmp}+T_0\text{Tmp(G)},$
$\quad T_3 \leftarrow T_2\text{Tmp}+T_1\text{Tmp(R)}, \ T_4 \leftarrow T_4\text{Tmp}+T_1\text{Tmp(E)};$
else if $(k_i, l_i, u_i, k_{i-1}, l_{i-1}, u_{i-1}) = (1, 0, 0, 0, 1, 0)$
$\quad T_0 \leftarrow T_1\text{Tmp}+T_0\text{Tmp(P)}, \ T_1 \leftarrow T_3\text{Tmp}+T_0\text{Tmp(A)}, \ T_2 \leftarrow T_4\text{Tmp}+T_0\text{Tmp(G)},$
$\quad T_3 \leftarrow T_3\text{Tmp}+T_1\text{Tmp(Q)}, \ T_4 \leftarrow T_4\text{Tmp}+T_1\text{Tmp(E)};$
else if $(k_i, l_i, u_i, k_{i-1}, l_{i-1}, u_{i-1}) = (1, 0, 0, 0, 1, 1)$
$\quad T_0 \leftarrow T_1\text{Tmp}+T_0\text{Tmp(P)}, \ T_1 \leftarrow T_2\text{Tmp}+T_0\text{Tmp(C)}, \ T_2 \leftarrow T_3\text{Tmp}+T_0\text{Tmp(A)},$
$\quad T_3 \leftarrow T_4\text{Tmp}+T_0\text{Tmp(G)}, \ T_4 \leftarrow T_4\text{Tmp}+T_1\text{Tmp(E)};$
else if $(k_i, l_i, u_i, k_{i-1}, l_{i-1}, u_{i-1}) = (1, 0, 0, 1, 0, 0)$
$\quad T_0 \leftarrow T_1\text{Tmp}+T_0\text{Tmp(P)}, \ T_1 \leftarrow 2T_1\text{Tmp}, \ T_2 \leftarrow T_2\text{Tmp}+T_1\text{Tmp(R)}, \ T_3 \leftarrow T_3\text{Tmp}+T_1\text{Tmp(Q)},$
$\quad T_4 \leftarrow T_4\text{Tmp}+T_1\text{Tmp(E)};$
else if $(k_i, l_i, u_i, k_{i-1}, l_{i-1}, u_{i-1}) = (1, 0, 0, 1, 0, 1)$
$\quad T_0 \leftarrow T_1\text{Tmp}+T_0\text{Tmp(P)}, \ T_1 \leftarrow T_2\text{Tmp}+T_0\text{Tmp(C)}, \ T_2 \leftarrow 2T_1\text{Tmp}, \ T_3 \leftarrow T_2\text{Tmp}+T_1\text{Tmp(R)},$
$\quad T_4 \leftarrow T_4\text{Tmp}+T_1\text{Tmp(E)};$
else if $(k_i, l_i, u_i, k_{i-1}, l_{i-1}, u_{i-1}) = (1, 0, 0, 1, 1, 0)$
$\quad T_0 \leftarrow T_1\text{Tmp}+T_0\text{Tmp(P)}, \ T_1 \leftarrow T_3\text{Tmp}+T_0\text{Tmp(A)}, \ T_2 \leftarrow 2T_1\text{Tmp}, \ T_3 \leftarrow T_3\text{Tmp}+T_1\text{Tmp(Q)},$
$\quad T_4 \leftarrow T_4\text{Tmp}+T_1\text{Tmp(E)};$
else if $(k_i, l_i, u_i, k_{i-1}, l_{i-1}, u_{i-1}) = (1, 0, 0, 1, 1, 1)$
$\quad T_0 \leftarrow T_2\text{Tmp}+T_0\text{Tmp(C)}, \ T_1 \leftarrow T_4\text{Tmp}+T_0\text{Tmp(G)}, \ T_2 \leftarrow T_2\text{Tmp}+T_1\text{Tmp(R)},$
$\quad T_3 \leftarrow T_3\text{Tmp}+T_1\text{Tmp(Q)}, \ T_4 \leftarrow T_4\text{Tmp}+T_1\text{Tmp(E)};$
else if $(k_i, l_i, u_i, k_{i-1}, l_{i-1}, u_{i-1}) = (1, 0, 1, 0, 0, 0)$
$\quad T_0 \leftarrow T_3\text{Tmp}+T_0\text{Tmp(C)}, \ T_1 \leftarrow T_3\text{Tmp}+T_1\text{Tmp(P)}, \ T_2 \leftarrow T_4\text{Tmp}+T_0\text{Tmp(G)},$
$\quad T_3 \leftarrow T_3\text{Tmp}+T_2\text{Tmp(R)}, \ T_4 \leftarrow T_4\text{Tmp}+T_2\text{Tmp(E)};$
else if $(k_i, l_i, u_i, k_{i-1}, l_{i-1}, u_{i-1}) = (1, 0, 1, 0, 0, 1)$
$\quad T_0 \leftarrow T_3\text{Tmp}+T_0\text{Tmp(C)}, \ T_1 \leftarrow T_3\text{Tmp}+T_1\text{Tmp(P)}, \ T_2 \leftarrow T_4\text{Tmp}+T_1\text{Tmp(A)}, \ T_3 \leftarrow 2T_3\text{Tmp},$
$\quad T_4 \leftarrow T_4\text{Tmp}+T_3\text{Tmp(Q)};$
else if $(k_i, l_i, u_i, k_{i-1}, l_{i-1}, u_{i-1}) = (1, 0, 1, 0, 1, 0)$
$\quad T_0 \leftarrow T_3\text{Tmp}+T_0\text{Tmp(C)}, \ T_1 \leftarrow T_4\text{Tmp}+T_0\text{Tmp(G)}, \ T_2 \leftarrow T_4\text{Tmp}+T_1\text{Tmp(A)},$
$\quad T_3 \leftarrow T_4\text{Tmp}+T_2\text{Tmp(E)}, \ T_4 \leftarrow T_4\text{Tmp}+T_3\text{Tmp(Q)};$
else if $(k_i, l_i, u_i, k_{i-1}, l_{i-1}, u_{i-1}) = (1, 0, 1, 0, 1, 1)$
$\quad T_0 \leftarrow T_3\text{Tmp}+T_0\text{Tmp(C)}, \ T_1 \leftarrow T_3\text{Tmp}+T_1\text{Tmp(P)}, \ T_2 \leftarrow T_4\text{Tmp}+T_0\text{Tmp(G)},$
$\quad T_3 \leftarrow T_4\text{Tmp}+T_1\text{Tmp(A)}, \ T_4 \leftarrow T_4\text{Tmp}+T_3\text{Tmp(Q)};$
else if $(k_i, l_i, u_i, k_{i-1}, l_{i-1}, u_{i-1}) = (1, 0, 1, 1, 0, 0)$

$T_0 \leftarrow T_3\mathtt{Tmp}+T_0\mathtt{Tmp(C)}$, $T_1 \leftarrow T_3\mathtt{Tmp}+T_2\mathtt{Tmp(R)}$, $T_2 \leftarrow 2T_3\mathtt{Tmp}$, $T_3 \leftarrow T_4\mathtt{Tmp}+T_2\mathtt{Tmp(E)}$,
$T_4 \leftarrow T_4\mathtt{Tmp}+T_3\mathtt{Tmp(Q)}$;
else if $(k_i,l_i,u_i,k_{i-1},l_{i-1},u_{i-1}) = (1,0,1,1,0,1)$
$T_0 \leftarrow T_3\mathtt{Tmp}+T_0\mathtt{Tmp(C)}$, $T_1 \leftarrow T_3\mathtt{Tmp}+T_1\mathtt{Tmp(P)}$, $T_2 \leftarrow T_3\mathtt{Tmp}+T_2\mathtt{Tmp(R)}$, $T_3 \leftarrow 2T_3\mathtt{Tmp}$,
$T_4 \leftarrow T_4\mathtt{Tmp}+T_3\mathtt{Tmp(Q)}$;
else if $(k_i,l_i,u_i,k_{i-1},l_{i-1},u_{i-1}) = (1,0,1,1,1,0)$
$T_0 \leftarrow T_3\mathtt{Tmp}+T_0\mathtt{Tmp(C)}$, $T_1 \leftarrow T_4\mathtt{Tmp}+T_0\mathtt{Tmp(G)}$, $T_2 \leftarrow T_3\mathtt{Tmp}+T_2\mathtt{Tmp(R)}$,
$T_3 \leftarrow T_4\mathtt{Tmp}+T_2\mathtt{Tmp(E)}$, $T_4 \leftarrow T_4\mathtt{Tmp}+T_3\mathtt{Tmp(Q)}$;
else if $(k_i,l_i,u_i,k_{i-1},l_{i-1},u_{i-1}) = (1,0,1,1,1,1)$
$T_0 \leftarrow T_3\mathtt{Tmp}+T_1\mathtt{Tmp(P)}$, $T_1 \leftarrow T_4\mathtt{Tmp}+T_1\mathtt{Tmp(A)}$, $T_2 \leftarrow 2T_3\mathtt{Tmp}$, $T_3 \leftarrow T_4\mathtt{Tmp}+T_2\mathtt{Tmp(E)}$,
$T_4 \leftarrow T_4\mathtt{Tmp}+T_3\mathtt{Tmp(Q)}$;
else if $(k_i,l_i,u_i,k_{i-1},l_{i-1},u_{i-1}) = (1,1,0,0,0,0)$
$T_0 \leftarrow T_3\mathtt{Tmp}+T_0\mathtt{Tmp(A)}$, $T_1 \leftarrow T_4\mathtt{Tmp}+T_0\mathtt{Tmp(G)}$, $T_2 \leftarrow T_3\mathtt{Tmp}+T_1\mathtt{Tmp(P)}$,
$T_3 \leftarrow T_3\mathtt{Tmp}+T_2\mathtt{Tmp(Q)}$, $T_4 \leftarrow 2T_3\mathtt{Tmp}$;
else if $(k_i,l_i,u_i,k_{i-1},l_{i-1},u_{i-1}) = (1,1,0,0,0,1)$
$T_0 \leftarrow T_3\mathtt{Tmp}+T_0\mathtt{Tmp(A)}$, $T_1 \leftarrow T_4\mathtt{Tmp}+T_0\mathtt{Tmp(G)}$, $T_2 \leftarrow T_4\mathtt{Tmp}+T_1\mathtt{Tmp(C)}$,
$T_3 \leftarrow T_4\mathtt{Tmp}+T_2\mathtt{Tmp(E)}$, $T_4 \leftarrow T_4\mathtt{Tmp}+T_3\mathtt{Tmp(R)}$;
else if $(k_i,l_i,u_i,k_{i-1},l_{i-1},u_{i-1}) = (1,1,0,0,1,0)$
$T_0 \leftarrow T_3\mathtt{Tmp}+T_0\mathtt{Tmp(A)}$, $T_1 \leftarrow T_3\mathtt{Tmp}+T_1\mathtt{Tmp(P)}$, $T_2 \leftarrow T_4\mathtt{Tmp}+T_1\mathtt{Tmp(C)}$, $T_3 \leftarrow 2T_3\mathtt{Tmp}$,
$T_4 \leftarrow T_4\mathtt{Tmp}+T_3\mathtt{Tmp(R)}$;
else if $(k_i,l_i,u_i,k_{i-1},l_{i-1},u_{i-1}) = (1,1,0,0,1,1)$
$T_0 \leftarrow T_3\mathtt{Tmp}+T_0\mathtt{Tmp(A)}$, $T_1 \leftarrow T_4\mathtt{Tmp}+T_0\mathtt{Tmp(G)}$, $T_2 \leftarrow T_3\mathtt{Tmp}+T_1\mathtt{Tmp(P)}$,
$T_3 \leftarrow T_4\mathtt{Tmp}+T_1\mathtt{Tmp(C)}$, $T_4 \leftarrow T_4\mathtt{Tmp}+T_3\mathtt{Tmp(R)}$;
else if $(k_i,l_i,u_i,k_{i-1},l_{i-1},u_{i-1}) = (1,1,0,1,0,0)$
$T_0 \leftarrow T_3\mathtt{Tmp}+T_0\mathtt{Tmp(A)}$, $T_1 \leftarrow T_3\mathtt{Tmp}+T_2\mathtt{Tmp(Q)}$, $T_2 \leftarrow T_4\mathtt{Tmp}+T_2\mathtt{Tmp(E)}$,
$T_3 \leftarrow 2T_3\mathtt{Tmp}$, $T_4 \leftarrow T_4\mathtt{Tmp}+T_3\mathtt{Tmp(R)}$;
else if $(k_i,l_i,u_i,k_{i-1},l_{i-1},u_{i-1}) = (1,1,0,1,0,1)$
$T_0 \leftarrow T_3\mathtt{Tmp}+T_0\mathtt{Tmp(A)}$, $T_1 \leftarrow T_4\mathtt{Tmp}+T_0\mathtt{Tmp(G)}$, $T_2 \leftarrow T_3\mathtt{Tmp}+T_2\mathtt{Tmp(Q)}$,
$T_3 \leftarrow T_4\mathtt{Tmp}+T_2\mathtt{Tmp(E)}$, $T_4 \leftarrow T_4\mathtt{Tmp}+T_3\mathtt{Tmp(R)}$;
else if $(k_i,l_i,u_i,k_{i-1},l_{i-1},u_{i-1}) = (1,1,0,1,1,0)$
$T_0 \leftarrow T_3\mathtt{Tmp}+T_0\mathtt{Tmp(A)}$, $T_1 \leftarrow T_3\mathtt{Tmp}+T_1\mathtt{Tmp(P)}$, $T_2 \leftarrow T_3\mathtt{Tmp}+T_2\mathtt{Tmp(Q)}$, $T_3 \leftarrow 2T_3\mathtt{Tmp}$,
$T_4 \leftarrow T_4\mathtt{Tmp}+T_3\mathtt{Tmp(R)}$;
else if $(k_i,l_i,u_i,k_{i-1},l_{i-1},u_{i-1}) = (1,1,0,1,1,1)$
$T_0 \leftarrow T_4\mathtt{Tmp}+T_0\mathtt{Tmp(G)}$, $T_1 \leftarrow T_4\mathtt{Tmp}+T1\mathtt{Tmp(C)}$, $T_2 \leftarrow T_4\mathtt{Tmp}+T_2\mathtt{Tmp(E)}$, $T_3 \leftarrow 2T_3\mathtt{Tmp}$,
$T_4 \leftarrow T_4\mathtt{Tmp}+T_3\mathtt{Tmp(R)}$;
else if $(k_i,l_i,u_i,k_{i-1},l_{i-1},u_{i-1}) = (1,1,1,0,0,0)$
$T_0 \leftarrow T_3\mathtt{Tmp}+T_0\mathtt{Tmp(H)}$, $T_1 \leftarrow T_4\mathtt{Tmp}+T_0\mathtt{Tmp(A)}$, $T_2 \leftarrow T_3\mathtt{Tmp}+T_1\mathtt{Tmp(D)}$,
$T_3 \leftarrow T_3\mathtt{Tmp}+T_2\mathtt{Tmp(F)}$, $T_4 \leftarrow T_4\mathtt{Tmp}+T_3\mathtt{Tmp(R)}$
else if $(k_i,l_i,u_i,k_{i-1},l_{i-1},u_{i-1}) = (1,1,1,0,0,1)$
$T_0 \leftarrow T_3\mathtt{Tmp}+T_0\mathtt{Tmp(H)}$, $T_1 \leftarrow T_4\mathtt{Tmp}+T_0\mathtt{Tmp(A)}$, $T_2 \leftarrow T_4\mathtt{Tmp}+T_1\mathtt{Tmp(P)}$,
$T_3 \leftarrow T_4\mathtt{Tmp}+T_2\mathtt{Tmp(Q)}$, $T_4 \leftarrow 2T_4\mathtt{Tmp}$;
else if $(k_i,l_i,u_i,k_{i-1},l_{i-1},u_{i-1}) = (1,1,1,0,1,0)$
$T_0 \leftarrow T_3\mathtt{Tmp}+T_0\mathtt{Tmp(H)}$, $T_1 \leftarrow T_3\mathtt{Tmp}+T_1\mathtt{Tmp(D)}$, $T_2 \leftarrow T_4\mathtt{Tmp}+T_1\mathtt{Tmp(P)}$,
$T_3 \leftarrow T_4\mathtt{Tmp}+T_3\mathtt{Tmp(R)}$, $T_4 \leftarrow 2T_4\mathtt{Tmp}$;
else if $(k_i,l_i,u_i,k_{i-1},l_{i-1},u_{i-1}) = (1,1,1,0,1,1)$
$T_0 \leftarrow T_3\mathtt{Tmp}+T_0\mathtt{Tmp(H)}$, $T_1 \leftarrow T_4\mathtt{Tmp}+T_0\mathtt{Tmp(A)}$, $T_2 \leftarrow T_3\mathtt{Tmp}+T_1\mathtt{Tmp(D)}$,
$T_3 \leftarrow T_4\mathtt{Tmp}+T_1\mathtt{Tmp(P)}$, $T_4 \leftarrow 2T_4\mathtt{Tmp}$;
else if $(k_i,l_i,u_i,k_{i-1},l_{i-1},u_{i-1}) = (1,1,1,1,0,0)$
$T_0 \leftarrow T_3\mathtt{Tmp}+T_0\mathtt{Tmp(H)}$, $T_1 \leftarrow T_3\mathtt{Tmp}+T_2\mathtt{Tmp(F)}$, $T_2 \leftarrow T_4\mathtt{Tmp}+T_2\mathtt{Tmp(Q)}$,
$T_3 \leftarrow T_4\mathtt{Tmp}+T_3\mathtt{Tmp(R)}$, $T_4 \leftarrow 2T_4\mathtt{Tmp}$;
else if $(k_i,l_i,u_i,k_{i-1},l_{i-1},u_{i-1}) = (1,1,1,1,0,1)$
$T_0 \leftarrow T_3\mathtt{Tmp}+T_0\mathtt{Tmp(H)}$, $T_1 \leftarrow T_4\mathtt{Tmp}+T_0\mathtt{Tmp(A)}$, $T_2 \leftarrow T_3\mathtt{Tmp}+T_2\mathtt{Tmp(F)}$,
$T_3 \leftarrow T_4\mathtt{Tmp}+T_2\mathtt{Tmp(Q)}$, $T_4 \leftarrow 2T_4\mathtt{Tmp}$;
else if $(k_i,l_i,u_i,k_{i-1},l_{i-1},u_{i-1}) = (1,1,1,1,1,0)$
$T_0 \leftarrow T_3\mathtt{Tmp}+T_0\mathtt{Tmp(H)}$, $T_1 \leftarrow T_3\mathtt{Tmp}+T_1\mathtt{Tmp(D)}$, $T_2 \leftarrow T_3\mathtt{Tmp}+T_2\mathtt{Tmp(F)}$,
$T_3 \leftarrow T_4\mathtt{Tmp}+T_3\mathtt{Tmp(R)}$, $T_4 \leftarrow 2T_4\mathtt{Tmp}$;
else if $(k_i,l_i,u_i,k_{i-1},l_{i-1},u_{i-1}) = (1,1,1,1,1,1)$
$T_0 \leftarrow T_4\mathtt{Tmp}+T_0\mathtt{Tmp(A)}$, $T_1 \leftarrow T_4\mathtt{Tmp}+T_1\mathtt{Tmp(P)}$, $T_2 \leftarrow T_4\mathtt{Tmp}+T_2\mathtt{Tmp(Q)}$,
$T_3 \leftarrow T_4\mathtt{Tmp}+T_3\mathtt{Tmp(R)}$, $T_4 \leftarrow 2T_4\mathtt{Tmp}$;
end for
[Finalize]

if $(k_0, l_0, u_0) = (0,0,0)$
    $W \leftarrow 2T_0$
else if $(k_0, l_0, u_0) = (0,0,1)$
    $W \leftarrow T_1 + T_0$ (R)
else if $(k_0, l_0, u_0) = (0,1,0)$
    $W \leftarrow T_1 + T_0$ (Q)

else if $(k_0, l_0, u_0) = (0,1,1)$
    $W \leftarrow T_3 + T_0$ (E)
else if $(k_0, l_0, u_0) = (1,0,0)$
    $W \leftarrow T_1 + T_0$ (P)
else if $(k_0, l_0, u_0) = (1,0,1)$
    $W \leftarrow T_3 + T_0$ (C)

else if $(k_0, l_0, u_0) = (1,1,0)$
    $W \leftarrow T_3 + T_0$ (A)
else if $(k_0, l_0, u_0) = (1,1,1)$
    $W \leftarrow T_3 + T_0$ (H)

Return x−coordinate of W by computing $x = X/Z$

References

1. Stinson, D.: Cryptography: Theory and Practice, 3rd edn. CRC Press, Boca Raton (2005)
2. Bellman, R., Straus, E.G.: Addition chains of vectors (problem 5125). Am. Math. Mon. **71**, 806–808 (1964)
3. ElGamal, T.: A public key cryptosystem and a signature scheme base on discrete logarithms. In: Blakley, G.R., Chaum, D. (eds.) CRYPTO 1984. LNCS, vol. 196, pp. 10–18. Springer, Heidelberg (1985)
4. Cohen, H., Frey, G.: Handbook of Elliptic and Hyperelliptic Curve Cryptography. CRC Press, Boca Raton (2006)
5. Solinas, J.A.: Low-weight binary representations for pairs of integers. Combinatorics and Optimization Research Report CORR 2001-41. University of Waterloo (2001)
6. Montgomery, P.L.: Speeding the pollard and elliptic curve methods of factorization. Math. Comput. **48**(177), 243–264 (1987)
7. Akishita, T.: Fast simultaneous scalar multiplication on elliptic curve with montgomery form. In: Vaudenay, S., Youssef, A.M. (eds.) SAC 2001. LNCS, vol. 2259, pp. 255–267. Springer, Heidelberg (2001)
8. Stam, M.: Speeding up subgroup cryptosystems. Ph.D. thesis, Technische Universiteit Eindhoven (2003)
9. Knuth, D.E.: The Art of Computer Programming. Seminumerical algorithms, vol. 2, 3rd edn. Pearson, London (1998)
10. Bernstein, D.J.: Differential Addition Chains (2006). http://cr.yp.to/ecdh/diffchain-20060219.pdf. Accessed 25 January 2015
11. Brown, D.R.L.: Multi-dimensional Montgomery ladders for elliptic curves (2006). http://eprint.iacr.org/2006/220. Accessed 25 January 2015
12. Brown, D.R.L.: Multi-dimensional Montgomery ladders for elliptic curves. Patent No. US8750500 B2 (2014). http://www.google.com/patents/US8750500
13. Montgomery, P.L.: Evaluating recurrences of form $X_{m+n} = f(x_m, X_n, X_{m-n})$ via Lucas chains (1992). https://cr.yp.to/bib/1992/montgomery-lucas.ps. Accessed 2 February 2016
14. Azarderakhsh, R., Karabina, K.: A New Double Point Multiplication Method and its Implementation on Binary Elliptic Curves with Endomorphisms. http://cacr.uwaterloo.ca/techreports/2012/cacr2012-24.pdf
15. Okeya, K., Sakurai, K.: Efficient elliptic curve cryptosystems from a scalar multiplication algorithm with recovery of the y-coordinate on a montgomery form elliptic curve. In: Ko, K., Naccache, D., Paar, C. (eds.) CHES 2001. LNCS, vol. 2162, pp. 126–141. Springer, Heidelberg (2001)

16. Brent, R., Zimmermann, P.: Modern Computer Arithmetic. Cambridge Monographs on Applied and Computational Mathematics. Cambridge University Press, Cambridge (2010)
17. Subramanya Rao, S.R.: A note on Schoenmakers' algorithm for multiexponentiation. In: Obaidat, M.S., Lorenz, P., Samarati, P. (eds.) Proceedings of International Conference on Security and Cryptography, SECRYPT 2015, pp. 384–391. SciTePress, Setúbal (2015)
18. Menezes, A., van Oorschot, P., Vanstone, S.: Handbook of Applied Cryptography. Taylor and Francis, London (1997)
19. Antipa, A., Brown, D., Gallant, R., Lambert, R., Struik, R., Vanstone, S.: Accelerated verification of ECDSA signatures. http://cacr.uwaterloo.ca/techreports/2005/cacr2005-28.pdf. Accessed 2 February 2016
20. Cheon, J.H., Yi, J.H.: Fast batch verification of multiple signatures. In: Okamoto, T., Wang, X. (eds.) PKC 2007. LNCS, vol. 4450, pp. 442–457. Springer, Heidelberg (2007)
21. Karati, S., Das, A., Roychoudhury, D.: Randomized batch verification of standard ECDSA signatures. In: Chakraborty, R.S., Matyas, V., Schaumont, P. (eds.) SPACE 2014. LNCS, vol. 8804, pp. 237–255. Springer, Heidelberg (2014)
22. Bernstein, D.J., Birkner, P., Lange, T., Peters, C.: Optimizing Double-Base Elliptic-Curve Single-Scalar Multiplication. https://cr.yp.to/antiforgery/doublebase-20071028.pdf. Accessed 2 February 2016
23. Dimitrov, V.S., Imbert, L., Mishra, P.K.: Efficient and Secure Elliptic Curve Point Multiplicaton Using Double-Base Chains. https://www.iacr.org/archive/asiacrypt2005/059/059.pdf. Accessed 2 February 2016
24. Dimitrov, V.S., Cooklev, T.: Hybrid algorithm for the computation of the matrix polynomial $I + A + \cdots + A^{n-1}$. IEEE Trans. Circ. Syst. **42**(7), 377–380 (1995)
25. Mishra, P.K., Dimitrov, V.S.: Efficient Quintuple Formuals for Elliptic Curves and Efficeint Scalar Multiplication Using Multibase Number Representation. https://eprint.iacr.org/2007/040.pdf. Accessed 2 February 2016
26. Giorgi, P., Imbert, L., Izard, T.: Optimizing elliptic curve scalar multiplications for small scalars. In: Mathematics for Signal and Information Processing, San Diego, CA, United States, p. 74440N (2009)
27. Longa, P., Miri, A.: New Multibase Non-Adjacent Form Scalar Multiplication and its applications to Elliptic Curve Cryptosystems. https://eprint.iacr.org/2008/052.pdf. Accessed 2 February 2016
28. Lopez, J., Dahab, R.: Fast multiplication on elliptic curves over $GF(2^m)$ without precomputation. In: Ko, K., Paar, C. (eds.) CHES 1999. LNCS, vol. 1717, pp. 316–327. Springer, Heidelberg (1999)
29. Fischer, W., Giraud, C., Knudsen, E.W., Seifert, J.-P.: Parallel scalar multiplication on general elliptic curves over $\mathbb{F}_p$ hedged against Non-Differential Side-Channel Attacks. http://eprint.iacr.org/2002/007.pdf. Accessed 2 February 2016
30. Brier, E., Joye, M.: Weierstrass elliptic curves and side-channel attacks. In: Naccache, D., Paillier, P. (eds.) PKC 2002. LNCS, vol. 2274, pp. 335–345. Springer, Heidelberg (2002)
31. Bernstein, D.J., Lange, T., Rezaeian Farashahi, R.: Binary edwards curves. In: Oswald, E., Rohatgi, P. (eds.) CHES 2008. LNCS, vol. 5154, pp. 244–265. Springer, Heidelberg (2008)
32. Justus, B., Loebenberger, D.: Differential addition in generalized edwards coordinates. In: Echizen, I., Kunihiro, N., Sasaki, R. (eds.) IWSEC 2010. LNCS, vol. 6434, pp. 316–325. Springer, Heidelberg (2010)

33. Devigne, J., Joye, M.: Binary huff curves. In: Kiayias, A. (ed.) CT-RSA 2011. LNCS, vol. 6558, pp. 340–355. Springer, Heidelberg (2011)

34. Hutter, M., Joye, M., Sierra, Y.: Memory-constrained implementations of elliptic curve cryptography in co-Z coordinate representation. In: Nitaj, A., Pointcheval, D. (eds.) AFRICACRYPT 2011. LNCS, vol. 6737, pp. 170–187. Springer, Heidelberg (2011)

35. Wu, H., Tang, C., Feng, R.: A new model of binary elliptic curves. In: Galbraith, S., Nandi, M. (eds.) INDOCRYPT 2012. LNCS, vol. 7668, pp. 399–411. Springer, Heidelberg (2012)

36. Farashahi, R.R., Joye, M.: Efficient arithmetic on hessian curves. In: Nguyen, P.Q., Pointcheval, D. (eds.) PKC 2010. LNCS, vol. 6056, pp. 243–260. Springer, Heidelberg (2010)

37. Abarzúa, R., Thériault, N.: Complete atomic blocks for elliptic curves in jacobian coordinates over prime fields. In: Hevia, A., Neven, G. (eds.) LatinCrypt 2012. LNCS, vol. 7533, pp. 37–55. Springer, Heidelberg (2012)

38. Longa, P., Miri, A.: Fast and flexible elliptic curves point arithmetic over prime fields. IEEE Trans. Comput. **57**(3), 289–302 (2008)

39. Bernstein, D.J.: Curve25519: New Diffie Hellman Speed Records. https://cr.yp.to/ecdh/curve25519-20060209.pdf. Accessed 2 February 2016

Secret-Key Cryptanalysis

Cryptanalysis of PRINCE with Minimal Data

Shahram Rasoolzadeh and Håvard Raddum[(✉)]

Simula Research Laboratory, Fornebu, Norway
haavardr@simula.no

Abstract. We investigate two attacks on the PRINCE block cipher in the most realistic scenario, when the attacker only has a minimal amount of known plaintext available. The first attack is called Accelerated Exhaustive Search, and is able to recover the key for up to the full 12-round PRINCE with a complexity slightly lower than the security claim given by the designers. The second attack is a meet-in-the-middle attack, where we show how to successfully attack 8- and 10-round PRINCE with only two known plaintext/ciphertext pairs. Both attacks take advantage of the fact that the two middle rounds in PRINCE are unkeyed, so guessing the state before the first middle round gives the state after the second round practically for free. These attacks are the fastest until now in the known plaintext scenario for the 8 and 10 reduced-round versions and the full 12-round of PRINCE.

Keywords: PRINCE · Lightweight cipher · Cryptanalysis · Exhaustive search · Meet-in-the-middle

1 Introduction

Designing ciphers that require only a minimum of resources in implementations is known as lightweight cryptography. Several lightweight block and stream ciphers have been proposed, and their design and analysis have been a very active area of research the last decade. PRINCE is a prominent example of a lightweight block cipher, and has received much attention since it was proposed in 2012.

One reason for the interest in cryptanalysis of cryptanalytic results in clearly defined settings. PRINCE' innovative structure has also attracted cryptanalysts to investigate this block cipher. This has resulted in a variety of cryptanalysis in different models, including single key, related key and physical attacks. As the designers did not claim any security in either related key or physical attack models, we focus on the normal single key mode.

Previous works on PRINCE in the single key setting include some attacks on the PRINCE$_{core}$ [3,4,7,8,11] or attacks with change on the original structure [5,8] or an attack in the multi-user case [9]. The attacks which investigate the original structure of PRINCE involves a variety of atttacks, such as integral [4,13], sieve-in-the-middle (SITM) [6], meet-in-the-middle (MITM) [7,12], differential [8,12], and time-memory or time-data-memory trade-off attacks [4,10]. All of these attacks except one exhaustive search-like attack in [4] are chosen

© Springer International Publishing Switzerland 2016
D. Pointcheval et al. (Eds.): AFRICACRYPT 2016, LNCS 9646, pp. 109–126, 2016.
DOI: 10.1007/978-3-319-31517-1_6

Table 1. Summary of cryptanalytic results on PRINCE

Mode	Rounds	Time	Data	Memory	Technique	Ref.
CP	4	2^{64}	2^4	2^4	Integral	[4]
	4	2^{11}	2^7	2^4	Integral	[13]
	4	2^{28}	$2^{5.58}$	2^4	Integral	[13]
	6	2^{64}	2^{16}	2^{16}	Integral	[4]
	6	2^{41}	$2^{18.58}$	2^{16}	Integral	[13]
	6	$2^{32.9}$	$2^{14.9}$	$\ll 2^{27}$	Differential/Logic	[12]
	6	$2^{33.7}$	2^{16}	$2^{31.9}$	MITM	[12]
	8	2^{60}	2^{53}	2^{30}	MITM	[7]
	8	$2^{50.7}$ *	2^{16}	$2^{84.9}$	MITM	[12]
	8	$2^{65.7}$ *	2^{16}	$2^{68.9}$	MITM	[12]
	10	$2^{60.62}$	$2^{57.94}$	$2^{61.52}$	Multiple Differential	[8]
	10	2^{68} *	2^{57}	2^{41}	MITM	[12]
KP	4	2^{43}	2^5	?	? **	[2]
	6	2^{101}	2^6	?	? **	[2]
	8	2^{124}	2	2^{20}	SITM	[6]
	8	$2^{122.74}$	2	negl.	Acc. Exh. Search	3.1
	8	$2^{109.34}$	2	$2^{65.03}$	MITM	4.1
	10	$2^{124.06}$	2	negl.	Acc. Exh. Search	3.2
	10	$2^{122.15}$	2	$2^{53.3}$	MITM	4.2
	12	$2^{125.47}$	2	negl.	Exh. Search	[4]
	12	$2^{125.14}$	2	negl	Acc. Exh. Search	3.3

* Online Time
** Attacks reported by Derbez, but not published yet.

plaintext attacks. There is also a known plaintext attack on a reduced-round version in [2] reported by Patrick Derbez but not published yet. A summary of these attacks are given in Table 1.

In this paper we investigate attacks where we assume the attacker only has a minimal amount of known plaintext available, typically only two known plaintext/ciphertext pairs. This is the most realistic scenario, so the results reported here should apply to most implementations. When we have so little data, integral attacks or attacks that rely on statistical biases can not be used, so we are left with algebraic attacks or guess/verify types of attacks. We will focus on the last type of attack, and look at two different attacks of the guess/verify kind that both of them are based on guessing the states right before and after two middle round of the cipher.

The first we call Accelerated Exhaustive Search, and as the name suggests in essence it is an exhaustive search for the key. However, we will show how to exploit certain properties and make several shortcuts when guessing, so the resulting complexity for this attack becomes significantly lower than in a straightforward exhaustive search. For the full 12-round PRINCE Accelerated Exhaustive Search has a complexity that is lower than the security claim given by the designers (about 1.8 times faster).

The second attack we investigate is a MITM attack. In contrast to [12], but similar to Accelerated Exhaustive Search, our MITM attack by breaking the whole cipher to two smaller sub-ciphers get a 2-dimensional MITM attack [16,17]. We show that the two dimensions can be treated in parallel in PRINCE due to the reflection property, so we can do matching in both sides at the same time and only need to build one big table of values to match instead of two. The main result of this part of the paper is that 10-round PRINCE can be efficiently attacked (with respect to designers' security claim) using only two known plaintexts. Moreover, we get a lower time/data trade-off value $T*D$ than the one reported in [12].

This paper is organized as follows. Section 2 presents a brief description of PRINCE. In Sect. 3 we outline the Accelerated Exhaustive Search attacks and Sect. 4 presents the MITM attacks. Finally, Sect. 5 concludes the paper.

2   PRINCE Block Cipher

PRINCE [1] is a lightweight block cipher with a block size of 64 bits and two keys that both have length 64 bits. It has an FX construction where one of the keys (K_0) are used for whitening and the other one (K_1) is used as a round key for the core of the structure (see Fig. 1). We denote the plaintext/ciphertext pair of PRINCE by P/C, and the corresponding input/output of the PRINCE$_{core}$ function by P'/C'. These variables are related through the following equations.

$$P' = P \oplus K_0, \tag{1}$$

$$C' = C \oplus K_0', \tag{2}$$

where K_0' is the following linear mapping of K_0

$$K_0' = L(K_0) = (K_0 \ggg 1) \oplus (K_0 \gg 63). \tag{3}$$

For any FX constructed block cipher with a linear mapping between the whitening keys (K_0/K_0'), having a known pair of P/C gives us some information about P'/C' of the core structure which is independent from the whitening keys. This is summarized in the following lemma.

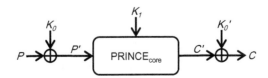

Fig. 1. PRINCE FX construction

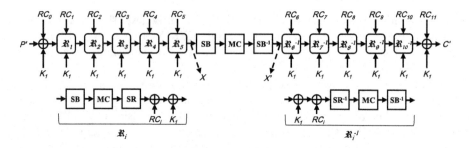

Fig. 2. PRINCE core

Lemma 1. *For a P/C pair and its corresponding P'/C' in the FX construction block cipher which uses a linear mapping L between whitening keys, the following equation holds:*

$$L(P') \oplus C' = L(P) \oplus C \qquad (4)$$

Proof. We can eliminate K_0 from (1) and (2), first by applying the $L(.)$ transformation to (1), an then summing up these two equations, which results in (4). So for each pair of P/C we can compute the value of $L(P') \oplus C'$, which is independent of K_0 and K_0'. □

The PRINCE_{core} is a block cipher of its own and similar to AES. It employs an involutive 12 rounds structure which, in the beginning, consists of two *xors* with the key and a round constant. This is followed by 5 forward rounds, a middle layer, 5 backward rounds and at the end, two more *xors* with a round constant and the key. Figure 2 shows the schematic view of the PRINCE_{core}.

The state can be defined as a 4×4 matrix like for AES, but in PRINCE the cells contain nibbles and not bytes. Each round of the PRINCE_{core} consists of 5 operations: S-box, matrix multiplication, shift row, round constant addition and key addition. These are described as follows.

- **S-box** (SB): Every nibble in the state is replaced using a 4-bit S-box.
- **Matrix Multiplication** (MC): The state is multiplied with an involutive 64×64 binary matrix. More precisely, this large matrix can be expressed as four 16×16 matrices where each of these mixes four nibbles in one column of the state.
- **Shift Row** (SR): It is exactly the same as the shift row operation in the AES. Row i of the state, with row 0 as the top row, is cyclically rotated i positions to the left.
- **Round Constant Addition** (RC): A bit-wise *xor*ing with a round constant RC_i , $i = 0, ..., 11$.
- **Key Addition** (AK): A bit-wise *xor*ing with the key K_1.

The middle rounds contain three layers, SB, MC, SB^{-1} which makes it an involutive keyless transformation. This transformation can also be separated into four smaller transformations, one for each column in the state.

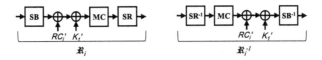

Fig. 3. Modified round function for PRINCE

In the backward rounds, the order of the operations are inverse of the forward rounds, and SB and SR are replaced with SB^{-1} and SR^{-1}. The round constants are also different, but related to the round constants in the forward rounds. The difference $RC_i \oplus RC_{11-i}, i = 0, ..., 11$ is always equal to the constant value $\alpha =$ 0xc0ac29b7c97c50dd.

As a result of this involutive structure of PRINCE$_{core}$, in implementations decryption can use the same circuit as encryption. In decryption mode the key only needs to be *xored* with α, i.e.

$$C' = PRINCE_{core}(P', K_1) \iff P' = PRINCE_{core}(C', K_1 \oplus \alpha). \quad (5)$$

This property is called α-reflection.

To ease the analysis in this paper we define an equivalent key for PRINCE$_{core}$, which is equal to

$$K_1' = SR^{-1}(MC(K_1)). \quad (6)$$

As will be shown in the next sections, using K_1' allows us to simplify the equations used. When we use this equivalent key, we position the AK layer between the SB and MC layers of the round to get an equivalent description of PRINCE$_{core}$ (see Fig. 3). Clearly, by recovering K_1' we can recover K_1.

Finally, as shown in Fig. 2, we denote the internal states exactly before and after the middle rounds by X and X', respectively. Given X, the value of X' can be computed directly, since there is no key involved between X and X'. By using the modified round function with K_1' instead of K_1, we can also expand these keyless rounds by two SR and two MC operations. The X/X' states will be used frequently in the attacks presented in the next sections.

3   Accelerated Exhaustive Key Search

In this section we will describe how to perform an accelerated exhaustive key search on PRINCE. Our way of doing this will be faster than a straight-forward exhaustive key search. By a straight-forward exhaustive key search we mean the attack where we guess a key, fully encrypt a known plaintext, and checks if it matches the given ciphertext. Our attack involves guessing the middle state X and compute the corresponding X'. Knowing a value for K_1' and X in PRINCE$_{core}$ we can easily compute P' and deduce K_0 from (1).

For a plaintext P and the corresponding ciphertext C we will guess the value of X/X' occurring for P and C. For each of the 2^{64} possible X/X'-values, we will search for candidates for K_1'. For each X/X'-value there will be one value of K_1'

in average that will produce P'- and C'-values that will match the given right-hand side in (4). The P'-value computed from this K_1' is then computed and used to deduce K_0. So for each X/X' guess we can expect one (K_0, K_1) candidate. This candidate for the full key can be tried on one other plaintext/ciphertext pair, and if it matches it should be the correct key.

At the outset this looks like an attack with complexity equal to exhaustive key search, but we will show below that the number of S-box look-ups needed to find the (K_0, K_1) candidates can be significantly smaller than in a straight-forward exhaustive key search. Similarly to the biclique attack on AES [14] we count the number of S-box applications we need to use in the attack, and evaluate how many full encryptions this amounts to by trading one round for 16 S-box look-ups. It is argued that a large majority of the time spent in an encryption is used on S-boxes so this trade-off should give rather accurate results.

Our analysis tries to minimise the number of S-box look-ups needed, and the results show that the full PRINCE can be attacked with complexity equal to $2^{125.14}$ encryptions using 2 known plaintexts. This is a little lower than the 2^{127-d}-claim made by the designers when using 2^d texts [1, p. 6].

3.1   Accelerated Exhaustive Search for 8-Round PRINCE

Assuming known X/X', the strategy is to guess on the values of the nibbles in K_1' in such a way that the total number of S-box look-ups for verifying/rejecting a guess becomes minimal. Figure 4 shows the guessing strategy, and which S-boxes that will be computed in the attack. In the following we explain what happens in Fig. 4, focusing on the $P' \leftrightarrow X$ part. Because of the reflective property of PRINCE, the exact same that is done in this part can be done in the $X' \leftrightarrow C'$ part.

Referring to Fig. 4, the nibbles of K_1' will be guessed in alphabetical order, starting with A. When A has been guessed, we have a fixed output of one S-box in round 3. We use one S-box look-up to find the corresponding input, and store this input value. Next, we guess on the value of B, and find the input value to the corresponding third round S-box. To compute these two input values for all the 2^8 possible values (A, B), we must do $2^4(1 + 2^4(1)) = 272$ S-box look-ups since we reuse the stored input for A. This is less than the 512 S-box applications we would have to do in a straight-forward exhaustive search on these two S-boxes. We continue with C and D, storing the input of the third round S-box for each guess.

After D has been guessed we have enough known nibbles between MC and SR in round 2 to go backwards through MC and find the input. At this point we have already guessed the A-value of K_1', so we can add this to the top left nibble and compute the input to the top left S-box in round 2. This is indicated with the state with a single D in this position. The total number of S-box look-ups needed for finding this nibble for all possible values of A, B, C, D is given by the expression

$$2^4(1 + 2^4(1 + 2^4(1 + 2^4(2)))) \tag{7}$$

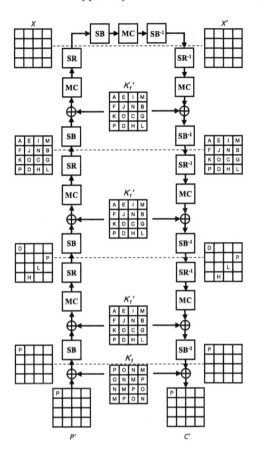

Fig. 4. Order of guessing K_1'-nibbles on 8-round PRINCE.

The 2 in the innermost bracket is because we do S-box look-ups in both round 3 and round 2. We continue in this way, guessing the values of $E - P$ in order, storing the inputs to S-boxes in rounds 2 and 3 whenever they can be computed. The letters in the states indicate which nibbles can be computed after which guess. Note that we do not evaluate all S-box inputs in round 2, only the four indicated with letters.

The number of S-box look-ups needed for computing all indicated nibbles on the $P' \leftrightarrow X$ side when cycling through all possible values of $A - P$ is given by

$$2^4(1 + 2^4(1 + 2^4(1 + 2^4(2 + 2^4(\dots(1 + 2^4(2))\dots))))), \tag{8}$$

where every fourth starting number in the brackets is a 2. This number should be multiplied with 2 to count all S-box look-ups for both $P' \leftrightarrow X$ and $X' \leftrightarrow C'$ sides.

Verifying a Guess: When we have made a full guess for K_1' and reached the bottom state in Fig. 4, we are in a position to verify whether the guess was correct or not. We first apply two S-box look-ups to find the top left nibble of $P' = (p_{63}, \ldots, p_0)$ and $C' = (c_{63}, \ldots, c_0)$. The four nibbles we have learned are $(p_{63}, p_{62}, p_{61}, p_{60})$ and $(c_{63}, c_{62}, c_{61}, c_{60})$. From the definition of L, the first nibble in $L(P') \oplus C'$ is $(p_0 \oplus c_{63}, p_{63} \oplus c_{62}, p_{62} \oplus c_{61}, p_{61} \oplus c_{60})$. Only p_0 is unknown, so we can check the current guess against three bits of the constant $L(P) \oplus C$. If our current guess matches the three bits we evaluate two more S-boxes to learn $(p_{59}, \ldots, p_{56})$ and $(c_{59}, \ldots, c_{56})$. We can now check $(p_{60} \oplus c_{59}, \ldots, p_{57} \oplus c_{56})$ against the constant $L(P) \oplus C$, a total of four new bits.

Continuing in this way, we evaluate the next pair of S-boxes only if the current guess has matched the given part of $L(P) \oplus C$ so far. Note that if the check matches in the first four nibbles, we have to calculate another four nibbles in the bottom state of Fig. 4 before applying the next pair of S-boxes for verification. In these cases we therefore have to apply a total of 10 S-box look-ups instead of 2.

With this analysis we can estimate the number of S-box look-ups needed to verify/reject a guess. This number is given as

$$2 \times (1 + 2^{-3} + 2^{-7} + 2^{-11} + 5 \times 2^{-15} + 2^{-19} + 2^{-23} + 2^{-27} + 5 \times 2^{-31} + \ldots + 2^{-63}) \quad (9)$$

Exploiting the α-Reflection Property: The fact that PRINCE is a reflection cipher can be exploited to reduce the amount of guessing. A given value x for the middle state X and a given value k for K_1' will determine particular values p' for P' and c' for C'. Let this be denoted as

$$(x, k) \rightarrow (p', c').$$

Because X and X' are related through an involution, if we chose x' for X, we will get x as a value for X'. PRINCE$_{core}$ is a reflection cipher where decryption is done by encrypting c' with $\alpha \oplus k$. We then know

$$(x', k + \alpha) \rightarrow (c', p'),$$

without having to evaluate all the S-boxes over again. So when we compute the first nibbles of P' and C' and check the first bits of $L(P') \oplus C'$, we can at the same time check $L(C') \oplus P'$. In other words, we check both (x, k) and $(x', k \oplus \alpha)$ at the same time and in this way cut the search space in half.

This can be implemented by enumerating the X-values as $x_i = i$, and do the guessing of the X-values in the natural order $x_0, x_1, \ldots$. We try all keys k for each x_i. When we reach an x_i such that $x_i' < x_i$, we simply skip this x_i because all values k (or rather, $k \oplus \alpha$) have been tried when we had x_i' as a value for X. In this way we only need to try 2^{63} values for the middle state X.

Complexity: When we check nibbles of both $L(P') \oplus C'$ and $L(C') \oplus P'$ for a match against $L(P) \oplus C$, the probability of getting a match which will invoke further S-box look-ups doubles. So the final expression for the number of S-box look-ups needed for verifying a guess becomes

$$2 \times (1 + 2^{-2} + 2^{-6} + 2^{-10} + 5 \times 2^{-14} + 2^{-18} + \ldots + 2^{-62}) = 2 \times 1.2669. \quad (10)$$

This number should be added to the innermost bracket in (8) to give the final expression for the number of S-box look-ups needed to do the accelerated exhaustive search for each X/X'-value:

$$2 \times \left[2^4(1 + 2^4(1 + 2^4(\ldots(1 + 2^4(2 + 1.2669))\ldots)))\right] = 2^{66.74}. \tag{11}$$

We will repeat this for each of 2^{63} values for X, bringing the grand total of S-box look-ups to $2^{129.74}$. Equating 16 S-box look-ups with one round of PRINCE and eight rounds for one encryption, this amounts to a complexity of $2^{122.74}$ encryptions to find the full key (K_0, K_1).

One thing we have glossed over so far in our analysis is the number of S-box look-ups needed to compute X' from X. This is 32 for each of the 2^{63} X-values, so we should add 2^{68} to the total above. The 2^{68} is a negligible addition, so the complexity remains at $2^{122.74}$.

3.2 Accelerated Exhaustive Search for 10-Round PRINCE

The accelerated exhaustive search on 10-round PRINCE is similar to the attack in the previous section, but there are a few differences. One difference is that we have to apply another 16 S-boxes on each of the $P' \leftrightarrow X$ and $X' \leftrightarrow C$ branches. Another is that we will guess the nibbles of K_1' in a different order than in the 8-round attack. The reason for this is to minimize the value of the expression similar to (8) that applies to the 10-round version. To minimize this value, we want the starting numbers in the brackets to be larger in the outer brackets, and smaller the further into the brackets we get. The order of guessing we have found that minimizes this value is shown in Fig. 5, where we also can see which nibbles that can be computed in the states after each guess.

The expression for counting the number of S-box look-ups will have 16 nested brackets, the first for guessing the A-nibble, the next for B, etc. up to P for the innermost bracket. The starting number in each bracket is the number of new S-box look-ups we can do, and store, after each guess. We must compute the full states at the input to rounds 3 and 4, but only 4 nibbles in the input to round 2, so these numbers will add up to 36. By counting the number of $A, B, C, \ldots, P$ nibbles in the cipher states in Fig. 5, the sequence of starting numbers in the brackets are

$$1, 1, 1, 2, 2, 2, 2, 1, 1, 3, 2, 2, 1, 4, 2, 9.$$

The cost of verifying/rejecting a guess is exactly the same as in the 8-round attack. Multiplying with 2 to cover both the $P' \leftrightarrow X$ and $X' \leftrightarrow C$ branches, the total number of S-box look-ups for doing the accelerated exhaustive search on 10-round PRINCE for one X-value is

$$2 \times \left[2^4(1 + 2^4(1 + 2^4(1 + \ldots + 2^4(2 + 2^4(9 + 1.2669))\ldots)))\right] = 2^{68.38} \tag{12}$$

Repeating this for all 2^{63} values of X, the total S-box look-ups in the whole attack will be $2^{131.38}$ that is equal to $2^{124.06}$ 10-round PRINCE encryptions.

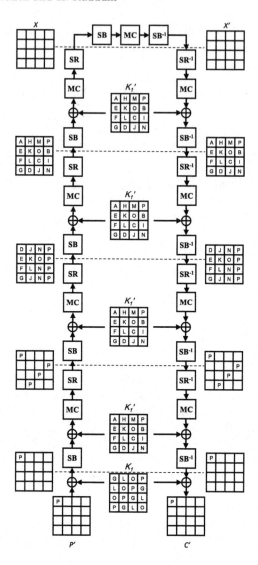

Fig. 5. Order of guessing K_1'-nibbles on 10-round PRINCE.

3.3   Accelerated Exhaustive Search for Full 12-Round PRINCE

For the full PRINCE we can guess the nibbles of K_1' in the same order as for the 10-round version. The expression for counting the total number of S-box look-ups for the $P' \leftrightarrow X$ branch is the same as in (12), except that we must add 16 to the innermost bracket. The total number of S-box look-ups is then

$$2 \times \left[2^4(1 + \ldots + 2^4(25 + 1.2669)\ldots) \right] = 2^{69.72} \tag{13}$$

Repeating for 2^{63} X-values amounts to $2^{132.72}$ S-box look-ups in the total attack, which is equal to $2^{125.14}$ PRINCE encryptions.

The attack uses only 2 known plaintexts, and the security claim given by the designers in this case is that the attacker must use an effort equivalent to 2^{126} PRINCE encryptions to find the secret key. Expecting to find the key half-way through the search, accelerated exhaustive search breaks this bound with an average complexity of $2^{124.14}$ encryptions for finding the secret key.

4   Meet-in-the-Middle Attack on PRINCE

In this section we will briefly introduce the Meet-in-the-Middle (MITM) attack and the technique of guessing only non-linearly involved key nibbles and then we explain the idea of how we do MITM cryptanalysis on PRINCE.

The basic MITM attack is a generic technique presented by Diffie and Hellman to cryptanalyse DES [15]. Despite the fact that this technique is arguably much less common than differential or linear attacks on block ciphers, there are some extensions and applications of this attack to specific primitives which are more successful than differential and linear attacks.

Let $E_{i,j}(S, K_f)$ denote the partial encryption of the state S, beginning from the start of round i and ending at the start of round j, where K_f is a particular sequence of subkeys corresponding to these $j-i$ rounds. Similarly, let $D_{j,i}(S, K_b)$ denote the partial decryption of S, beginning from the start of round j and ending at the start of round i, where K_b is the sequence of subkeys corresponding to these $j - i$ rounds. The main idea of a MITM attack is that the subkeys in both parts of the cipher can be guessed separately. First, the attacker guesses K_f and computes $E_{0,r}(P, K_f)$ for a known plaintext P. Next, he guesses K_b and computes $D_{R,r}(C, K_b)$ for the corresponding ciphertext. If

$$E_{0,r}(P, K_f) = D_{R,r}(C, K_b), \tag{14}$$

then the guessed values for K_f and K_b are candidates for representing the correct secret key.

Linearly Involved Key Bits: In the technique of guessing only non-linearly involved key bits, we do not guess the key bits which are only *xor*ed to the bits we use for matching. For example, in the MITM attack we can write:

$$\begin{cases} E_{0,r}(P, K_f) = E'_{0,r}(P, K'_f) \oplus L_f K''_f, \\ D_{R,r}(C, K_b) = D'_{R,r}(C, K'_b) \oplus L_b K''_b. \end{cases} \tag{15}$$

Here K'_f and K'_b are subsets of K_f and K_b such that $E_{0,r}(P, K_f)$ and $D_{R,r}(C, K_b)$ are non-linearly dependent on them. L_f and L_b are binary matrices, only *xor*ing some bits of K''_f or K''_b to the state bits.

When the key schedule is linear and K'_f and K'_b together determine the user-selected key, we can always find two binary matrices L'_f and L'_b of full rank which satisfy

$$L'_f \cdot K'_f \oplus L'_b \cdot K'_b = L_f \cdot K''_f \oplus L_b \cdot K''_b. \tag{16}$$

That is, K_f'' and K_b'' can be expressed as linear combinations of K_f' and K_b' bits. Instead of checking equation (14) for the whole K_f and K_b, we can then check for

$$E_{0,r}'(P, K_f') \oplus L_f' \cdot K_f' = D_{R,r}'(C, K_b') \oplus L_b' \cdot K_b', \qquad (17)$$

where the left and right hand sides can be calculated by K_f' and K_b' in the forward and backward sides of a MITM attack, respectively. This technique enables us to reduce the number of guessed key bits in each side of a MITM attack.

MITM on PRINCE: For PRINCE, because of the reflection property, knowing the value of X, we can break the whole $R = 2r + 2$-round structure into two equal sub-ciphers with r rounds and a linearly related key. Assume we know the values of X/X' for a P/C pair. Then we have two equations for the same sub-ciphers:

$$\begin{cases} F(P \oplus K_0, K_1) = X, \\ F(C \oplus K_0', K_1 \oplus \alpha) = X', \end{cases} \qquad (18)$$

where $F(S, K)$ denotes the encryption function of state S under key of K, for r forward rounds of the PRINCE$_{core}$ structure.

As finding a MITM matching for r rounds is easier than for $2r+2$ rounds, our idea is that for a known P/C pair we guess a value of X/X' and break the cipher into two smaller sub-ciphers and do a MITM attack on each of the sub-ciphers in parallel. From the P/C sides, we will guess some bits of K_0 (denoted by K_w) and some nibbles of K_1 (denoted by K_s). For a guessed value of K_w and K_s we will calculate m bits from a state in the middle of each r-round sub-ciphers ($2m$ bits in total) and save them in a table. From the middle of the whole structure we will guess values of X/X' and some nibbles of K_1' (denoted by K_m) that allows us to calculate the same $2m$ bits in the middle of each r-round sub-cipher. Then we can check equality of m bits in each sub-cipher as defined by (18).

As both K_s and K_m are derived from K_1, they may have some common information bits, which we denote as K_c. Guessing values of K_c before any other values will help us to reduce the complexity of the attack. The algorithm of the attack is described in Algorithm 1, where $E_{0,i}(.,.)$ and $D_{r,i}(.,.)$ denotes partially encryption or decryption functions for the r-round sub-cipher defined by $F(.,.)$ in (18).

This extension of the MITM attack may be considered as a multidimensional (MD) MITM attack [16–18], because we break the whole cipher into two sub-ciphers by guessing a full state in the middle. On the other hand, here we do two MITM matchings in parallel with each other, while in a MD MITM attack matchings happens serially, one after another.

The data complexity of the attack is 2 known plaintexts. The memory complexity of the attack is storing the table $\mathcal{T}$ which will cost $2^{|K_w|+|K_s-K_c|}$ words of memory.

Algorithm 1. MITM attack on PRINCE

---

for $k_c \in K_c$ **do**
    for $k_w \in K_w$ **do**
        for $k_s \in (K_s - K_c)$ **do**
            Compute $v_1 = E_{0,i}(P, (k_w, k_c, k_s))$ and $v_2 = E_{0,i}(C, (k_w, k_c, k_s))$;
            Store (k_w, k_s) into a table $\mathcal{T}$ indexed by (v_1, v_2);
        end for
    end for
    for $X \in \mathbb{F}_2^{64}$ **do**
        Compute X';
        for $k_m \in (K_m - K_c)$ **do**
            Compute $v_1' = D_{r,i}(X, (k_c, k_m))$ and $v_2' = D_{r,i}(X', (k_c, k_m))$;
            Find $(k_w, k_s) = \mathcal{T}[v_1', v_2']$ (if it exists);
            Verify/reject the candidate (k_w, k_s, k_m, k_c) against other state bits;
            if (k_w, k_s, k_m, k_c) fits all other state bits **then**
                Check it on another known plaintext/ciphertext pair;
                if (k_w, k_s, k_m, k_c) fits the second plaintext/ciphertext pair **then**
                    Return (k_w, k_s, k_m, k_c) as correct key;
                end if
            end if
        end for
    end for
end for

4.1   MITM Attack to 8-Round PRINCE

Guessing the value of X/X' will break 8-round PRINCE into two 3-round sub-ciphers. From the P/C sides, we guess 48 bits in the three leftmost columns of K_0 and K_0' (equal to 49 bits of K_0) and the three leftmost columns of K_1. From the X/X' sides we guess all nibbles of K_1' except nibbles on the secondary diagonal of the state. These 48 bits of K_s and 48 bits of K_m have 32 common information bits denoted by K_c. After guessing the value of the 32 K_c bits, only 16 bits from each set of keys remain for guessing.

Figure 6 shows the procedure of the attack. From the P/C sides we will be able to calculate 9 nibbles of the states before the MC layer in the second and seventh rounds (Gray/White squares are related to computed/un-computed nibbles). From the X/X' sides we can calculate 12 nibbles of the state before the MC layer in the second and seventh rounds, so we can do a matching on 2×9 common nibbles of these states.

In the second and seventh rounds the AK layer is not shown in Fig. 6. As K_s and K_m determine all 64 bits of K_1, we can include it using the technique of equation (17).

Attack Procedure: The attack follows Algorithm 1. In the first stage of the attack we create a table $\mathcal{T}$ from the P/C sides. In the table $\mathcal{T}$ we should save the value of $49 + 16 = 65$ bits (k_w, k_s) indexed by the 72 bits of (v_1, v_2) from the two states before the MC layer. Only a fraction of $2^{65-72} = 2^{-7}$ indexes in $\mathcal{T}$

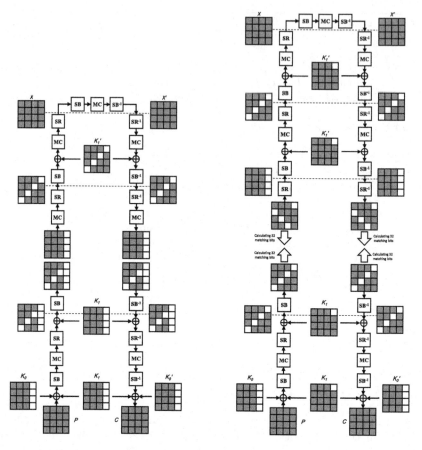

Fig. 6. MITM cryptanalysis of 8-round PRINCE

Fig. 7. MITM cryptanalysis of 10-round PRINCE

will then be filled so the storage in the table would be larger than it needs to be. Instead we save the 65 bits of (k_w, k_s) and the 7 last bits of (v_1, v_2) in the cell indexed by the 65 first bits of (v_1, v_2). Then we expect each cell in $\mathcal{T}$ to contain one value.

In the second stage of the attack we evaluate 64 bits of (v'_1, v'_2) for the guessed values of (X, k_m, k_c) and pick up the content in $\mathcal{T}$ for the index of the 65 first bits of (v'_1, v'_2). First we check whether the 7 last bits of (v'_1, v'_2) are equal to the 7 last bits of (v_1, v_2) or not. If they match we have a candidate for $(k_w, k_s, k_m, k_c) = (k_w, K_1)$, 113 bits of the key, and the middle values X/X'.

Using the values of X/X' and K_1 we calculate the nibbles which we did not evaluate (white squares in Fig. 6), coming from X/X' to plaintext and ciphertext sides. By evaluating them we can do a matching for equality of 3 nibbles at the output of the first round and 3 nibbles of the input to the eighth round. If these 24 bits match, we will evaluate P' and C' which will allow us to find a unique value for the 15 un-guessed bits of K_0 and also another $32 - 15 = 17$ matching bits. If this matching happens we will have only one candidate for (K_0, K_1) that with probability of 2^{-64} is the correct key. Using another pair of plaintext and ciphertext will verify whether it is the correct key or not.

Complexity: The time complexity of this attack will be dominated by the computation time of the second stage. In the second stage, for each guess of 32 bits k_c, 16 bits k_m and 64 bits X, we must calculate X' from X (32 S-box calls) and 12 nibbles in the middle of second and seventh rounds (12 calls for each). Trading 16 S-box look-ups with one round of PRINCE, this is equal to

$$2^{32+16+64} \times (32 + 2 \times 12) \times \frac{1}{16} \times \frac{1}{8} = 7 \times 2^{108} \qquad (19)$$

encryptions of 8-round PRINCE.

As X is the last parameter to guess in the second stage, instead of guessing the whole 64 bits we can guess it one nibble at the time (the same technique used for accelerated key search in Sect. 3) to reduce the time complexity by a fraction of about 0.362. So the final complexities of the attack will be $2^{109.34}$ encryptions for time and 2^{65} 72-bit blocks for memory.

4.2 MITM Attack to 10-Round PRINCE

Guessing the value of X/X' will break 10-round PRINCE into two 4-round subciphers. We guess the same 49 bits of K_0 and K_0' as for the 8-round attack, and all nibbles of K_1 except the one on top of the rightmost column. From the X/X' sides we guess all nibbles of K_1' except the one on top of the rightmost column. These 60 bits of K_s and 60 bits of K_m have 56 common information bits denoted by K_c. After guessing the value of the 56 K_c bits, only 4 bits from each set of keys remain to be guessed.

As Fig. 7 shows, from the P/C sides we are able to calculate 12 nibbles of the states before the MC operation in the second and ninth rounds. From the X/X' sides, we can also calculate 12 nibbles of the states on the other side of the MC layers in the second and ninth rounds. Exactly one nibble in each column is unknown in each state.

Matching Through MC: We have two partially known states on both sides of the MC operation, and we can match these in a similar way to what is done in [19]. Let one column of bits in the input to MC be (x, a, b, c) with output (y, d, e, f), where only the x and y bits are unknown. As will become clear, the exact positions of the unknown bits do not matter. The MC operation on this column gives 4 linear equations in the input and output:

$$\begin{aligned}
l_0(x, a, b, c) + y &= 0 \\
l_1(x, a, b, c) + d &= 0 \\
l_2(x, a, b, c) + e &= 0 \\
l_3(x, a, b, c) + f &= 0
\end{aligned} \qquad (20)$$

We can eliminate the unknown x and y variables and get two linear equations in the known $a, \ldots, f$:

$$\begin{aligned} l'_0(a, b, c) + l'''_0(d, e, f) &= 0 \\ l'_1(a, b, c) + l'''_1(d, e, f) &= 0 \end{aligned} \tag{21}$$

Also we will have unique equations for x and y:

$$\begin{aligned} x &= l'_2(a, b, c) + l''_2(d, e, f) \\ y &= l'_3(a, b, c) + l''_3(d, e, f) \end{aligned} \tag{22}$$

Coming from the plaintext side we define the two bits of v_1 relating to this column to be (l'_0, l'_1), and from the X side we define the corresponding two bits of v'_1 to be (l''_0, l'''_1). Besides these two values, from the plaintext side we will save l'_3 to compute the unknown value of y later, after matching. From the X side we can evaluate the value of y by computing and adding l'''_3 to l'_3. Repeating for the other 15 columns gives a total of 32 bits in v_1 and v'_1 that can be used for matching, plus four l'_3-values.

The same procedure is done in the ninth round, so all together we get 64-bit values for (v_1, v_2) and (v'_1, v'_2) that can be used for matching as described in Algorithm 1. In addition we get 32 stored bits for evaluating the unknown values in the state after the MC layer in both of the second and ninth rounds.

Attack Procedure: The attack follows the same procedure as before, but with different numbers of guessed bits. In the table $\mathcal{T}$ we will save 53 bits of (k_w, k_s) and the 11 last bits of (v_1, v_2), indexed by the 53 first bits of (v_1, v_2). In the second stage, we compute (v'_1, v'_2) from the X/X' sides and do the matching with the corresponding value in $\mathcal{T}$. We then have a candidate value for (k_w, K_1) of 113 bits and known X/X' values. Using the saved l'_3-values we can then compute the rest of K_0 and the un-evaluated nibbles of the states and verify the current guess for correctness.

There will be one (K_0, K_1) candidate surviving for each X/X' guess, which has to be verified against another plaintext/ciphertext pair.

Complexity: The number of bits to guess in the first stage is 15 bits less than the number of bits to guess in the second stage, so the time complexity of this attack will be dominated by the computation time of the second stage. In the second stage, for each guess of 56 bits k_c, 4 bits k_m and 64 bit X, we must calculate X' from X (32 S-box calls) and 12 nibbles in the middle of second and ninth rounds (24 calls for each). Trading 16 S-box look-ups with one round of PRINCE, this is equal to

$$2^{56+4+64} \times (32 + 2 \times 24) \times \frac{1}{16} \times \frac{1}{10} = 2^{123} \tag{23}$$

encryptions of 10-round PRINCE.

Here again, instead of guessing the whole 64 bits of X we can guess it nibble-wise to reduce time complexity by a fraction of 0.553. The final complexities will then be $2^{122.15}$ for time, and 2^{53} 96-bit blocks for memory.

5   Conclusions

In this paper we have shown that PRINCE can be efficiently attacked with respect to the security bound 2^{127-d}, even when having only a minimal amount of known plaintext available. To our knowledge, accelerated exhaustive search is the first reported attack on the full PRINCE with complexity lower than the claim given by the designers.

The practice of counting the number of S-box look-ups needed in an attack and translating this into number-of-encryptions complexity has already been established. It is clear that evaluating S-boxes takes the majority of the time in implementations, but it would be good to get more accurate numbers on exactly how large percentage of the time is spent on S-box look-ups and how much is spent on the linear layers. We have been in contact with some of the designers of PRINCE, and know they are working on producing these numbers. This will give a better scientific foundation for estimating the time complexity, and should allow to report very accurate numbers for this with confidence.

We have also implemented a new meet-in-the-middle attack to PRINCE where we can successfully attack 8 and 10 rounds, again with only two known plaintext/ciphertext pairs. Although meet-in-the-middle attacks have a big memory complexity, it shows that only having a minimum of known plaintext available, 8- and 10-round PRINCE can still be attacked efficiently using a meet-in-the-middle approach. As far as we know, these attacks for the 8 and 10 reduced-round versions are the fastest until now in the known plaintext scenario.

References

1. Borghoff, J., et al.: PRINCE – A low-latency block cipher for pervasive computing applications. In: Wang, X., Sako, K. (eds.) ASIACRYPT 2012. LNCS, vol. 7658, pp. 208–225. Springer, Heidelberg (2012)
2. The PRINCE Teamd: PRINCE Challenge. https://www.emsec.rub.dde/research/research_startseite/prince-challenge/
3. Abed, F., List, E., Lucks, S.: On the security of the core of PRINCE against biclique and differential cryptanalysis. IACR Cryptology ePrint Archive, Report /712, 2012 (2012)
4. Jean, J., Nikolić, I., Peyrin, T., Wang, L., Wu, S.: Security analysis of PRINCE. In: Moriai, S. (ed.) FSE 2013. LNCS, vol. 8424, pp. 92–111. Springer, Heidelberg (2014)
5. Soleimany, H., Blondeau, C., Yu, X., Wu, W., Nyberg, K., Zhang, H., Zhang, L., Wang, Y.: Reflection cryptanalysis of PRINCE-like ciphers. In: Moriai, S. (ed.) FSE 2013. LNCS, vol. 8424, pp. 71–91. Springer, Heidelberg (2014)
6. Canteaut, A., Naya-Plasencia, M., Vayssière, B.: Sieve-in-the-middle: Improved MITM attacks. In: Canetti, R., Garay, J.A. (eds.) CRYPTO 2013, Part I. LNCS, vol. 8042, pp. 222–240. Springer, Heidelberg (2013)
7. Li, L., Jia, K., Wang, X.: Improved meet-in-the-middle attacks on AES-192 and PRINCE, IACR Cryptology ePrint Archive, Report /573, 2013 (2013)
8. Canteaut, A., Fuhr, T., Gilbert, H., Naya-Plasencia, M., Reinhard, J.-R.: Multiple differential cryptanalysis of round-reduced PRINCE. In: Cid, C., Rechberger, C. (eds.) FSE 2014. LNCS, vol. 8540, pp. 591–610. Springer, Heidelberg (2015)

9. Fouque, P.-A., Joux, A., Mavromati, C.: Multi-user collisions: Applications to discrete logarithm, even-mansour and PRINCE. In: Sarkar, P., Iwata, T. (eds.) ASIACRYPT 2014. LNCS, vol. 8873, pp. 420–438. Springer, Heidelberg (2014)

10. Dinur, I.: Cryptanalytic time-memory-data tradeoffs for FX-constructions with applications to PRINCE and PRIDE. In: Oswald, E., Fischlin, M. (eds.) EUROCRYPT 2015. LNCS, vol. 9056, pp. 231–253. Springer, Heidelberg (2015)

11. Zhao, G., Sun, B., Li, C., Su, J.: Truncated differential cryptanalysis of PRINCE. Secur. Commun. Netw. **8**, 2875–2887 (2015). Wiley

12. Derbez, P., Perrin, L.: Meet-in-the-middle attacks and structural analysis of round-reduced PRINCE. In: Leander, G. (ed.) FSE 2015. LNCS, vol. 9054, pp. 190–216. Springer, Heidelberg (2015)

13. Morawiecki, P.: Practical attacks on the round-reduced PRINCE. IACR Cryptology ePrint Archive, Report /245, 2015 (2015)

14. Bogdanov, A., Khovratovich, D., Rechberger, C.: Biclique cryptanalysis of the Full AES. In: Lee, D.H., Wang, X. (eds.) ASIACRYPT 2011. LNCS, vol. 7073, pp. 344–371. Springer, Heidelberg (2011)

15. Diffie, W., Hellman, M.: Exhaustive cryptanalysis of the NBS data encryption standard. IEEE Comput. Soc. Press **10**(6), 74–84 (1977)

16. Zhu, B., Gong, G.: Multidimensional meet-in-the-middle attack and its applications to KATAN32/48/64. Cryptography and Communications **6**, 313–333 (2014). Springer

17. Boztaş, Ö., Karakoç, F., Çoban, M.: Multidimensional meet-in-the-middle attacks on reduced-round TWINE-128. In: Avoine, G., Kara, O. (eds.) LightSec 2013. LNCS, vol. 8162, pp. 55–67. Springer, Heidelberg (2013)

18. Rasoolzadeh, S., Raddum, H.: Multidimensional meet in the middle cryptanalysis of KATAN. IACR Cryptology ePrint Archive, Report /077, 2016 (2016)

19. Sasaki, Y.: Meet-in-the-middle preimage attacks on AES hashing modes and an application to whirlpool. In: Joux, A. (ed.) FSE 2011. LNCS, vol. 6733, pp. 378–396. Springer, Heidelberg (2011)

Authentication Key Recovery on Galois/Counter Mode (GCM)

John Mattsson[(✉)] and Magnus Westerlund

Ericsson Research, Stockholm, Sweden
{john.mattsson,magnus.westerlund}@ericsson.com

Abstract. GCM is used in a vast amount of security protocols and is quickly becoming the de facto mode of operation for block ciphers due to its exceptional performance. In this paper we analyze the NIST standardized version (SP 800-38D) of GCM, and in particular the use of short tag lengths. We show that feedback of successful or unsuccessful forgery attempt is almost always possible, contradicting the NIST assumptions for short tags. We also provide a complexity estimation of Ferguson's authentication key recovery method on short tags, and suggest several novel improvements to Fergusons's attacks that significantly reduce the security level for short tags. We show that for many truncated tag sizes; the security levels are far below, not only the current NIST requirement of 112-bit security, but also the old NIST requirement of 80-bit security. We therefore strongly recommend NIST to revise SP 800-38D.

Keywords: Secret-key cryptography · Message Authentication Codes · Block ciphers · Cryptanalysis · Galois/Counter Mode · GCM · Authentication key recovery · AES-GCM · Suite B · IPsec · ESP · SRTP · Re-forgery

1 Introduction

Galois/Counter Mode (GCM) [1] is quickly becoming the de facto mode of operation for block ciphers. GCM is included in the NSA Suite B set of cryptographic algorithms [2], and AES-GCM is the benchmark algorithm for the AEAD competition CAESAR [3]. Together with Galois Message Authentication Code (GMAC), GCM is used in a vast amount of security protocols:

- Many protocols such as IPsec [4], TLS [5], SSH [6], JOSE [7], 802.1AE (MACsec) [8], 802.11ad (WiGig) [9], 802.11ac (Wi-Fi) [10], P1619.1 (data storage) [11], Fibre Channel [12], and SRTP [13,14][1] only allow 128-bit tags.
- The exceptions are IPsec [15] that allows 64, 96, and 128 bit tags, CMS [16] that allows 96, 104, 112, 120, and 128 bit tags, NFC-SEC [17,18] that only allows 96 bit tags, and QUIC [19] that only allows 96 bit tags.

---

[1] The Internet Drafts specifying the use of GCM in SRTP did originally allow also 64-bit and 96-bit tags, but this was removed after the publication of this paper on the Cryptology ePrint Archive and the discussion of this paper on the IETF AVTCORE mailing list.

© Springer International Publishing Switzerland 2016
D. Pointcheval et al. (Eds.): AFRICACRYPT 2016, LNCS 9646, pp. 127–143, 2016.
DOI: 10.1007/978-3-319-31517-1_7

GCM is also used in several cryptography APIs:

- W3C Web Cryptography API [20] and Oracle Java SE [21] allow 32, 64, 96, 104, 112, 120, and 128 bit tags. PKCS #11 [22] allows tags of any length between 0 and 128 bits, and for Microsoft Cryptography API [23] we could not find any information on allowed tag lengths.

The popularity is very well deserved, GCM has exceptional performance and proven security, it is online and fully parallelizable, and it is efficient in both hardware and software, especially on new processors with dedicated AES-GCM instructions. Weaknesses of the GCM decryption function were described by Ferguson [24], which showed that the forgery probability is not 2^{-t}, and that feedback on successful forgeries allows an attacker to recover the authentication key H. As a note, the fact that the substitution probability decreases as message length increases was already known [25]. The results in this paper rely heavily on Ferguson's attack [24] and do not violate the provable security given in the original version of GCM [24]. The version standardized by NIST [1] makes normative changes to short tag lengths (32 and 64 bits) aimed to improve security, but NIST does not provide any estimated security levels given by these changes. The complexity of Ferguson's authentication key recovery method for the NIST approved short tags has not previously been analyzed.

Our results:

- In Sect. 3.1 we describe how to extend Fergusons's method for message forgery and authentication key recovery method [24] to use associated data, which is needed to apply the attack to IPv6 Jumbograms. We then describe an improvement that reduces the effective tag lengths for re-forgeries, derive a formula for the effective tag lengths, and use this improved method to calculate the probabilities for multiple message forgeries.
- In Sect. 3.2 we use these probabilities to calculate the complexity for authentication key recovery using Ferguson's method for the NIST approved short tag lengths (32 and 64 bits) showing that NIST seems to have chosen the parameters for 64 bit tags to get 80-bit security against Ferguson's attack.
- In Sect. 3.3 we suggest several novel improvements to Fergusons's method that significantly reduces the security levels for short tags, in one case the already low complexity is reduced from $2^{81.0}$ to $2^{70.0}$. We show that independently of the encryption key size, the security levels (i.e. the effective key lengths) are only 62–67 bits for 32-bit tags, and 70–75 bits for 64-bit tags. For these tag sizes, the security levels are far below, not only the current NIST requirement of 112-bit security, but also the old NIST requirement of 80-bit security. The results are applicable to both GCM and GMAC.
- In Sect 3.4 we show that feedback of successful or unsuccessful forgery attempt is almost always possible, contradicting the NIST assumptions for short tags. This illustrates that the key recovery attacks are practical and that the NIST assumption of no feedback is not valid for reasonable protocols and deployments. This is true especially for SRTP, which NIST claims meet the guidelines for use of short tags.

We strongly recommend NIST to revise SP 800-38D [1] so that the security levels of all allowed options are clearly stated, that short tags are removed, and that it is explained why any options offering less than 112-bit security against online attacks are acceptable.

We do however fully recommend GCM for usage with 128-bit tags, especially with AES-128. In fact we believe that with its excellent performance and proven security, it should be the first choice for everybody wanting an AEAD algorithm.

2   Preliminaries

2.1   Galois/Counter Mode (GCM)

Galois/Counter Mode (GCM) is an Authenticated Encryption with Associated Data (AEAD) mode of operation for block ciphers with a block size of 128 bits. It was designed by McGrew and Viega [26,27] and is standardized in NIST SP 800-38D [1] and ISO/IEC 19772:2009 [28]. The analysis in this paper is based on [1]. GCM combines the well-known counter mode of encryption with the Galois mode of authentication, which is based on universal hashing. The Galois mode of authentication makes use of the function $\text{GHASH}_H(A, C)$, which uses multiplications in $\text{GF}(2^{128})$ that can easily be parallelized. The 128-bit authentication tag is defined as

$$\text{Tag} = E_K(N) \oplus \text{GHASH}_H(A, C), \tag{1}$$

where K is the encryption key, N is the nonce, $H = E_K(0^{128})$ is the authentication key (the encryption of 128 zero bits), A is the associated data (to be authenticated but not encrypted), and C is the ciphertext. The output of the authenticated decryption function is either the plaintext P or the special error code *FAIL*. Explicit weaknesses of the GCM functions have been discussed by Ferguson [24], Joux [29], Handschuh and Preneel [30], Saarinen [31], Procter and Cid [32], and Abdelraheem et al. [33]. An extensive evaluation of GCM was done by Rogaway [34].

Galois Message Authentication Code (GMAC) is an authentication-only variant of GCM. It can be seen as a special case of GCM where the ciphertext C is the empty string. We refer to [1] for the full specification of GCM and GMAC.

2.2   Authentication Weaknesses in GCM

During the NIST standardization of AES-GCM, Fergusson [24] demonstrated through a concrete attack that due to the linear behavior of the GCM authentication function, the forgery probability is not 2^{-t}, and that feedback on successful forgeries allows an attacker to recover the authentication key H.

Fergusson considers the case when there is no associated data and the attacker tries to change the ciphertext C without changing the tag. Let C_i be block i of C, where the blocks are numbered so that C_1 encodes the length of the ciphertext. The tag can now be written as

$$\text{Tag} = E_K(N) \oplus \sum_{i \geq 1} C_i \cdot H^i. \tag{2}$$

The attacker does not change the number of blocks in C and only changes blocks in C where i is a power two. Let C' be the modified ciphertext and define the error polynomial E as

$$E = \sum_{i \geq 0} D_i \cdot H^{2^i} = \sum_{i \geq 0} (C_{2^i} - C'_{2^i}) \cdot H^{2^i}, \tag{3}$$

where $D_i = (C_{2^i} - C'_{2^i})$. Fergusson shows that the error polynomial E is a linear function of H and that the attacker can force a number of bits in E to zero. If the length of C is at least $2^l - 1$ blocks and not a multiple of 16, the attacker has $128l$ free variables and can in the first forgery force $e_0 = l$ bits of the error polynomial E to zero. The effective tag length for the first forgery is therefore $t_0 = t - l$.

Fergusson then shows that feedback of successful forgery of a message with effective tag length t_n allows recovery of t_n additional bits of the authentication key H. The effective tag length for each succeeding forgery is therefore decreasing until the attacker has full knowledge of H and can forge all subsequent tags with probability 1. As the attack is dominated by the complexity of finding the first forgery, full authentication key recovery requires approximately $2^{t_0} = 2^{-l} \cdot 2^t$ forgery attempts. As pointed out by McGrew and Viega in [35], Fergusson's attack does not break the security guarantees of GCM; it proves that the bounds in [27] are tight.

2.3   NIST Standardized Version of GCM

The NIST standard SP 800-38D [1] specifies that the 128-bit authentication tag may be truncated to 96, 104, 112, or 120 bits. For tag lengths of at least 96 bits, the maximum combined length of A and C is $L = 2^{57}$ blocks and the maximum number of invocations q of the authenticated decryption function is unlimited. For certain applications the tag may be truncated to 32 or 64 bits, and for these tag lengths, L and q are more restricted. In Appendix B of SP 800-38D [1], NIST summarizes some particulars of the GCM authentication function that were pointed out by Ferguson [24]:

– For t-bit tags, the forgery probability is not the ideal 2^{-t} but instead $2^l \cdot 2^{-t}$ where 2^l is the length in blocks of the largest message (A and C) processed by the authenticated encryption function.
– Each successful forgery enables the adversary to recover parts of the authentication key H and increases the probability of subsequent forgeries.

Table 1. NIST requirements on the usage of GCM with short tags.

t	32						64					
L	2^1	2^2	2^3	2^4	2^5	2^6	2^{11}	2^{13}	2^{15}	2^{17}	2^{19}	2^{21}
q	2^{22}	2^{20}	2^{18}	2^{15}	2^{13}	2^{11}	2^{32}	2^{29}	2^{26}	2^{23}	2^{20}	2^{17}

NIST then draws the conclusion that the following additional requirement shall apply to short tags:

1. There should not be feedback of whether a forgery attempt is successful or unsuccessful.
2. The maximum combined length L of A and C shall be heavily restricted.
3. The maximum number of invocations q of the authenticated decryption function shall be restricted.

The details of requirement 2 and 3 are listed in Table 1. Unfortunately, NIST does not give any motivations for the specific choice of parameters, or for that matter the security levels they were assumed to give. In Sect. 3.4 we show that requirement 1 on feedback is not realistic and that feedback is almost always possible when security protocols like IPsec or SRTP are used. In Sect. 3.3 we show that with our improvements to Fergusson's attack, requirement 2 and 3 has smaller effect than expected.

3 Our Results

3.1 Use of Associated Data and Lowered Effective Tag Length

As mentioned above, Fergusson [24] demonstrated through a concrete attack that due to the linear behavior of the GCM authentication function, the forgery probability for t-bit tags is not 2^{-t}. The tag and message lengths must therefore be chosen so that the forgery probability $L \cdot 2^{-t}$ is acceptable. We do not find this overly problematic and our view is that complexity is a better and more natural measure of forgery resistance. For an ideal MAC, the data complexity to perform a single forgery is $2^0 \cdot 2^t = 2^t$. For GCM, the data complexity to perform a single forgery is $2^l \cdot 2^{t-l} = 2^t$. The fact that each successful forgery enables the adversary to recover parts of the authentication key H and increases the probability of subsequent forgeries is more problematic.

Reading [24], it is not trivial to understand or calculate the effective tag lengths for re-forgeries. In this section we extend Fergusson's method to use associated data in addition to ciphertext. This extension is needed to apply Ferguson's attack to IPv6 Jumbograms. We then suggest an improvement to Ferguson's method, derive a formula for the effective tag lengths, and apply this formula to the NIST approved tag and message lengths. These effective tag lengths are then used in Sect. 3.2 to evaluate the data complexity of Ferguson's method for authentication key recovery.

Extension to Use Associated Data. The attacker tries to change the associated data A and the ciphertext C without changing the tag. The attacker does not change the number of blocks in A and C. Let A' be the modified associated data, let C' be the modified ciphertext, and define B and B' as

$$
\begin{aligned}
B &= A \,||\, 0^{128-v} \,||\, C \,||\, 0^{128-u} \,||\, \text{len}(A) \,||\, \text{len}(C)\,, \\
B' &= A' \,||\, 0^{128-v} \,||\, C' \,||\, 0^{128-u} \,||\, \text{len}(A) \,||\, \text{len}(C)\,,
\end{aligned}
\tag{4}
$$

where v is the bit length of the final block of A and u is the bit length of the final block of C. Let B_i be block i of B, where we number the blocks in the same order as Ferguson, i.e. $B_1 = \text{len}(A) \,||\, \text{len}(C)$. We can now define the error polynomial E as

$$
E = \sum_{i \geq 0} D_i \cdot H^{2^i} = \sum_{i \geq 0} (B_{2^i} - B'_{2^i}) \cdot H^{2^i}\,,
\tag{5}
$$

where $D_i = (B_{2^i} - B'_{2^i})$.

Effective Tag Length. Let t_n be the effective tag length after n successful forgeries (with feedback). Following the procedures in [24] and assuming that:

- The byte length of A or C is not a multiple of 16. This implies that the attacker can modify the length encoding in D_0.
- The combined length of A and C is at least $2^l - 1$ blocks.

With these assumptions, the attacker has $128l$ free variables and can in the first forgery force $e_0 = l$ bits of the error polynomial E to zero. The effective tag length is therefore $t_0 = t - l$. In subsequent forgeries the attacker knows more bits of the authentication key H and can force even more bits of the error polynomial E to zero. Feedback of successful forgery of a message with effective tag length t_n allows recovery of t_n additional bits of the authentication key H. After n successful forgeries, the attacker knows h_n bits of H and can force e_n bits of the error polynomial E to zero where

$$
h_n = \sum_{j=0}^{n-1} t_j \quad \text{and} \quad e_n = \left\lfloor \frac{128l}{128 - h_n} \right\rfloor .
\tag{6}
$$

Following [24], the effective tag length is $t_n = t - e_n$ until the attacker knows all 128 bits of H ($h_n \geq 128$) or when the attacker can force more then t bits of the error polynomial to zero ($e_n \geq t$), in which case the effective tag size is zero.

Exhaustive Search Improvement. We notice that when $128 - h_n < t - e_n$, the effective tag length can be reduced by doing exhaustive search on the $128 - h_n$ unknown bits of H instead of doing exhaustive search on the $t - e_n$ bits of the tag that could not be forced to zero. With this improvement, the effective tag size is

$$
t_n = \max\left(0,\, \min(t - e_n, 128 - h_n)\right) .
\tag{7}
$$

Table 2. Effective tag lengths for the NIST approved tag and message lengths.

t	32						64						96			104			112			120			128		
L	2^1	2^2	2^3	2^4	2^5	2^6	2^{11}	2^{13}	2^{15}	2^{17}	2^{19}	2^{21}	2^{12}	2^{28}	2^{57}	2^{12}	2^{28}	2^{57}	2^{12}	2^{28}	2^{57}	2^{12}	2^{28}	2^{57}	2^{12}	2^{28}	2^{57}
t_0	31	30	29	28	27	26	53	51	49	47	45	43	84	68	39	92	76	47	100	84	55	108	92	63	116	100	71
t_1	31	30	29	27	26	25	46	43	40	38	35	33	44	37	15	36	36	14	28	31	13	20	21	8	0	0	0
t_2	31	29	27	25	24	23	16	16	15	14	14	13	0	0	0	0	0	0	0	0	0	0	0	0	0	0	0
t_3	29	26	24	22	20	18	0	0	0	0	0	0	0	0	0	0	0	0	0	0	0	0	0	0	0	0	0
t_4	6	13	12	13	12	11	0	0	0	0	0	0	0	0	0	0	0	0	0	0	0	0	0	0	0	0	0
t_5	0	0	0	0	0	2	0	0	0	0	0	0	0	0	0	0	0	0	0	0	0	0	0	0	0	0	0

This improvement significantly reduces some of the effective tag lengths, but has negligible effect on the authentication key recovery complexities in the coming sections. The result of applying the improved formula (7) to the NIST approved tag and maximum message lengths, as well as the maximum message lengths of 2^{12} and 2^{28} blocks imposed by IPv4 and IPv6 is shown in Table 2.

While the values t_0 might look short, the complexity of performing a single forgery is still the expected 2^t. If a tag length of $t = 128$ is used with an encryption key of length 128 bits, performing a single forgery is as hard as recovering the encryption key, hardly a weakness.

The effective tag lengths in Table 2 are calculated with the greedy algorithm used by Ferguson. Using the suggestions we propose in Sect. 3.3, it is possible to decrease the effective tag length of later forgeries by increasing the effective tag length of earlier forgeries.

3.2   Complexity of Ferguson's Authentication Key Recovery Method

The discussions [35,36] after Ferguson's paper [24] focused mostly on multiple forgeries and authentication key recovery after nonce collisions in the encryption function, i.e. the forbidden attack later discussed by Joux [29]. We think the most important aspect of Ferguson's paper is the full recovery of the authentication key H after successful forgeries to the decryption function. While we agree with McGrew and Viega that the expected complexity to perform multiple forgeries is unclear, the expected complexity against key recovery is very clear. The complexity of performing full key recovery is expected to be 2^k where k is the stated security level. Unless stated otherwise, k is expected to be equal to the key length. In e.g. HMAC-SHA-256 the complexity for key recovery is believed to be 2^{256}, unless the authentication key is derived from a smaller key. In GCM, the authentication key is always 128 bits, which means that the security level against authentication key recovery is never more than 128 bits, even if block ciphers with larger key sizes like AES-192 or AES-256 are used. Other AEAD schemes like CCM and OCB give a security level equal to the encryption key size. This shortcoming is not mentioned in [1,27,34].

An important detail mentioned in [29] but not in [1,24] is that as the authentication tag depends on $E_K(N)$, authentication key recovery in GCM does not mean that the attacker can independently create new messages. If the length of N is fixed, knowledge of the authentication key H enables an attacker to modify a valid message by freely choosing A and C, but not N. Assuming known-plaintext, an attacker can freely chose A and P, where P is the plaintext. Still, we would expect a security level of no less than the encryption key length against authentication key recovery attacks. In [33] Abdelraheem et al. show that if a GCM implementation supports variable nonce lengths and the attacker has knowledge of H, slide universal forgeries using twisted polynomials enable an attacker to choose N as well.

Complexity Without Query Restrictions. Assuming a maximum combined length of $L = 2^l$ blocks, the effective tag length is $t_0 = t - l$, and the data complexity (measured in blocks) of performing the first forgery is $2^l \cdot 2^{t-l} = 2^t$. As the complexity of Ferguson's authentication key recovery method is dominated by the complexity of the first forgery, this is also the data complexity c for full authentication key recovery

$$c \approx 2^t . \tag{8}$$

Hence, without restrictions on q and irrespective of encryption key length, the security level of GCM against full authentication key recovery is only equal to the tag length t. This shortcoming is not mentioned in [1,34].

Complexity with Query Restrictions. The complexity of Fergusson's key recovery method with restrictions q and L has not previously been analyzed. In this section we derive the complexities for the NIST approved tag and maximum message lengths. Let p_n be the probability that an attacker succeeds with n forgeries in q attempts and let $l = \log_2 L$. We can now calculate the complexity c of authentication key recovery with Ferguson's method as

$$c \approx q \cdot 2^l / p_n , \tag{9}$$

where n is the number of forgeries needed to recover the full authentication key. Limiting the maximum number of invocations q of the decryption function so that $2^{t_0} \gg q \gg 2^{t_1}$ does not increase the complexity of authentication key recovery. The data complexity is $q \cdot 2^l$ and the probability that the attacker succeeds with one forgery in q attempts is $p_1 \approx q \cdot 2^{-t_0}$, resulting in the same total complexity of $q \cdot 2^l / p_1 = 2^l / 2^{l-t} = 2^t$.

Restricting q so that $2^{t_1} \gg q$ does however increase the complexity of Ferguson's method. Let $\phi_i = 2^{-t_i}$. The probability that the first successful forgery will occur on forgery attempt f is approximately $\phi_0 (1 - \phi_0)^{f-1}$ and the probability of a second forgery is approximately $\phi_1 (q - f)$. The probability p_2 that an attacker succeeds with two forgeries in q attempts is therefore:

$$p_2 \approx \sum_{f=1}^{q} \phi_0 (1 - \phi_0)^{f-1} \cdot \phi_1 (q - f) = \frac{\phi_0 \phi_1}{2} q^2 + \mathcal{O}\left(\frac{\phi_0^2 \phi_1}{6} q^3\right) . \tag{10}$$

We used SageMath to calculate the Taylor series and then collected the leading terms for the domain $\phi_0, \phi_1 \ll q^{-1}$. McGrew and Viega prove a formula similar to (10) in [36], but do not calculate further values. With the above approximation for p_2 we can approximate p_3 using that the probability of a second and third forgery is approximately $\phi_1 \phi_2 (q-f)^2/2$, and with p_n we can approximate p_{n+1}, etc.[2]

$$p_3 \approx \sum_{f=1}^{q} \phi_0 (1-\phi_0)^{f-1} \cdot \frac{\phi_1 \phi_2}{2} (q-f)^2 = \frac{\phi_0 \phi_1 \phi_2}{6} q^3 + \mathcal{O}\left(\frac{\phi_0^2 \phi_1 \phi_2}{24} q^4\right),$$

$$p_4 \approx \sum_{f=1}^{q} \phi_0 (1-\phi_0)^{f-1} \cdot \frac{\phi_1 \phi_2 \phi_3}{6} (q-f)^3 = \frac{q^4}{4!} \prod_{j=0}^{3} \phi_j + \mathcal{O}\left(\frac{\phi_0 q^5}{5!} \prod_{j=0}^{3} \phi_j\right),$$

$$p_5 \approx \sum_{f=1}^{q} \phi_0 (1-\phi_0)^{f-1} \cdot \frac{\phi_1 \phi_2 \phi_3 \phi_4}{24} (q-f)^4 = \frac{q^5}{5!} \prod_{j=0}^{4} \phi_j + \mathcal{O}\left(\frac{\phi_0 q^6}{6!} \prod_{j=0}^{4} \phi_j\right),$$

$$\tag{11}$$

Complexity for the NIST Tag and Message Lengths. With the approximations for p_1, p_2, p_3, p_4, p_5 we can calculate the complexity of authentication key recovery with Ferguson's method. Table 3 shows the complexities achieved by applying (9), (10), and (11) to the NIST approved tag and maximum message lengths. The grey coloring shows the t_n that was used in the calculation In a few cases the domain assumption does not hold as $2^{t_n} \approx q$. In these cases we have chosen n to overestimate rather than underestimate the complexity. Note that the complexities for authentication key recovery are independent of the encryption key length.

Our analysis show that with Ferguson's method the security levels for 32-bit tags are below the old NIST requirement of 80-bit security (that was in place in 2007 when [1] was published), while 64-bit tags are just on the border. In fact, NIST seems to have chosen the parameters for 64 bit tags to get 80-bit security against Ferguson's attack.

Only 112, 120, and 128 bit tags fulfill the current NIST requirement of 112-bit security. Unfortunately, NIST does not give any motivations for the exact restrictions they put on 32 and 64 bit tags, or for that matter the security levels they were assumed to give.

3.3 Our Improved Method for Authentication Key Recovery

In this section we propose several novel improvements to Ferguson's method for authentication key recovery. These improvements significantly reduce the security levels for short tags.

---

[2] The calculations below lead us to the hypothesis that $p_n \approx \frac{q^n}{n!} \prod_{j=0}^{n-1} \phi_j + \mathcal{O}\left(\frac{\phi_0 q^{n+1}}{(n+1)!} \prod_{j=0}^{n-1} \phi_j\right)$. This is however something that we do not use and that we do not prove, but by dividing q into n intervals, it is easy to prove that $p_n \geq \frac{q^n}{n!} \prod_{j=0}^{n-1} \phi_j$.

Table 3. Data complexity with Ferguson's method for full authentication key recovery.

t	32						64						96			104			112			120			128		
L	2^1	2^2	2^3	2^4	2^5	2^6	2^{11}	2^{13}	2^{15}	2^{17}	2^{19}	2^{21}	2^{12}	2^{28}	2^{57}	2^{12}	2^{28}	2^{57}	2^{12}	2^{28}	2^{57}	2^{12}	2^{28}	2^{57}	2^{12}	2^{28}	2^{57}
q	2^{22}	2^{20}	2^{18}	2^{15}	2^{13}	2^{11}	2^{32}	2^{29}	2^{26}	2^{23}	2^{20}	2^{17}	∞	∞	∞	∞	∞	∞	∞	∞	∞	∞	∞	∞	∞	∞	∞
l_0	31	30	29	28	27	26	53	51	49	47	45	43	84	68	39	92	76	47	100	84	55	108	92	63	116	100	71
t_1	31	30	29	27	26	25	46	43	40	38	35	33	44	37	15	36	36	14	28	31	13	20	21	8	0	0	0
t_2	31	29	27	25	24	23	16	16	15	14	14	13	0	0	0	0	0	0	0	0	0	0	0	0	0	0	0
t_3	29	26	24	22	20	18	0	0	0	0	0	0	0	0	0	0	0	0	0	0	0	0	0	0	0	0	0
t_4	6	13	12	13	12	11	0	0	0	0	0	0	0	0	0	0	0	0	0	0	0	0	0	0	0	0	0
c	$2^{61.6}$	$2^{61.6}$	$2^{62.6}$	$2^{65.6}$	$2^{68.9}$	$2^{71.9}$	$2^{79.0}$	$2^{79.0}$	$2^{79.0}$	$2^{80.0}$	$2^{80.0}$	$2^{81.0}$	$2^{96.0}$	$2^{96.0}$	$2^{96.0}$	2^{104}	2^{104}	2^{104}	2^{112}	2^{112}	2^{112}	2^{120}	2^{120}	2^{120}	2^{128}	2^{128}	2^{128}

- The attacker may choose to modify a message with a message length that is smaller than the maximum message length L.
- After each successful forgery, the attacker may choose to modify a different message.
- The attacker may choose to modify messages with different lengths 2^{l_0}, 2^{l_1}, 2^{l_2}, ...

Let the length of the first message be 2^{l_0} and let $l = \max(l_1, l_2, \dots)$. The probability that the attacker does not achieve a single successful forgery in q attempts is $(1 - \phi_0)^q$ in which case the attacker sends $q2^{l_0}$ blocks of data. The probability that the first successful forgery will occur on forgery attempt f is approximately $\phi_0(1 - \phi_0)^{f-1}$ in which case the attacker sends at most $f2^{l_0} + (q - f)2^l$ blocks of data. The average number of blocks w sent by the attacker is therefore bounded by:

$$w \le (1 - \phi_0)^q \cdot q2^{l_0} + \sum_{f=1}^{q} \phi_0(1 - \phi_0)^{f-1} \cdot \left(q2^l - f(2^l - 2^{l_0})\right)$$

$$= q2^{l_0} + \frac{1}{2}\phi_0 q^2(2^l - 2^{l_0}) + \mathcal{O}(\phi_0^2 q^3 2^{l_0}). \tag{12}$$

We used SageMath to calculate the Taylor series and then collected the leading terms for the domain $\phi_0 \ll q^{-1}$. Using this improved method, the data complexity c of authentication key recovery is

$$c \approx q \cdot 2^{l_0}/p_n. \tag{13}$$

64-Bit Tags. Let $l_0 = 0$ and $l = \log_2 L$. For 64-bit tags, the effective tag lengths are $t_0 = 64$, $t_1 = 64 - 2l$, and the complexity is

$$c_{64} \approx q \cdot 2^{l_0}/p_2 = 2^{t_0+t_1+1}/q = 2^{129}/L^2 q. \tag{14}$$

By applying (14) to the column ($L = 2^{21}$, $q = 2^{17}$), the already low complexity is reduced from $2^{81.0}$ to $2^{70.0}$. It seems infeasible to increase the security level to 112 bits, as this would either restrict the message length too much or make deployments vulnerable to denial-of-service attacks.

Table 4. Data complexity with our improved method for full authentication key recovery.

t	32						64						96			104			112			120			128		
L	2^1	2^2	2^3	2^4	2^5	2^6	2^{11}	2^{13}	2^{15}	2^{17}	2^{19}	2^{21}	2^{12}	2^{28}	2^{57}	2^{12}	2^{28}	2^{57}	2^{12}	2^{28}	2^{57}	2^{12}	2^{28}	2^{57}	2^{12}	2^{28}	2^{57}
q	2^{22}	2^{20}	2^{18}	2^{15}	2^{13}	2^{11}	2^{32}	2^{29}	2^{26}	2^{23}	2^{20}	2^{17}	∞	∞	∞	∞	∞	∞	∞	∞	∞	∞	∞	∞	∞	∞	∞
t_0	32	32	32	32	32	32	64	64	64	64	64	64	96	96	96	104	104	104	112	112	112	120	120	120	128	128	128
t_1	31	30	28	27	26	24	42	38	34	30	26	22	32	0	0	24	0	0	16	0	0	0	0	0	0	0	0
t_2	31	29	27	25	23	22	0	0	0	0	0	0	0	0	0	0	0	0	0	0	0	0	0	0	0	0	0
t_3	29	26	23	21	19	17	0	0	0	0	0	0	0	0	0	0	0	0	0	0	0	0	0	0	0	0	0
t_4	5	9	11	10	10	9	0	0	0	0	0	0	0	0	0	0	0	0	0	0	0	0	0	0	0	0	0
c	$2^{61.6}$	$2^{61.6}$	$2^{60.6}$	$2^{64.6}$	$2^{65.9}$	$2^{66.9}$	$2^{75.0}$	$2^{74.0}$	$2^{73.0}$	$2^{72.0}$	$2^{71.0}$	$2^{70.0}$	$2^{96.0}$	$2^{96.0}$	$2^{96.0}$	2^{104}	2^{104}	2^{104}	2^{112}	2^{112}	2^{112}	2^{120}	2^{120}	2^{120}	2^{128}	2^{128}	2^{128}

Table 4 shows the complexities achieved by applying our improved method 13(13) with $l_0 = 0$ and $l = \log_2 L$ to the NIST approved tag and maximum message lengths. This significantly reduces the data complexities of authentication key recovery for short tags. With our improved method, the security levels are 62–67 bits for 32-bit tags and 70–75 bits for 64-bit tags; this is below the old NIST requirement of 80-bit security and far below the current NIST requirement of 112-bit security.

3.4   Analysis of the Use of GCM in Security Protocols

We show that neither IPsec nor SRTP fulfills the NIST requirements for short tags. The specification of the use of GCM with 64 bit tags in IPsec [15] was published shortly after Fergusson's paper [24] and does not refer to the NIST specification [1]. The RFC [13] and Internet Draft [14] specifying the use of GCM in SRTP do no longer allow the use of truncated tags, but the NIST specification mentions SRTP as an example of a protocol fulfilling the guidelines for short tags. Two of these guidelines are:

- There should not be feedback of whether a forgery attempt is successful or unsuccessful.
- The AAD within packets should be limited to the necessary header information.

Analysis of GCM Usage in IPsec ESP. RFC 4106 [15] specifies the use of GCM with 64, 96, and 128 bit tags. The specification does not discuss the problems with short tags and does not require implementations to restrict the maximum message length L or the maximum number of invocations q of the authenticated decryption function. While ESP limits the AAD to necessary header information and silently discards datagrams that fail the integrity check, ESP does not silently discard datagrams that passes the integrity check and information leakage regarding the integrity of individual packets is therefore possible in many deployments.

- If any request-response protocol is sent over an IPsec protected path, an attacker can attempt forgery by modifying a datagram containing a request (e.g. HTTP GET). If integrity fails, the IPsec implementation will silently discard the datagram. If the datagram passes the integrity check, a response (e.g. 200 OK) will be sent. The datagram containing the response will also be encrypted, but assuming small amounts of other traffic (the attacker may e.g. block certain traffic) the attacker can see that a response was sent and conclude that the forgery was successful.
- If tunnel mode is used, the attacker may modify the inner destination IP address so that the packet in case of a successful forgery is routed to the adversary himself/herself.

If multicast is used [37], the attacker may attempt forgery towards several instances of the GCM decryption function in parallel, and the maximum number of invocations q of the decryption function would need to be calculated over all instances of the decryption function. Theoretically this could be done with synchronization, but in practice the only solution would be to restrict the number of invocations of each instance to q/r where r is the total number of receivers. This makes q/r impractically small and makes the system vulnerable to denial-of-service attacks.

IPsec ESP with GCM and 64-bit tags offers 64 bits of security against online authentication key recovery and IPsec ESP with GCM and 96-bit tags offers 96 bits of security. A probable attack could be detected by an intrusion detection system by identifying a large number of messages only differencing in blocks B_i where i is a power two according to (4).

Analysis of GCM Usage in SRTP. The Real-time Transport Protocol (RTP) [38] is a network protocol for transmitting real-time data, such as audio, video, and text. RTP is used in conjunction with the RTP Control Protocol (RTCP) to specify quality of service feedback and synchronization between media streams. The Secure Real-time Transport Protocol (SRTP) [39] provides encryption, message authenticity, and replay protection to RTP and RTCP. While RTP and SRTP are standardized in RFC 3550 [38] and RFC 3711 [39], there are numerous extensions to both protocols. In Appendix C of [1], NIST makes the statement:

> "An example of a protocol that meets these guidelines is Secure Real-time Transport Protocol carrying Voice over Internet Protocol, running over User Datagram Protocol".

This is not a correct statement and SRTP does in fact violate both of the guidelines mentioned before.

- The AAD is not at all limited. In RTP, the associated data consists of the RTP header, which is not limited as e.g. the header in the TLS record layer. The RTP header is extensible with proprietary header extensions carrying any type of information. In RTCP, the scope of the AAD depends on the encryption flag E. If the encryption flag is '1', the AAD data is limited to necessary header information, but if the encryption flag is '0', the AAD consists of the *entire* RTCP packet.
- RTCP receiver reports (RR) provide a wealth of information that can be used to determine the integrity of individual forged RTP packages, e.g. SSRC of the source, cumulative number of packets lost, extended highest sequence number received, last sender report (SR) timestamp, and delay since last SR. The RTCP extension for port mapping [40] is even worse as it echoes back the 64-bit nonce received in the request.
- RTP Rapid Synchronisation [41] is used; a forged Rapid Resynchronisation Request results in a RTP header extension with sync information sent from the sender.
- If the RTP header extension Client-to-Mixer Audio Level Indication [42] is used, a forged RTP packet with a high audio level will result in the Multipoint Control Unit (MCU) forwarding the SSRC. As the SSRC is not encrypted, this is easily detected by the attacker.

Even if encryption of RTCP is mandated and specific RTP header extensions and RTCP packets types are forbidden, an attacker may still in many cases determine whether a forgery was successful by looking at the length of packets. Either by looking at the length of RTCP packets from the sender or by looking at the length of RTP packets forwarded by an MCU.

A further problem with SRTP and GCM is that SRTP is very often used in one-to-many scenarios. The maximum number of invocations of each instance of the authenticated decryption function would have to be restricted to q/r, where q is the maximum total number of invocations of the authenticated decryption function, and r is the total number of receivers, including any late joiners.

All in all, SRTP does absolutely not meet the NIST guidelines for usage of GCM with short tags.

Summary. While many protocols silently discard packets with failed integrity check, very few are totally silent when the integrity check is valid. Even if the security protocol itself does not provide feedback, the higher level messages protected by the security protocol likely do. We believe that feedback of successful or unsuccessful forgery attempt is almost always possible. The NIST guideline is therefore unrealistic, and the authentication key recovery attacks practically possible. Analyzing the possibility of information feedback from successful forgeries is not trivial and the NIST statement regarding SRTP is obviously incorrect. We strongly recommend NIST to remove short tags from SP 800-38D [1].

4   Conclusions

The security levels of GCM and GMAC against authentication key recovery are for many tag sizes far below, not only the current NIST requirement of 112-bit security, but also the old NIST requirement of 80-bit security. With our improved authentication key recovery method, the security levels are 62–67 bits for the NIST approved usage of 32-bit tags and 70–75 bits for the NIST approved usage of 64-bit tags. For larger tags the security levels are as previously known t bits for t-bit tags where $t = 96, 104, 112, 120$, or 128. It seems infeasible to increase the low security levels to 112 bits, as this would either restrict the message length too much or make deployments vulnerable to denial-of-service attacks.

We note that as the authentication key is always 128 bits, the security level against authentication key recovery is never more than 128 bits, even if block ciphers with larger key sizes like AES-192 or AES-256 are used. Other AEAD schemes like CCM and OCB give a security level equal to the encryption key size.

One might argue that it is acceptable to allow a lower security level against authentication key recovery than encryption key recovery, especially if authentication key recovery requires online access to the hopefully short-lived GCM instances. With this arguing, 96-bit tags could be acceptable, even if they only offer 96 bits of security against online authentication key recovery. We do not take a stance on this, but note that the current NIST requirements in NIST SP 800-57 Part 3 [43] states that the authentication key strength shall be equal or greater than 112 bits and that less than 112 bits of security shall not be used.

NIST states that implementations should not provide feedback on the integrity of individual packets and then nevertheless heavily restricts the number of invocations of the decryption function. We have illustrated that feedback on the integrity of individual packets is almost always possible. The NIST guideline is therefore unrealistic, and the authentication key recovery attacks practically possible. Analyzing the possibility of information feedback of successful forgeries is not trivial and the NIST statement regarding SRTP is obviously incorrect. We therefore strongly recommend NIST to remove short tags from SP 800-38D [1].

Furthermore, we recommend that such analysis is never left to the user, and we strongly recommend against standardizing any cryptographic algorithms that relies on the assumption of no information feedback from successful forgeries.

We strongly recommend NIST to make a revise SP 800-38D [1] so that the security levels of all allowed options are clearly stated, that short tags are removed, and that it is explained why any options offering less than 112-bit security against online attacks are acceptable. We do however fully recommend GCM for usage with 128-bit tags, especially with AES-128. In fact we believe that with its excellent performance and proven security, it should be the first choice for everybody wanting an AEAD algorithm. We note that the design choices causing the security problems with truncated tags are also responsible for the excellent performance of GCM.

References

1. NIST SP 800–38D.: Recommendations for Block Cipher Modes of Operation: Galois/Counter Mode (GCM) and GMAC, November 2007. http://csrc.nist.gov/publications/nistpubs/800-38D/SP-800-38D.pdf
2. NSA: Suite B Cryptography. https://www.nsa.gov/ia/programs/suiteb_cryptography/
3. CAESAR: Competition for Authenticated Encryption: Security, Applicability, and Robustness. http://competitions.cr.yp.to/caesar.html
4. IETF RFC 4543.: The Use of Galois Message Authentication Code (GMAC) in IPsec ESP and AH, May 2006. https://tools.ietf.org/html/rfc4543
5. IETF RFC 5288: AES Galois Counter Mode (GCM) Cipher Suites for TLS, August 2008. https://tools.ietf.org/html/rfc5288
6. IETF RFC 5647.: AES Galois Counter Mode for the Secure Shell Transport Layer Protocol, August 2009. https://tools.ietf.org/html/rfc5647
7. IETF RFC 7518.: JSON Web Algorithms (JWA), May 2015. https://tools.ietf.org/html/rfc7518
8. IEEE 802.1AE-2006.: Media Access Control (MAC) Security, August 2006. http://standards.ieee.org/getieee802/download/802.1AE-2006.pdf
9. IEEE 802.11ad-2012.: Part 11: Wireless LAN Medium Access Control (MAC) and Physical Layer (PHY) Specifications - Amendment 3: Enhancements for Very High Throughput in the 60 GHz Band, October 2012 . http://standards.ieee.org/getieee802/download/802.11ad-2012.pdf
10. IEEE 802.11ac-2013.: Part 11: Wireless LAN Medium Access Control (MAC) and Physical Layer (PHY) Specifications - Amendment 4: Enhancements for Very High Throughput for Operation in Bands below 6 GHz, December 2013. http://standards.ieee.org/getieee802/download/802.11ac-2013.pdf
11. IEEE 1619.1-2007.: IEEE Standard for Cryptographic Protection of Data on Block-Oriented Storage Devices, May 2008
12. ANSI INCITS 496–2012.: Information technology - Fibre Channel Security Protocol 2 (FC-SP-2)
13. IETF RFC 7714.: AES-GCM Authenticated Encryption in Secure RTP (SRTP), December 2015. https://tools.ietf.org/html/rfc7714
14. Kim, W., Lee, J., Park, J., Kwon, D.: The ARIA Algorithm and Its Use with the Secure Real-time Transport Protocol (SRTP). (IETF work in progress), November 2015. https://tools.ietf.org/html/draft-ietf-avtcore-aria-srtp-09
15. IETF RFC 4106.: The Use of Galois/Counter Mode (GCM) in IPsec Encapsulating Security Payload (ESP), June 2005. https://tools.ietf.org/html/rfc4106
16. IETF RFC 5084.: Using AES-CCM and AES-GCM Authenticated Encryption in the Cryptographic Message Syntax (CMS), November 2007. https://tools.ietf.org/html/rfc5084
17. ECMA-409.: NFC-SEC-02: NFC-SEC Cryptography Standard using ECDH-256 and AES-GCM, December 2014. http://www.ecma-international.org/publications/files/ECMA-ST/ECMA-409.pdf
18. ECMA-411.: NFC-SEC-04: NFC-SEC Entity Authentication and Key Agreement using Symmetric Cryptography, December 2014. http://www.ecma-international.org/publications/files/ECMA-ST/ECMA-411.pdf
19. Langley, A., Chang, W.T.: QUIC Crypto, July 2015. https://docs.google.com/document/d/1g5nIXAIkN_Y-7XJW5K45IblHd_L2f5LTaDUDwvZ5L6g/edit

20. W3C.: Web Cryptography API, December 2014. http://www.w3.org/TR/WebCryptoAPI/
21. Oracle: Java Platform, Standard 8th edn. API Specification. https://docs.oracle.com/javase/8/docs/api/index.html
22. OASIS: PKCS #11 Cryptographic Token Interface Current Mechanisms Specification Version 2.40, September 2014. http://docs.oasis-open.org/pkcs11/pkcs11-curr/v2.40/cs01/pkcs11-curr-v2.40-cs01.pdf
23. Microsoft: Cryptography API: Next Generation. https://msdn.microsoft.com/en-us/library/windows/desktop/aa376210
24. Ferguson.: Authentication weaknesses in GCM, May 2005. http://csrc.nist.gov/groups/ST/toolkit/BCM/documents/comments/CWC-GCM/Ferguson2.pdf
25. Kabatianskii, G., Smeets, B., Johansson, T.: On the cardinality of systematic authentication codes via error-correcting codes. IEEE Trans. Inf. Theory **42**(2), 566–578 (1996)
26. McGrew, D.A., Viega, J.: The Galois/Counter Mode of Operation (GCM), May 2005. http://csrc.nist.gov/groups/ST/toolkit/BCM/documents/proposedmodes/gcm/gcm-revised-spec.pdf
27. McGrew, D.A., Viega, J.: The Security and Performance of the Galois/Counter Mode of Operation, October 2004. http://eprint.iacr.org/2004/193.pdf
28. ISO, IEC 9772: 2009.: Information technology - Security techniques - Authenticated encryption, July 2008. http://www.iso.org/iso/home/store/catalogue_tc/catalogue_detail.htm?csnumber=46345
29. Joux.: Authentication Failures in NIST version of GCM (2006). http://csrc.nist.gov/groups/ST/toolkit/BCM/documents/comments/800-38_Series-Drafts/GCM/Joux_comments.pdf
30. Handschuh, H., Preneel, B.: Key-recovery attacks on universal hash function based MAC algorithms. In: Wagner, D. (ed.) CRYPTO 2008. LNCS, vol. 5157, pp. 144–161. Springer, Heidelberg (2008). http://www.cosic.esat.kuleuven.be/publications/article-1150.pdf
31. Saarinen.: GCM, GHASH and Weak Keys (2011). http://www.iacr.org/archive/fse2012/75490220/75490220.pdf
32. Procter, G., Cid, C.: On weak keys and forgery attacks against polynomial-based MAC schemes. In: Moriai, S. (ed.) FSE 2013. LNCS, vol. 8424, pp. 287–304. Springer, Heidelberg (2014). https://eprint.iacr.org/2013/144.pdf
33. Abdelraheem, M.A., Beelen, P., Bogdanov, A., Tischhauser, E.: Twisted polynomials and forgery attacks on GCM. In: Oswald, E., Fischlin, M. (eds.) EUROCRYPT 2015. LNCS, vol. 9056, pp. 762–786. Springer, Heidelberg (2015). https://eprint.iacr.org/2015/1224.pdf
34. CRYPTREC TR No. 2012.: Evaluation of Some Blockcipher Modes of Operation, February 2011. http://www.cryptrec.go.jp/estimation/techrep_id2012_2.pdf
35. McGrew, D.A., Viega, J.: GCM Update, May 2005, http://csrc.nist.gov/groups/ST/toolkit/BCM/documents/comments/CWC-GCM/gcm-update.pdf
36. McGrew, D.A., Fluhrer, S.R.: Multiple forgery attacks against Message Authentication Codes, May 2005. http://csrc.nist.gov/groups/ST/toolkit/BCM/documents/comments/CWC-GCM/multi-forge-01.pdf
37. IETF RFC 5374.: Multicast Extensions to the Security Architecture for the Internet Protocol, November 2008. https://tools.ietf.org/html/rfc5374
38. IETF RFC 3550.: RTP: A Transport Protocol for Real-Time Applications, July 2003. https://tools.ietf.org/html/rfc3550
39. IETF RFC 3711.: The Secure Real-time Transport Protocol (SRTP), March 2004. https://tools.ietf.org/html/rfc3711

40. IETF RFC 6284.: Port Mapping between Unicast and Multicast RTP Sessions, June 2011. https://tools.ietf.org/html/rfc6284
41. IETF RFC 6051.: Rapid Synchronisation of RTP Flows, November 2010. https:// tools.ietf.org/html/rfc6051
42. IETF RFC 6464.: A Real-time Transport Protocol (RTP) Header Extension for Client-to-Mixer Audio Level Indication, December 2011. https://tools.ietf.org/ html/rfc6464
43. NIST SP 800–57 Part 3-Rev.1.: Recommendation for Key Management: Part 3 - Application-Specific Key Management Guidance, January 2015. http://nvlpubs. nist.gov/nistpubs/SpecialPublications/NIST.SP.800-57Pt3r1.pdf

Efficient Implementations

Extreme Pipelining Towards the Best Area-Performance Trade-Off in Hardware

Stjepan Picek[1(✉)], Dominik Sisejkovic[2], Domagoj Jakobovic[2], Lejla Batina[3], Bohan Yang[1], Danilo Sijacic[1], and Nele Mentens[1]

[1] KU Leuven ESAT/COSIC and IMinds,
Kasteelpark Arenberg 10, 3001 Leuven-Heverlee, Belgium
stjepan@computer.org
[2] Faculty of Electrical Engineering and Computing,
University of Zagreb, Zagreb, Croatia
[3] Digital Security Group, ICIS, Radboud University Nijmegen,
Nijmegen, The Netherlands

Abstract. This paper presents a novel framework for the automatic pipelining of AES S-boxes using composite field representations. The framework is capable of finding positions to insert flip-flops in an almost optimal way, resulting in S-boxes with an almost optimal critical path. Our novel method is using memetic algorithms and is shown to be fast, reliable and successful. We demonstrate our framework for composite field S-boxes using a polynomial and a normal basis, respectively. Our results prove that this method should be consulted when an optimal solution is of interest. Besides experimental results with the new memetic algorithms, we also discuss the ideal model of a circuit, which can be used when assessing the quality of the obtained solutions. We emphasize that this method can be used for any circuit of interest and not only for AES S-boxes.

Keywords: Real-time cryptography · Pipelining · AES S-box · Optimization · Memetic algorithm

1 Introduction

Implementations of cryptographic primitives present constant challenges in today's security applications. On the one side, embedded security relies on multiple trade-offs in terms of constraints on area, timing, power and energy and at the same time requires implementations to be secure against side-channel adversaries. On the other side, various high-speed implementations in high-bandwidth servers aim at ever faster versions of algorithms without a substantial increase in resources.

Considering block ciphers like AES that are commonly used for bulk encryption applications, a clear preference is often given to the counter mode of operation as it is parallelizable and hence suitable for high throughput, which is required by applications such as VPN setup, IPSec, etc. It may appear that

© Springer International Publishing Switzerland 2016
D. Pointcheval et al. (Eds.): AFRICACRYPT 2016, LNCS 9646, pp. 147–166, 2016.
DOI: 10.1007/978-3-319-31517-1_8

pipelining and parallelism are the terms that do not go well with constrained platforms but it is less certain where one should draw the line defining embedded security devices. For example, ARM has recently announced its next generation ARM Cortex-A72 processor to be used for mobile phones that is based on the 64-bit ARM v8-A architecture. ARM claims that the new chip delivers as much as 50 times the performance compared to processors from just five years ago and that it is at the same time 75 % more energy efficient than the previous generation.

The situation is even more unclear with hardware modules. Basically, applications that require hardware implementations such as RFID tags and smart cards are often developed for unique purposes and tailored towards a specific scenario. It may be the case that high speed is of utmost importance even though the application is embedded. It is fair to say that techniques that boost the performance in hardware such as pipelining and parallelism remain important for efficient implementations.

In this work we focus on pipelining and more precisely, we look for the optimal way to put registers (flip-flops) such that we reduce the critical path substantially. Naturally, at the same moment we do not want to pay for it too much with area or power overhead. Our goal is to develop a novel framework that could be useful for hardware designers and in general, implementers. To this end, we use memetic algorithms as a known approach in the Evolutionary Computation (EC) area. We demonstrate our approach on composite field S-boxes, because they result in circuits with a high number of gates and a high number of unbalanced paths. We elaborate on our ideas and contributions in the remainder of this paper.

Motivation and Contributions. The goal of this work is to derive a framework that is applicable to real-world scenarios. The authors of [1] give a proof of concept where they succeed in pipelining an AES S-box with an improved throughput as a result. However, to come up with a generic and at the same time optimal strategy, significant improvements in the choice of algorithm and the optimization function are necessary. Therefore, the main difference with our work is that we use more powerful search algorithm as well as improved evaluation mechanisms. Although maybe at a first glance those differences do not seem important, they are crucial in the transition from a proof of concept work that was not able to produce optimal results, to our framework that produces much better results in a smaller amount of time.

More specifically, our main contributions are:

1. Development of a new optimization algorithm that is able to produce correct solutions with a high certainty. Since we use heuristics, we cannot guarantee that all obtained solutions will be correct. Nevertheless, the experimental results in this paper did not yield any incorrect solution.
2. Improvement of the evaluation process that enables one to obtain results relatively fast. The evaluation process consists of testing whether all paths have the same number of flip-flops.
3. Extensive tests showing the suitability of our approach.

4. Building a whole system that accepts as an input the netlist and outputs a ready to be simulated netlist with inserted flip-flops.
5. Pipelined S-boxes with an optimized critical path compared to related work.

Besides those main contributions, we have a few more things to report on. Firstly, we have conducted all necessary experiments with several optimization techniques to find the best one. Furthermore, we have developed a tool that enables us to test a circuit in order to a priori determine what kind of results are expected. For this purpose we experimented with several different representations of the problem, in order to find the optimal one. Next, we present a framework that is capable to decompose a network (i.e. a circuit) into several subnetworks. Finally, we introduce the notion of the Ideal Circuit Model that helps us to evaluate the quality of our solutions. We give more details on all the aspects of this work below.

The remainder of this paper is organized as follows. In Sect. 2, we give the necessary information about AES and the methods for implementing S-boxes in hardware. Furthermore, we give the basic circuit terminology that we follow in this work. We continue in Sect. 3, where we present related work from the cryptographic, the design automation and the evolutionary computation perspective. In Sect. 4, we give an extensive description of our framework. To justify the model we use, we also present several other options with their advantages and drawbacks. Here, we also present the Ideal Circuit Model, an abstraction that helps us to assess the quality of the obtained solutions. Section 5 gives results of our EC experiments as well as the results of the synthesis process. Furthermore, we give a short discussion on the relevance of those results as well as some guidelines for future work. Finally, in Sect. 6 we conclude this study.

2   Preliminaries

In this section, we give the necessary background information for following this work. First, we define the network related terminology we use and then we shortly describe the AES cipher.

2.1   Circuit Terminology

Retiming represents a technique that transforms the circuit by moving registers from one location in the circuit to another in such a way that the functional behavior of the circuit as a whole is preserved [2]. Retiming can optimize several objectives [3]:

- minperiod - minimizes the clock frequency of a circuit,
- minarea - minimizes the number of registers in a circuit, and
- constrained minarea - minimizes the number of registers in a circuit subject to a maximum constraint on the clock period.

Pipelining is a system design technique that increases the performance of a system by partitioning a complex combinatorial circuit into a number of circuits. The pipelined circuit has a reduced critical path and could be operated on a higher working frequency [4]. Pipelining can be regarded as a special case of a minperiod objective [2].

A piece of hardware that implements the functionality of a Boolean function is called a logic gate. A circuit (network) is a set of interconnected logic gates. Networks are commonly described using netlists, which contain information about the types of logic gates employed, as well as their interconnections. Therefore, within a netlist logic gates can be perceived as network nodes.

When an output of a logic gate A, contained within a circuit, is connected to an input of a logic gate B we say that the gate A drives (the input of) gate B. Inputs of a circuit are inputs of logic gates within the circuit that are not driven by any of the logic gates of the circuit. Outputs of a circuit are outputs of logic gates that do not drive any of the gates of the circuit.

A path is a unique combination of nodes connecting a single input to a single output. Each node in the path (logic gate) introduces a delay corresponding to the time required for the signal to propagate from node inputs to node outputs. The number of paths denotes the number of different possible paths through the circuit from an input to an output. The delay of a path is equal to the sum of the delays of all its nodes. The path with the largest delay is called the critical path; for it limits the rate at which circuit-inputs may be changed (system clock frequency).

2.2 Standard Cells and Delays

A standard-cell design approach is based on using pre-made logic gates – called cells – that implement a variety of combinatorial and sequential functions. For further information about standard cells and delays, we refer the readers to [5].

In an effort to have results that are possible to compare with those from previous work, we use the same standard cell library, namely the UMC 0.13 μm low-leakage (LL) standard cell library [6]. For versatility we provide results for all operating conditions of this library, as well results for the UMC 0.13 μm high-speed (HS) standard cell library.

2.3 AES Cipher

As already stated, the target for the pipelining in this work is the S-box as used in the AES cipher. Furthermore, we experiment with both polynomial and normal basis. In accordance with that, here we give the necessary details about the AES cipher, and various ways of implementing the S-box. The AES cipher is a symmetric 128-bit block cipher [7]. To obtain a ciphertext, the plaintext needs to pass a number of round transformations. The number of rounds depends on the length of the key and is 10 rounds for a 128-bit key, 12 rounds for a 192-bit key and 14 rounds for a 256-bit key. Each round has a unique key that is calculated from the initial key. The operations in the AES cipher are on a

4×4 byte array, called the state. Those operations are *AddRoundKey*, *SubBytes*, *ShiftRows* and *MixColumns*. All the rounds consist of the same set of operations, except that there is an additional *AddRoundKey* operation before the first round, and the last round does not have the *MixColumns* operation. With regards to the whole AES cipher, there are a variety of objectives for the implementation. As a result, there exist approaches that e.g. maximize the throughput [8], minimize the circuitry [9] or minimize power consumption [10].

The nonlinear layer (*SubBytes* operation) of the AES cipher is the substitution layer realized with S-boxes. This *SubBytes* operation replaces each byte of the input and involves an inversion in the Galois field $GF(2^8)$. This calculation is not easy and therefore there are several approaches to this problem. In the rest of this paper, we only consider AES that has 10 rounds in order to simplify the considerations. Since the S-box is a 8-input 8-output lookup table (LUT), a memory to store the S-box would have a size of 256 bytes. To reduce the circuitry, Rijmen suggested to calculate the inverse of the Galois field by using subfield arithmetic [11]. This idea was further extended by work of Satoh et al. who suggested to use the tower field approach [9]. Works of Canright [12] and Mentens et al. [13] showed that the most compact solutions rely on composite field arithmetic. For details about the polynomial and normal bases, we refer the readers to [12,13]. For details about tower fields, we refer the readers to [14].

3   Related Work

In the next section, we briefly summarize several important works on hardware implementations and design automation as well as on evolutionary computation techniques for applications in cryptology. First, we list seminal works dealing with the retiming problem. Leiserson and Saxe presented the retiming technique that is able to minimize the area or maximize the clock frequency without changing the functionality of a circuit as a whole [2]. Furthermore, they showed that the problem of determining a retimed circuit with a minimum number of registers is solvable in polynomial time. Maheshwari and Sapatnekar presented an approach for the minarea retiming problem that is able to handle large circuits [15]. Narendra and Rudell discuss implementation issues arising when implementing retiming algorithms and they give a number of experimental results for circuits of various sizes [3]. Münzner and Hemme presented an algorithm that converts combinational circuits into pipelined data paths where the first step is to use timing requirements to find parts of the circuit where the register placement is possible. The second step utilizes a modified maxperiod algorithm to position a minimal number of registers [16]. However, we note that this algorithm does not guarantee to find the global optimum of flip-flops. For more information about retiming, we refer the readers to [17].

Next, we present related works that concentrate on cryptographic hardware implementations where the design choice is similar to ours. The focus is on implementations that use composite field arithmetic to boost compactness or speed. Satoh et al. were the first to take advantage of the composite field $GF(((2^2)^2)^2)$

for low area implementations, which results in the most compact S-box at the time with a gate complexity of 5.4 kgates [9]. This paper has triggered many related works looking into one or the other tower field approach.

Similarly, Wolkerstorfer et al. use arithmetic in $GF((2^4)^2)$ to achieve an implementation with a gate count comparable to the one presented by Satoh et al. (5.7 kgates) [18]. An additional goal was to make the best out of reusing hardware area for both encryption and decryption. Mentens et al. experiment with the choice of polynomials and representations to optimize the S-box on compactness for polynomial bases [13]. The main result proves that one can make better choices with different irreducible polynomials and representations of elements in this special type of tower field. Canright picked up on this work, applying the ideas to normal bases [12]. Systematically exploring all the possibilities, he deduced the smallest S-box at the time, a result that held up for almost a decade. Only recently Moradi et al. have published the most compact AES implementation of a size of only 2.4 kgates [19]. This result is obtained by focusing on AES encryption and squeezing the area on all the design layers.

Macchetti and Bertoni [20] describe an ASIC implementation for the same composite field $\mathbb{F}((2^4)^2)$ as Wolkerstorfer et al., but looking into a different representation. We mention here just a handful of the most influential papers, but it is obvious that the plethora of implementation options of AES has contributed to a huge amount of results that vary from exploiting one or the other design alternative. Looking into high-speed implementations, Hodjat and Verbauwhede describe an ASIC implementation for the same composite field $GF((2^4)^2)$ as Wolkerstorfer [21]. Their approach was to perform an area-throughput trade-off by fully pipelining the architecture and also optimizing the key-schedule implementation. The same authors also consider a pipelined AES implementation on an FPGA [22]. In [23,24], Boyar and Peralta presented a technique to improve the implementation of the AES S-Box. Their result provides different tradeoffs between the implementation area and the logic depth.

From the Evolutionary Computation perspective, we can find a number of papers that explore various applications that could be of interest in cryptology, the most prominent ones being the evolution of Boolean functions and S-boxes [25–27]. However, here we list only a few works that have clear connections with the problem we describe. Yagain and Vijayakrishna present a framework for the retiming problem when considering DSP blocks [28]. They experiment with the multi-objective genetic algorithm and report as a main advantage of their approach a number of viable solutions instead of one.

Batina et al. conduct the first experiments in trying to evolve the AES S-box in the form of a combinatorial circuit with the goal of increased throughput [1]. We point to this paper as a proof of concept, which is also our starting point and we present a complete novel framework that can be used in real-world security systems. However, we note that the results presented in that paper are worse than even those obtained by manually inserting flip-flops in the design phase as given in Sect. 5.1.

4   The Optimization Framework

In this section, we start by defining an ideal circuit that can be pipelined into circuits of the same size. In an ideal circuit, each part of the circuit in between the pipelining stages has an equal delay. Furthermore, it is always possible to divide the longest path of such a circuit into $n + 1$ partitions of exactly the same size where n is the number of flip-flop layers one adds.

As an example, consider a circuit that has a critical path equal to $1\,000$ ns. After inserting one layer of flip-flops on all necessary positions, the critical path would equal 500 ns (we disregard the delay of flip-flops). Such ideal model can help us when evaluating the quality of obtained solutions and guide us towards the best possible (optimal) solution. Naturally, it is hard to expect that a realistic circuit can be divided so perfectly. Therefore, we aim that the best possible solution should be as close to the ideal solution as possible. Next, we define the maximal number of flip-flops that can be added to a circuit.

Definition 1. *The upper bound of the number of flip-flop layers is equal to the number of cells that can be added to the shortest path connecting the input to the output of the circuit.*

4.1   The Choice of the Optimization Procedure

Similarly to the approach from [1], we regard this as an optimization problem: pipelining of a combinatorial circuit in a way that minimizes the critical path of a circuit while retaining its correctness, can be viewed as an optimization problem.

To be able to run the optimization, we introduce the notion of a correct solution.

Definition 2. *A correct solution is represented by any circuit with flip-flops in which there is the same number of flip-flops on every path connecting any input to any output.*

It is obvious that, to be able to pipeline the signal, there has to be at least one flip-flop on each path; but for the solution to be correct, that number must be the same for each path.

Since we established that we regard pipelining as an optimization problem, next we discuss which algorithm to use. We regard this problem as a black box scenario and therefore we assume no specific knowledge about the circuit. If we start with an initial circuit that has no flip-flops and then randomly add a certain number of flip-flops, it would be highly unrealistic to expect correct solutions. Therefore, we decide to use heuristics. Heuristics are algorithms that find good solutions on a large-size problem instance. Alternatively, heuristics can be defined as parts of an optimization algorithm. There, heuristics use the information currently gathered by the algorithm to help decide which solution candidate should be tested next or how the next solution can be produced [29]. Heuristic algorithms can be divided into specific heuristics and metaheuristics.

Specific heuristics are methods that are tailor-made to solve a specific problem and therefore not appropriate here (since we are not aware of any tailor-made heuristic algorithm for this problem). Metaheuristics are general-purpose algorithms that can be applied to solve almost any optimization problem. To classify metaheuristics, one can follow many criteria, but we divide it into single-solution based metaheuristics and population based heuristics [30]. Single-solution based metaheuristics manipulate and transform a single solution during the search as in the case of algorithms like local search or simulated annealing. In contrast, population based metaheuristics work on a population of solutions. On the basis of the aforementioned classification, we decide to use population based metaheuristics, and more precisely evolutionary algorithms (EAs).

We experiment with three different evolutionary algorithms, namely, Genetic Algorithms (GAs) [31], Evolution Strategy (ES) [32] and Genetic Annealing (GAn) [33]. First, in order to conduct the experiments we need to define the representation of the problem as well as the objective function. We use the same objective function as in [1] for an easier comparison of the results. The goal is the **minimization** of the following equation:

$$fitness = max\_delay\_time + (1,000 * number\_invalid\_paths). \qquad (1)$$

In the previous equation, the second term acts as a penalty for solutions that are not correct. In other words, we allow the incorrect (infeasible) solutions while searching the solution space, but guide the search towards correct solutions. Here we presume that the user specifies the target number of flip-flop layers $n >= 1$ to be inserted. Consequently, the number of invalid paths presents the number of paths that contain a different number of flip-flops. After a solution is obtained, we simulate it in the Synopsys tool as described in Sect. 5.4.

Next, we discuss how to encode the solution of the problem. We use the same representation as in [1] where for a position with no flip-flops, we write 0 and for a position with an inserted flip-flop, we write 1.

We developed a tool that translates a netlist into a bitstring representation that can be used in the optimization algorithm. The same tool returns the solution back into the netlist format after the flip-flops are inserted. The tool itself is written in the Java programming language, but the implementation details are of secondary importance so they are not presented here.

However, the question is what is a possible insertion position? The most general option is to allow an insertion of a flip-flop to every input of every cell in the circuit, which we denote as *input-based* encoding. Thus, a potential solution is represented as a string of bits with a length equal to the product of the number of cells and their inputs. This length may be denoted with S. Since each bit may be independently set to either one or zero, the size of the search space is 2^S. We have shown experimentally that in general only a very small fraction of this space represents correct solutions. Naturally, one can suggest to encode the solution in a way where each cell represents one possible insertion position. Therefore, in this kind of encoding we do not put flip-flops on each input of a cell, but on the output of a cell (*output-based* encoding). In this way we are able to reduce the solution length and size of the search space significantly. However, this also results in the fact that some correct solutions, which can be obtained with the first encoding, cannot be represented using the second one.

4.2   Genetic Algorithms

Genetic algorithms (GAs) are probabilistic algorithms whose search methods model some natural phenomena: genetic inheritance and survival of the fittest [30, 31]. GAs are a subclass of evolutionary algorithms where the elements of the search space are arrays of elementary types like strings of bits, integers, floating-point values and permutations [29, 31]. Usual variation operators in the GA are mutation and crossover [31]. In the context of optimization, exploration (diversification) means finding new points in previously unexplored areas of the search space, which is achieved by mutation in GAs. Exploitation (intensification) represents the process of improving and combining the traits of known solutions which is why crossover is used [29]. For an optimization algorithm to be successful, it needs to have a good balance between those two notions to avoid a too fast convergence to a local optimum from one side, but also a too long operation time from the other side. For further information about GAs, we refer to [31]. After the initial round of experiments, the results have shown that GAs outperform by far the ES and GAn algorithms. Therefore, in the rest of the paper we consider only GAs in our experiments.

4.3   Design of the Optimization Algorithm

As noted, in our experiments we use GAs in order to find suitable locations for the insertion of flip-flops. However, it is easy to notice that a GA on itself is often not enough. Recall our fitness function where we penalize each incorrect path. The smaller the number of incorrect paths, the better the solution. Consider the situation where a GA produces a solution that is not correct, but has only a small number of incorrect paths. Mutation will help to explore new search space areas, but in general will not help to correct a slightly incorrect solution. We noticed that often solutions are incorrect, but we need only a small change to make them correct. To amend this disadvantage of GAs, we add a local search (LS) algorithm that tries to correct almost-correct paths. Since now we combine GAs and local search, we deviate from evolutionary algorithms, and instead go to the evolutionary computation area. Such a combination of algorithms is called a memetic algorithm (MA). Memetic Algorithms (MAs) represent a synergy between evolutionary algorithms (or any other population-based algorithms) and local improvement algorithms [29]. Most MAs can be interpreted as search strategies in which a population of solutions cooperate and compete [34]. Next, we give the pseudocode for our optimization algorithm in Algorithms 1, 2, 3 and 4.

Algorithm 1 represents the main part of our framework and is a somewhat customized version of a genetic algorithm.

Algorithm 1. Greedy Hibrid SSGA.

P = createInitPopulation(POP_SIZE)
evaluate(P)
while not termination **do**
  if LS **then**
    (I1, I2) = getTwoBestFrom(P)
    for all individual from (I1, I2) **do**
      I = GreedyLocalSearch(individual)
      if fitness(I) better than fitness(bestOf(I1, I2)) **then**
        switch I with worst from P
      end if
    end for
  end if
  repeat
    randomly add k individuals to the tournament
    select the worst one in tournament
    (R1, R2) = randomly select two parents from the remaining ones in the tournament
    D = randomCrx(R1, R2)
    evaluate(D)
    replace the worst in P with D
  until POP_SIZE times
end while

The LS algorithm presented in Algorithm 2 helps us to locate correct solutions that are close to those obtained by the GA.

Algorithm 2. Greedy Local Search.

Require: iteration = 0
  repeat
    N(I(iteration)) = Neighborhood(I);
    I(iteration + 1) = getBestOf(N(I(iteration)))
    LocalOp(I(iteration + 1))
    iteration = iteration + 1
  until MAX_ITER times

Next, the Neighborhood algorithm is used to generate a population of solutions that are within Hamming distance of the current solution. Here, by Hamming distance we mean the number of positions (flip-flops) that differ in the two solutions. The Neighborhood algorithm is given in Algorithm 3.

Finally, the LocalOp algorithm is used to compare the quality of solutions generated by the local greedy search algorithm and is presented in Algorithm 4.

Algorithm 3. Neighborhood.

Require: iteration = 0
  while N_SIZE > iteration **do**
    create_individual at Hamming distance d from individual
  end while

Algorithm 4. LocalOp.

---

  for all bit postion i in bitsOf(I) **do**
    oldFitness = fitness(I);
    flip bit on position i in bitsOf(I);
    evaluate(I);
    if fitness(I) worse than oldFitness **then**
      flip bit on position i in bitsOf(I);
    end if
  end for

Common Parameters. To be able to assess the effectiveness of the optimization algorithm, and compare the alternatives, we need to define parameter values
for each algorithm variant. Since the observed algorithms are stochastic, their
performance must be evaluated on the basis of repeated runs; therefore, the number of independent runs for each setting in our experiments is 100. The other
common parameters are selected on the basis of tuning experiments and include
the population size, which is set to 50. The tournament size k in the tournament
selection is equal to 3. The tournament selection works by randomly choosing
3 individuals and then removing the worst one. From the remaining two individuals, one new solution is created via the crossover operation. The mutation
probability is set to 0.01 per individual where we make a choice on the basis of a
small set of tuning experiments that showed this was the best result on average.
Local search is called every fourth generation, with a maximum of 6 iterations
for local search. The neighborhood size is 35 and the Hamming distance is 10.
Furthermore, we display all common parameters in Table 1.

4.4   Circuit Decomposition

Here, we briefly discuss the additional functionality that our framework incorpo-
rates. It allows to decompose a network on several levels, i.e. subnetworks divided
by flip-flops. Each of those subnetworks realize a part of the functionality of the
complete network and it is possible to pipeline only a subnetwork. We call this
procedure network decomposition. However, it is important to state that it is
not always possible to pipeline a subnetwork (or even a network). Therefore,
with regards to Definition 1, we offer the following definition:

Definition 3. *It is possible to add flip-flops only to those subnetworks that do
not contain cells with direct inputs to the network.*

Table 1. Common parameters.

Parameter	Parameter Value
Number of runs	100
Tournament size	3
Population size	50
Stopping criterion	Stagnation in 10 generations
Mutation rate	0.01
LS rate	4
Max iteration for LS	6
Neighborhood size for LS	35
Hamming distance in LS	10

5   Experimental Results

In this section, we first introduce the results obtained with two different methods where the focus is on those results obtained with the optimization algorithm. We perform static timing analysis (STA) on pre-layout netlists synthesized using Synopsys Design Compiler, which is used to report the area of the designs, while we use PrimeTime - a golden timing signoff solution and environment by Synopsys - to perform STA.

In all of our experiments we are using the smallest D-Flip-Flop (DFF) cells from the appropriate libraries (DFFCLD and DFFCHD) for driving the inputs. Furthermore, we use the load of these cells for all outputs. This models the placement of the combinatorial network between two registers. The same DFF cells are used for the pipeline registers.

Lastly, in order to depict the impact of the proposed method on purely combinatorial networks, we do not include the setup times of sequential elements in the presented results.

5.1   Introducing Flip-Flops Manually in the Design Phase

We established that randomly setting flip-flops cannot result in a correct network when working with such complex networks as given here. However, what about inserting flip-flops in the design phase? In this way, we avoid working with netlists, but rather with an abstraction of a network that is much easier to comprehend. Furthermore, this approach is the dominant one when considering practical applications [4]. As an example, here we take the AES S-box in polynomial basis and then we insert flip-flops into the inverse8 part. This is represented in Fig. 1a and b. Flip-flops are depicted as "fd" cells in the latter figure. We note that the tool itself changes the network when adding the cells in the design phase.

(a) Top view of AES S-box in polynomial basis.

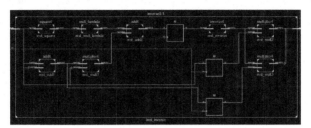

(b) Zoom into inverse8 with added FFs.

Fig. 1. Example of inserting FFs in design phase.

5.2 Results for the Optimization Algorithm

In this section, we present the best results we obtained with our memetic algorithm. Alongside, we give basic statistics on the netlists without inserted flip-flops in Table 2. For an example of full statistics, we point the readers to Appendix A. In order to ease the comparison, we calculated the delays for cells as in [1] where the values are obtained as averages for all possible combinations for each element. To model the delay of a flip-flop, we can use any value from the library as long as it is the same for the whole circuit, and here we work with a D-FF with a single output and no clear signal that has an average delay time of 320.35 ps.

In Table 3, we give the best obtained results for our algorithm. If written only Polynomial, it means that the flip-flops are inserted to the input of a cell,

Table 2. Statistics of the preliminary S-box design.

Basis	# of cells	# of inputs	# of paths	Critical path (ps)
Polynomial	165	432	8 023 409	3 884.52
Normal	181	497	139 221 044	4 685.724

Table 3. Best solutions.

Basis	Layers	Critical path (ps)	# of added FFs
Polynomial	2	2 065.7435	64
Polynomial, out	2	3 075.6087	11
Normal	2	2 508.8050	73

Table 4. Obtained number of correct solutions (%).

Basis	Layers	1.5 - 2	2 - 2.5	2.5 - 3	3 - 3.5	3.5 - 4	4 - 4.5	4.5 - 5
Polynomial	2	-	13.04	53.26	32.6	1.08	-	-
Polynomial, out	2	-	-	-	80	20	-	-
Normal	2	-	-	10.52	-	5.26	36.84	47.37

when flip-flops are added to the output of a cell we denote it with Polynomial, out.

Finally, in Table 4, we give the percentage value of times that each correct solution reached a certain critical delay time.

5.3   The Performance of the Memetic Algorithm

After discussing the successfulness of our approach in the previous section, here we discuss its reliability and speed. As already stated, those objectives are what we believe to be the differentiation of a proof of concept from the real-world framework. For all results we use PCs with Intel i5-3470 CPU with 3.2 GHz, 6 Gb of RAM and 64-bit Windows 7 OS. To obtain the following statistics, we run every setup 100 times. We consider the algorithm successful if it generates at least one correct solution. The rationale behind this is supported by the fact that every stochastic optimization algorithm is meant to be run at least several times (in other words, it is meaningless to run a stochastic algorithm on a given problem only once).

When adding one level of flip-flops to the S-box in polynomial representation where flip-flops are positioned on the input and with 100 000 evaluations, we obtain a successfulness of 93 %. When running the same setup, but with flip-flops positioned on the output of cells (output-based), the successfulness drops to only 36.8 %. When working with S-boxes in normal basis with flip-flops based on the inputs of the cells, the successfulness equals 91.6 %. To summarize the previous results, we can conclude that our algorithm is reliable since it has a reasonably high success rate. Next, we discuss the speed of our approach based on the speed of evaluation. Here, an evaluation is the whole process of obtaining a new individual and examining its fitness. Since it is clear that the evaluation process depends on the number of paths, it is easy to see that the evaluation of a solution in polynomial representation will be faster than the one in normal representation since it has a smaller number of paths as given in Table 2. A single evaluation of a polynomial representation solution lasts around 100 ms and of a normal representation solution around 120 ms. However, 10 evaluations last 800 ms, and 100 evaluations last 8 000 ms. We observe that more evaluations are comparably faster since in Java implementation we have a "warm up" phase due to the optimizations and JIT compilation. Finally, on average, our algorithm needs 150 generations to find a solution which amounts to 7 500 evaluations on average. When accounting for the speed of evaluation, we see that our approach needs on average 12.5 min to output a correct solution with an improved critical path.

5.4 Static Timing Analysis Results

The critical paths of the synthesized netlists are evaluated using the following design constraints. Firstly, for both libraries that are used we are using the enG10k wire load model. Secondly, we perform STA for all available operating conditions in order to take into account the available driving powers of combinatorial networks. Thirdly, we assume the combinatorial networks are driven by the smallest DFFs. Consistently, all outputs are loaded with the same DFFs. The setting used for STA is depicted in Fig. 2.

In order to depict the impact of the proposed method on purely combinatorial networks, we do not include the setup times of the sequential elements in the presented results. Lastly, due to the fact that DFFs typically have a smaller load of input pins—while providing stronger drives—than the combinatorial elements used in the initial network, the sum of delays through both networks is smaller than the delay of the original network.

In Table 5 we give the critical paths of the networks for all observed cases. The last column gives information on the ratio between two critical paths, including the rising edge setup times of the DFFs used for the model.

5.5 Discussion and Future Work

The results presented in this work show that our methodology is capable of finding almost optimal positions for adding flip-flops. However, we must also ask the question if it is worth while? Although our framework is capable of generating good results relatively fast, this is still significantly slower than what is the case when adding flip-flops manually in the design phase. Therefore, the answer to the previous question depends on the setting. If we have a setting where we require a critical path that is as small as possible and where we can afford the cost of added flip-flops, this methodology represents a valuable resource. Otherwise, the total cost versus the benefit is much less favorable. Furthermore, as main advantage of our approach compared with the retiming technique is the possibility to divide

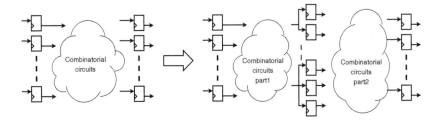

Fig. 2. STA setting.

Table 5. Critical paths of original and pipelined networks.

Basis	Library	Operating Conditions	C. Path (ns)	Pipeline C. Path (ns)	Ratio (%)	With setup (%)
Poly	UMC0.13LL	BCCOM	5.91	2.87	48.89	49.92
Poly	UMC0.13LL	TCCOM	9.83	4.78	49.03	49.45
Poly	UMC0.13LL	WCCOM	17.09	8.36	48.84	49.39
Poly	UMC0.13HS	BCCOM	2.35	1.19	50.64	53.78
Poly	UMC0.13HS	TCCOM	3.64	1.85	50.82	52.89
Poly	UMC0.13HS	WCCOM	6.30	3.20	50.79	52.01
Norm	UMC0.13LL	BCCOM	6.28	3.07	49.61	50.16
Norm	UMC0.13LL	TCCOM	10.36	5.08	49.38	49.81
Norm	UMC0.13LL	WCCOM	18.10	8.84	48.92	49.29
Norm	UMC0.13HS	BCCOM	2.58	1.28	49.61	52.55
Norm	UMC0.13HS	TCCOM	4.01	1.98	49.38	51.32
Norm	UMC0.13HS	WCCOM	6.98	3.45	49.43	50.56

the circuit in parts of almost the same critical path size and therefore obtaining an optimal solution. The same often cannot be said for the retiming technique due to the optimization towards a minimal number of registers.

We believe our approach can be coupled with the retiming technique to provide even better results (i.e. our critical path, but with a smaller number of registers). We emphasize that although we work here on S-boxes realized in tower fields, there is nothing stopping us to use this method with any other kind of combinatorial circuit. Naturally, the smaller the critical delay, the smaller the benefit of pipelining. In any case, pipelining has a big impact on the efficiency of certain modes of operation. For fully exploiting the power of the AES instructions, one needs a small delay in the mode of operation and that has the unfortunate side effect that the "better modes of operation" such as CBC are much less applied and one tends to do counter mode (fully parallelizable). In our future work we want to extend this research to the whole AES round. The results showed here suggest our technique should be regarded as a viable option when looking for optimal pipelining. However, the final verdict must be done only after a whole cipher round is examined. Besides that, we plan to further improve the local search part of the algorithm since its efficiency has an extreme impact on the efficiency of the whole algorithm. On top of that, one interesting research avenue would be to combine our algorithm with techniques for finding ASAP (as-soon-as-possible) and ALAP (as-late-as-possible) locations [16] for flip-flops which could result in a decrease of the search space size for our optimization algorithm. Finally, it is worth mentioning that the results presented in this paper are pre-layout results. We are aware that the outcome might change when post-layout results are used, as also mentioned in [35, 36].

6   Conclusion

In this paper we present a framework that is able to pipeline combinatorial circuits. To show its performance, we experimented with the AES S-box realized with tower fields in both polynomial and normal representation. The obtained results show our approach is highly competitive when the goal is to minimize the critical path. Furthermore, our results can be regarded as the best possible since they divide the circuit into two equal parts. The method presented in [1], as well as the method dominantly used today (Sect. 5.1) give worse results. Naturally, the methodology used in this work can be used in other applications besides cryptography when the goal is to decrease the critical path as much as possible and where each nanosecond makes a difference.

Acknowledgments. This work has been supported in part by the Croatian Science Foundation under the project IP-2014-09-4882. In addition, this work was supported in part by the Research Council KU Leuven (C16/15/058) and IOF project EDA-DSE (HB/13/020). D. Sijacic is supported by the Marie Curie-Sklodowska research fellowship, within the ECRYPT-NET framework.

A   Appendix

Here, we give an example of the results for our statistical tool for a circuit of interest.

AES S-box Polynomial Basis

```
---  Network report [start]  ---
File: sbox_poli.txt
Num of paths: 8023409
Max path length: 3848.862013890002
Max possible layers: 4 (3 flip-flops)
Max possible num of flip-flops on max path: 31
Solution (BitString) size: 432
Network path delay statistics:
 [0-500>: 2
 [500-1000>: 2164
 [1000-1500>: 149944
 [1500-2000>: 2026442
 [2000-2500>: 3580150
 [2500-3000>: 1899675
 [3000-3500>: 361708
 [3500-4000>: 3324
 [4000-4500>: 0
 [4500-5000>: 0
---  Network report [end]  ---
```

In Fig. 3, we give a graphical representation of the AES S-box in polynomial basis. Blue lines depict internal nodes and red lines direct inputs.

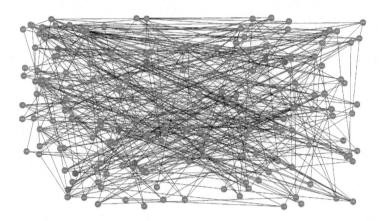

Fig. 3. Graphical representation of the S-box in polynomial basis(Color figure online).

References

1. Batina, L., Jakobovic, D., Mentens, N., Picek, S., Piedra, A.D.L., Sisejkovic, D.: S-box pipelining using genetic algorithms for high-throughput AES implementations: how fast can we go?. In: Proceedings of the Progress in Cryptology - INDOCRYpPT 2014–15th International Conference on Cryptology in India, New Delhi, India, December 14–17, 2014, pp. 322–337 (2014)
2. Leiserson, C.E., Saxe, J.B.: Retiming synchronous circuitry. Algorithmica **6**(1), 5–35 (1991)
3. Shenoy, N., Rudell, R.: Efficient implementation of retiming. In: Kuehlmann, A. (ed.) The Best of ICCAD, pp. 615–630. Springer, New York (2003)
4. Lin, M.B.: Introduction to VLSI Systems: A Logic, Circuit, and System Perspective. CRC Press, Boca Raton (2011)
5. Tillich, S., Feldhofer, M., Großschädl, J.: Area, delay, and power characteristics of standard-cell implementations of the AES S-box. In: Vassiliadis, S., Wong, S., Hämäläinen, T.D. (eds.) SAMOS 2006. LNCS, vol. 4017, pp. 457–466. Springer, Heidelberg (2006)
6. Corp., F.T.: Faraday Cell Library 0.13 μm Standard Cell (2004)
7. Daemen, J., Rijmen, V.: The Design of Rijndael. Springer-Verlag New York Inc, Secaucus (2002)
8. Morioka, S., Satoh, A.: A 10 GBPS full-aes crypto design with a twisted-BDD S-box architecture. In: Proceedings of 2002 IEEE International Conference on Computer Design: VLSI in Computers and Processors, pp. 98–103(2002)
9. Satoh, A., Morioka, S., Takano, K., Munetoh, S.: A compact rijndael hardware architecture with S-box optimization. In: Boyd, C. (ed.) ASIACRYPT 2001. LNCS, vol. 2248, pp. 239–254. Springer, Heidelberg (2001)
10. Morioka, S., Satoh, A.: An optimized S-box circuit architecture for low power aes design. In: Kaliski Jr., B.S., Koç, Ç.K., Paar, C. (eds.) CHES 2002. LNCS, vol. 2523, pp. 172–186. Springer, Heidelberg (2003)
11. Rijmen, V.: Efficient Implementation of the Rijndael S-box
12. Canright, D.: A very compact S-box for AES. In: Rao, J.R., Sunar, B. (eds.) CHES 2005. LNCS, vol. 3659, pp. 441–455. Springer, Heidelberg (2005)

13. Mentens, N., Batina, L., Preneel, B., Verbauwhede, I.: A systematic evaluation of compact hardware implementations for the rijndael S-box. In: Menezes, A. (ed.) CT-RSA 2005. LNCS, vol. 3376, pp. 323–333. Springer, Heidelberg (2005)
14. Paar, C.: Efficient VLSI architectures for bit parallel computation in Galios [Galois] fields. VDI-Verlag (1994)
15. Maheshwari, N., Sapatnekar, S.: Efficient retiming of large circuits. IEEE Trans. Very Large Scale Integr. VLSI Syst. 6(1), 74–83 (1998)
16. Münzer, A., Hemme, G.: Converting combinational circuits into pipelined data paths. In: 1991 IEEE International Conference on Computer-Aided Design, ICCAD 1991, Digest of Technical Papers, pp. 368–371, November 1991
17. Jiang, J.H., Brayton, R.: Retiming and resynthesis: a complexity perspective. IEEE Trans. Comput. Aided Des. Integr. Circuits Syst. 25(12), 2674–2686 (2006)
18. Wolkerstorfer, J., Oswald, E., Lamberger, M.: An ASIC implementation of the AES SBoxes. In: Preneel, B. (ed.) CT-RSA 2002. LNCS, vol. 2271, pp. 67–78. Springer, Heidelberg (2002)
19. Moradi, A., Poschmann, A., Ling, S., Paar, C., Wang, H.: Pushing the limits: a very compact and a threshold implementation of AES. In: Paterson, K.G. (ed.) EUROCRYPT 2011. LNCS, vol. 6632, pp. 69–88. Springer, Heidelberg (2011)
20. Bertoni, G., Breveglieri, L., Fragneto, P., Macchetti, M., Marchesin, S.: Efficient software implementation of AES on 32-bit platforms. In: Kaliski Jr., B.S., Koç, Ç.K., Paar, C. (eds.) CHES 2002. LNCS, vol. 2523, pp. 159–171. Springer, Heidelberg (2003)
21. Hodjat, A., Verbauwhede, I.: Area-throughput trade-offs for fully pipelined 30 to 70 gbits/s AES processors. IEEE Trans. Comput. 55(4), 366–372 (2006)
22. Hodjat, A., Verbauwhede, I.: A 21.54 Gbits/s fully pipelined AES processor on FPGA. In: 12th Annual IEEE Symposium on Field-Programmable Custom Computing Machines, FCCM 2004, pp. 308–309, April 2004
23. Boyar, J., Peralta, R.: A small depth-16 circuit for the AES S-box. In: Gritzalis, D., Furnell, S., Theoharidou, M. (eds.) SEC 2012. IFIP AICT, vol. 376, pp. 287–298. Springer, Heidelberg (2012)
24. Boyar, J., Peralta, R.: A new combinational logic minimization technique with applications to cryptology. In: Festa, P. (ed.) SEA 2010. LNCS, vol. 6049, pp. 178–189. Springer, Heidelberg (2010)
25. Clark, J.A., Jacob, J.L., Stepney, S., Maitra, S., Millan, W.L.: Evolving boolean functions satisfying multiple criteria. In: Menezes, A., Sarkar, P. (eds.) INDOCRYPT 2002. LNCS, vol. 2551, pp. 246–259. Springer, Heidelberg (2002)
26. Burnett, L., Carter, G., Dawson, E., Millan, W.L.: Efficient methods for generating MARS-like S-boxes. In: Schneier, B. (ed.) FSE 2000. LNCS, vol. 1978, pp. 300–314. Springer, Heidelberg (2001)
27. Picek, S., Papagiannopoulos, K., Ege, B., Batina, L., Jakobovic, D.: Confused by confusion: systematic evaluation of DPA resistance of various S-boxes. In: Proceedings of Progress in Cryptology - INDOCRYpPT 2014–15th International Conference on Cryptology in India, New Delhi, India, December 14–17, pp. 374–390 (2014)
28. Yagain, D., Vijayakrishna, A.: A novel framework for retiming using evolutionary computation for high level synthesis of digital filters. Swarm Evol. Comput. 20, 37–47 (2015)
29. Weise, T.: Global Optimization Algorithms - Theory and Application, 2 edn. Self-Published (2009). http://www.it-weise.de/
30. Talbi, E.G.: Metaheuristics: From Design to Implementation. Wiley Publishing, Hoboken (2009)

31. Eiben, A.E., Smith, J.E.: Introduction to Evolutionary Computing. Springer-Verlag, Heidelberg (2003)
32. Beyer, H.G., Schwefel, H.P.: Evolution Strategies a comprehensive introduction. Natural Comput. 1(1), 3–52 (2002)
33. Yao, X.: Optimization by genetic annealing. In: Proceedings of 2nd Australian Conference on Neural Networks, pp. 94–97 (1991)
34. Glover, F.W., Kochenberger, G.A. (eds.): Handbook of Metaheuristics. International Series in Operations Research & Management Science, vol. 114, 1st edn. Springer, Heideelberg (2003)
35. Standaert, F.-X., Rouvroy, G., Quisquater, J.-J., Legat, J.-D.: Efficient implementation of rijndael encryption in reconfigurable hardware: improvements and design tradeoffs. In: Walter, C.D., Koç, Ç.K., Paar, C. (eds.) CHES 2003. LNCS, vol. 2779, pp. 334–350. Springer, Heidelberg (2003)
36. Kerckhof, S., Durvaux, F., Hocquet, C., Bol, D., Standaert, F.-X.: Towards green cryptography: a comparison of lightweight ciphers from the energy viewpoint. In: Prouff, E., Schaumont, P. (eds.) CHES 2012. LNCS, vol. 7428, pp. 390–407. Springer, Heidelberg (2012)

A Deeper Understanding of the XOR Count Distribution in the Context of Lightweight Cryptography

Sumanta Sarkar[1(✉)] and Siang Meng Sim[2]

[1] TCS Innovation Labs, Hyderabad, India
sumanta.sarkar1@tcs.com
[2] Nanyang Technological University, Singapore, Singapore
ssim011@e.ntu.edu.sg

Abstract. In this paper, we study the behavior of the XOR count distributions under different bases of finite field. XOR count of a field element is a simplified metric to estimate the hardware implementation cost to compute the finite field multiplication of an element. It is an important criterion in the design of lightweight cryptographic primitives, typically to estimate the efficiency of the diffusion layer in a block cipher. Although several works have been done to find lightweight MDS diffusion matrices, to the best of our knowledge, none has considered finding lightweight diffusion matrices under other bases of finite field apart from the conventional polynomial basis. The main challenge for considering different bases for lightweight diffusion matrix is that the number of bases grows exponentially as the dimension of a finite field increases, causing it to be infeasible to check all possible bases. Through analyzing the XOR count distributions and the relationship between the XOR count distributions under different bases, we find that when all possible bases for a finite field are considered, the collection of the XOR count distribution is invariant to the choice of the irreducible polynomial of the same degree. In addition, we can partition the set of bases into equivalence classes, where the XOR count distribution is invariant in an equivalence class, thus when changing bases within an equivalence class, the XOR count of a diffusion matrix will be the same. This significantly reduces the number of bases to check as we only need to check one representative from each equivalence class for lightweight diffusion matrices. The empirical evidence from our investigation says that the bases which are in the equivalence class of the polynomial basis are the recommended choices for constructing lightweight MDS diffusion matrices.

Keywords: Lightweight cryptography · Finite field multiplication · Basis of finite field · XOR count · MDS matrices · Diffusion layer

1 Introduction

In today's world *Internet of Things (IoT)* is a buzzword. The devices that are involved in IoT are equipped with very limited power and memory. The standard cryptographic primitives often do not suit in these devices. Thus to cater

© Springer International Publishing Switzerland 2016
D. Pointcheval et al. (Eds.): AFRICACRYPT 2016, LNCS 9646, pp. 167–182, 2016.
DOI: 10.1007/978-3-319-31517-1_9

the security requirement of IoT, the so-called topic *lightweight cryptography* has emerged.

Lightweight cryptography is mostly based on symmetric-key cryptography. Examples of lightweight ciphers include eSTREAM finalists Grain v1 [7], MICKEY 2.0 [1], and Trivium [11]. On the other hand, the block ciphers CLEFIA [9], PRESENT [2] have already been included in the ISO standardization project of lightweight cryptography ISO/IEC 29192. The block cipher PRINCE [3] is another block cipher that is lightweight, and after its arrival in 2012, it has generated a lot of interest in the community.

There are two important cryptographic criteria of a block cipher, and other cryptographic primitives such as hash functions that are based on block ciphers— confusion and diffusion. The confusion layer makes the relation between key and ciphertext as complex as possible, and on the other hand the diffusion layer spreads the plaintext statistics through the ciphertext. A popular choice for constructing the diffusion layer is to use maximum distance separable (MDS) matrices, for instance AES [4] and LED [6] use MDS matrix to achieve the maximum diffusion power. However, having MDS matrix in a lightweight cipher is a real challenge for the designers as MDS matrices tend to have high implementation cost. To quantify the hardware cost of the diffusion layer, a metric to estimate the cost for implementing the coefficients of the diffusion matrix is required.

Before [8], a common belief was that field elements with low Hamming weight tends to be lightweight. For instance, one of the rationales for the choice of AES diffusion matrix coefficients was its simplicity and low Hamming weight. However, there was no clear implication of how low Hamming weight elements would result in lightweight implementation. In 2014, the authors of [8] proposed to look at the number of XORs required to compute the multiplication of a fixed field elements. As a result, they found MDS diffusion matrices that required lesser XORs to implement than the AES diffusion matrix and yet with higher total Hamming weight. In 2015, the authors of [10] extended the search for lightweight diffusion matrices, with special focus on involutory (self-inverse) MDS matrices, over other finite fields defined by other irreducible polynomials besides the irreducible polynomial used for AES diffusion matrix. Besides finding new lightweight diffusion matrices, the authors proposed that the choice of irreducible polynomial to construct lightweight matrices should not be dependent on the Hamming weight of the polynomial, but the high standard deviation of the XOR count distribution. Although all possible irreducible polynomials for generating finite fields have been studied, the choice of the basis has not been considered.

In symmetric-key cryptography, the conventional choice of basis is the polynomial basis. However, there are many other choices of basis, for instance a normal basis, which is commonly used in elliptic curve cryptography. These new choices of basis give rise to new sets of XOR count distributions. Hence a natural question is whether there exist even lighter MDS diffusion matrices when we consider different bases besides the polynomial basis, which is the main motivation of this work. However, extending the search for lightweight matrices to other bases brings about a new challenge—the number of bases grows exponentially as the dimension of the finite field increases. Perhaps this is one reason that little work in any aspect of cryptography has looked into the different choices of bases.

Contributions. In this paper we deeply study the distribution of XOR count of field elements and characterize how sensitive they are to the change of basis. Prior to this work, little work has been done on analyzing different finite field bases in the cryptographical aspect. In Sect. 2, after giving a brief introduction to finite field and its bases, we describe how to compute the XOR count of a field element and the XOR count of a diffusion matrix. In Sect. 3.2, we analyze the distribution of XOR counts and show that the mean of the XOR count distribution is invariant of the irreducible polynomial and basis. In addition, we prove that the collection of XOR count distributions is the same for any irreducible polynomial of the same degree. This implies that we only need to consider XOR count distributions under one irreducible polynomial. In Sect. 3.3, we show that there are bases that generate similar XOR count distributions, which means that there are "redundant bases", and we can reduce the number of bases to consider when we search for lightweight diffusion matrices. In Sect. 4, we formally define the equivalence relation between bases whose XOR count distributions are invariant, and propose the concept of equivalence classes of bases. Since it is sufficient to search for lightweight MDS diffusion matrices under one representative basis from each equivalence class, this significantly reduces the number of bases to consider. In Sect. 5, we describe the algorithms for finding all equivalence classes of bases, and searching lightweight MDS and involutory MDS diffusion matrices under the representative bases. Although we do not find new lighter (involutory) MDS diffusion matrices, our empirical evidence shows that the polynomial basis, and its equivalent bases, are the recommended choice of bases for constructing lightweight MDS diffusion matrices.

2   Preliminary

In this section, first we give a short recap on finite field and its bases. Next, we describe how the XOR count of a field element and XOR count of a diffusion matrix under some irreducible polynomial are computed.

2.1   Finite Field

We denote by $GF(2^n)$ the finite field with 2^n elements, $n \geq 1$. The addition $+$ over $GF(2^n)$ will be used in this paper with ambiguity, however implication will be clear from the context. The exclusive-or (XOR) sign $\oplus$ will sometimes be used to mean addition modulo 2.

The extension field $GF(2^n)$ of $GF(2)$ is constructed using an irreducible polynomial of degree n. Let $GF(2^n)/p(X)$ denote the field having the underlying irreducible polynomial $p(X)$ of degree n^1. Note that for any other irreducible polynomial $q(X)$ of degree n, the two fields $\mathrm{GF}(2^n)/p(X)$ and $\mathrm{GF}(2^n)/q(X)$ are isomorphic. Throughout the paper we will be using the notation $\mathrm{GF}(2^n)/p(X)$ only when we need to mention $p(X)$ explicitly.

The number of irreducible polynomial of degree n over $\mathrm{GF}(2)$, denoted as $M_n(2)$, is given by the following formula,

$$M_n(2) = \frac{1}{n} \sum_{d|n} \mu(d) 2^{\frac{n}{d}}, \tag{1}$$

where $\mu(d)$ is the Möbius function [5].

2.2   Bases of a Finite Field

Let α denote a primitive element of $\mathrm{GF}(2^n)$, then any nonzero element in the finite field can be expressed as α^i. Given $\mathrm{GF}(2^n)$, consider a set of elements in the field $\mathcal{B} = \{\alpha^{r_0}, \alpha^{r_1}, ..., \alpha^{r_{n-1}}\}$, where r_i's are non-negative integers. If the $\mathrm{GF}(2)$-linear combinations of elements of $\mathcal{B}$ span the entire field, we call this as a basis, that is through the basis, we identify $\mathrm{GF}(2^n)$ with the vector space $\mathrm{GF}(2)^n$. Sometimes we denote the basis as $\{\alpha^{r_i}\}_{i=0}^{n-1} = \{\alpha^{r_0}, \alpha^{r_1}, ..., \alpha^{r_{n-1}}\}$.

The number of bases for a given finite field $\mathrm{GF}(2^n)$ is given as

$$\frac{1}{n!} \prod_{s=0}^{n-1} (2^n - 2^s). \tag{2}$$

Conventionally, we use the polynomial basis $\{\alpha^0, \alpha^1, ..., \alpha^{n-1}\}$, but there are many other bases such as a normal basis, which is of the form $\{\alpha^i, \alpha^{2i}, ..., \alpha^{2^{n-1}i}\}^2$, where integer $i > 0$.

2.3   XOR Count of Finite Field Elements and Diffusion Matrices

MDS matrices are popular choice for building the diffusion layer of a block cipher. Towards the construction of lightweight diffusion layer, it is required that the total operations needed to execute the diffusion layer on an input vector (the product of the matrix and a vector) should also be low. In this paper, we consider XOR count as the metric for lightweightness of matrices as done in [8,10].

In practice, a finite field element is represented by its corresponding vector space element by choosing some basis. Then to realize a product of two finite field elements we need to express the product in terms of the basis elements, where the coefficients are linear functions of coordinates of the two elements.

---

[1] This notation should not be confused with the finite field notation $\mathrm{GF}(2)[X]/(P)$, where (P) is an ideal generated by irreducible polynomial P. Nevertheless, both notations refer to the same thing. i.e., $\mathrm{GF}(2^n)/p(X) = \mathrm{GF}(2)[X]/(P)$.

[2] This is a necessary condition for a normal basis, not every i forms a basis.

Definition 1. *The XOR count of an element θ in the field* $\mathrm{GF}(2^n)$ *is the number of XORs required to implement the multiplication of θ with an arbitrary element β. We name the set of XOR counts of all the elements of* $\mathrm{GF}(2^n)$ *as the XOR count distribution.*

For example, consider $\mathrm{GF}(2^3)/(X^3 + X + 1)$ and a basis $\{1, \alpha, \alpha^2\}$. Consider the multiplication of $\alpha^4 = \alpha + \alpha^2$ with an arbitrary element $\beta = b_0 + b_1\alpha + b_2\alpha^2$, where $b_i \in \{0, 1\}$

$$(b_0 + b_1\alpha + b_2\alpha^2)(\alpha + \alpha^2) = (b_1 + b_2) + (b_0 + b_1)\alpha + (b_0 + b_1 + b_2)\alpha^2.$$

In other words, the product of the α^4 and β is of the form

$$(b_1 \oplus b_2, b_0 \oplus b_1, b_0 \oplus b_1 \oplus b_2),$$

in which there are 4 XORs[3]. Therefore, the XOR count of the element α^4 is 4. It is trivial to check that the zero element will have XOR count 0. Since the coefficients of MDS diffusion matrices must be nonzero, in the XOR count distribution we will not mention the XOR count of the zero element. One may also check that for this basis, identity element also has XOR count 0.

We observe that the XOR count distribution of a field may differ as per the choice of basis. For example, consider $\mathrm{GF}(2^3)/(X^3 + X + 1)$ and enumerate the nonzero field elements as $\{\alpha^i\}_{i=0}^6$. For the basis $\{1, \alpha, \alpha^2\}$, the XOR count distribution is $\{0, 1, 2, 4, 4, 3, 1\}$. However, if we consider the normal basis $\{\alpha^3, \alpha^6, \alpha^{12}\}^4$, then the XOR count distribution is $\{0, 3, 3, 2, 3, 2, 2\}$.

The XOR count of one row of a diffusion matrix can be computed using the following formula given in [8]:

$$\text{XOR count of one row} = \sum_{i=1}^{k} \gamma_i + (\ell - 1) \cdot n,$$

where γ_i is the XOR count of the i-th entry in the row of the matrix, k being the order of the diffusion matrix, ℓ is the number of nonzero coefficients in the row and n is the dimension of the finite field. For example, the first row of the AES diffusion matrix being $(1, 1, 2, 3)$ over the field $\mathrm{GF}(2^8)/(X^8 + X^4 + X^3 + X + 1)$, so the XOR count for the first row is $(0 + 0 + 3 + 11) + 3 \times 8 = 38$. Note that for MDS matrices, all coefficients are nonzero thus we can assume $\ell = k$. Since the latter term of the formula is dependent of the dimension of the finite field and order of the MDS matrix, it will be a fixed constant for a given finite field and order of the MDS matrix. Hence, we are only interested in the sum of the XOR count of the coefficients.

In this paper, sometime we describe a diffusion matrix with relatively lower XOR counts as a lightweight matrices.

---

[3] We acknowledge that common terms in the expression could be computed just once and reused to save some XOR count. However, that would require additional cycle and extra memory cost which would very likely to outweigh the cost saved for the XOR count.

[4] Note that the element α^{12} can also be written as α^5 as the finite field multiplication of primitive element has a cycle of length 7.

3 XOR Count Distribution

In this section, we first give a special property of the XOR count distribution under normal bases. Next in Sect. 3.2, we analyze the XOR count distribution and its relation between different irreducible polynomials. We show that any choice of the irreducible polynomial generates the same collection of XOR count distributions when all bases are considered. Lastly in Sect. 3.3, we study the similarity of the XOR count distribution under different bases. This is the building block for constructing the equivalence classes of bases in Sect. 4.

3.1 XOR Count Distribution Under Normal Bases

We give an interesting property of the XOR count regarding normal bases. First, it is known that the binary representation of an element α^{2i} is a shift rotation of the binary representation for α^i under a normal basis. This is a nice feature in the context of hardware implementation.

Proposition 1. *Under a normal basis, α^i of $\mathrm{GF}(2^n)$ has the same XOR count as α^{2i}.*

Proof. Without loss of generality, let the normal basis be $\{\alpha, \alpha^2, ..., \alpha^{2^{n-1}}\}$, an element α^i can be expressed as a polynomial $\alpha^i = a_0\alpha + a_1\alpha^2 + ... + a_{n-1}\alpha^{2^{n-1}}$, while the square of the element has a shift in the coefficient, $\alpha^{2i} = a_{n-1}\alpha + a_0\alpha^2 + ... + a_{n-2}\alpha^{2^{n-1}}$.

For any arbitrary element $b_0\alpha + b_1\alpha^2 + ... + b_{n-1}\alpha^{2^{n-1}}$, the XOR count of α^{2i} can be computed as

$$(a_{n-1}\alpha + a_0\alpha^2 + ... + a_{n-2}\alpha^{2^{n-1}})(b_0\alpha + b_1\alpha^2 + ... + b_{n-1}\alpha^{2^{n-1}})$$
$$= \left((a_0\alpha + a_1\alpha^2 + ... + a_{n-1}\alpha^{2^{n-1}})(b_1\alpha + b_2\alpha^2 + ... + b_0\alpha^{2^{n-1}})\right)^2.$$

Since squaring is simply a shift in the binary representation, the number of XORs in $\left(\alpha^i(b_1\alpha + b_2\alpha^2 + ... + b_0\alpha^{2^{n-1}})\right)^2$ is the same as that of $\alpha^{2i}(b_1\alpha + b_2\alpha^2 + ... + b_0\alpha^{2^{n-1}})$. Furthermore, the number of XORs in $\alpha^{2i}(b_1\alpha + b_2\alpha^2 + ... + b_0\alpha^{2^{n-1}})$ is the same as that of $\alpha^{2i}(b_0\alpha + b_1\alpha^2 + ... + b_{n-1}\alpha^{2^{n-1}})$ as $\{b_0, ..., b_{n-1}\}$ is simply a permutation of $\{b_1, ..., b_0\}$. Hence, the XOR count of α^{2i} is the same as the XOR count of α^i. □

Thus there will be several repetitions in the XOR count distributions when normal basis is considered. As one can see from the example in the previous section that the elements α, α^2 and α^4 have the same XOR count 3 while α^3, α^6 and α^5 have the same XOR count 2.

3.2   XOR Count Spectrum

For a field element θ, we can define a matrix such that the XOR count of the product with an arbitrary element b can be computed directly from that matrix.

Let $\{1, \alpha\}$ be a basis of $GF(2^2)$. For a fixed element $a_0 + a_1\alpha$ of $GF(2^2)$ the multiplication with an arbitrary element $b_0 + b_1\alpha$ will give

$$(b_0 + b_1\alpha)(a_0 + a_1\alpha) = b_0a_0 + b_1a_1 + (b_0a_1 + (a_0 + a_1)b_1)\alpha.$$

In vector notation, this product is actually $(b_0a_0 \oplus b_1a_1, b_0a_1 \oplus (a_0 \oplus a_1)b_1$, which can be written as a matrix product

$$\begin{pmatrix} a_0 & a_1 \\ a_1 & a_0 \oplus a_1 \end{pmatrix} \times \begin{pmatrix} b_0 \\ b_1 \end{pmatrix}.$$

Clearly if there are k_i 1's in the i-th row, then there will be $k_i - 1$ XORs of b_i's in the i-th coordinate of the product.

In general if $\{\alpha^{r_1}, \ldots, \alpha^{r_n}\}$ is a basis of $GF(2^n)/p(X)$, the product of a fixed element $\theta = a_0\alpha^{r_1} + \ldots + a_{n-1}\alpha^{r_n}$ and an arbitrary element $b = b_0\alpha^{r_1} + \ldots + b_{n-1}\alpha^{r_n}$ can be expressed as a multiplication matrix M_θ and $(b_0, \ldots, b_{n-1})$, where

$$M_\theta = \begin{bmatrix} L_{0,0}(a_0, \ldots, a_{n-1}) & \cdots & L_{0,n-1}(a_0, \ldots, a_{n-1}) \\ L_{1,0}(a_0, \ldots, a_{n-1}) & \cdots & L_{1,n-1}(a_0, \ldots, a_{n-1}) \\ \vdots & \ddots & \vdots \\ L_{n-1,0}(a_0, \ldots, a_{n-1}) & \cdots & L_{n-1,n-1}(a_0, \ldots, a_{n-1}) \end{bmatrix},$$

note that each $L_{i,j}(a_0, \ldots, a_{n-1})$ is some $GF(2)$-linear combination of $\{a_0, \ldots, a_{n-1}\}$. As said before if there are k_i 1's in row i, the total number of XORs needed is $\sum_{i=1}^{n}(k_i - 1)$.

It is to be noted that the matrix M_θ is invertible, since $\theta^{-1}\theta b = b$, equivalently $M_\theta^{-1} M_\theta \times [b_0, \ldots, b_{n-1}]^T$ should give $[b_0, \ldots, b_{n-1}]^T$. This fact is used to determine the following property of the matrix M_θ.

We call an n-tuple binary vector nonzero if at least one coordinate of it is nonzero.

Lemma 1. *The collection of the row vectors taken from any fixed row of all the matrices M_θ for all nonzero θ, is in bijection with the set of nonzero n-tuple binary vectors.*

Proof. It is clear that every row of the matrix M_θ of a nonzero element θ is nonzero n-tuple binary vectors, else M_θ is not invertible. Consequently, for each row i, row vectors are pairwise distinct for all such matrices. Suppose not, let θ_1 and θ_2 be distinct elements with the same binary vector in row i. Then $\theta_1 + \theta_2$ is another nonzero element with zeroes in row i which contradicts that nonzero elements are invertible. $\square$

Proposition 2. *The total XOR count of the elements in $GF(2^n)$ is $n\sum_{i=2}^{n}\binom{n}{i}(i - 1)$, and it is invariant of the choice of irreducible polynomial and basis.*

Proof. By Lemma 1, the row i of nonzero multiplication matrices is in bijection with the set of nonzero n-tuple binary vectors over $GF(2)$. Hence, summing the number of XORs for the row i of all elements is $\sum_{i=2}^{n} \binom{n}{i}(i-1)$. Since there are n rows, we have $n \sum_{i=2}^{n} \binom{n}{i}(i-1)$. $\square$

This proposition shows that there is no clear advantage in choosing some particular irreducible polynomial and basis over another.

As the example in Sect. 2.3 shows that XOR count distribution may change under different basis, therefore, one may think that varying over all possible bases, the set of XOR count distributions might be different for $GF(2^n)/p(X)$ and $GF(2^n)/q(X)$. However, our analysis shows that for a basis $\mathcal{B}$ in $GF(2^n)/p(X)$ there will be a basis $\mathcal{B}'$ in $GF(2^n)/q(X)$ such that XOR count distribution of $GF(2^n)/p(X)$ under $\mathcal{B}$ will be equal to that of $GF(2^n)/q(X)$ under $\mathcal{B}'$. The proof is as follows.

For brevity, we call the set of all XOR count distributions for all possible bases as the *XOR count spectrum*.

Lemma 2. *Let* $\psi : GF(2^n)/p(X) \rightarrow GF(2^n)/q(X)$ *an isomorphism between these two finite fields. If* $\{\alpha_0, \ldots, \alpha_{n-1}\}$ *is a basis of* $GF(2^n)/p(X)$, *then the set* $\{\psi(\alpha_0), \ldots, \psi(\alpha_{n-1})\}$ *is a basis of* $GF(2^n)/q(X)$.

Theorem 1. *The XOR count spectrum of* $GF(2^n)/p(X)$ *and* $GF(2^n)/q(X)$ *are the same.*

Proof. We show that for a basis of $GF(2^n)/p(X)$, there is a basis of $GF(2^n)/q(X)$, where XOR count distribution will be the same. Let α and β be the primitive elements of $GF(2^n)/p(X)$ and $GF(2^n)/q(X)$ respectively. Suppose $\{\alpha_0, \ldots, \alpha_{n-1}\}$ is a basis of $GF(2^n)/p(X)$. Consider an arbitrary element of $GF(2^n)/p(X)$ as $b_0\alpha_0 + \ldots + b_{n-1}\alpha_{n-1}$ and multiply with the element α^i

$$\alpha^i(b_0\alpha_0 + \ldots + b_{n-1}\alpha_{n-1}) = L_0\alpha_0 + \ldots + L_{n-1}\alpha_{n-1}, \qquad (3)$$

where L_i's are some linear combinations of $\{b_0, \ldots, b_{n-1}\}$. If in the linear combination L_i there are c_i XORs, then XOR count is $\sum_{i=0}^{n-1} c_i$. Notice that the value of each $L_i \in \{0, 1\}$.

Apply ψ on both sides of (3), and we get

$$\psi(\alpha)^i(b_0\psi(\alpha_0) + \ldots + b_{n-1}\psi(\alpha_{n-1})) = L_0\psi(\alpha_0) + \ldots + L_{n-1}\psi(\alpha_{n-1}).$$

From Lemma 2, we know that $\{\psi(\alpha_0), \ldots, \psi(\alpha_{n-1})\}$ is a basis of $GF(2^n)/q(X)$, and from the above we get that there is $\psi(\alpha)^i$ in $GF(2^n)/q(X)$ such that its XOR count under $\{\psi(\alpha_0), \ldots, \psi(\alpha_{n-1})\}$ is $\sum_{i=0}^{n-1} c_i$.

Thus the XOR count spectrum obtained for $GF(2^n)/p(X)$ will be the same for $GF(2^n)/q(X)$. $\square$

Therefore, we see that there is no gain in considering $GF(2^n)$ under different irreducible polynomials, as this will not generate any new XOR count spectrum. Hence for the rest of the paper, we omit the irreducible polynomial of the corresponding field unless necessary.

3.3   Bases with Similar XOR Count Distributions

Let us now check if there is any similar XOR count distribution within the XOR count spectrum, more precisely saying that we would like to see if a given finite field $\mathrm{GF}(2^n)/p(X)$, there are bases whose corresponding XOR count distributions are equal (up to a permutation). In the following we present the results.

Lemma 3. *If* $\{\alpha^{r_0}, \ldots, \alpha^{r_{n-1}}\}$ *is a basis of* $\mathrm{GF}(2^n)$, *then* $\{\alpha^{r_0+1}, \ldots, \alpha^{r_{n-1}+1}\}$ *is also a basis of* $\mathrm{GF}(2^n)$.

Proposition 3. *Given a finite field* $\mathrm{GF}(2^n)$ *and bases* $\mathcal{B} = \{\alpha^{r_0}, \ldots, \alpha^{r_{n-1}}\}$ *and* $\mathcal{B}^{+t} = \{\alpha^{r_0+t}, \ldots, \alpha^{r_{n-1}+t}\}$, *for integer* $t > 0$, *the XOR count distribution of* $\mathrm{GF}(2^n)$ *under these bases are exactly the same.*

Proof. For simplicity we prove it for $t = 1$, the rest follows by induction. For an arbitrary element $b = b_0\alpha^{r_0} + \ldots + b_{n-1}\alpha^{r_{n-1}}$, we can express the multiplication with α^j under $\mathcal{B}$ as

$$\alpha^j(b_0\alpha^{r_0} + \ldots + b_{n-1}\alpha^{r_{n-1}}) = L_0\alpha^{r_0} + \ldots + L_{n-1}\alpha^{r_{n-1}},$$

where L_i's are some linear combinations of $\{b_0, \ldots, b_{n-1}\}$. Suppose c_i is the number of XORs in L_i, then XOR count of α^j under $\mathcal{B}$ is $\sum_{i=0}^{n-1} c_i$.

On the other hand, the multiplication with α^j under $\mathcal{B}^{+1}$ can be expressed as

$$\begin{aligned}
\alpha^j(b_0\alpha^{r_0+1} + \ldots + b_{n-1}\alpha^{r_{n-1}+1}) &= \alpha^j(b_0\alpha^{r_0} + \ldots + b_{n-1}\alpha^{r_{n-1}})\alpha \\
&= (L_0\alpha^{r_0} + \ldots + L_{n-1}\alpha^{r_{n-1}})\alpha \\
&= L_0\alpha^{r_0+1} + \ldots + L_{n-1}\alpha^{r_{n-1}+1}.
\end{aligned}$$

Clearly the XOR count in this case is $\sum_{i=0}^{n-1} c_i$ too.

Therefore, the XOR count distribution of $\mathrm{GF}(2^n)$ under $\{\alpha^{r_0}, \ldots, \alpha^{r_{n-1}}\}$ and $\{\alpha^{r_0+1}, \ldots, \alpha^{r_{n-1}+1}\}$ are exactly the same. ☐

Next we find that there is another set of bases where the corresponding XOR count distributions are the same up to a permutation.

Lemma 4. *If* $\{\alpha^{r_0}, \ldots, \alpha^{r_{n-1}}\}$ *is a basis of* $\mathrm{GF}(2^n)$, *then* $\{\alpha^{2r_0}, \ldots, \alpha^{2r_{n-1}}\}$ *is also a basis of* $\mathrm{GF}(2^n)$.

Proposition 4. *Given a finite field* $\mathrm{GF}(2^n)$, *the XOR count distribution under the bases* $\mathcal{B} = \{\alpha^{r_0}, \ldots, \alpha^{r_{n-1}}\}$ *and* $\mathcal{B}^{\times 2^s} = \{\alpha^{2^s r_0}, \ldots, \alpha^{2^s r_{n-1}}\}$, *for integer* $s > 0$, *are the same up to a permutation.*

Proof. For simplicity, we prove for $s = 1$, the rest will follow by induction.

For an arbitrary element $b = b_0\alpha^{r_0}, \ldots, b_{n-1}\alpha^{r_{n-1}}$, we can express the multiplication with α^j under $\mathcal{B}$ as

$$\alpha^j(b_0\alpha^{r_0} + \ldots + b_{n-1}\alpha^{r_{n-1}}) = L_0\alpha^{r_0} + \ldots + L_{n-1}\alpha^{r_{n-1}}, \qquad (4)$$

where L_i's are linear combinations of $\{b_0, \ldots, b_{n-1}\}$. If c_i is the number of XORs in L_i, then the XOR count of α^j under $\mathcal{B}$ is $\sum_{i=0}^{n-1} c_i$.

To compute the XOR count for α^{2i} under $\mathcal{B}^{\times 2}$, we square (4) to obtain

$$\alpha^{2j}(b_0\alpha^{2r_0} + \ldots + b_{n-1}\alpha^{2r_{n-1}}) = L_0\alpha^{2r_0} + \ldots + L_{n-1}\alpha^{2r_{n-1}}. \tag{5}$$

Clearly the XOR count obtained from (5) is also $\sum_{i=0}^{n-1} c_i$. Since $\gcd(2, 2^n - 1) = 1$, the mapping from α^j under $\mathcal{B}$ to α^{2i} under $\mathcal{B}^{\times 2}$ is bijection. Therefore, XOR count distribution under $\mathcal{B}$ and $\mathcal{B}^{\times 2}$ are just permutation of each other. $\qquad\square$

4   Equivalence Classes of Bases

In the previous section, we have seen the similarities in some of the XOR count distributions generated by different bases. In this section, we formally introduce the equivalence relation between bases whose XOR count distributions produce the lightest MDS matrix with the same XOR count. Using this equivalence relation, we construct the equivalence classes of bases.

From Proposition 3, it is clear that for any MDS diffusion matrix $M = [\beta_{i,j}]_{k \times k}$ has the same XOR count both under $\mathcal{B}$ and $\mathcal{B}^{+t}$. As for the other type of basis $\mathcal{B}^{\times 2^s}$, by Proposition 4, we know that the XOR count of M under $\mathcal{B}$ will match with that of another matrix $M' = [\beta_{i,j}^2]_{k \times k}$ under $\mathcal{B}^{\times 2}$, however, it is unclear if M' is also an MDS matrix. Thus, we need the following lemma.

Lemma 5. *Suppose $M = [\beta_{i,j}]_{k \times k}$ is an MDS matrix over $\mathrm{GF}(2^n)$, then $M' = [\beta_{i,j}^2]_{k \times k}$ is also an MDS matrix.*

Proof. It is known that all square submatrices of an MDS matrix have nonzero determinants. Since $\mathrm{GF}(2^n)$ has characteristic 2, the determinants of the submatrices of M' are square of the determinants of the corresponding submatrices of M, which are also nonzero. $\qquad\square$

With Lemma 5, it is now clear that M' is also MDS. Therefore, we can say that by Propositions 3 and 4, the XOR count distributions of $\mathrm{GF}(2^n)$ under $\mathcal{B}$, $\mathcal{B}^{+t}$ and $\mathcal{B}^{\times 2^s}$ are invariant for the MDS matrices over finite fields under these bases. Because for every MDS matrix with some XOR count found under $\mathcal{B}$, there will be another MDS matrix having the same XOR count under $\mathcal{B}^{+t}$ and $\mathcal{B}^{\times 2^s}$, and vice versa. With that said, we can partition the set of all bases of $\mathrm{GF}(2^n)$ into distinct equivalence classes.

Definition 2. *The bases $\mathcal{B} = \{\alpha^{r_i}\}_{i=0}^{n-1}$ and $\mathcal{B}' = \{\alpha^{u_i}\}_{i=0}^{n-1}$ of $\mathrm{GF}(2^n)$ are equivalent if $u_i = (2^s r_i + t) \bmod 2^n - 1$ for some $s \geq 0$ and $t \geq 0$. The collection of these equivalent bases forms an equivalence class of bases.*

With these equivalence classes of bases, it is sufficient to consider one basis representative from each equivalence class in order to find one of the lightest MDS matrices over all possible bases.

Next, we analyze the cardinality of the equivalence classes. Interestingly, the bases are not uniformly partitioned into equivalence classes. For instance, for $GF(2^3)$, there are 28 bases and only 2 equivalence classes, where one consists of 21 bases while the other has 7. This complicates the counting of the number of equivalence classes for a given field dimension. Therefore, instead of finding the exact cardinality of the equivalence class, we give a bound to it.

Lemma 6. *The cardinality of any equivalence classes of bases of $GF(2^n)$ is a multiple of $2^n - 1$.*

Proof. Consider basis of the form $\mathcal{B}^{+t} = \{\alpha^{r_i+t}\}_{i=0}^{n-1}$, for positive integer t, which is in the equivalence class of $\mathcal{B} = \{\alpha^{r_i}\}_{i=0}^{n-1}$. Then the proof is immediate if we can show that the smallest positive integer t that satisfies $\mathcal{B}^{+t} = \mathcal{B}$ is $t = 2^n - 1$.

Since $\alpha^{2^n-1} = 1$, it is clear that $\mathcal{B}^{+t} = \mathcal{B}$ when $t = 2^n - 1$. Suppose there exists $t_0 < 2^n - 1$ such that $\mathcal{B}^{+t_0} = \mathcal{B}$, taking the summation of the elements in the basis, we have

$$\alpha^{r_0+t_0} + \ldots + \alpha^{r_{n-1}+t_0} = \alpha^{r_0} + \ldots + \alpha^{r_{n-1}}.$$

Since $\{\alpha^{r_i}\}_{i=0}^{n-1}$ is a basis, the summation, $\sum_{i=0}^{n-1} \alpha^{r_i}$, is nonzero and invertible. Hence we can simplify the equation and obtain $\alpha^{t_0} = 1$, which is a contradiction that α is a primitive element of the finite field.   □

Theorem 2. *A lower bound and upper bound of the cardinality of any equivalence classes of bases of $GF(2^n)$ is $2^n - 1$ and $n \cdot 2^n - 1$ respectively.*

Proof. From Lemma 6, we know that the lower bound of the cardinality of equivalence class is $2^n - 1$. Since $\alpha^{2^n} = \alpha$, it is clear that $\mathcal{B}^{\times 2^s} = \mathcal{B}$ when $s = n$. Therefore, the largest possible cardinality is $n \cdot 2^n - 1$, when these n sets of bases, $\{\mathcal{B}^{+t}\}_{t=0}^{2^n-1}$, $\{(\mathcal{B}^{\times 2})^{+t}\}_{t=0}^{2^n-1}$, $\{(\mathcal{B}^{\times 2^2})^{+t}\}_{t=0}^{2^n-1}$, ..., $\{(\mathcal{B}^{\times 2^{n-1}})^{+t}\}_{t=0}^{2^n-1}$, belong to the same equivalence class and are pairwise distinct.   □

Lastly, we show that every equivalence class contains one certain kind of basis. This allows us to find a representative basis from each equivalence classes more efficiently.

Proposition 5. *Every equivalence class always contains a basis of the form $\{1, \alpha^{u_1} \ldots, \alpha^{u_{n-1}}\}$.*

Proof. Given a basis $\{\alpha^{r_0}, \alpha^{r_1} \ldots, \alpha^{r_{n-1}}\}$ from an equivalence class, consider $t = 2^n - 1 - r_0$, then the equivalent basis $\{\alpha^{r_i+t}\}_{i=0}^{n-1} = \{1, \ldots, \alpha^{2^n-1-r_0+r_{n-1}}\}$ also belongs to the same equivalence class.   □

5   Search Algorithms and Results

In this section, we first present our strategy to find all the equivalence classes of bases, then it is sufficient for us to apply our search on one representative of each equivalence classes for lightweight (in terms of low XOR count) MDS diffusion matrices. Next, we adopt the similar strategy as described in [10, Sect. 5.2] and extend the search to different bases for lightweight (involutory) MDS Hadamard matrices of order 4.

5.1   Enumerating Equivalence Classes

Search Algorithm. By Proposition 5, we know that the representative of an equivalence class is of the form $\{1, \alpha^{u_1}, \ldots, \alpha^{u_{n-1}}\}$. Therefore, all we need is to start from $\{1, \alpha^{u_1}, \ldots, \alpha^{u_{n-1}}\}$, if it is a basis then we can generate the equivalence class by

$$\{1, \alpha^{u_1}, \ldots, \alpha^{u_{n-1}}\} \rightarrow \{1, \alpha^{2^s u_1 + t}, \ldots, \alpha^{2^s u_{n-1} + t}\}, s \geq 0, t \geq 0.$$

This way we need to test $\binom{2^n - 2}{n-1}$ possible basis representatives in the worst case. The pseudocode for enumerating the equivalence class is presented in Appendix A.

Results. Due to memory issue, we are unable to compute the exact number of equivalence classes of bases for $n \geq 6$. However, by Theorem 2, we are able to estimate the number of equivalence classes as we know the lower bound and upper bound of the cardinality of an equivalence class to be $2^n - 1$ and $n \cdot 2^n - 1$ respectively.

In Table 1, the second column shows the number of irreducible polynomials for each dimension which can be computed from (1). By Theorem 1, we only need to consider one arbitrary irreducible polynomial. The total number of bases which can be computed using (2) is given in the third column, while the number of equivalence classes of bases for each dimension is given in the last column.

5.2   Finding Lightweight (involutory) MDS Matrices Under Different Bases

Search Algorithm. The authors of [10] analyzed the structure of Hadamard matrices and presented the equivalence classes of Hadamard matrices and a simplified check for MDS property on Hadamard matrices. In this paper, we focus on Hadamard matrices of order 4 as 4×4 matrices are commonly used in diffusion layer of a block cipher, for instance in AES. In addition, involutory MDS Hadamard matrices can be easily constructed, as a Hadamard matrix is

Table 1. Number of equivalence classes of bases

Dimension of finite field	Number of irreducible polynomial	Number of bases	Number of equivalence classes
$n = 3$	2	28	2
$n = 4$	3	840	16
$n = 5$	6	83328	540
$n = 6$	7	$2^{24.74}$	$2^{16.18} \sim 2^{18.76}$
$n = 7$	18	$2^{34.92}$	$2^{25.12} \sim 2^{27.93}$
$n = 8$	30	$2^{46.91}$	$2^{35.92} \sim 2^{38.92}$

involution iff the XOR-sum of the first row is 1. Based on these results, we only need to choose a set of lightweight coefficients and test for MDS, the arrangement of the entries is invariant as there is only one equivalence class of Hadamard matrices for order 4 [10].

From Sect. 4, we see that bases within an equivalence class of bases have the same (w.r.t. XOR count) collection of MDS matrices. Hence, it is sufficient to check one representative from each equivalence class. To search for lightweight MDS matrices over a given basis, we set some threshold value as the upper bound for the total XOR count of the coefficients. If the sum of the XOR count of the candidate is lower than the threshold, then we check if it forms an MDS Hadamard matrix. In order to search for lightweight involutory MDS Hadamard matrices, an additional condition that the XOR-sum of the candidates equals to 1 is required. For $GF(2^4)$ and $GF(2^8)$, we set the threshold value to be the XOR count of the lightest MDS matrices found in [10]. For other order of finite fields, we set the threshold value to some arbitrary large value. The threshold value will be updated whenever we find a new (involutory) MDS Hadamard matrix with lesser total XOR count. The pseudocode for finding lightweight (involutory) MDS Hadamard matrices is presented in Appendix B.

Results. For $n = 3, 4, 5$, we search through all the equivalence classes of bases. For $n = 8$, we consider the polynomial and normal bases because these are the two most commonly used bases. The outcome is that the lightest MDS and involutory MDS Hadamard matrices are found for the bases that belong to the equivalence class containing the polynomial basis. And naturally for $n = 4, 8$, the XOR count of the lightest MDS and involutory MDS Hadamard matrices match with the results from [10].

5.3   Recommended Choice of Basis

Although we do not find MDS diffusion matrices with XOR count lesser than the existing ones, it is interesting to see that the lightest diffusion matrices are found under the polynomial basis. From Proposition 2, it seems that there is no clear implication that one basis is strictly better than another, as the mean XOR count is the same for any basis. However, the XOR count distribution may vary for different bases, that is quantified by the standard deviation. A high standard deviation implies that the distribution of XOR count is far apart from the mean, thus there will be more elements with relatively lower/higher XOR count. As pointed out in [10], in general the order of the finite field is much larger than the order of the diffusion matrix, since only a few elements of the finite field are used, there is a better chance of finding lightweight diffusion matrix under XOR distributions with higher standard deviation.

To illustrate this concept, consider taking the two XOR count distributions, $D_1 = \{0, 1, 2, 4, 4, 3, 1\}$ and $D_2 = \{0, 3, 3, 2, 3, 2, 2\}$, from Sect. 2.3 as an example. One can observe that the standard deviations of D_1 and D_2 are 1.57 and 1.07 respectively. Suppose we want to construct an MDS matrix of order 2, we need to

Table 2. Highest standard deviation of various bases

Dimension of finite field	Polynomial basis	Other basis
$n = 3$	1.46	0.99
$n = 4$	2.68	1.71
$n = 5$	4.09	3.55
Dimension of finite field	Polynomial basis	Normal basis
$n = 8$	7.53	4.48

pick 2 distinct nonzero elements. Under D_1, we can pick 2 elements corresponding to XOR count 0 and 1, which is lower than any choice that we make under D_2. The main reason being that the XOR counts in D_2 are much closer to the mean, while under D_1 we are able to pick elements with relatively lower XOR count and check if they form an MDS matrix. Therefore, we look into the standard deviation of the XOR count distribution of the bases.

By computing the standard deviation for all representation bases of the equivalence classes of bases, we observe that the standard deviation of the polynomial bases are significantly larger than the highest standard deviation of the non-polynomial bases. The results are summarized in Table 2.

For any finite field $GF(2^n)$, we conjecture that the XOR distribution under a polynomial basis tends to have a higher standard deviation as compared to other bases. Therefore, we think that considering the polynomial basis, or its equivalent bases, is the preferable choice for finding lightweight diffusion matrices.

5.4   Conclusion

In this paper, we study the behavior of the XOR count distribution under different bases and irreducible polynomials. We show that for all irreducible polynomials, the XOR count spectrum is the same. Hence, we only need to consider one irreducible polynomial when all bases are considered. Under a fixed irreducible polynomial, the bases can be partitioned into equivalence classes, where the XOR count distribution is invariant under these bases. In addition, we provide a search algorithm for finding all the equivalence classes of bases. Using these equivalence classes of bases, we complete the search for lightweight MDS and involutory MDS Hadamard matrix of order 4 for finite field dimension $n = 3, 4, 5$. Our result suggests that the bases from the equivalence class of polynomial basis are the recommended choice for constructing lightweight MDS diffusion matrices.

Acknowledgements. The authors would like to thank Thomas Peyrin for his valuable comments. The second author is supported by Singapore National Research Foundation Fellowship 2012 (NRF-NRFF2012-06).

A Pseudocode for finding equivalent bases of $GF(2^n)$

Algorithm 1. Finding equivalent bases for $GF(2^n)$.

INPUT: $GF(2^n)$ generated by a primitive element α, $\mathbf{S} = \emptyset$.

OUTPUT: $\mathbf{B}$ the set of basis representatives of distinct equivalence classes of bases.

  set $\mathbf{B} = \emptyset$ and $\texttt{counter} = 0$

  for each set $\{(0, i_1, \ldots, i_{n-1}) : i_j \in [1, 2^n - 2]\}$ chosen from $\binom{2^n - 2}{n-1}$ possible combinations **do**

    generate $E = \{\alpha^{2^s i_1 + t \bmod 2^n - 1}, \ldots, \alpha^{2^s i_n + t \bmod 2^n - 1} : s \in [0, n-1], t \in [0, 2^n - 2]\}$

    store every member of E in $\mathbf{S}$ that has 1 and is new to $\mathbf{S}$, and update $\texttt{counter}{+}{+}$

    if $\{1, \alpha^{i_1}, \ldots, \alpha^{i_{n-1}}\}$ is a basis **then**

      store $\{1, \alpha^{i_1}, \ldots, \alpha^{i_{n-1}}\}$ in $\mathbf{B}$

      if $\texttt{counter} = \binom{2^n - 2}{n-1}$ **then**

        return $\mathbf{B}$ as the set of bases that are representatives to all distinct equivalence classes

      end if

    end if

  end for

B Pseudocode for Finding Lightweight (involutory) MDS Hadamard Matrices over $GF(2^n)$

Algorithm 2. Finding lightweight (involutory) MDS Hadamard matrices for $GF(2^n)$.

INPUT: `MDS_threshold`, `IMDS_threshold`, nonzero elements of $GF(2^n)$, XOR count of the field elements.

OUTPUT: XOR count of the lightest MDS and involutory MDS Hadamard matrices of order 4.

  sort the elements in ascending order according to their XOR counts

  for each set S of 4 elements chosen from $\binom{2^n - 1}{4}$ possible combinations **do**

    if XOR-sum of elements $= 1$ **then**

      if sum of XOR count $<$ `IMDS_threshold` **then**

        construct Hadamard matrix H from S

        if H is MDS **then**

          update `IMDS_threshold` $=$ sum of XOR count

        end if

      end if

    else if sum of XOR count $<$ `MDS_threshold` **then**

      construct Hadamard matrix H from S

      if H is MDS **then**

        update `MDS_threshold` $=$ sum of XOR count

      end if

    end if

  end for

  return `MDS_threshold` and `IMDS_threshold`

References

1. Babbage, S., Dodd, M.: The stream cipher MICKEY 2.0 (2006). http://www.ecrypt.eu.org/stream/mickeypf.html
2. Bogdanov, A.A., Knudsen, L.R., Leander, G., Paar, C., Poschmann, A., Robshaw, M., Seurin, Y., Vikkelsoe, C.: PRESENT: an ultra-lightweight block cipher. In: Paillier, P., Verbauwhede, I. (eds.) CHES 2007. LNCS, vol. 4727, pp. 450–466. Springer, Heidelberg (2007)
3. Borghoff, J., Canteaut, A., Güneysu, T., Kavun, E.B., Kneževic, M., Knudsen, L.R., Leander, G., Nikov, V., Paar, C., Rechberger, C., Rombouts, P., Thomsen, S.S., Yalçin, T.: PRINCE - a low-latency block cipher for pervasive computing applications (Full version). Cryptology ePrint Archive, Report /529 (2012). http://eprint.iacr.org/
4. Daemen, J., Rijmen, V.: The Design of Rijndael: AES - The Advanced Encryption Standard. Information Security and Cryptography. Springer, Heidelberg (2002)
5. Gehring, F.W., Halmos, P.R. (eds.): Introduction to Analytic Number Theory. Undergraduate Texts in Mathematics. Springer, New York (1976)
6. Guo, J., Peyrin, T., Poschmann, A., Robshaw, M.: The LED block cipher. In: Preneel, B., Takagi, T. (eds.) CHES 2011. LNCS, vol. 6917, pp. 326–341. Springer, Heidelberg (2011)
7. Hell, M., Johansson, T., Meier, W.: Grain : a stream cipher for constrained environments. Int. J. Wire. Mob. Comput. 2(1), 86–93 (2007)
8. Khoo, K., Peyrin, T., Poschmann, A.Y., Yap, H.: FOAM: searching for hardware-optimal SPN structures and components with a fair comparison. In: Batina, L., Robshaw, M. (eds.) CHES 2014. LNCS, vol. 8731, pp. 433–450. Springer, Heidelberg (2014)
9. Shirai, T., Shibutani, K., Akishita, T., Moriai, S., Iwata, T.: The 128-bit blockcipher CLEFIA (extended abstract). In: Biryukov, A. (ed.) FSE 2007. LNCS, vol. 4593, pp. 181–195. Springer, Heidelberg (2007)
10. Sim, S.M., Khoo, K., Oggier, F., Peyrin, T.: Lightweight MDS involution matrices. In: Leander, G. (ed.) FSE 2015. LNCS, vol. 9054, pp. 471–493. Springer, Heidelberg (2015)
11. Tian, Y., Chen, G., Li, J.: On the design of Trivium. Cryptology ePrint Archive, Report /431 (2009). http://eprint.iacr.org/

Secure Protocols

Prover-Efficient Commit-and-Prove Zero-Knowledge SNARKs

Helger Lipmaa[(✉)]

Institute of Computer Science, University of Tartu, Tartu, Estonia
helger.lipmaa@gmail.com

Abstract. Zk-SNARKs (succinct non-interactive zero-knowledge arguments of knowledge) are needed in many applications. Unfortunately, all previous zk-SNARKs for interesting languages are either inefficient for the prover, or are non-adaptive and based on a commitment scheme that depends both on the prover's input and on the language, i.e., they are not commit-and-prove (CaP) SNARKs. We propose a proof-friendly extractable commitment scheme, and use it to construct prover-efficient adaptive CaP succinct zk-SNARKs for different languages, that can all reuse committed data. In new zk-SNARKs, the prover computation is dominated by a linear number of cryptographic operations. We use batch-verification to decrease the verifier's computation; importantly, batch-verification can be used also in QAP-based zk-SNARKs.

Keywords: Batch verification · Commit-and-prove · CRS · NIZK · Numerical NP-complete languages · Range proof · SUBSET-SUM · zk-SNARK

1 Introduction

Recently, there has been a significant surge of activity in studying succinct non-interactive zero knowledge (NIZK) arguments of knowledge (also known as zk-SNARKs) [3–6,12,13,17,19,23,24,28]. The prover of a zk-SNARK outputs a short (ideally, a small number of group elements) argument π that is used to convince many different verifiers in the truth of the same claim without leaking any side information. The verifiers can verify independently the correctness of π, without communicating with the prover. The argument must be efficiently verifiable. Constructing the argument can be less efficient, since it is only done once. Still, prover-efficiency is important, e.g., in a situation where a single server has to create many arguments to different clients or other servers.

Many known zk-SNARKs are non-adaptive, meaning that the common reference string, CRS, can depend on the concrete instance of the language (e.g., the circuit in the case of CIRCUIT-SAT). In an adaptive zk-SNARK, the CRS is independent on the instance and thus can be reused many times. This distinction is important, since generation and distribution of the CRS must be

© Springer International Publishing Switzerland 2016
D. Pointcheval et al. (Eds.): AFRICACRYPT 2016, LNCS 9646, pp. 185–206, 2016.
DOI: 10.1007/978-3-319-31517-1_10

done securely. The most efficient known *non-adaptive* zk-SNARKs for NP-complete languages from [17] are based on either Quadratic Arithmetic Programs (QAP, for arithmetic CIRCUIT-SAT) or Quadratic Span Programs (QSP, for Boolean CIRCUIT-SAT). There, the prover computation is dominated by $\Theta(n)$ cryptographic operations (see the full version [26] for a clarification on cryptographic/non-cryptographic operations), where n is the number of the gates. QAP, QSP [17,24] and other related approaches like SSP [13] have the same asymptotic complexity.

QSP-based CIRCUIT-SAT SNARK can be made adaptive by using universal circuits [33]. Then, the CRS depends on the construction of universal circuit and not on the concrete input circuit itself. However, since the size of a universal circuit is $\Theta(n \log n)$, the prover computation in resulting adaptive zk-SNARKs is $\Theta(n \log^2 n)$ non-cryptographic operations and $\Theta(n \log n)$ cryptographic operations. (In the case of QAP-based arithmetic CIRCUIT-SAT SNARK, one has to use universal arithmetic circuits [30] that have an even larger size $\Theta(r^4 n)$, where r is the degree of the polynomial computed by the arithmetic circuit. Thus, we will mostly give a comparison to the QSP-based approach.)

Since Valiant's universal circuits incur a large constant $c = 19$ in the $\Theta(\cdot)$ expression, a common approach [21,31] is to use universal circuits with the overhead of $\Theta(\log^2 n)$ but with a smaller constant $c = 1/2$ in $\Theta(\cdot)$. The prover computation in the resulting adaptive zk-SNARKs is $\Theta(n \log^3 n)$ non-cryptographic operations and $\Theta(n \log^2 n)$ cryptographic operations.[1]

Another important drawback of the QSP/QAP-based SNARKs is that they use a circuit-dependent commitment scheme. To use the same input data in multiple sub-SNARKs, one needs to construct a single large circuit that implements all sub-SNARKs, making the SNARK and the resulting *new* commitment scheme more complicated. In particular, these SNARKs are not commit-and-prove (CaP [9,20]) SNARKs. We recall that in CaP SNARKs, a commitment scheme C is fixed first, and the statement consists of commitments of the witness using C; see Sect. 2. Hence, a CaP commitment scheme is *instance-independent*. In addition, one would like the commitment scheme to be *language-independent*, enabling one to first commit to the data and only then to decide in what applications (e.g., verifiable computation of a later fixed function) to use it.

See Table 1 for a brief comparison of the efficiency of proposed adaptive zk-SNARKs for NP-complete languages. SUBSET-SUM is here brought as an example of a wider family of languages; it can be replaced everywhere say with PARTITION or KNAPSACK, see the full version [26]. Here, $N = r_3^{-1}(n) = o(n 2^{2\sqrt{2 \log_2 n}})$, where $r_3(n)$ is the density of the largest progression-free set in $\{1, \ldots, n\}$. According to the current knowledge, $r_3^{-1}(n)$ is comparable to (or only slightly smaller than) n^2 for $n < 2^{12}$; this makes all known CaP SNARKs [15,19,23] arguably impractical unless n is really small. In all cases, the verifier's computation is dominated by either $\Theta(n)$ cryptographic or $\Theta(n \log n)$

---

[1] Recently, [12] proposed an independent methodology to improve the prover's computational complexity in QAP-based arguments. However, [12] does not spell out their achieved prover's computational complexity.

Table 1. Prover-efficiency of known *adaptive* zk-SNARKs for NP-complete languages. Here, n is the number of the gates (in the case of CIRCUIT-SAT) and the number of the integers (in the case of SUBSET-SUM). Green background denotes the best known asymptotic complexity of the *concrete* NP-complete language w.r.t. to the concrete parameter. The solutions marked with * use proof bootstrapping from [12]

Paper	Language	Prover computation		|CRS|
		non-crypt. op.	crypt. op.	
Not CaP-s				
QAP, QSP ([14, 19, 27])	CIRCUIT-SAT	$\Theta(n \log^2 n)$	$\Theta(n \log n)$	$\Theta(n)$
CaP-s				
Gro10 ([21])	CIRCUIT-SAT	$\Theta(n^2)$	$\Theta(n^2)$	$\Theta(n^2)$
Lip12 ([26])	CIRCUIT-SAT	$\Theta(n^2)$	$\Theta(N)$	$\Theta(N)$
Lip14 + Lip12 ([26, 28])*	CIRCUIT-SAT	$\Theta(N \log^2 n)$	$\Theta(N \log n)$	$\Theta(N \log n)$
Lip14 + current paper ([28])*	CIRCUIT-SAT	$\Theta(n \log^2 n)$	$\Theta(n \log n)$	$\Theta(n \log n)$
FLZ13 ([16])	SUBSET-SUM	$\Theta(N \log n)$	$\Theta(N)$	$\Theta(N)$
Current paper	SUBSET-SUM	$\Theta(n \log n)$	$\Theta(n)$	$\Theta(n)$

non-cryptographic operations (with the verifier's online computation usually being $\Theta(1)$), and the communication consists of a small constant number of group elements.[2] Given all above, it is natural to ask the following question:

The Main Question of This Paper: *Is it possible to construct* adaptive CaP *zk-SNARKs for NP-complete languages where the prover computation is dominated by a linear number of cryptographic operations?*

We answer the "main question" positively by improving on Groth's modular approach [19]. Using the modular approach allows us to modularize the security analysis, first proving the security of underlying building blocks (the product and the shift SNARKs), and then composing them to construct master SNARKs for even NP-complete languages. The security of master SNARKs follows easily from the security of the basic SNARKs. We also use batch verification to speed up verification of almost all known SNARKs.

All new SNARKs use the same commitment scheme, the interpolating commitment scheme. Hence, one can reuse their input data to construct CaP zk-SNARKs for different unrelated languages, chosen only after the commitment was done. Thus, one can first commit to some data, and only later decide in which application and to what end to use it. Importantly, by using CaP zk-SNARKs, one can guarantee that all such applications use exactly the same data.

---

[2] We emphasize that CIRCUIT-SAT is *not* our focus; the lines corresponding to CIRCUIT-SAT are provided only for the sake of comparison. One can use proof boot-strapping [12] to decrease the length of the resulting CIRCUIT-SAT argument from $\Theta(\log n)$, as stated in [25], to $\Theta(1)$; we omit further discussion.

The resulting SNARKs are not only commit-and-prove, but also very efficient, and often more efficient than any previously known SNARKs. The new CaP SNARKs have prover-computation dominated by $\Theta(n)$ cryptographic operations, with the constant in $\Theta(\cdot)$ being reasonably small. Importantly, we propose the most efficient known succinct range SNARK. Since the resulting zk-SNARKs are sufficiently different from QAP-based zk-SNARKs, we hope that our methodology by itself is of independent interest. Up to the current paper, Groth's modular approach has resulted in significantly less efficient zk-SNARKs than the QSP/QAP-based approach.

In Sect. 3, we construct a new natural extractable trapdoor commitment scheme (the interpolating commitment scheme). Here, commitment to $\boldsymbol{a} \in \mathbb{Z}_p^n$, where n is a power of 2, is a short garbled and randomized version $g_1^{L_a(\chi)}(g_1^{\chi^n-1})^r$ of the Lagrange interpolating polynomial $L_a(X)$ of $\boldsymbol{a}$, for a random secret key χ, together with a knowledge component. This commitment scheme is arguably a very natural one, and in particular its design is not influenced by the desire to tailor it to one concrete application. Nevertheless, as we will see, using it improves the efficiency of many constructions while allowing to reuse many existing results.

The new CaP zk-SNARKs are based on the interpolating commitment scheme and two CaP witness-indistinguishable SNARKs: a product SNARK (given commitments to vectors $\boldsymbol{a}$, $\boldsymbol{b}$, $\boldsymbol{c}$, it holds that $c_i = a_i b_i$; see [15,19,23]), and a shift SNARK (given commitments to $\boldsymbol{a}$, $\boldsymbol{b}$, it holds that $\boldsymbol{a}$ is a coordinate-wise shift of $\boldsymbol{b}$; see [15]). One can construct an adaptive CIRCUIT-SAT CaP zk-SNARK from $\Theta(\log n)$ product and shift SNARKs [19,25], or adaptive CaP zk-SNARKs for NP-complete languages like SUBSET-SUM (and a similar CaP range SNARK) by using a constant number of product and shift SNARKs [15].

In Sect. 4, we propose a CaP product SNARK, that is an argument of knowledge under a computational and a knowledge (needed solely to achieve extractability of the commitment scheme) assumption. Its prover computation is dominated by $\Theta(n \log n)$ non-cryptographic and $\Theta(n)$ cryptographic operations. This can be compared to $r_3^{-1}(n)$ non-cryptographic operations in [15]. The speed-up is mainly due to the use of the interpolating commitment scheme.

In Sect. 5, we propose a variant of the CaP shift SNARK of [15], secure when combined with the interpolating commitment scheme. We prove that this SNARK is an adaptive argument of knowledge under a computational and a knowledge assumption. It only requires the prover to perform $\Theta(n)$ cryptographic and non-cryptographic operations.

Product and shift SNARKs are already very powerful by itself. E.g., a prover can commit to her input vector $\boldsymbol{a}$. Then, after agreeing with the verifier on a concrete application, she can commit to a different yet related input vector (that say consists of certain permuted subset of $\boldsymbol{a}$'s coefficients), and then use the basic SNARKs to prove that this was done correctly. Here, she may use the permutation SNARK [25] that consists of $O(\log n)$ product and shift SNARKs. Finally, she can use another, application-specific, SNARK (e.g., a range SNARK) to prove that the new committed input vector has been correctly formed.

In Sect. 6, we describe a modular adaptive CaP zk-SNARK, motivated by [15], for the NP-complete language, SUBSET-SUM. (SUBSET-SUM was chosen by us mainly due to the simplicity of the SNARK; the rest of the paper considers more applications.) This SNARK consists of three commitments, one application of the shift SNARK, and three applications of the product SNARK. It is a zk-SNARK given that the commitment scheme, the shift SNARK, and the product SNARK are secure. Its prover computation is strongly dominated by $\Theta(n)$ cryptographic operations, where n is the instance size, the number of integers. More precisely, the prover has to perform only nine ($\approx n$)-wide multi-exponentiations, which makes the SNARK efficient not only asymptotically (to compare, the size of Valiant's arithmetic circuit has constant 19, and this constant has to be multiplied by the overhead of non-adaptive QSP/QAP/SSP-based solutions). Thus, we answer positively to the stated main question of the current paper. Moreover, the prover computation is highly parallelizable, while the *online* verifier computation is dominated by 17 pairings (this number will be decreased later).

In Sect. 7, we propose a new CaP range zk-SNARK that the committed value belongs to a range $[L \ldots H]$. This SNARK looks very similar to the SUBSET-SUM SNARK, but with the integer set S of the SUBSET-SUM language depending solely on the range length. Since here the prover has a committed input, the simulation of the range SNARK is slightly more complicated than of the SUBSET-SUM SNARK. Its prover-computation is similarly dominated by $\Theta(n)$ cryptographic operations, where this time $n := \lceil \log_2(H - L) \rceil$. Differently from the SUBSET-SUM SNARK, the verifier computation is dominated only by $\Theta(1)$ cryptographic operations, more precisely, by 19 pairings (also this number will be decreased later). Importantly, this SNARK is computationally more efficient than any of the existing *succinct* range SNARKs either in the standard model (i.e., random oracle-less) or in the random oracle model. E.g., the prover computation in [22] is $\Theta(n^2)$ under the Extended Riemann Hypothesis, and the prover computation in [15] is $\Theta(r^{-3}(n) \log r^{-3}(n))$. It is also significantly simpler than the range SNARKs of [11,15], mostly since we do not have to consider different trade-offs between computation and communication.

In the full version [26], we outline how to use the new basic SNARKs to construct efficient zk-SNARKs for several other NP-complete languages like Boolean and arithmetic CIRCUIT-SAT, TWO-PROCESSOR SCHEDULING, SUBSET-PRODUCT, PARTITION, and KNAPSACK [16]. Table 1 includes the complexity of SUBSET-SUM and CIRCUIT-SAT, the complexity of most other SNARKs is similar to that of SUBSET-SUM zk-SNARK. It is an interesting open problem why some NP-complete languages like SUBSET-SUM have more efficient zk-SNARKs in the modular approach (equivalently, why their verification can be performed more efficiently in the parallel machine model that consists of Hadamard product and shift) than languages like CIRCUIT-SAT. We note that [14] used recently some of the ideas from the current paper to construct an efficient shuffle argument. However, they did not use product or shift arguments.

In the full version [26], we show that by using batch-verification [2], one can decrease the verifier's computation of all presented SNARKs. In particular, one can decrease the verifier's computation in the new Range SNARK from 19 pairings to 8 pairings, one 4-way multi-exponentiation in $\mathbb{G}_1$, two 3-way multi-exponentiations in $\mathbb{G}_1$, one 2-way multi-exponentiation in $\mathbb{G}_1$, three exponentiations in $\mathbb{G}_1$, and one 3-way multi-exponentiation in $\mathbb{G}_2$. Since one exponentiation is much cheaper than one pairing [8] and one m-way multi-exponentiation is much cheaper than m exponentiations [29,32], this results in a significant win for the verifier. A similar technique can be used to also speed up other SNARKs; a good example here is the CIRCUIT-SAT argument from [25] that uses $\Theta(\log n)$ product and shift arguments. To compare, in Pinocchio [28] and Geppetto [12], the verifier has to execute 11 pairings; however, batch-verification can also be used to decrease this to 8 pairings and a small number of (multi-)exponentiations.

Finally, all resulting SNARKs work on data that has been committed to by using the interpolating commitment scheme. This means that one can repeatedly reuse committed data to compose different zk-SNARKs (e.g., to show that we know a satisfying input to a circuit, where the first coefficient belongs to a certain range). This is not possible with the known QSP/QAP-based zk-SNARKs where one would have to construct a single circuit of possibly considerable size, say n'. Moreover, in the QSP/QAP-based SNARKs, one has to commit to the vector, the length of which is equal to the total length of the input and witness (e.g., n' is the number of wires in the case of CIRCUIT-SAT). By using a modular solution, one can instead execute several zk-SNARKs with smaller values of the input and witness size; this can make the SNARK more prover-efficient since the number of non-cryptographic operations is superlinear. This emphasizes another benefit of the modular approach: one can choose the value n, the length of the vectors, accordingly to the desired tradeoff, so that larger n results in faster verifier computation, while smaller n results in faster prover computation. We are not aware of such a tradeoff in the case of the QSP/QAP-based approach.

We provide some additional discussion (about the relation between n and then input length, and about possible QSP/QAP-based solutions) in the full version [26]. Due to the lack of space, many proofs and details are only given in the full version [26]. We note that an early version of this paper, [26], was published in May 2014 and thus predates [12]. The published version differs from this early version mainly by exposition, and the use of proof bootstrapping (from [12]) and batching.

2   Preliminaries

By default, all vectors have dimension n. Let $\boldsymbol{a} \circ \boldsymbol{b}$ denote the Hadamard (i.e., element-wise) product of two vectors, with $(\boldsymbol{a} \circ \boldsymbol{b})_i = a_i b_i$. We say that $\boldsymbol{a}$ is a *shift-right-by-z* of $\boldsymbol{b}$, $\boldsymbol{a} = \boldsymbol{b} \gg z$, iff $(a_n, \ldots, a_1) = (0, \ldots, 0, b_n, \ldots, b_{1+z})$. For a tuple of polynomials $\mathcal{F} \subseteq \mathbb{Z}_p[X, Y_1, \ldots, Y_{m-1}]$, define $Y_m \mathcal{F} = (Y_m \cdot f(X, Y_1, \ldots, Y_{m-1}))_{f \in \mathcal{F}} \subseteq \mathbb{Z}_p[X, Y_1, \ldots, Y_m]$. For a tuple of polynomials $\mathcal{F}$ that have the same domain, denote $h^{\mathcal{F}(\boldsymbol{a})} := (h^{f(\boldsymbol{a})})_{f \in \mathcal{F}}$. For a group $\mathbb{G}$, let $\mathbb{G}^*$ be the set of its invertible elements. Since the direct product $\mathbb{G}_1 \times \ldots \times \mathbb{G}_m$ of groups is also a group, we use notation like $(g_1, g_2)^c = (g_1^c, g_2^c) \in \mathbb{G}_1 \times \mathbb{G}_2$ without prior definition. Let κ be the security parameter. We denote $f(\kappa) \approx_\kappa g(\kappa)$ if $|f(\kappa) - g(\kappa)|$ is negligible in κ.

On input 1^κ, a *bilinear map generator* BP returns $\mathsf{gk} = (p, \mathbb{G}_1, \mathbb{G}_2, \mathbb{G}_T, \hat{e})$, where $\mathbb{G}_1$, $\mathbb{G}_2$ and $\mathbb{G}_T$ are three multiplicative cyclic groups of prime order p (with $\log p = \Omega(\kappa)$), and $\hat{e}$ is an efficient bilinear map $\hat{e}: \mathbb{G}_1 \times \mathbb{G}_2 \to \mathbb{G}_T$ that satisfies in particular the following two properties, where g_1 (resp., g_2) is an arbitrary generator of $\mathbb{G}_1$ (resp., $\mathbb{G}_2$): (i) $\hat{e}(g_1, g_2) \neq 1$, and (ii) $\hat{e}(g_1^a, g_2^b) = \hat{e}(g_1, g_2)^{ab}$. Thus, if $\hat{e}(g_1^a, g_2^b) = \hat{e}(g_1^c, g_2^d)$ then $ab \equiv cd \pmod{p}$. We also give BP another input, n (intuitively, the input length), and allow p to depend on n. We assume that all algorithms that handle group elements verify by default that their inputs belong to corresponding groups and reject if they do not. In the case of many practically relevant pairings, arithmetic in (say) $\mathbb{G}_1$ is considerably cheaper than in $\mathbb{G}_2$; hence, we count separately exponentiations in both groups.

For $\kappa = 128$, the current recommendation is to use an optimal (asymmetric) Ate pairing over Barreto-Naehrig curves [1]. In that case, at security level of $\kappa = 128$, an element of $\mathbb{G}_1/\mathbb{G}_2/\mathbb{G}_T$ can be represented in respectively $256/512/3072$ bits. To speed up interpolation, we will additionally need the existence of the n-th, where n is a power of 2, primitive root of unity modulo p (under this condition, one can interpolate in time $\Theta(n \log n)$, otherwise, interpolation takes time $\Theta(n \log^2 n)$). For this, it suffices that $(n + 1) \mid (p - 1)$ (recall that p is the elliptic curve group order). Fortunately, given κ and a practically relevant value of n, one can easily find a Barreto-Naehrig curve such that $(n + 1) \mid (p - 1)$ holds; such an observation was made also in [5]. For example, if $\kappa = 128$ and $n = 2^{10}$, one can use Algorithm 1 of [1] to find an elliptic curve group of prime order $N(x_0)$ over a finite field of prime order $P(-x_0)$ for $x_0 = 1753449050$, where $P(x) = 36x^4 + 36x^3 + 24x^2 + 6x + 1$, $T(x) = 6x^2 + 1$, and $N(x) = P(x) + 1 - T(x)$. One can then use the curve $E : y^2 = x^3 + 6$.

In proof bootstrapping [12], one needs an additional elliptic curve group $\tilde{E}$ over a finite field of order $N(x_0)$ (see [12] for additional details). Such elliptic curve group can be found by using the Cocks-Pinch method; note that $\tilde{E}$ has somewhat less efficient arithmetic than E.

The security of the new commitment scheme and of the new SNARKs depends on the following q-type assumptions, variants of which have been used in many previous papers. The assumptions are parameterized but non-interactive in the sense that q is related to the parameters of the language (most generally, to the input length) and not to the number of the adversarial queries. All known (to us) adaptive zk-SNARKs are based on q-type assumptions about BP.

Let $d(n) \in poly(n)$ be a function. Then, BP is

- $d(n)$-*PDL (Power Discrete Logarithm) secure* if for any $n \in poly(\kappa)$ and any non-uniform probabilistic polynomial-time (NUPPT) adversary A, $\Pr[\mathsf{gk} \leftarrow \mathsf{BP}(1^\kappa, n), (g_1, g_2, \chi) \leftarrow_r \mathbb{G}_1^* \times \mathbb{G}_2^* \times \mathbb{Z}_p : \mathsf{A}(\mathsf{gk}; ((g_1, g_2)^{\chi^i})_{i=0}^{d(n)}) = \chi] \approx_\kappa 0$.
- n-*TSDH (Target Strong Diffie-Hellman) secure* if for any $n \in poly(\kappa)$ and any NUPPT adversary A, $\Pr[\mathsf{gk} \leftarrow \mathsf{BP}(1^\kappa, n), (g_1, g_2, \chi) \leftarrow_r \mathbb{G}_1^* \times \mathbb{G}_2^* \times \mathbb{Z}_p : \mathsf{A}(\mathsf{gk}; ((g_1, g_2)^{\chi^i})_{i=0}^n) = (r, \hat{e}(g_1, g_2)^{1/(\chi-r)})] \approx_\kappa 0$.

For algorithms A and X_{A}, we write $(y; y') \leftarrow (\mathsf{A}||X_{\mathsf{A}})(\chi)$ if A on input χ outputs y, and X_{A} on the same input (including the random tape of A) outputs y'. We will need knowledge assumptions w.r.t. several knowledge secrets γ_i. Let m be the number of different knowledge secrets in any concrete SNARK. Let $\mathcal{F} = (P_i)_{i=0}^n$ be a tuple of univariate polynomials, and $\mathcal{G}_1$ (resp., $\mathcal{G}_2$) be a tuple of univariate (resp., m-variate) polynomials. Let $i \in [1 .. m]$. Then, BP is $(\mathcal{F}, \mathcal{G}_1, \mathcal{G}_2, i)$-*PKE (Power Knowledge of Exponent)* secure if for any NUPPT adversary A there exists an NUPPT extractor X_{A}, such that

$$\Pr \left[\begin{array}{l} \mathsf{gk} \leftarrow \mathsf{BP}(1^\kappa, n), (g_1, g_2, \chi, \gamma) \leftarrow_r \mathbb{G}_1^* \times \mathbb{G}_2^* \times \mathbb{Z}_p \times \mathbb{Z}_p^m, \\ \gamma_{-i} \leftarrow (\gamma_1, \ldots, \gamma_{i-1}, \gamma_{i+1}, \ldots, \gamma_m), \mathsf{aux} \leftarrow (g_1^{\mathcal{G}_1(x)}, g_2^{\mathcal{G}_2(x, \gamma_{-i})}), \\ (h_1, h_2; (a_i)_{i=0}^n) \leftarrow (\mathsf{A}||X_{\mathsf{A}})(\mathsf{gk}; (g_1, g_2^{\gamma_i})^{\mathcal{F}(x)}, \mathsf{aux}) : \\ \hat{e}(h_1, g_2^{\gamma_i}) = \hat{e}(g_1, h_2) \wedge h_1 \neq g_1^{\sum_{i=0}^n a_i P_i(x)} \end{array} \right] \approx_\kappa 0.$$

Here, aux can be seen as the common auxiliary input to A and X_{A} that is generated by using benign auxiliary input generation. If $\mathcal{F} = (X^i)_{i=0}^d$ for some $d = d(n)$, then we replace the first argument in $(\mathcal{F}, \ldots)$-PKE with d. If $m = 1$, then we omit the last argument i in $(\mathcal{F}, \ldots, i)$-PKE. While knowledge assumptions are non-falsifiable, we recall that non-falsifiable assumptions are needed to design succincts SNARKs for interesting languages [18].

By generalizing [7,19,23], one can show that the TSDH, PDL, and PKE assumptions hold in the generic bilinear group model.

Within this paper, $m \leq 2$, and hence we denote γ_1 just by γ, and γ_2 by δ.

An extractable trapdoor commitment scheme in the CRS model consists of two efficient algorithms $\mathsf{G}_{\mathsf{com}}$ (that outputs a CRS ck and a trapdoor) and, (that, given ck, a message m and a randomizer r, outputs a commitment $\mathsf{C}_{\mathsf{ck}}(m; r)$), and must satisfy the following security properties.

Computational Binding: without access to the trapdoor, it is intractable to open a commitment to two different messages.

Trapdoor: given access to the original message, the randomizer and the trapdoor, one can open the commitment to any other message.

Perfect Hiding: commitments of any two messages have the same distribution.

Extractability: given access to the CRS, the commitment, and the random coins of the committer, one can open the commitment to the committed message.

See, e.g., [19] for formal definitions. In the context of the current paper, the message is a vector from $\mathbb{Z}_p^n$. We denote the randomizer space by $\mathfrak{R}$.

Let $\mathcal{R} = \{(u, w)\}$ be an efficiently verifiable relation with $|w| = \mathrm{poly}(|u|)$. Here, u is a statement, and w is a witness. Let $\mathcal{L} = \{u : \exists w, (u, w) \in \mathcal{R}\}$ be an **NP**-language. Let $n = |u|$ be the input length. For fixed n, we have a relation $\mathcal{R}_n$ and a language $\mathcal{L}_n$.

Following [9,20], we will define commit-and-prove (CaP) argument systems. Intuitively, a CaP non-interactive zero knowledge argument system for $\mathcal{R}$ allows to create a common reference string (CRS) crs, commit to some values w_i

(say, $u_i = C_{ck}(w_i; r_i)$, where ck is a part of crs), and then prove that a subset $u := (u_{i_j}, w_{i_j}, r_{i_j})_{j=1}^{\ell_m(n)}$ (for publicly known indices i_j) satisfies that u_{i_j} is a commitment of w_{i_j} with randomizer r_{i_j}, and that $(w_{i_j}) \in \mathcal{R}$.

Differently from most of the previous work (but see also [12]), our CaP argument systems will use computationally binding trapdoor commitment schemes. This means that without their openings, commitments $u_i = C_{ck}(a_i; r_i)$ themselves do not define a valid relation, since u_i can be a commitment to any a_i', given a suitable r_i'. Rather, we define a new relation $\mathcal{R}_{ck} := \{(\boldsymbol{u}, \boldsymbol{w}, \boldsymbol{r}) : (\forall i, u_i = C_{ck}(w_i; r_i)) \wedge \boldsymbol{w} \in \mathcal{R}\}$, and construct argument systems for $\mathcal{R}_{ck}$.

Within this subsection, we let vectors $\boldsymbol{u}$, $\boldsymbol{w}$, and $\boldsymbol{r}$ be of dimension $\ell_m(n)$ for some polynomial $\ell_m(n)$. However, we allow committed messages w_i themselves to be vectors of dimension n. Thus, $\ell_m(n)$ is usually very small. In some argument systems (like the SUBSET-SUM SNARK in Sect. 6), also the argument will include some commitments. In such cases, technically speaking, $\boldsymbol{w}$ and $\boldsymbol{r}$ are of higher dimension than $\boldsymbol{u}$. To simplify notation, we will ignore this issue.

A *commit-and-prove non-interactive zero-knowledge argument system* [9, 20] Π for $\mathcal{R}$ consists of an ($\mathcal{R}$-independent) trapdoor commitment scheme $\Gamma = (G_{com}, C)$ and of a non-interactive zero-knowledge argument system (G, P, V), that are combined as follows: 1. the CRS generator G (that, in particular, invokes $(ck, td_C) \leftarrow G_{com}(1^\kappa, n)$) outputs (crs $= (crs_p, crs_v), td) \leftarrow G(1^\kappa, n)$, where both crs_p and crs_v include ck, and td includes td_C. 2. the prover P produces an argument π, $\pi \leftarrow P(crs_p; \boldsymbol{u}; \boldsymbol{w}, \boldsymbol{r})$, where presumably $u_i = C_{ck}(w_i; r_i)$. 3. the verifier V, $V(crs_v; \boldsymbol{u}, \pi)$, outputs either 1 (accept) or 0 (reject). [(i)] Now, Π is *perfectly complete*, if for all $n = \text{poly}(\kappa)$, $\Pr[(crs, td) \leftarrow G(1^\kappa, n), (\boldsymbol{u}, \boldsymbol{w}, \boldsymbol{r}) \leftarrow \mathcal{R}_{ck,n} : V(crs_v; \boldsymbol{u}, P(crs_p; \boldsymbol{u}, \boldsymbol{w}, \boldsymbol{r})) = 1] = 1$.

Since Γ is computationally binding and trapdoor (and hence u_i can be commitments to *any* messages), soundness of the CaP argument systems only makes sense together with the argument of knowledge property.

Let $b(X)$ be a non-negative polynomial. Π is a *(b-bounded-auxiliary-input) argument of knowledge* for $\mathcal{R}$, if for all $n = \text{poly}(\kappa)$ and every NUPPT A, there exists an NUPPT extractor X_A, such that for every auxiliary input aux $\in \{0, 1\}^{b(\kappa)}$, $\Pr[(crs, td) \leftarrow G(1^\kappa, n), ((\boldsymbol{u}, \pi); \boldsymbol{w}, \boldsymbol{r}) \leftarrow (A\|X_A)(crs; aux) : (\boldsymbol{u}, \boldsymbol{w}, \boldsymbol{r}) \notin \mathcal{R}_{ck,n} \wedge V(crs_v; \boldsymbol{u}, \pi) = 1] \approx_\kappa 0$. As in the definition of PKE, we can restrict the definition of an argument of knowledge to benign auxiliary information generators, where aux is known to come from; we omit further discussion.

Π is *perfectly witness-indistinguishable*, if for all $n = \text{poly}(\kappa)$, it holds that if $(crs, td) \in G(1^\kappa, n)$ and $((\boldsymbol{u}; \boldsymbol{w}, \boldsymbol{r}), (\boldsymbol{u}; \boldsymbol{w}', \boldsymbol{r}')) \in \mathcal{R}^2_{ck,n}$ with $r_i, r_i' \leftarrow_r \mathfrak{R}$, then the distributions $P(crs_p; \boldsymbol{u}; \boldsymbol{w}, \boldsymbol{r})$ and $P(crs_p; \boldsymbol{u}; \boldsymbol{w}', \boldsymbol{r}')$ are equal. Note that a witness-indistinguishable argument system does not have to have a trapdoor.

Π is *perfectly composable zero-knowledge*, if there exists a probabilistic poly-time simulator S, s.t. for all stateful NUPPT adversaries A and $n = \text{poly}(\kappa)$, $\Pr[(crs, td) \leftarrow G(1^\kappa, n), (\boldsymbol{u}, \boldsymbol{w}, \boldsymbol{r}) \leftarrow A(crs), \pi \leftarrow P(crs_p; \boldsymbol{u}; \boldsymbol{w}, \boldsymbol{r}) : (\boldsymbol{u}, \boldsymbol{w}, \boldsymbol{r}) \in \mathcal{R}_{ck,n} \wedge A(\pi) = 1] = \Pr[(crs, td) \leftarrow G(1^\kappa, n), (\boldsymbol{u}, \boldsymbol{w}, \boldsymbol{r}) \leftarrow A(crs), \pi \leftarrow S(crs; \boldsymbol{u}, td) : (\boldsymbol{u}, \boldsymbol{w}, \boldsymbol{r}) \in \mathcal{R}_{ck,n} \wedge A(\pi) = 1]$. Here, the prover and the simulator use the same CRS, and thus we have *same-string zero knowledge*. Same-string statistical zero knowledge allows to use the same CRS an unbounded number of times.

An argument system that satisfies above requirements is known as *adaptive*. An argument system where the CRS depends on the statement is often called *non-adaptive*. It is not surprising that non-adaptive SNARKs can be much more efficient than adaptive SNARKs.

A non-interactive argument system is *succinct* if the output length of P and the running time of V are polylogarithmic in the P's input length (and polynomial in the security parameter). A succinct non-interactive argument of knowledge is usually called *SNARK*. A zero-knowledge SNARK is abbreviated to *zk-SNARK*.

3   New Extractable Trapdoor Commitment Scheme

We now define a new extractable trapdoor commitment scheme. It uses the following polynomials. Assume n is a power of two, and let ω be the n-th primitive root of unity modulo p. Then,

- $Z(X) := \prod_{i=1}^{n}(X - \omega^{i-1}) = X^n - 1$ is the unique degree n monic polynomial, such that $Z(\omega^{i-1}) = 0$ for all $i \in [1 .. n]$.
- $\ell_i(X) := \prod_{j \neq i}((X - \omega^{j-1})/(\omega^{i-1} - \omega^{j-1}))$, the *ith Lagrange basis polynomial*, is the unique degree $n-1$ polynomial, such that $\ell_i(\omega^{i-1}) = 1$ and $\ell_i(\omega^{j-1}) = 0$ for $j \neq i$.

Clearly, $L_a(X) = \sum_{i=1}^{n} a_i \ell_i(X)$ is the interpolating polynomial of $\boldsymbol{a}$ at points ω^{i-1}, with $L_a(\omega^{i-1}) = a_i$, and can thus be computed by executing an inverse Fast Fourier Transform. Moreover, $(\ell_i(\omega^{j-1}))_{j=1}^{n} = \boldsymbol{e}_i$ (the ith unit vector) and $(Z(\omega^{j-1}))_{j=1}^{n} = \boldsymbol{0}_n$. Thus, $Z(X)$ and $(\ell_i(X))_{i=1}^{n}$ are $n+1$ linearly independent degree $\leq n$ polynomials, and hence $\mathcal{F}_{\mathsf{C}} := (Z(X), (\ell_i(X))_{i=1}^{n})$ is a basis of such polynomials. Clearly, $Z^{-1}(0) = \{j : Z(j) = 0\} = \{\omega^{i-1}\}_{i=1}^{n}$.

Definition 1 (Interpolating Commitment Scheme). *Let* $n = \mathrm{poly}(\kappa)$, $n > 0$, *be a power of two. First,* $\mathsf{G}_{\mathsf{com}}(1^\kappa, n)$ *sets* $\mathsf{gk} \leftarrow \mathsf{BP}(1^\kappa, n)$, *picks* $g_1 \leftarrow_r \mathbb{G}_1^*$, $g_2 \leftarrow_r \mathbb{G}_2^*$, *and then outputs the CRS* $\mathsf{ck} \leftarrow (\mathsf{gk}; (g_1^{f(\chi)}, g_2^{\gamma f(\chi)})_{f \in \mathcal{F}_{\mathsf{C}}})$ *for* $\chi \leftarrow_r \mathbb{Z}_p \backslash Z^{-1}(0)$ *and* $\gamma \leftarrow_r \mathbb{Z}_p^*$. *The trapdoor is equal to* χ.

The commitment of $\boldsymbol{a} \in \mathbb{Z}_p^n$, *given a randomizer* $r \leftarrow_r \mathbb{Z}_p$, *is* $\mathsf{C}_{\mathsf{ck}}(\boldsymbol{a}; r) := (g_1^{Z(\chi)}, g_2^{\gamma Z(\chi)})^r \cdot \prod_{i=1}^{n}(g_1^{\ell_i(\chi)}, g_2^{\gamma \ell_i(\chi)})^{a_i} \in \mathbb{G}_1 \times \mathbb{G}_2$, *i.e.,* $\mathsf{C}_{\mathsf{ck}}(\boldsymbol{a}; r) := (g_1, g_2^\gamma)^{r(\chi^n - 1) + L_a(\chi)}$. *The validity of a commitment* (A_1, A_2^γ) *is checked by verifying that* $\hat{e}(A_1, g_2^{\gamma Z(\chi)}) = \hat{e}(g_1^{Z(\chi)}, A_2^\gamma)$. *To open a commitment, the committer sends* $(\boldsymbol{a}, r)$ *to the verifier.*

The condition $Z(\chi) \neq 0$ is needed in Theorem 1 to get perfect hiding and the trapdoor property. The condition $\gamma \neq 0$ is only needed in Theorem 5 to get perfect zero knowledge. Also, (a function of) γ is a part of the trapdoor in the range SNARK of Sect. 7.

Clearly, $\log_{g_1} A_1 = \log_{g_2^\gamma} A_2^\gamma = rZ(\chi) + \sum_{i=1}^{n} a_i \ell_i(\chi)$. The second element, A_2^γ, of the commitment is known as the knowledge component.

Theorem 1. *The interpolating commitment scheme is perfectly hiding and trap-door. If* BP *is n-PDL secure, then it is computationally binding. If* BP *is* $(n, \emptyset, \emptyset)$-*PKE secure, then it is extractable.*

Proof. PERFECT HIDING: since $Z(\chi) \neq 0$, then $rZ(\chi)$ (and thus also $\log_{g_1} A_1$) is uniformly random in $\mathbb{Z}_p$. Hence, (A_1, A_2^γ) is a uniformly random element of the multiplicative subgroup $\langle (g_1, g_2^\gamma) \rangle \subset \mathbb{G}_1^* \times \mathbb{G}_2^*$ generated by (g_1, g_2^γ), indepen-dently of the committed value. TRAPDOOR: given χ, $\boldsymbol{a}$, r, $\boldsymbol{a}^*$, and $c = \mathsf{C}_{\mathsf{ck}}(\boldsymbol{a}; r)$, we compute r^* s.t. $(r^* - r)Z(\chi) + \sum_{i=1}^{n}(a_i^* - a_i)\ell_i(\chi) = 0$. This is possible since $Z(\chi) \neq 0$. Clearly, $c = \mathsf{C}_{\mathsf{ck}}(\boldsymbol{a}^*; r^*)$. EXTRACTABILITY: clear from the statement.

COMPUTATIONAL BINDING: assume that there exists an adversary A_{C} that outputs $(\boldsymbol{a}, r_a)$ and $(\boldsymbol{b}, r_b)$ with $(\boldsymbol{a}, r_a) \neq (\boldsymbol{b}, r_b)$, s.t. the polynomial $d(X) :=$ $(r_a Z(X) + \sum_{i=1}^{n} a_i \ell_i(X)) - (r_b Z(X) + \sum_{i=1}^{n} b_i \ell_i(X))$ has a root at χ.

Construct now the following adversary A_{pdl} that breaks the PDL assumption. Given an n-PDL challenge, since $\mathcal{F}_{\mathsf{C}}$ consists of degree $\leq n$ polynomials, A_{pdl} can compute a valid ck from (a distribution that is statistically close to) the correct distribution. He sends ck to A_{C}. If A_{C} is successful, then $d(X) \in \mathbb{Z}_p[X]$ is a non-trivial degree-$\leq n$ polynomial. Since the coefficients of d are known, A_{pdl} can use an efficient polynomial factorization algorithm to compute all roots r_i of $d(X)$. One of these roots has to be equal to χ. A_{pdl} can establish which one by comparing each (say) $g_1^{\ell_1(r_i)}$ to the element $g_1^{\ell_1(\chi)}$ given in the CRS. Clearly, $g_1^{\ell_1(r_i)}$ is computed from g_1 (which can be computed, given the CRS, since $1 \in \mathrm{span}(\mathcal{F}_{\mathsf{C}})$), the coefficients of $\ell_1(X)$, and r_i. A_{pdl} has the same success probability as A_{C}, while her running time is dominated by that of A_{C} plus the time to factor a degree-$\leq n$ polynomial. $\qquad \square$

Theorem 1 also holds when instead of $Z(X)$ and $\ell_i(X)$ one uses any $n + 1$ linearly independent low-degree polynomials (say) $P_0(X)$ and $P_i(X)$. Given the statement of Theorem 1, this choice of the concrete polynomials is very natural: $\ell_i(X)$ interpolate linearly independent vectors (and thus are linearly indepen-dent; in fact, they constitute a basis), and the choice to interpolate unit vectors is the conceptually clearest way of choosing $P_i(X)$. Another natural choice of independent polynomials is to set $P_i(X) = X^i$ as in [19], but that choice has resulted in much less efficient (CaP) SNARKs.

In the full version [26] we show how to use batch-verification techniques to speed up simultaneous validity verification of many commitments.

4   New Product SNARK

Assume the use of the interpolating commitment scheme. In a *CaP product SNARK* [19], the prover aims to convince the verifier that she knows how to open three commitments (A, A^γ), (B, B^γ), and (C, C^γ) to vectors $\boldsymbol{a}$, $\boldsymbol{b}$ and $\boldsymbol{c}$ (together with the used randomizers), such that $\boldsymbol{a} \circ \boldsymbol{b} = \boldsymbol{c}$. Thus,

$$
\mathcal{R}_{\mathsf{ck},n}^\times := \left\{ \begin{array}{l} (u_\times, w_\times, r_\times) : u_\times = ((A_1, A_2^\gamma), (B_1, B_2^\gamma), (C_1, C_2^\gamma)) \wedge \\ w_\times = (\boldsymbol{a}, \boldsymbol{b}, \boldsymbol{c}) \wedge r_\times = (r_a, r_b, r_c) \wedge (A_1, A_2^\gamma) = \mathsf{C}_{\mathsf{ck}}(\boldsymbol{a}; r_a) \wedge \\ (B_1, B_2^\gamma) = \mathsf{C}_{\mathsf{ck}}(\boldsymbol{b}; r_b) \wedge (C_1, C_2^\gamma) = \mathsf{C}_{\mathsf{ck}}(\boldsymbol{c}; r_c) \wedge \boldsymbol{a} \circ \boldsymbol{b} = \boldsymbol{c} \end{array} \right\} .
$$

Next, we propose an efficient CaP product SNARK. For this, we need Lemma 1.

Lemma 1. *Let $A(X)$, $B(X)$ and $C(X)$ be polynomials with $A(\omega^{i-1}) = a_i$, $B(\omega^{i-1}) = b_i$ and $C(\omega^{i-1}) = c_i$, $\forall i \in [1..n]$. Let $Q(X) = A(X)B(X) - C(X)$. Assume that (i) $A(X), B(X), C(X) \in \mathrm{span}\{\ell_i(X)\}_{i=1}^n$, and (ii) there exists a degree $n - 2$ polynomial $\pi(X)$, s.t. $\pi(X) = Q(X)/Z(X)$. Then $\boldsymbol{a} \circ \boldsymbol{b} = \boldsymbol{c}$.*

Proof. From (i) it follows that $A(X) = L_a(X)$, $B(X) = L_b(X)$, and $C(X) = L_c(X)$, and thus $Q(\omega^{i-1}) = a_i b_i - c_i$ for all $i \in [1..n]$. But (ii) iff $Z(X) \mid Q(X)$, which holds iff $Q(X)$ evaluates to 0 at all n values ω^{i-1}. Thus, $\boldsymbol{a} \circ \boldsymbol{b} = \boldsymbol{c}$. Finally, if (i) holds then $\deg Q(X) = 2n - 2$ and thus $\deg \pi(X) = n - 2$. □

If privacy and succinctness are not needed, one can think of the product argument being equal to $\pi(X)$. We achieve privacy by picking $r_a, r_b, r_c \leftarrow_r \mathbb{Z}_p$, and defining $Q_{wi}(X) := (L_a(X) + r_a Z(X)) (L_b(X) + r_b Z(X)) - (L_c(X) + r_c Z(X))$. Here, the new addends of type $r_a Z(X)$ guarantee hiding. On the other hand, $Q_{wi}(X)$ remains divisible by $Z(X)$ iff $\boldsymbol{c} = \boldsymbol{a} \circ \boldsymbol{b}$. Thus, $\boldsymbol{a} \circ \boldsymbol{b} = \boldsymbol{c}$ iff

(i') $Q_{wi}(X)$ can be expressed as $Q_{wi}(X) = A(X)B(X) - C(X)$ for some polynomials $A(X)$, $B(X)$ and $C(X)$ that belong to the span of $\mathcal{F}_C$, and
(ii') there exists a polynomial $\pi_{wi}(X)$, such that

$$\pi_{wi}(X) = Q_{wi}(X)/Z(X). \tag{1}$$

The degree of $Q_{wi}(X)$ is $2n$, thus, if $\pi_{wi}(X)$ exists, then it has degree n.

However, $|\pi_{wi}(X)|$ is not sublinear in n. To minimize communication, we let the prover transfer a "garbled" evaluation of $\pi_{wi}(X)$ at a random secret point χ. More precisely, the prover computes $\pi_\chi := g_1^{\pi_{wi}(\chi)}$, using the values $g_1^{\chi^i}$ (given in the CRS) and the coefficients π_i of $\pi_{wi}(X) = \sum_{i=0}^n \pi_i X^i$, as follows:

$$\pi_\chi := g_1^{\pi_{wi}(\chi)} \leftarrow \prod_{i=0}^n (g_1^{\chi^i})^{\pi_i}. \tag{2}$$

Similarly, instead of (say) $L_a(X) + r_a Z(X)$, the verifier has the succinct interpolating commitment $C_{ck}(\boldsymbol{a}; r_a) = (g_1, g_2^\gamma)^{L_a(\chi) + r_a Z(\chi)}$ of $\boldsymbol{a}$.

We now give a full description of the new product SNARK Π_χ, given the interpolating commitment scheme (G_{com}, C) and the following tuple of algorithms, (G_χ, P_χ, V_χ). Note that $C_{ck}(\boldsymbol{1}_n; 0) = (g_1, g_2^\gamma)$.

CRS Generation: $G_\chi(1^\kappa, n)$: Let $gk \leftarrow BP(1^\kappa)$, $(g_1, g_2, \chi, \gamma) \leftarrow_r \mathbb{G}_1^* \times \mathbb{G}_2^* \times \mathbb{Z}_p^2$ with $Z(\chi) \neq 0$ and $\gamma \neq 0$. Let $crs_p = ck \leftarrow (gk; (g_1, g_2^\gamma)^{\mathcal{F}_C(\chi)})$ and $crs_v \leftarrow (gk; g_2^{\gamma Z(\chi)})$. Output $crs_\chi = (crs_p, crs_v)$.
Common Input: $u_\chi = ((A_1, A_2^\gamma), (B_1, B_2^\gamma), (C_1, C_2^\gamma))$.
Proving: $P_\chi(crs_p; u_\chi; w_\chi = (\boldsymbol{a}, \boldsymbol{b}, \boldsymbol{c}), r_\chi = (r_a, r_b, r_c))$: Compute $\pi_{wi}(X) = \sum_{i=0}^n \pi_i X^i$ as in Eq. (1) and π_χ as in Eq. (2). Output π_χ.
Verification: $V_\chi(crs_v; u_\chi; \pi_\chi)$: accept if $\hat{e}(A_1, B_2^\gamma) = \hat{e}(g_1, C_2^\gamma) \cdot \hat{e}(\pi_\chi, g_2^{\gamma Z(\chi)})$.

Since one can recompute it from ck, inclusion of $g_2^{\gamma Z(\chi)}$ in the CRS is only needed to speed up the verification. Here as in the shift SNARK of Sect. 5, validity of the commitments will be verified in the master SNARK. This is since the master SNARKs use some of the commitments in several sub-SNARKs, while it suffices to verify the validity of every commitment only once.

To obtain an argument of knowledge, we use knowledge assumptions in all following proofs. This SNARK is not zero-knowledge since the possible simulator gets three commitments as inputs but not their openings; to create an accepting argument the simulator must at least know how to open the commitment $(A_1B_1/C_1, A_2^\gamma B_2^\gamma/C_2^\gamma)$ to $\boldsymbol{a} \circ \boldsymbol{b} - \boldsymbol{c}$. It is witness-indistinguishable, and this suffices for the SUBSET-SUM and other master SNARKs to be zero-knowledge.

Theorem 2. $\Pi_\times$ *is perfectly complete and witness-indistinguishable. If the input consists of valid commitments, and* BP *is* n-TSDH *and* $(n, \emptyset, \emptyset)$-PKE *secure, then* $\Pi_\times$ *is an* $(\Theta(n)$-*bounded-auxiliary-input) adaptive argument of knowledge.*

Proof. PERFECT COMPLETENESS: follows from the discussion in the beginning of this section. PERFECT WITNESS-INDISTINGUISHABILITY: since the argument $\pi_\times$ that satisfies the verification equations is unique, all witnesses result in the same argument, and thus this argument is witness-indistinguishable.

ARGUMENT OF KNOWLEDGE: Assume that A_{aok} is an adversary that, given $\mathsf{crs}_\times$, returns $(u_\times, \pi)$ such that $\mathsf{V}_\times(\mathsf{crs}_v; u_\times, \pi) = 1$. Assume that the PKE assumption holds, and let X_A be the extractor that returns openings of the commitments in $u_\times$, i.e., $(\boldsymbol{a}, r_a)$, $(\boldsymbol{b}, r_b)$, and $(\boldsymbol{c}, r_c)$. We now claim that X_A is also the extractor needed to achieve the argument of knowledge property.

Assume that this is not the case. We construct an adversary A_{tsdh} against n-TSDH. Given an n-TSDH challenge $ch = (\mathsf{gk}, ((g_1, g_2)^{\chi^i})_{i=0}^n)$, A_{tsdh} first generates $\gamma \leftarrow_r \mathbb{Z}_p^*$, and then computes (this is possible since $\mathcal{F}_\mathsf{C}$ consists of degree $\leq n$ polynomials) and sends $\mathsf{crs}_\times$ to A_{aok}. Assume $(\mathsf{A}_{aok} || X_A)(\mathsf{crs}_\times)$ returns $((u_\times = ((A_1, A_2^\gamma), (B_1, B_2^\gamma), (C_1, C_2^\gamma)), \pi), (w_\times = (\boldsymbol{a}, \boldsymbol{b}, \boldsymbol{c}), r_\times = (r_a, r_b, r_c)))$, s.t. $u_i = \mathsf{C}_{\mathsf{ck}}(w_i; r_i)$ but $(u_\times, w_\times, r_\times) \notin \mathcal{R}_{\mathsf{ck}, n}^\times$. Since the openings are correct, $\boldsymbol{a} \circ \boldsymbol{b} \neq \boldsymbol{c}$ but π is accepting. According to Lemma 1, thus $Z(X) \nmid Q_{wi}(X)$.

Since $Z(X) \nmid Q_{wi}(X)$, then for some $i \in [1..n]$, $(X - \omega^{i-1}) \nmid Q_{wi}(X)$. Write $Q_{wi}(X) = q(X)(X - \omega^{i-1}) + r$ for $r \in \mathbb{Z}_p^*$. Clearly, $\deg q(X) \leq 2n - 1$. Moreover, we write $q(X) = q_1(X)Z(X) + q_2(X)$ with $\deg q_i(X) \leq n - 1$. Since the verification succeeds, $\hat{e}(g_1, g_2^\gamma)^{Q_{wi}(\chi)} = \hat{e}(\pi_\times, g_2^{\gamma Z(\chi)})$, or $\hat{e}(g_1, g_2^\gamma)^{q(\chi)(\chi - \omega^{i-1}) + r} = \hat{e}(\pi_\times, g_2^{\gamma Z(\chi)})$, or $\hat{e}(g_1, g_2^\gamma)^{q(\chi) + r/(\chi - \omega^{i-1})} = \hat{e}(\pi_\times, g_2^{\gamma Z(\chi)/(\chi - \omega^{i-1})})$, or $\hat{e}(g_1, g_2^\gamma)^{1/(\chi - \omega^{i-1})} = (\hat{e}(\pi_\times, g_2^{\gamma Z(\chi)/(\chi - \omega^{i-1})})/\hat{e}(g_1^{q(\chi)}, g_2^\gamma))^{r^{-1}}$.

Now, $\hat{e}(g_1^{q(\chi)}, g_2^\gamma) = \hat{e}(g_1^{q_1(\chi)}, g_2^{\gamma Z(\chi)})\hat{e}(g_1^{q_2(\chi)}, g_2^\gamma)$, and thus it can be efficiently computed from $((g_1^{\chi^i})_{i=0}^{n-1}, g_2^\gamma, g_2^{\gamma Z(\chi)}) \subset \mathsf{crs}$. Moreover, $Z(X)/(X - \omega^{i-1}) = \ell_i(X) \cdot \prod_{j \neq i}(\omega^{i-1} - \omega^{j-1})$, and thus $g_2^{\gamma Z(\chi)/(\chi - \omega^{i-1})}$ can be computed from $g_2^{\gamma \ell_i(\chi)}$ by using generic group operations. Hence, $\hat{e}(g_1, g_2^\gamma)^{1/(\chi - \omega^{i-1})}$ can be computed from $((g_1^{\chi^i})_{i=0}^{n-1}, g_2^\gamma, g_2^{\gamma Z(\chi)}, (g_2^{\gamma \ell_i(\chi)})_{i=1}^n)$ (that can be computed from ch), by using generic group operations. Thus, the adversary has computed $(r = \omega^{i-1}, \hat{e}(g_1, g_2)^{1/(\chi - r)})$, for $r \neq \chi$. Since A_{tsdh} knows $\gamma \neq 0$, he can finally compute $(r, \hat{e}(g_1, g_2)^{1/(\chi - r)})$, and thus break the n-TSDH assumption.

Hence, the argument of knowledge property follows.   □

We remark that the product SNARK (but not the shift SNARK of Sect. 5) can be seen as a QAP-based SNARK [17], namely for the relation $\boldsymbol{a} \circ \boldsymbol{b} - \boldsymbol{c}$. (Constructing a QAP-based shift SNARK is possible, but results in using different polynomials and thus in a different commitment scheme.)

The prover computation is dominated by the following: (i) one $(n + 1)$-wide multi-exponentiation in $\mathbb{G}_1$. By using the Pippenger's multi-exponentiation algorithm for *large* n this means approximately $n + 1$ bilinear-group multiplications, see [29]. For small values of n, one can use the algorithm by Straus [32]; then one has to execute $\Theta(n/\log n)$ bilinear-group exponentiations. (ii) three polynomial interpolations, one polynomial multiplication, and one polynomial division to compute the coefficients of the polynomial $\pi_{wi}(X)$. Since polynomial division can be implemented as 2 polynomial multiplications (by using pre-computation and storing some extra information in the CRS, [24]), this part is dominated by two inverse FFT-s and three polynomial multiplications.

The verifier computation is dominated by 3 pairings. (We will count the cost of validity verifications separately in the master SNARKs.) In the special case $C_1 = A_1$ (e.g., in the *Boolean SNARK*, where we need to prove that $\boldsymbol{a} \circ \boldsymbol{a} = \boldsymbol{a}$, or in the *restriction SNARK* [19], where we need to prove that $\boldsymbol{a} \circ \boldsymbol{b} = \boldsymbol{a}$ for a *public* Boolean vector $\boldsymbol{b}$), the verification equation can be simplified to $\hat{e}(A_1, B_2^\gamma / g_2^\gamma) = \hat{e}(\pi_\times, g_2^{\gamma Z(\chi)})$, which saves one more pairing. In the full version [26], we will describe a batch-verification technique that allows to speed up simultaneous verification of several product SNARKs.

Excluding gk, the prover CRS together with ck consists of $2(n + 1)$ group elements, while the verifier CRS consists of 1 group element. The CRS can be computed in time $\Theta(n)$, by using an algorithm from [3].

5   New Shift SNARK

In a *shift-right-by-z* SNARK [15] (shift SNARK, for short), the prover aims to convince the verifier that for 2 commitments (A, A^γ) and (B, B^γ), he knows how to open them as $(A, A^\gamma) = \mathsf{C}_{\mathsf{ck}}(\boldsymbol{a}; r_a)$ and $(B, B^\gamma) = \mathsf{C}_{\mathsf{ck}}(\boldsymbol{b}; r_b)$, s.t. $\boldsymbol{a} = \boldsymbol{b} \gg z$. I.e., $a_i = b_{i+z}$ for $i \in [1 .. n - z]$ and $a_i = 0$ for $i \in [n - z + 1 .. n]$. Thus,

$$\mathcal{R}_{\mathsf{ck},n}^{\mathsf{rsft}} := \left\{ \begin{array}{l} (u_\times, w_\times, r_\times) : u_\times = ((A_1, A_2^\gamma), (B_1, B_2^\gamma)) \wedge w_\times = (\boldsymbol{a}, \boldsymbol{b}) \wedge \\ r_\times = (r_a, r_b) \wedge (A_1, A_2^\gamma) = \mathsf{C}_{\mathsf{ck}}(\boldsymbol{a}; r_a) \wedge \\ (B_1, B_2^\gamma) = \mathsf{C}_{\mathsf{ck}}(\boldsymbol{b}; r_b) \wedge (\boldsymbol{a} = \boldsymbol{b} \gg z) \end{array} \right\}.$$

An efficient shift SNARK was described in [15]. We now reconstruct this SNARK so that it can be used together with the interpolating commitment scheme. We can do it since the shift SNARK of [15] is *almost* independent of the commitment scheme. We also slightly optimize the resulting SNARK; in particular, the verifier has to execute one less pairing compared to [15].

Our strategy of constructing a shift SNARK follows the strategy of [19,23]. We start with a concrete verification equation that also contains the argument, that we denote by π_1. We write the discrete logarithm of π_1 (that follows from this equation) as $F_\pi(\chi) + F_{con}(\chi)$, where χ is a secret key, and $F_\pi(X)$ and $F_{con}(X)$ are two polynomials. The first polynomial, $F_\pi(X)$, is identically zero iff the prover is honest. Since the spans of certain two polynomial sets do not intersect, this results in an efficient adaptive shift SNARK that is an argument of knowledge under (two) PKE assumptions.

Now, for a non-zero polynomial $Z^*(X)$ to be defined later, consider the verification equation $\hat{e}(A_1, g_2^{\gamma Z^*(\chi)})/\hat{e}(B_1\pi_1, g_2^\gamma) = 1$ (due to the properties of pairing, this is equivalent to verifying that $\pi_1 = A_1^{Z^*(\chi)}/B_1$), with (A_1, A_2^γ) and (B_1, B_2^γ) being interpolating commitments to $\boldsymbol{a}$ and $\boldsymbol{b}$, and $\pi_1 = g_1^{\pi(\chi)}$ for some polynomial $\pi(X)$. Denote $r(X) := (r_a Z^*(X) - r_b)Z(X)$. Taking a discrete logarithm of the verification equation, we get that $\pi(X) = (r_a Z(X) + \sum_{i=1}^n a_i\ell_i(X))Z^*(X) - (r_b Z(X) + \sum_{i=1}^n b_i\ell_i(X)) = Z^*(X)\sum_{i=1}^n a_i\ell_i(X) - \sum_{i=1}^n b_i\ell_i(X) + r(X) = \left(\sum_{i=1}^{n-z} a_i\ell_i(X) + \sum_{i=n-z+1}^n a_i\ell_i(X)\right)Z^*(X) + r(X) - \sum_{i=1}^{n-z} b_{i+z}\ell_{i+z}(X) - \sum_{i=1}^z b_i\ell_i(X)$. Hence, $\pi(X) = F_\pi(X) + F_{con}(X)$, where

$$F_\pi(X) = \left(\sum_{i=1}^{n-z}(a_i - b_{i+z})\ell_i(X) + \sum_{i=n-z+1}^n a_i\ell_i(X)\right) \cdot Z^*(X),$$

$$F_{con}(X) = \left(\sum_{i=z+1}^n b_i(\ell_{i-z}(X)Z^*(X) - \ell_i(X)) - \sum_{i=1}^z b_i\ell_i(X)\right) + r(X).$$

Clearly, the prover is honest iff $F_\pi(X) = 0$, which holds iff $\pi(X) = F_{con}(X)$, i.e., $\pi(X)$ belongs to the span of $\mathcal{F}_{z-\mathsf{rsft}} := (\ell_{i-z}(X)Z^*(X) - \ell_i(X))_{i=z+1}^n, (\ell_i(X))_{i=1}^z, Z(X)Z^*(X), Z(X))$. For the shift SNARK to be an argument of knowledge, we need that

(i) $(\ell_i(X)Z^*(X))_{i=1}^n$ is linearly independent, and
(ii) $F_\pi(X) \cap \mathrm{span}(\mathcal{F}_{z-\mathsf{rsft}}) = \emptyset$.

Together, (i) and (ii) guarantee that from $\pi(X) \in \mathrm{span}(\mathcal{F}_{z-\mathsf{rsft}})$ it follows that $\boldsymbol{a}$ is a shift of $\boldsymbol{b}$.

We guarantee that $\pi(X) \in \mathrm{span}(\mathcal{F}_{z-\mathsf{rsft}})$ by a knowledge assumption (w.r.t. another knowledge secret δ); for this we will also show that $\mathcal{F}_{z-\mathsf{rsft}}$ is linearly independent. As in the case of the product SNARK, we also need that (A_1, A_2^γ) and (B_1, B_2^γ) are actually commitments of n-dimensional vectors (w.r.t. γ), i.e., we rely on two PKE assumptions.

Denote $\mathcal{F}_\pi := \{\ell_i(X)Z^*(X)\}_{i=1}^n$. For a certain choice of $Z^*(X)$, both (i) and (ii) follow from the next lemma.

Lemma 2. *Let* $Z^*(X) = Z(X)^2$. *Then* $\mathcal{F}_\pi \cup \mathcal{F}_{z-\mathsf{rsft}}$ *is linearly independent.*

Proof. Assume that there exist $\boldsymbol{a} \in \mathbb{Z}_p^n$, $\boldsymbol{b} \in \mathbb{Z}_p^n$, $c \in \mathbb{Z}_p$, and $d \in \mathbb{Z}_p$, s.t. $f(X) := \sum_{i=1}^n a_i\ell_i(X)Z^*(X) + \sum_{i=z+1}^n b_i(\ell_{i-z}(X)Z^*(X) - \ell_i(X)) -$

$\sum_{i=1}^{z} b_i \ell_i(X) + cZ(X)Z^*(X) + dZ(X) = 0$. But then also $f(\omega^{j-1}) = 0$, for $j \in [1 .. n]$. Thus, due to the definition of $\ell_i(X)$ and $Z(X)$, $\sum_{i=1}^{n} b_i e_i = \mathbf{0}_n$ which is only possible if $b_i = 0$ for all $i \in [1 .. n]$. Thus also $f'(X) := f(X)/Z(X) = \sum_{i=1}^{n} a_i \ell_i(X)Z^*(X)/Z(X) + cZ^*(X) + d = 0$. But then also $f'(\omega^{j-1}) = 0$ for $j \in [1 .. n]$. Hence, $cZ^*(\omega^{j-1}) + d = d = 0$. Finally, $f''(X) := f(X)/Z^*(X) = \sum_{i=1}^{n} a_i \ell_i(X) + cZ(X) = 0$, and from $f''(\omega^{j-1}) = 0$ for $j \in [1 .. n]$, we get $\mathbf{a} = \mathbf{0}_n$. Thus also $c = 0$. This finishes the proof. $\square$

Since the argument of knowledge property of the new shift SNARK relies on $\pi(X)$ belonging to a certain span, similarly to [15], we will use an additional knowledge assumption. That is, it is necessary that there exists an extractor that outputs a witness that $\pi(X) = F_{con}(X)$ belongs to the span of $\mathcal{F}_{z-\mathsf{rsft}}$.

Similarly to the product SNARK, the shift SNARK does not contain $\pi(X) = F_{con}(X)$, but the value $\pi_{\mathsf{rsft}} = (g_1, g_2^\delta)^{\pi(\chi)}$ for random χ and δ (necessary due to the use of the second PKE assumption), computed as

$$\pi_{\mathsf{rsft}} \leftarrow (\pi_1, \pi_2^\delta) = (g_1, g_2^\delta)^{\pi(\chi)}$$
$$= \prod_{i=z+1}^{n}((g_1, g_2^\delta)^{\ell_{i-z}(\chi)Z^*(\chi)-\ell_i(\chi)})^{b_i} \cdot \prod_{i=1}^{z}((g_1, g_2^\delta)^{\ell_i(\chi)})^{-b_i}. \quad (3)$$
$$((g_1, g_2^\delta)^{Z(\chi)Z^*(\chi)})^{r_a} \cdot ((g_1, g_2^\delta)^{Z(\chi)})^{-r_b}.$$

We are now ready to state the new shift-right-by-z SNARK Π_{rsft}. It consists of the interpolating commitment scheme and of the following three algorithms:

CRS Generation: $\mathsf{G}_{\mathsf{rsft}}(1^\kappa, n)$: Let $Z^*(X) = Z(X)^2$. Let $\mathsf{gk} \leftarrow \mathsf{BP}(1^\kappa)$, $(g_1, g_2, \chi, \gamma, \delta) \leftarrow \mathbb{G}_1^* \times \mathbb{G}_2^* \times \mathbb{Z}_p^3$, s.t. $Z(\chi) \neq 0$, $\gamma \neq 0$. Set $\mathsf{ck} \leftarrow (\mathsf{gk}; (g_1, g_2^\gamma)^{\mathcal{F}_\mathsf{c}(\chi)})$, $\mathsf{crs}_p \leftarrow (\mathsf{gk}; (g_1, g_2^\delta)^{\mathcal{F}_{z-\mathsf{rsft}}(\chi)})$, $\mathsf{crs}_v \leftarrow (\mathsf{gk}; (g_1, g_2^\delta)^{Z(\chi)}, g_2^{\delta Z(\chi)Z^*(\chi)})$. Return $\mathsf{crs}_{\mathsf{rsft}} = (\mathsf{ck}, \mathsf{crs}_p, \mathsf{crs}_v)$.
Common Input: $u_{\mathsf{rsft}} = ((A_1, A_2^\gamma), (B_1, B_2^\gamma))$.
Proving: $\mathsf{P}_{\mathsf{rsft}}(\mathsf{crs}_p; u_{\mathsf{rsft}}; w_{\mathsf{rsft}} = (\mathbf{a}, \mathbf{b}), r_{\mathsf{rsft}} = (r_a, r_b))$: return $\pi_{\mathsf{rsft}} \leftarrow (\pi_1, \pi_2^\delta)$ from Eq. (3).
Verification: $\mathsf{V}_{\mathsf{rsft}}(\mathsf{crs}_v; u_{\mathsf{rsft}}; \pi_{\mathsf{rsft}} = (\pi_1, \pi_2^\delta))$: accept if $\hat{e}(\pi_1, g_2^{\delta Z(\chi)}) = \hat{e}(g_1^{Z(\chi)}, \pi_2^\delta)$ and $\hat{e}(B_1\pi_1, g_2^{\delta Z(\chi)}) = \hat{e}(A_1, g_2^{\delta Z(\chi)Z^*(\chi)})$.

Since crs_v can be recomputed from $\mathsf{ck} \cup \mathsf{crs}_p$, then clearly it suffices to take CRS to be $\mathsf{crs}_{\mathsf{rsft}} = (\mathsf{gk}; g_1^{\mathcal{F}_\mathsf{c}(\chi) \cup \mathcal{F}_{z-\mathsf{rsft}}(\chi)}, g_2^{\gamma \mathcal{F}_\mathsf{c}(\chi) \cup \delta \mathcal{F}_{z-\mathsf{rsft}}(\chi)})$.

Theorem 3. *Let* $Z^*(X) = Z(X)^2$, $y = \deg(Z(X)Z^*(X)) = 3n$. Π_{rsft} *is perfectly complete and witness-indistinguishable. If the input consists of valid commitments, and* BP *is* y-PDL, $(n, \mathcal{F}_{z-\mathsf{rsft}}, Y_2\mathcal{F}_{z-\mathsf{rsft}}, 1)$-PKE, *and* $(\mathcal{F}_{z-\mathsf{rsft}}, \mathcal{F}_\mathsf{c}, Y_1\mathcal{F}_\mathsf{c}, 2)$-PKE *secure, then* Π_{rsft} *is an* $(\Theta(n)$-*bounded-auxiliary-input) adaptive argument of knowledge.*

The prover computation is dominated by two $(n+2)$-wide multi-exponentiations (one in $\mathbb{G}_1$ and one in $\mathbb{G}_2$); there is no need for polynomial interpolation, multiplication or division. The communication is 2 group elements. The verifier computation is dominated by 4 pairings. In the full version [26], we describe a batch-verification technique that allows to speed up simultaneous verification of several shift SNARKs. Apart from gk, the prover CRS and ck together contain $4n + 6$ group elements, and the verifier CRS contains 3 group elements.

A shift-left-by-z (necessary in [25] to construct a permutation SNARK) SNARK can be constructed similarly. A rotation-left/right-by-z SNARK (one committed vector is a *rotation* of another committed vector) requires only small modifications, see [15].

6   New Subset-Sum SNARK

For fixed n and $p = n^{\omega(1)}$, the NP-complete language SUBSET-SUM over $\mathbb{Z}_p$ is defined as the language $\mathcal{L}_n^{\text{SUBSET-SUM}}$ of tuples $(\boldsymbol{S} = (S_1, \ldots, S_n), s)$, with $S_i, s \in \mathbb{Z}_p$, such that there exists a vector $\boldsymbol{b} \in \{0,1\}^n$ with $\sum_{i=1}^n S_i b_i = s$ in $\mathbb{Z}_p$. SUBSET-SUM can be solved in pseudo-polynomial time $O(pn)$ by using dynamic programming. In the current paper, since $n = \kappa^{o(1)}$ and $p = 2^{O(\kappa)}$, pn is not polynomial in the input size $n \log_2 p$.

In a SUBSET-SUM SNARK, the prover aims to convince the verifier that he knows how to open commitment (B_1, B_2^γ) to a vector $\boldsymbol{b} \in \{0,1\}^n$, such that $\sum_{i=1}^n S_i b_i = s$. We show that by using the new product and shift SNARKs, one can design a prover-efficient adaptive SUBSET-SUM zk-SNARK Π_{ssum}. We emphasize that SUBSET-SUM is just one of the languages for which we can construct an efficient zk-SNARK; Sect. 7 and the full version [26] have more examples.

First, we use the interpolating commitment scheme. The CRS generation G_{ssum} invokes CRS generations of the commitment scheme, the product SNARK and the shift SNARK, sharing the same gk, g_1, g_2, γ, and trapdoor $\mathsf{td} = \chi$ between the different invocations. (Since here the argument must be zero knowledge, it needs a trapdoor.) Thus, $\mathsf{crs}_{\text{ssum}} = \mathsf{crs}_{\text{rsft}}$ for $z = 1$.

Let $\boldsymbol{e}_i$ be the ith unit vector. The prover's actions are depicted by Fig. 1 (a precise explanation of this SNARK will be given in the completeness proof in Theorem 4). This SNARK, even without taking into account the differences in the product and shift SNARKs, is both simpler and moth efficient than the

---

Let $\boldsymbol{b} \in \{0,1\}^n$ be such that $\sum_{i=1}^n S_i b_i = s$.
Let (B_1, B_2^γ) be a commitment to $\boldsymbol{b}$.
Construct a product argument π_1 to show that $\boldsymbol{b} \circ \boldsymbol{b} = \boldsymbol{b}$.
Let (C_1, C_2^γ) be a commitment to $\boldsymbol{c} \leftarrow \boldsymbol{S} \circ \boldsymbol{b}$.
Construct a product argument π_2 to show that $\boldsymbol{c} = \boldsymbol{S} \circ \boldsymbol{b}$.
Let (D_1, D_2^γ) be a commitment to $\boldsymbol{d}$, where $d_i = \sum_{j \geq i} c_j$.
Construct a shift-right-by-1 argument $(\pi_{31}, \pi_{32}^\delta)$ to show that $\boldsymbol{d} = (\boldsymbol{d} - \boldsymbol{c}) \gg 1$.
Construct a product argument π_4 to show that $\boldsymbol{e}_1 \circ (\boldsymbol{d} - s\boldsymbol{e}_1) = \boldsymbol{0}_n$.
Output $\pi_{\text{ssum}} = (B_1, B_2^\gamma, C_1, C_2^\gamma, D_1, D_2^\gamma, \pi_1, \pi_2, \pi_{31}, \pi_{32}^\delta, \pi_4)$.

Fig. 1. The new SUBSET-SUM SNARK Π_{ssum} (prover's operations)

SUBSET-SUM SNARK presented in [15] where one needed an additional step of proving that $\boldsymbol{b} \neq \boldsymbol{O}_n$.

We remark that the vector $\boldsymbol{d}$, with $d_i = \sum_{j \geq i} c_j$, is called either a *vector scan*, an *all-prefix-sums*, or a *prefix-sum* of $\boldsymbol{c}$, and $(\pi_{31}, \pi_{32}^\delta)$ can be thought of as a *scan SNARK* [15] that $\boldsymbol{d}$ is a correct scan of $\boldsymbol{c}$.

After receiving π_{ssum}, the verifier computes $S_1' \leftarrow \prod_i (g_1^{\ell_i(\chi)})^{S_i}$ as the first half of a commitment to $\boldsymbol{S}$, and then performs the following verifications: (i) Three commitment validations: $\hat{e}(B_1, g_2^\gamma) = \hat{e}(g_1, B_2^\gamma)$, $\hat{e}(C_1, g_2^\gamma) = \hat{e}(g_1, C_2^\gamma)$, $\hat{e}(D_1, g_2^\gamma) = \hat{e}(g_1, D_2^\gamma)$. (ii) Three product argument verifications: $\hat{e}(B_1/g_1, B_2^\gamma) = \hat{e}(\pi_1, g_2^{\gamma Z(\chi)})$, $\hat{e}(S_1', B_2^\gamma) = \hat{e}(g_1, C_2^\gamma) \cdot \hat{e}(\pi_2, g_2^{\gamma Z(\chi)})$, $\hat{e}(g_1^{\ell_1(\chi)}, D_2^\gamma/(g_2^{\ell_1(\chi)})^s) = \hat{e}(\pi_4, g_2^{\gamma Z(\chi)})$. (iii) One shift argument verification, consisting of two equality tests: $\hat{e}(\pi_{31}, g_2^{\delta Z(\chi)}) = \hat{e}(g_1^{Z(\chi)}, \pi_{32}^\delta)$, $\hat{e}(D_1/C_1\pi_{31}, g_2^{\delta Z(\chi)}) = \hat{e}(D_1, g_2^{\delta Z(\chi) Z^*(\chi)})$.

Theorem 4. Π_{ssum} *is perfectly complete and perfectly composable zero-knowledge. It is an ($\Theta(n)$-bounded-auxiliary-input) adaptive argument of knowledge if* BP *satisfies n-TSDH and the same assumptions as in Theorem 3 (for $z = 1$).*

The prover computation is dominated by three commitments and the application of 3 product SNARKs and 1 shift SNARK, i.e., by $\Theta(n \log n)$ non-cryptographic operations and $\Theta(n)$ cryptographic operations. The latter is dominated by nine ($\approx n$)-wide multi-exponentiations (2 in commitments to $\boldsymbol{c}$ and $\boldsymbol{d}$ and in the shift argument, and 1 in each product argument), 7 in $\mathbb{G}_1$ and 4 in $\mathbb{G}_2$. The argument size is constant (11 group elements), and the verifier computation is dominated by *offline* computation of two $(n+1)$-wide multi-exponentiations (needed to once commit to $\boldsymbol{S}$) and *online* computation of 17 pairings (3 pairings to verify π_2, 2 pairings to verify each of the other product arguments, 4 pairings to verify the shift argument, and 6 pairings to verify the validity of 3 commitments). In the full version [26], we will describe a batch-verification technique that allows to speed up on-line part of the verification of the SUBSET-SUM SNARK.

As always, multi-exponentiation can be sped up by using algorithms from [29, 32]; it can also be highly parallelized, potentially resulting in very fast parallel implementations of the zk-SNARK.

7   New Range SNARK

In a *range SNARK*, given public range $[L\,..\,H]$, the prover aims to convince the verifier that he knows how to open commitment (A_1, A_2^γ) to a value $a \in [L\,..\,H]$. That is, that the common input (A_1, A_2^γ) is a commitment to vector $\boldsymbol{a}$ with $a_1 = a$ and $a_i = 0$ for $i > 1$.

We first remark that instead of the range $[L\,..\,H]$, one can consider the range $[0\,..\,H-L]$, and then use the homomorphic properties of the commitment scheme to add L to the committed value. Hence, we will just assume that the range is equal to $[0\,..\,H]$ for some $H \geq 1$. Moreover, the efficiency of the following SNARK depends on the range length.

The new range SNARK Π_{rng} is very similar to Π_{ssum}, except that one has to additionally commit to a value $a \in [0..H]$, use a specific sparse $\boldsymbol{S}$ with $S_i = \lfloor (H + 2^{i-1})/2^i \rfloor$ [10,27], and prove that $a = \sum_{i=1}^{n} S_i b_i$ for the committed a. Since $\boldsymbol{S} = (S_i)_{i=1}^{n}$ does not depend on the instance (i.e., on a), the verifier computation is $\Theta(1)$. On the other hand, since the commitment to a is given as an input to the prover (and not created by prover as part of the argument), Π_{rng} has a more complex simulation strategy, with one more element in the trapdoor.

Let $n = \lfloor \log_2 H \rfloor + 1$. Define $S_i = \lfloor (H + 2^{i-1})/2^i \rfloor$ for $i \in [1..n]$ and $\boldsymbol{S} = (S_i)$. We again use the interpolating commitment scheme. To prove that $a \in [0..H]$, we do the following.

The CRS generation $\mathsf{G}_{\mathsf{rng}}$ invokes the CRS generations of the commitment scheme, the product SNARK and the shift SNARK, sharing the same gk and trapdoor $\mathsf{td} = (\chi, \delta/\gamma)$ between the different invocations. In this case, the trapdoor has to include δ/γ (which is well defined, since $\gamma \neq 0$) since the simulator does not know how to open (A_1, A_2^{γ}); see the proof of Theorem 5 for more details. We note that the trapdoor only has to contain δ/γ, and not γ and δ separately. The CRS also contains the first half of a commitment $S_1' \leftarrow \prod(g_1^{\ell_i(\chi)})^{S_i}$ to $\boldsymbol{S}$, needed for a later efficient verification of the argument π_2. Clearly, the CRS can be computed efficiently from $\mathsf{crs}_{\mathsf{rsft}}$ (for $z = 1$).

---

1   Let $a = \sum_{i=1}^{n} S_i b_i$ for $b_i \in \{0, 1\}$.
    Let (B_1, B_2^{γ}) be a commitment to $\boldsymbol{b}$.
    Construct a product argument π_1 to show that $\boldsymbol{b} = \boldsymbol{b} \circ \boldsymbol{b}$.
    Let (C_1, C_2^{γ}) be a commitment to $\boldsymbol{c} \leftarrow \boldsymbol{S} \circ \boldsymbol{b}$.
    Construct a product argument π_2 to show that $\boldsymbol{c} = \boldsymbol{S} \circ \boldsymbol{b}$.
    Let (D_1, D_2^{γ}) be a commitment to $\boldsymbol{d}$, where $d_i = \sum_{j \geq i} c_i$.
    Construct a shift argument $(\pi_{31}, \pi_{32}^{\delta})$ to show that $\boldsymbol{d} = (\boldsymbol{d} - \boldsymbol{c}) \gg 1$.
2   Construct a product argument π_4 to show that $\boldsymbol{e}_1 \circ (\boldsymbol{d} - \boldsymbol{a}) = \boldsymbol{0}_n$.
    Output $\pi_{\mathsf{rng}} = (B_1, B_2^{\gamma}, C_1, C_2^{\gamma}, D_1, D_2^{\gamma}, \pi_1, \pi_2, \pi_{31}, \pi_{32}^{\delta}, \pi_4)$.

Fig. 2. The new range argument Π_{rng}

The prover's actions on input (A_1, A_2^{γ}) are depicted by Fig. 2 (further explanations are given in the concise completeness proof in Theorem 5). The only differences, compared to the prover computation of Π_{ssum}, are the computation of b_i on step 1, and of π_4 on step 2. After receiving π_{rng}, the verifier performs the following checks: (i) Four commitment validations: $\hat{e}(A_1, g_2^{\gamma}) = \hat{e}(g_1, A_2^{\gamma})$, $\hat{e}(B_1, g_2^{\gamma}) = \hat{e}(g_1, B_2^{\gamma})$, $\hat{e}(C_1, g_2^{\gamma}) = \hat{e}(g_1, C_2^{\gamma})$, $\hat{e}(D_1, g_2^{\gamma}) = \hat{e}(g_1, D_2^{\gamma})$. (ii) Three product argument verifications: $\hat{e}(B_1/g_1, B_2^{\gamma}) = \hat{e}(\pi_1, g_2^{\gamma Z(\chi)})$, $\hat{e}(S_1', B_2^{\gamma}) = \hat{e}(g_1, C_2^{\gamma}) \cdot \hat{e}(\pi_2, g_2^{\gamma Z(\chi)})$, $\hat{e}(g_1^{\ell_1(\chi)}, D_2^{\gamma}/A_2^{\gamma}) = \hat{e}(\pi_4, g_2^{\gamma Z(\chi)})$. (iii) One shift argument verification, consisting of two equality tests: $\hat{e}(\pi_{31}, g_2^{\delta Z(\chi)}) = \hat{e}(g_1^{Z(\chi)}, \pi_{32}^{\delta})$, $\hat{e}(D_1/C_1 \pi_{31}, g_2^{\delta Z(\chi)}) = \hat{e}(D_1, g_2^{\delta Z(\chi) Z^*(\chi)})$.

Theorem 5. Π_{rng} *is perfectly complete and composable zero-knowledge. If* BP *satisfies* n*-TSDH and the assumptions of Theorem 3, then* Π_{rng} *is an adaptive* $(\Theta(n)$*-bounded-auxiliary-input) argument of knowledge.*

The prover computation is dominated by three commitments and the application of three product arguments and one shift argument, that is, by $\Theta(n \log n)$ non-cryptographic operations and $\Theta(n)$ cryptographic operations. The latter is dominated by nine $(\approx n)$-wide multi-exponentiations (2 in commitments to c and d and in the shift argument, and 1 in each product argument), seven in $\mathbb{G}_1$ and four in $\mathbb{G}_2$. The argument size is constant (11 group elements), and the verifier computation is dominated by 19 pairings (3 pairings to verify π_2, 2 pairings to verify each of the other product arguments, 4 pairings to verify the shift argument, and 8 pairings to verify the validity of 4 commitments). In this case, since the verifier does not have to commit to S, the verifier computation is dominated by $\Theta(1)$ cryptographic operations.

The new range SNARK is significantly more computation-efficient for the prover than the previous range SNARKs [11,15] that have prover computation $\Theta(r_3^{-1}(n) \log n)$. Π_{rng} has better communication (11 versus 31 group elements in [15]), and verification complexity (19 versus 65 pairings in [15]). Moreover, Π_{rng} is also simpler: since the prover computation is quasi-linear, we do not have to consider various trade-offs (though they are still available) between computation and communication as in [11,15]. In the full version [26], we will use batch verification to further speed up the verification of the Range SNARK.

Acknowledgments. We would like to thank Diego Aranha, Paulo Barreto, Markulf Kohlweiss, and Prastudy Fauzi for useful comments. This work was supported by the European Union's Horizon 2020 research and innovation programme under grant agreement No. 653497 (project PANORAMIX), and the Estonian Research Council.

References

1. Barreto, P.S.L.M., Naehrig, M.: Pairing-friendly elliptic curves of prime order. In: Preneel, B., Tavares, S. (eds.) SAC 2005. LNCS, vol. 3897, pp. 319–331. Springer, Heidelberg (2006)
2. Bellare, M., Garay, J.A., Rabin, T.: Batch verification with applications to cryptography and checking. In: Lucchesi, C.L., Moura, A.V. (eds.) LATIN 1998. LNCS, vol. 1380, pp. 170–191. Springer, Heidelberg (1998)
3. Ben-Sasson, E., Chiesa, A., Genkin, D., Tromer, E., Virza, M.: SNARKs for C: verifying program executions succinctly and in zero knowledge. In: Canetti, R., Garay, J.A. (eds.) CRYPTO 2013, Part II. LNCS, vol. 8043, pp. 90–108. Springer, Heidelberg (2013)
4. Ben-Sasson, E., Chiesa, A., Tromer, E., Virza, M.: Scalable zero knowledge via cycles of elliptic curves. In: Garay, J.A., Gennaro, R. (eds.) CRYPTO 2014, Part II. LNCS, vol. 8617, pp. 276–294. Springer, Heidelberg (2014)
5. Ben-Sasson, E., Chiesa, A., Tromer, E., Virza, M.: Succinct non-interactive zero knowledge for a von Neumann architecture. In: USENIX, pp. 781–796 (2014)

6. Bitansky, N., Chiesa, A., Ishai, Y., Ostrovsky, R., Paneth, O.: Succinct non-interactive arguments via linear interactive proofs. In: Sahai, A. (ed.) TCC 2013. LNCS, vol. 7785, pp. 315–333. Springer, Heidelberg (2013)
7. Boneh, D., Boyen, X.: Short signatures without random oracles and the SDH assumption in bilinear groups. J. Cryptol. **21**(2), 149–177 (2008)
8. Bos, J.W., Costello, C., Naehrig, M.: Exponentiating in pairing groups. In: Lange, T., Lauter, K., Lisoněk, P. (eds.) SAC 2013. LNCS, vol. 8282, pp. 438–455. Springer, Heidelberg (2014)
9. Canetti, R., Lindell, Y., Ostrovsky, R., Sahai, A.: Universally composable two-party and multi-party secure computation. In: STOC, pp. 494–503 (2002)
10. Chaabouni, R., Lipmaa, H., Shelat, A.: Additive combinatorics and discrete logarithm based range protocols. In: Steinfeld, R., Hawkes, P. (eds.) ACISP 2010. LNCS, vol. 6168, pp. 336–351. Springer, Heidelberg (2010)
11. Chaabouni, R., Lipmaa, H., Zhang, B.: A non-interactive range proof with constant communication. In: Keromytis, A.D. (ed.) FC 2012. LNCS, vol. 7397, pp. 179–199. Springer, Heidelberg (2012)
12. Costello, C., Fournet, C., Howell, J., Kohlweiss, M., Kreuter, B., Naehrig, M., Parno, B., Zahur, S.: Geppetto: versatile verifiable computation. In: IEEE SP, pp. 253–270 (2015)
13. Danezis, G., Fournet, C., Groth, J., Kohlweiss, M.: Square span programs with applications to succinct NIZK arguments. In: Sarkar, P., Iwata, T. (eds.) ASIACRYPT 2014. LNCS, vol. 8873, pp. 532–550. Springer, Heidelberg (2014)
14. Fauzi, P., Lipmaa, H.: Efficient culpably sound NIZK shuffle argument without random oracles. CT-RSA 2016. LNCS, vol. 9610. Springer, switzerland (2016)
15. Fauzi, P., Lipmaa, H., Zhang, B.: Efficient modular NIZK arguments from shift and product. In: Abdalla, M., Nita-Rotaru, C., Dahab, R. (eds.) CANS 2013. LNCS, vol. 8257, pp. 92–121. Springer, Heidelberg (2013)
16. Garey, M.R., Johnson, D.S.: Computers and Intractability: A Guide to the Theory of NP-Completeness. Series of Books in the Mathematical Sciences. W.H. Freeman, New York (1979)
17. Gennaro, R., Gentry, C., Parno, B., Raykova, M.: Quadratic span programs and succinct NIZKs without PCPs. In: Johansson, T., Nguyen, P.Q. (eds.) EUROCRYPT 2013. LNCS, vol. 7881, pp. 626–645. Springer, Heidelberg (2013)
18. Gentry, C., Wichs, D.: Separating succinct non-interactive arguments from all falsifiable assumptions. In: STOC, pp. 99–108 (2011)
19. Groth, J.: Short pairing-based non-interactive zero-knowledge arguments. In: Abe, M. (ed.) ASIACRYPT 2010. LNCS, vol. 6477, pp. 321–340. Springer, Heidelberg (2010)
20. Kilian, J.: Uses of randomness in algorithms and protocols. Ph.D. thesis, Massachusetts Institute of Technology, USA (1989)
21. Kolesnikov, V., Schneider, T.: A practical universal circuit construction and secure evaluation of private functions. In: Tsudik, G. (ed.) FC 2008. LNCS, vol. 5143, pp. 83–97. Springer, Heidelberg (2008)
22. Lipmaa, H.: On diophantine complexity and statistical zero-knowledge arguments. In: Laih, C.-S. (ed.) ASIACRYPT 2003. LNCS, vol. 2894, pp. 398–415. Springer, Heidelberg (2003)
23. Lipmaa, H.: Progression-free sets and sublinear pairing-based non-interactive zero-knowledge arguments. In: Cramer, R. (ed.) TCC 2012. LNCS, vol. 7194, pp. 169–189. Springer, Heidelberg (2012)

24. Lipmaa, H.: Succinct non-interactive zero knowledge arguments from span programs and linear error-correcting codes. In: Sako, K., Sarkar, P. (eds.) ASIACRYPT 2013, Part I. LNCS, vol. 8269, pp. 41–60. Springer, Heidelberg (2013)
25. Lipmaa, H.: Efficient NIZK arguments via parallel verification of benes networks. In: Abdalla, M., De Prisco, R. (eds.) SCN 2014. LNCS, vol. 8642, pp. 416–434. Springer, Heidelberg (2014)
26. Lipmaa, H.: Prover-efficient commit-and-prove zero-knowledge SNARKs. TR 2014/396, IACR (2014). http://eprint.iacr.org/2014/396
27. Lipmaa, H., Asokan, N., Niemi, V.: Secure vickrey auctions without threshold trust. FC 2002. LNCS, vol. 2357, pp. 87–101. Springer, Heidelberg (2002)
28. Parno, B., Gentry, C., Howell, J., Raykova, M.: Pinocchio: nearly practical verifiable computation. In: IEEE SP, pp. 238–252 (2013)
29. Pippenger, N.: On the evaluation of powers and monomials. SIAM J. Comput. **9**(2), 230–250 (1980)
30. Raz, R.: Elusive functions and lower bounds for arithmetic circuits. Theor. Comput. **6**(1), 135–177 (2010)
31. Sadeghi, A.-R., Schneider, T.: Generalized universal circuits for secure evaluation of private functions with application to data classification. In: Lee, P.J., Cheon, J.H. (eds.) ICISC 2008. LNCS, vol. 5461, pp. 336–353. Springer, Heidelberg (2009)
32. Straus, E.G.: Addition chains of vectors. Amer. Math. Mon. **70**, 806–808 (1964)
33. Valiant, L.G.: Universal circuits (Preliminary report). In: STOC, pp. 196–203 (1976)

On the Security of the (F)HMQV Protocol

Augustin P. Sarr[1][✉] and Philippe Elbaz–Vincent[2]

[1] Laboratoire ACCA, Université Gaston Berger de Saint–Louis,
Saint Louis, Senegal
aug.sarr@gmail.com
[2] Institut Fourier – CNRS, Université Grenoble Alpes, Grenoble, France

Abstract. The HMQV protocol is under consideration for IEEE P1363 standardization. We provide a complementary analysis of the HMQV protocol. Namely, we point a Key Compromise Impersonation (KCI) attack showing that the two and three pass HMQV protocols cannot achieve their security goals. Next, we revisit the FHMQV building blocks, design and security arguments; we clarify the security and efficiency separation between HMQV and FHMQV, showing the advantages of FHMQV over HMQV.

Keywords: Authenticated key exchange · FHMQV · HMQV · KCI Attack · Security model

1 Introduction

Designing authenticated key agreement protocols is a notoriously subtle task. Bellare and Rogaway proposed a new approach for the analysis of key agreement protocols [3], which was later refined in many security models, including, and among others the CK [6], eCK [19] and seCK [28] models.

The HMQV protocol [16], inspired by the famous MQV [20] protocol, was shown secure in a variant of the CK model, termed here CK_{HMQV}. HMQV was designed to resist a variety of attacks and was shown to provably achieve its security attributes. Among others, Krawczyk was able to show that HMQV remains secure even if public keys are not tested to be of correct order ($\mathcal{G}$–tests). As the computational cost of these tests may be significant, avoiding them may induce a significant efficiency improvement. With this efficiency improvement, HMQV was proposed for standardization in P1363 [17]. The HMQV P1363 submission states that the tests to ensure ephemeral keys to be of correct order "are required only in settings where ephemeral exponents are more vulnerable to attack than long–term secrets. In all other cases, i.e., where ephemeral and long–term secrets are equally protected, HMQV can safely skip these tests, thus providing superior performance especially when the cofactor is large" [17, p. 2].

A.P. Sarr—Partially supported by the CEA–MITIC.
P. Elbaz–Vincent—Partially supported by the LabEx PERSYVAL–Lab (ANR–11–LABX–0025–01).

D. Pointcheval et al. (Eds.): AFRICACRYPT 2016, LNCS 9646, pp. 207–224, 2016.
DOI: 10.1007/978-3-319-31517-1_11

In [22,23], some attacks against HMQV are proposed to recover the victim's static private key; the attacks can be launched in the case the static and ephemeral keys are not tested to be of correct order. Even if the attack against the one–pass HMQV is realistic, the attacks proposed against the two and three pass variants seem less realistic, as the attacker needs to learn some ephemeral private keys from the target victim. The work [24] delves further in the effects of omitting public key validation in HMQV, and some new attacks are presented in the cases static public keys only or ephemeral public keys only are tested to be of correct order; however the attacks are proposed in groups which are not used in practice. In [7], Chalkias, *et al.* explore KCI against the One Pass (H)MQV protocols and show that these protocols are vulnerable to KCI attacks. In [26,27] Sarr *et al.* explore the consequences of secret exponent leakage in a HMQV session. They show that (partial) leakage on ephemeral secret exponents lead to impersonation and man–in–the–middle attacks. Basing on theses findings they propose the FHMQV protocol they show to confine the effects of such leakages.

In this paper, we investigate the effects of omitting ephemeral key validation in the HMQV protocol. We show that the (two and three pass) HMQV protocol(s) are vulnerable to KCI, unless further restrictions are considered in the underlying group. Namely, in the case the group keys are supposed to belong is a subgroup of a DSA group $GF(q)$, with $(q-1)$ divisible by a sufficiently "large" integer, without $\mathcal{G}$–tests, HMQV is vulnerable to a KCI; our attack invalidates the HMQV resistance to KCI, stated in [16, Theorem 18 and Lemma 21]. A main feature of the KCI attack we present is that it requires the entity to be impersonated to omit ephemeral key validation *only once*.

Besides, we re–examine the FHMQV building blocks, showing that contrary to what is suggested in [21] changing the interaction order has no effect on the building blocks security. We clarify also the separation between FHMQV and HMQV, showing the security and efficiency improvements in FHMQV.

This paper is organized as follows. In Sect. 2, we revisit the HMQV protocol, pointing a KCI attack. In Sect. 3 we revisit the FXCR scheme, showing that its security is not dependent to interaction order. The FDCR scheme is revised in Sect. 4. In Sect. 5 we clarify the separation between FHMQV and HMQV.

The following notations are used in this paper H is λ bit hash function, where λ is the security parameter; $\bar{H}$ is a $l = \lambda/2$ bits hash function. $\mathcal{G}$ is a multiplicatively written group, $\mathcal{G}^*$ is the set of non–identity elements in $\mathcal{G}$. If n is an integer, $|n|$ denotes its bit–length; we refer to the length of a list $\mathcal{L}$ by $|\mathcal{L}|$. The symbol $\in_R$ stands for "chosen uniformly at random in". For two bit strings m_1 and m_2, $m_1 \| m_2$ denotes the concatenation of m_1 and m_2. If $x_1, x_2, \cdots, x_k$ are objects belonging to different structures (group, bit–string, etc.) $(x_1, x_2, \cdots, x_k)$ denotes a representation such that each element can be univocally parsed.

2   Key Compromise Impersonation Against HMQV

A protocol is said to be vulnerable to KCI impersonation, if an attacker who learns the long term secret of a party, say $\hat{A}$, is able to impersonate another party, say $\hat{B}$, to $\hat{A}$. When a protocol is vulnerable to such attacks, a static key leakage may lead to harms that go far beyond the sole ability to impersonate the static key's owner. For instance, in the case $\hat{A}$ is a bank client and $\hat{B}$ a bank server, the attacker may impersonate the server to the client to collect more sensitive information (such as a credit card number or a security code, for instance). As another example, $\hat{B}$ may be a trusted software update server, in this case, the attacker may impersonate the server to $\hat{A}$ to make him/her install a malicious software (spyware, worm, virus, etc.), and gain much more sensitive information, such as passwords, credit card numbers, etc. KCI resilience is then an important security attribute, particularly for protocols intended to be standardized, such as HMQV.

In this section, we present a KCI against the three–pass HMQV protocol. About prime–order tests, we show that without these tests HMQV is vulnerable to KCI, unless further restrictions are specified about the underlying group. This shows also that the HMQV KCI resilience stated in [16, Theorem 18 on p. 40 and Lemma 21 on p. 41] does not hold.

Let q be a prime, p a prime dividing $(q-1)$, and G an element of $GF(q)$ of order p. Let $\hat{A}$ and $\hat{B}$ be two parties with respective static key pairs $(a, A = G^a)$, $(b, B = G^b)$ with $a, b \in \{1, \cdots, p-1\}$. An execution of the three–pass HMQV between $\hat{A}$ and $\hat{B}$ is as in Protocol 1; if any verification fails the execution aborts.

Protocol 1. Three Pass HMQV Key Exchange

I) The initiator $\hat{A}$ does the following:
    a) Choose $x \in_R \{1, \cdots, p-1\}$ and compute $X = G^x$.
    b) Send $(\hat{A}, \hat{B}, X)$ to $\hat{B}$.

II) At receipt of $(\hat{A}, \hat{B}, X)$, $\hat{B}$ does the following:
    a) Verify that $X \in GF(q) \setminus \{0, 1\}$.
    b) Choose $y \in_R \{1, \cdots, p-1\}$ and compute $Y = G^y$.
    c) Compute $d = \bar{H}(X, \hat{B})$ and $e = \bar{H}(Y, \hat{A})$.
    d) Compute $s_B = y + eb \bmod p$, $\sigma_B = (XA^d)^{s_B}$, $K = H(\sigma_B, 1)$ and $K_m = H(\sigma_B, 0)$.
    e) Send $\left(\hat{B}, \hat{A}, Y, MAC_{K_m}(\text{"1"})\right)$ to $\hat{A}$.

III) At receipt of $\left(\hat{B}, \hat{A}, Y, MAC_{K_m}(\text{"1"})\right)$, $\hat{A}$ does the following:
    a) Verify that $Y \in GF(q) \setminus \{0, 1\}$.
    b) Compute $d = \bar{H}(X, \hat{B})$ and $e = \bar{H}(Y, \hat{A})$.
    c) Compute $s_A = x + da \bmod p$, $\sigma_A = (YB^e)^{s_A}$, $K = H(\sigma_B, 1)$ and $K_m = H(\sigma_B, 0)$.
    d) Validate $MAC_{K_m}(\text{"1"})$.
    e) Send $\left(\hat{A}, \hat{B}, X, Y, MAC_{K_m}(\text{"0"})\right)$ to $\hat{B}$.

IV) At receipt of $\left(\hat{A}, \hat{B}, X, Y, MAC_{K_m}(\text{"0"})\right)$, $\hat{B}$ validates $MAC_{K_m}(\text{"0"})$.

V) The shared session key is K.

The HMQV protocol is shown secure in a variant of the CK model, the CK$_{\text{HMQV}}$ model [16] (see [8] for a comparison between the CK, CK$_{\text{HMQV}}$, and eCK models).

Suppose that q and p are primes such that $p \mid (q-1)$. Let G' be a primitive element in $GF(q)$; the element $G = G'^{(q-1)/p}$ has order p, and generates a group $\mathcal{G}$ of order p. For concreteness, suppose in addition that $(|q|, |p|) \in \{(1024, 160), (2048, 224), (3072, 256)\}$. The complexity of the Number Field Sieve for prime field discrete logarithm[1] [13,29,30] is

$$L_q[1/3, \sqrt[3]{64/9} + o(1)] \approx \exp\left(\left(\sqrt[3]{64/9} + o(1)\right)(\ln(q))^{1/3}(\ln\ln(q))^{2/3}\right).$$

Hence, omitting the term $o(1)$ we have, $L_q[1/3, \sqrt[3]{64/9}] \geqslant 2^{87}$ when $|q| = 1024$ and $L_q[1/3, \sqrt[3]{64/9}] \geqslant 2^{117}$ when $|q| = 2048$ and $L_q[1/3, \sqrt[3]{64/9}] \geqslant 2^{139}$ when $|q| = 3072$. So, we have $L_q[1/3, \sqrt[3]{64/9}] \geqslant p^{1/2}$; the complexity of the DLP on $\mathcal{G}$ reduces then to that of the generic attacks.

Let $\lambda = |p|$, $t = \lambda/3$ and suppose in addition, that $2^t|(q-1)$. Primes satisfying these conditions can be efficiently found using the following process[2]: (i) choose a prime p such that $|p| = \lambda$, and (ii) set $\alpha = 2^t \cdot p$; then (iii) try to find an integer s with bit–length $(|q| - |\alpha|)$, such that $q = s \cdot \alpha + 1$ is prime. By the theorem of Dirichlet on primes in arithmetic progression [11], we know that an infinity of primes in the form $s \cdot \alpha + 1$ exist; moreover the interval $[2^{|q|}, 1.048 \cdot 2^{|q|}]$ contains *at least* one prime from the progression [10]. Hence the interval $[2^{|q|}, 2^{|q|+1}]$ contains *at least* 14 of such primes (which is very pessimistic).

An example of such primes for $(|q|, |p|) = (3072, 256)$ is
$q_{3072} = 2^{91} \cdot 3^7 \cdot 5^7 \cdot 11^8 \cdot 17^3 \cdot 37^6 \cdot 67^2 \cdot 131^4 \cdot 257^4 \cdot 521^4 \cdot 1031^5 \cdot 2053 \cdot$
$\qquad 4099^7 \cdot 8209^4 \cdot 16411^5 \cdot 32771^4 \cdot 65537^4 \cdot 131101 \cdot 262147^5 \cdot 524309^7 \cdot$
$\qquad 1048583^5 \cdot 2097169^5 \cdot 4194319^2 \cdot 8388617 \cdot 16777259^4 \cdot 33554467^6 \cdot$
$\qquad 67108879^2 \cdot 134217757^5 \cdot 268435459^3 \cdot 536870923^8 \cdot 1073741827^5 \cdot$
$\qquad 2147483659^6 \cdot 4294967311^5 \cdot 8589934609^4 \cdot 17179869209^7 \cdot p_{256} + 1$, with
$p_{256} = 578960446186580977117854925043439539266349923328202820197287\backslash$
$\qquad 92003956564820063$.

Following the KCI scenario considered in [16, pp. 40–42], suppose that $\hat{A}$ and $\hat{B}$ are two honest parties, and $\mathcal{A}$ an attacker which knows $\hat{A}$'s static key a, and aims to impersonate $\hat{B}$ to $\hat{A}$. Suppose that $\hat{B}$ chooses his/her ephemeral keys in $\mathcal{G}^*$ as prescribed.

The attacker can proceed as follows: (i) using an invalid ephemeral public key, he/she learns the ephemeral secret exponent s_B at $\hat{B}$ in a three pass HMQV[3] session, as described, in Attack 2, and (ii) using s_B, the attacker impersonates indefinitely $\hat{B}$ to $\hat{A}$.

Attack 2. Online stage of an Ephemeral Secret Exponent Recovering

1) Compute $X = G'^{(q-1)/2^t}$.
2) Send $(\hat{A}, \hat{B}, X)$ to $\hat{B}$ to initiate a three–pass HMQV session.
3) Intercept $\hat{B}$'s response to $\hat{A}$, $(\hat{B}, \hat{A}, Y, \mathsf{tag}_{\hat{B}} = \mathrm{MAC}_{K_m}(\text{"1"}))$ and halt.

---

[1] This is to date the best sieving algorithm for discrete logarithm over a prime field.
[2] It takes few seconds on a i7–4790K to find such primes.
[3] To launch this phase in the two–pass HMQV, the attacker has simply to wait, for instance, that $\hat{B}$ uses the key to authenticate some value he/she knows.

In a three–pass HMQV session, the key used at the responder for MACing is $K_m = H(\sigma, 0)$ with $\sigma = (XA^d)^{s_B}$ wherein $s_B = y + eb \mod p$, with $d = \bar{H}(X, \hat{B})$ and $e = \bar{H}(Y, \hat{A})$. As

$$\sigma = (XA^d)^{s_B} = X^{s_B} (YB^e)^{da},$$

the attacker computes $\sigma_0 = (YB^e)^{da}$ and $\sigma_{1_i} = X^i$ and $K'_i = H(\sigma_{1_i}, \sigma_0, 0)$, for $i = 1, 2, 3, \cdots, 2^t$ until $\mathrm{MAC}_{K'_i}(\text{"1"}) = \mathrm{tag}_{\hat{B}}$. By this exhaustive search, the attacker finds the t least significant bits of s_B. Then, using the relation

$$\sigma_0 = (YB^e)^{da} = (A^d)^{s_B},$$

the attacker recovers the remaining bits of s_B (recall that $t = \lambda/3$) using $\mathcal{O}\left(2^{(\lambda-t)/2}\right) = \mathcal{O}(2^t)$ operations [12, §B]. And then, the whole offline stage requires $\mathcal{O}(2^t)$ operations. The rough computational cost of the attack for different values of λ, in the case $t = \lambda/3$, are given hereunder.

Value of λ	Rough computational cost
160	2^{54}
224	2^{75}
256	2^{86}

As a concrete example, for $\lambda = 224$ (recall that $\lambda = |p|$) we have $t = 75$, then recovering s_B requires roughly 2^{75} operations, which is far from the 2^{112} operations required for the discrete logarithm problem, and not out of reach of our computational capabilities [14,18].

From a knowledge of s_B and the ephemeral public key Y generated by $\hat{B}$, the attacker can indefinitely impersonate $\hat{B}$ to $\hat{A}$, in both the two and three pass HMQV variants [26]. We stress that the attacker cannot recover $\hat{B}$'s static private key from s_B; this shows that for any primes p and q such that p divides $(q-1)$, 2^t divides $(q-1)$ and $\max\{2^{(|p|-t)/2}, 2^t\}$ operations are not out of reach, omitting ephemeral key validation *only once* is sufficient for an effective KCI attack. As *the attacker never learns an ephemeral private key*, this invalidates the claim that public key validation is required in the HMQV protocol "only in settings where ephemeral exponents are more vulnerable to attack than long–term secrets" [17]. Also, the "minimal requirement for a secure key–exchange ... that the attacker, not knowing the private key of a party $\hat{B}$, should not be able to impersonate $\hat{B}$" [16, p. 18] is not achieved.

About the Factorization of $q - 1$. We presented our attack in the case a sufficiently large power of 2 divides $(q - 1)$, however the attack can be launched as long as $(q - 1)$ divisible by a "sufficiently large" integer. We stress that in real word settings, to avoid "sieving" attacks [25], q is chosen to be much larger than p; for instance the NIST recommends [1] the following pairs for $(|q|, |p|)$: (1024,160), (2048,224), and (3072,256). Hence for real word domain parameters, it is likely that q has a factor of bit–length $\approx |p|/3$. If M is a divisor of $q-1$ with bit–length $\lambda/3$ (recall that $\lambda = |p|$), the element $X = G'^{(q-1)/M}$ has order M, and can be used as outgoing ephemeral key in the online stage (Attack 2). By exhaustive search, $\beta_1 = s_B \mod M$ can be found using M operations. Then, as $s_B = y + eb \mod p = \beta_2 M + \beta_1 \mod p$, with $|\beta_2| \leqslant 2\lambda/3$, from the relation $\sigma_0 = (YB^e)^{da} = (A^d)^{s_B} = (A^d)^{\beta_2 M + \beta_1}$, one obtains $\sigma_0 A^{-d\beta_1} = (A^{dM})^{\beta_2}$; the remaining part β_2 can then be recovered using $\mathcal{O}(2^{\lambda/3})$ operations [12]. The attack can then be launched for any divisor M of $q-1$ of bit–length t such that $\mathcal{O}(\max\{2^{(\lambda-t)/2}, 2^t\})$ operations are not out of reach.

2.1   On the HMQV Security Reduction

The KCI attack is totally well grounded in the CK_{HMQV} model; a natural question is then how can it co–exist with the HMQV security reduction.

The attack is rooted in the interpretation of the XCR security reduction in the analysis of the DCR scheme. In fact, a DCR signature is an XCR signature by $\hat{B}$ (resp. $\hat{A}$) with the challenge XA^d (resp. YB^e). As the DCR reduction uses the XCR reduction [16, pp. 20–25], wherein challenges are supposed to belong to $\mathcal{G}^*$, it becomes a requirement that both XA^d and YB^e belong to $\mathcal{G}^*$. Hence, when KCI is considered, namely, when a is known to the attacker, the security reduction leads to $CDH(X, B)$. Unfortunately, when $X \notin \mathcal{G}^*$, there is no guarantee that computing $CDH(X, B)$ is hard. As the core of the HMQV protocol is the DCR scheme, it then becomes also a requirement that ephemeral keys be tested for membership in $\mathcal{G}^*$. This point was missed in the analysis of the HMQV protocol and explains the co–existence of the attack and the security reduction in [16].

We stress that contrary to the DCR and XCR schemes, the FDCR signature of $\hat{A}$ and $\hat{B}$ on messages m_1, m_2 and challenges X and Y is not a FXCR signature of $\hat{A}$ (resp. $\hat{B}$) on the message m_2 (resp. m_1) and challenge YB^e (resp. XA^d) [27]. Also, the attack does not apply to protocols that mandate ephemeral key validation, such as MQV [20] and FHMQV [27].

Nonetheless, in the case of MQV, when ephemeral keys are not validated, the attack can be launched. In this case, as the ephemeral secret exponents $s_A = x + (\tilde{X} \mod 2^l)a$ and $s_B = y + (\tilde{Y} \mod 2^l)b$, where $\tilde{X}$ is the integer representation of X, are not tied to the peer's identity, the attacker can not only impersonate $\hat{B}$ to $\hat{A}$, but to *any* party. Moreover, there is no need for the attacker to learn an honest party's static key, the attacker can *use his/her own static key together with an invalid ephemeral key*.

In the case of FHMQV, which is resilient to ephemeral secret exponent leakage, we do not know how the attack can be launched. However, if ephemeral keys are not validated and ephemeral private key leakage is considered at the victim $\hat{B}$, the attacker can disclose the victim's static private key, in both MQV, HMQV and FHMQV.

3   FXCR Security in the Reversed Interaction Order

In this section, we revisit the FXCR scheme [26], clarifying its advantages over
the XCR scheme. We show also that the recent critics from [21] about the FXCR
security reduction (which is the main ingredient in FHMQV security arguments)
are erroneous. As already reported in [26], even if the ephemeral keys are tested
for membership in $\mathcal{G}^*$, the HMQV protocol is sensitive to partial leakage of
the ephemeral exponents s_A and s_B. This observation lead to the design of the
FXCR scheme.

Definition 1 (FXCR Scheme). *Let $\hat{B}$ be an entity with public key $B \in \mathcal{G}^*$.*
$\hat{B}$'s signature on a challenge X provided by a verifier $\hat{A}$ together with a mes-
sage m is $FSig_{\hat{B}} = (Y, X^{y+\bar{H}(Y,X,m)b})$. The verifier accepts a pair (Y, σ) as a
valid signature if $\left(Y B^{\bar{H}(Y,X,m)}\right)^x = \sigma$.

From [26], it is shown that no efficient attacker, even if given the secret exponent s_B
at each signature generation can forge a valid FXCR signature, unless with negligible probability. The authors of [21][4] consider a reversed interaction order between
the signer and the verifier and claim that the FXCR security reduction is flawed,
as the simulation becomes invalid in this case. Namely, if the challenge is provided
to the signer after it generates Y, the security reduction does not hold.

Strictly speaking changing the interaction order defines another signature
scheme; and the security reduction may become inapplicable for the new scheme.
Furthermore, even if the interaction order is changed, all the security attributes
claimed in [26] about the FXCR scheme remain valid; and contrary to what is
suggested in [21], no additional Gap DH assumption is required. We still denote
the variant of the signature scheme obtained by changing the interaction order by
FXCR and consider a signer $\hat{B}$ and a verifier $\hat{A}$ interacting as described in Figure 3.

Figure 3. FXCR Interactions for Signature Generation

---

1) At signature request with a message m, $\hat{B}$ generates $Y \in \mathcal{G}^*$ and provides $\hat{A}$
   with (m, Y, B).
2) $\hat{A}$ chooses $x \in_R \{1, \cdots, p-1\}$ and provides $\hat{B}$ with $(m, X = G^x, Y, B)$.
3) $\hat{B}$ verifies that $X \in \mathcal{G}^*$. If the verification succeeds, it provides $\hat{A}$ with
   $(m, X, Y, B, \sigma, s_B)$ wherein $\sigma = X^{s_B}$ and $s_B = y + \bar{H}(Y, X, m)b$.
4) The verifier accepts $\hat{B}$'s signature as valid if $\left(Y B^{\bar{H}(Y,X,m)}\right)^x = \sigma$.

We stress that the verifier is provided also with the secret exponent s_B; this
models total secret exponent leakage in each signature generation.

Definition 2 (FXCR Security). *The FXCR scheme is said to be secure, if no*
efficient attacker can succeed in the game in Figure 4 with non–negligible probability.

---

[4] Their abstract starts with "HMQV is one of the most efficient (provably secure)
    authenticated key–exchange ˜protocols based on public–key cryptography, and is
    widely standardized." To date, we are not aware of any standardization body which
    has already adopted the HMQV protocol.

Figure 4. The FXCR Security Game

---

1) The attacker $\mathcal{A}$ is given a public key B, a challenge X_0, together with a signing oracle as described in Figure 7, and also a hashing oracle.
2) The attacker halts with output $(0,0,0,0,0)$ to indicate a failure, or a quintuple $(m_0, X_0, Y_0, B, \sigma_0)$ such that
   a) (Y_0, σ_0) is a valid signature with respect to the public key B and message–challenge pair (m_0, X_0), and
   b) (Y_0, σ_0) is a fresh signature, $i.\ e.$, (Y_0, σ_0) was never generated by $\hat{B}$ on signature request on (m_0, X_0).

Notice that contrary to [16, Sect. 4.1], which requires that "the pair (Y_0, m_0) did not appear in any of the responses of $\hat{B}$", we use the minimal requirement that (Y_0, σ_0) was not generated by the signer on the message–challenge pair (m_0, X_0).

Theorem 1. *Under the Computational Diffie–Hellman (CDH) assumption in $\mathcal{G}$ and the Random Oracle (RO) model, the FXCR scheme is secure in the sense of Definition 2.*

Proof. As the attacker is supposed to be polynomial in $|p|$, let P be a polynomial and $T = P(|p|)$ an upper bound on the number of digest queries on messages with format (Y, Z, m) with $Y, Z \in \mathcal{G}^*$, the attacker issues after it receives (m, Y, B) from the signing oracle (step 1 of Figure 3) and before it provides the signing oracle with its challenge (m, X, Y, B) (step 2 of Figure 3). Also, we suppose that the number of signature queries the attacker issues is upper bounded by $L = Q(|p|)$ for some polynomial Q.

To lighten the presentation, we suppose that the attacker behaves as follows. First, for each signature generation, the attacker issues exactly $T = P(|p|)$ digest queries on messages with format (Y, Z, m), with $Z \in \mathcal{G}^*$, after he/she receives (m, Y, B) from the signing oracle, and before he/she provides the signing oracle with the challenge (m, X, Y, B). The attacker may discard digest values with no interest. Second, among the digest queries the attacker issues, one query is on (Y, X, m), where X is the challenge to be submitted to the signing oracle. Third, the attacker never submits to the hashing oracle the same message twice (the attacker can keep track of his/her previous digest queries).

We stress that the attacker we consider remains polynomial in $|p|$ and from any efficient attacker $\mathcal{A}'$ one can derive an efficient attacker $\mathcal{A}$ which behaves as described and succeeds with the same probability than $\mathcal{A}'$. The attacker $\mathcal{A}$ behaves exactly as $\mathcal{A}'$ except that for signature generation, after he/she receives (m, Y, B) from the signing oracle and before he/she provides $\hat{B}$ with the challenge X, he/she ensures that T digest queries on messages with format (Y, Z, m), including one query on (Y, X, m), are issued. $\mathcal{A}$ ignores the digest values $\mathcal{A}'$ does not issue; he/she remains polynomial and has the same success probability than $\mathcal{A}'$.

The attacker's interactions with the signing and hashing oracles are summarized in Figure 5; without loss of generality, we omit digest queries of other kinds the attacker may issue between consecutive steps of Figure 5.

Figure 5. Modified queries for Signature Generation

1) For the j–the signature query, activate the signing oracle with a message m_j to obtain (m_j, Y_j, B).
2) Generate a challenge $X_j \in \mathcal{G}^*$ and issue T digest queries on messages with format $(Y_j, Z_{j,i}, m_j)_{i \in \{1, \cdots, T\}}$ with one of the $Z_{j,i}$'s being the challenge X_j to be submitted to the signing oracle.
3) Provide the signing oracle with (m_j, X_j, Y_j, B).
4) And receive the signature $(m_j, X_j, Y_j, B, \sigma_j, s_{j,B})$ from the signing oracle.

Let $\Pr(\mathrm{Succ}_\mathcal{A})$ denote the probability that $\mathcal{A}$ succeeds in the FXCR security game, and $\mathbf{V} = \{1, \cdots, T\}^L$ be the set of L–uples of elements of $\{1, \cdots, T\}$. We denote by V the random variable that takes values in $\mathbf{V}$ and describes the digest queries at step 2 of Figure 5 wherein the attacker provides the hashing oracle with the message (Y_j, X_j, m_j) (X_j being the challenge to be submitted to the signing oracle). Namely, for $v = (v_1, \cdots, v_L) \in \mathbf{V}$, we denote by $\Pr(V = v)$ the probability that for all $j \in \{1, \cdots, L\}$, the v_j–th digest query at step 2 in the j–th signature generation, Z_{j,v_j} equals the challenge X_j; i. e. $\Pr(V = v)$ denotes the probability that the attacker provides the signing oracle with challenge Z_{1,v_1} in the first signature query, and Z_{2,v_2} as a challenge in the second signature query, and so forth. For $v \in \mathbf{V}$, we say that v is *possible* if $\Pr(V = v)$ is non–zero and denote by $\mathrm{Poss}(\mathbf{V})$ the subset of $\mathbf{V}$ consisting of possible v's. The probability of success of the adversary $\mathcal{A}$ is

$$
\begin{aligned}
\Pr(\mathrm{Succ}_\mathcal{A}) &= \sum_{v \in \mathrm{Poss}(\mathbf{V})} \Pr(\mathrm{Succ}_\mathcal{A} \cap V = v) \\
&= \sum_{v \in \mathrm{Poss}(\mathbf{V})} \Pr(\mathrm{Succ}_\mathcal{A} \mid V = v)\Pr(V = v) \\
&\leqslant \sum_{v \in \mathrm{Poss}(\mathbf{V})} \left(\max_{v \in \mathrm{Poss}(\mathbf{V})} \Pr(\mathrm{Succ}_\mathcal{A} \mid V = v) \right) \Pr(V = v) \\
&\leqslant \max_{v \in \mathrm{Poss}(\mathbf{V})} \Pr(\mathrm{Succ}_\mathcal{A} \mid V = v). \tag{1}
\end{aligned}
$$

It then suffices to show that for all $v \in \mathrm{Poss}(\mathbf{V})$, $\Pr(\mathrm{Succ}_\mathcal{A} \mid V = v)$ is negligible. Suppose there is v such that $\Pr(\mathrm{Succ}_\mathcal{A} \mid V = v)$ is non–negligible. Using $\mathcal{A}$, we show the existence of an efficient CDH solver $\mathcal{S}$ which succeeds with non–negligible probability. The solver works as in Figure 6.

The simulator is efficient, moreover it provides a consistent simulation; and under the RO model, this simulated environment is indistinguishable from a real one. The probability that the attacker provides a valid forgery without issuing $\bar{H}(Y_0, X_0, m_0)$ is 2^{-l}. Hence, in this simulation, the attacker succeeds with non–negligible probability. From the General Forking Lemma [2], the probability the attacker succeeds in the simulation and in the repeat experiment is

$$
\Pr(\mathrm{Succ}_2) \geqslant \Pr(\mathrm{Succ}_\mathcal{A} \mid V = v) \left(\frac{\Pr(\mathrm{Succ}_\mathcal{A} \mid V = v)}{q} - 2^{-l} \right),
$$

Figure 6. A CDH solver $\mathcal{S}$ from $\mathcal{A}$

Run of $\mathcal{A}$:

1) When $\mathcal{S}$ is activated with a message m_j, it does as follows:
   a) Choose $s_{j,B} \in_R \{1, \cdots, p-1\}$, $e_j \in_R \{0,1\}^l$, set $Y_j = G^{s_{j,B}} B^{e_j^{-1}}$.
   b) Create an empty list $\mathcal{L}_{e_j, Y_j, s_{j,B}, m_j}$.
   c) Provide the attacker with (m_j, Y_j, B).
2) At digest query on a message which does not have format (Y, Z, m), the simulator $\mathcal{S}$ responds with $e \in_R \{0,1\}^l$.
3) At digest query on a message with format (Y, Z, m), $\mathcal{S}$ does as follows:
   a) If $\mathcal{A}$ was provided with (m_j, Y_j, B) (Step 2 of Figure 3) such that $m_j = m$ and $Y_j = Y$ and if $|\mathcal{L}_{e_j, Y_j, m_j}| = v_j - 1$, $\mathcal{S}$ provides the attacker with e_j, and appends Z to $\mathcal{L}_{e_j, Y_j, s_{j,B}, m_j}$.
   b) Otherwise, $\mathcal{S}$ responds with $e \in_R \{0,1\}^l$, and if $m_j = m$ and $Y_j = Y$, it appends Z to $\mathcal{L}_{e_j, Y_j, m_j}$.
4) When $\mathcal{A}$ provides $\mathcal{S}$ with (m_j, X_j, Y_j, B), $\mathcal{S}$ responds with $(m_j, X_j, Y_j, B, G^{s_{j,B}}, s_{j,B})$. Notice that this is consistent with the digest simulation at steps 2 and 3.
5) At $\mathcal{A}$'s halt, $\mathcal{S}$ verifies that $\mathcal{A}$'s output is different from $(0,0,0,0,0)$ and satisfies the following conditions; if not $\mathcal{S}$ aborts.
   - $Y_0 \in \mathcal{G}^*$ and $\bar{H}(Y_0, X_0, m_0)$ was issued from $\bar{H}$.
   - The signature (Y_0, σ_0) was not returned by $\hat{B}$ on query (m_0, X_0).

Repeat: $\mathcal{S}$ executes a new run of $\mathcal{A}$, using the same input and coins; and answering to all digest queries before $\bar{H}(Y_0, X_0, m_0)$ with the same values as in the previous run. The new query of $\bar{H}(Y_0, X_0, m_0)$ and subsequent queries to $\bar{H}$ are answered with new random values.

Output: If $\mathcal{A}$ outputs a second signature $(m_0, X_0, Y_0, B, \sigma_0')$ satisfying conditions of step 5, with a hash value $\bar{H}(Y_0, X_0, m_0)_2 = e_0' \neq e_0 = \bar{H}(Y_0, X_0, m_0)_1$ then $\mathcal{S}$ outputs $\left(\sigma_0/\sigma_0'\right)^{(e_0 - e_0')^{-1}}$ as a guess for $\mathrm{CDH}(B, X_0)$.

---

where q is the number of digest queries the attacker issues, which is non–negligible, unless $\Pr(\mathrm{Succ}_\mathcal{A} \mid V = v)$ is negligible. Moreover, if the repeat experiment succeeds, the digest values e_0 and e_0' are different with probability $1 - 2^{-l}$, and then the computation $\left(\sigma_0/\sigma_0'\right)^{(e_0 - e_0')^{-1}}$, leads to $\mathrm{CDH}(X_0, B)$ with non–negligible probability, contradicting then the CDH assumption. Hence, under the RO model and the CDH assumption, for all $v \in \mathbf{V}$, $\Pr(\mathrm{Succ}_\mathcal{A} \mid V = v)$ is negligible; using (1), we conclude that $\Pr(\mathrm{Succ}_\mathcal{A})$ is negligible. $\qquad \square$

This shows that the FXCR CDH–based security reduction holds not only in what the authors of [21] calls a "regular interaction order", but also if the interaction order is reversed.

4 FDCR Security in the Reversed Interaction Order

As for the FXCR scheme, we show here that the security of the FDCR scheme totally holds, in the reversed interaction order, wherein the signer provides his/her ephemeral public key before receiving a challenge from the verifier.

Definition 3 (FDCR Scheme). *Let $\hat{A}$ and $\hat{B}$ be entities with respective public keys A and B in $\mathcal{G}^*$. The FDCR signature of $\hat{A}$ and $\hat{B}$ on challenge–message pairs (X, m_1) and (Y, m_2) provided respectively by $\hat{A}$ and $\hat{B}$, with $X, Y \in \mathcal{G}^*$ is*

$$FDSig_{\hat{A},\hat{B}}(m_1, m_2, X, Y) = (XA^d)^{y+eb} = (YB^e)^{x+da},$$

where $d = \bar{H}(X, Y, m_1, m_2)$ and $e = \bar{H}(Y, X, m_1, m_2)$.

To show the FDCR security in the reversed interaction order, we consider a verifier interacting with a signer $\hat{B}$ as described in Figure 7, and the game in Figure 8.

Figure 7. FDCR interactions for Signature Generation

1) The verifier $\hat{A}$ provides $\hat{B}$ with (m_1, A).
2) The signer $\hat{B}$ responds with (m_1, m_2, Y, A, B), with $Y \in \mathcal{G}^*$.
3) The verifier chooses $x \in \{1, \cdots, p - 1\}$ and provides $\hat{B}$ with $(m_1, m_2, X = G^x, Y, A, B)$.
4) The signer verifies that $X \in \mathcal{G}^*$, and provides $\hat{A}$ with $(m_1, m_2, X, Y, A, B, \sigma)$ wherein $\sigma = (XA^d)^{y+eb}$ with $d = \bar{H}(X, Y, m_1, m_2)$ and $e = \bar{H}(Y, X, m_1, m_2)$.
5) The verifier accepts $\hat{B}$'s signature as valid if $(YB^e)^{x+da} = \sigma$.

Figure 8. The FDCR Security Game

1) The attacker $\mathcal{A}$ is given a key pair (A, a) and a message–challenge pair (X_0, m_{1_0}); $\mathcal{A}$ is also given access to a hashing oracle, and is allowed to interact with a signing oracle as described in Figure 7.
2) The attacker halts with output $(0, 0, 0, 0, 0, 0, 0)$ to indicate a failure, or a septuple $(m_{1_0}, m_{2_0}, X_0, Y_0, A, B, \sigma_0)$ such that
   a) σ_0 is a valid FDCR signature on messages m_{1_0}, m_{2_0} and challenges X_0, Y_0 with respect to the public keys A and B.
   b) σ_0 is a fresh, *i. e.*, it was not generated as a signature on message–challenge pairs (m'_1, X_0), (m'_2, Y_0) such that $m'_1 \| m'_2 = m_{1_0} \| m_{2_0}$.

Definition 4 (FDCR Security). *The FDCR scheme is said to be secure if no efficient attacker can succeed in the game in Figure 8 with non–negligible probability.*

Theorem 2. *Under the RO model and the CDH assumption, the FDCR scheme is secure in the sense of Definition 4.*

Proof. To lighten the presentation, as in Theorem 1, we suppose that the attacker issues $L = Q(|p|)$ signature queries and $T = P(|p|)$ digest queries on messages with format (Z_1, Z_2, m_1, m_2), with $Z_1, Z_2 \in \mathcal{G}$, between the steps 2 and 3 of Figure 7. We suppose also that the attacker issues a digest query on (Y, X, m_1, m_2), before providing its challenge X to the signer; and also that he/she does not issue the same digest query twice. We stress that the attacker remains polynomial, and may discard the digest values of no interest. We summarize the queries for a signature generation in Figure 9.

Figure 9. Attacker's queries for FDCR Signature Generation

---

1) For the j–the signature query, $\mathcal{A}$ activates the signing oracle with (m_{1_j}, A).
2) The signer provides the attacker with $(m_{1_j}, m_{2_j} Y_j, A, B)$ with $Y_j \in \mathcal{G}^*$.
3) $\mathcal{A}$ generates $X_j \in \mathcal{G}^*$ and issues T digest queries on messages with format $(Y_j, Z_{j,i}, m_{1_j}, m_{2,j})_{i \in \{1, \cdots, T\}}$ with one query on $(Y_j, X_j, m_{1_j}, m_{2,j})$.
4) The attacker provides the signing oracle with $(m_{1_j}, m_{2_j}, X_j, Y_j, A, B)$.
5) And receives $(m_{1_j}, m_{2_j}, X_j, Y_j, A, B, \sigma_j)$ from the signing oracle.

We still denote $\{1, \cdots, T\}^L$ by $\mathbf{V}$, for $v = (v_1, \cdots, v_L) \in \mathbf{V}$, we denote by $\Pr(V = v)$ the probability that for all $j \in \{1, \cdots, L\}$, for the j–th signature generation, the attacker issues a digest query on $(Y_j, X_j, m_{1_j}, m_{2,j})$ at the v_j–th digest query in step 2. The notations $\mathrm{Poss}(\mathbf{V})$ from the proof of Theorem 1 is used again. Conditioning on V, we still obtain

$$\Pr(\mathrm{Succ}_{\mathcal{A}}) \leqslant \max_{v \in \mathrm{Poss}(\mathbf{V})} \Pr(\mathrm{Succ}_{\mathcal{A}} \mid V = v).$$

Suppose that there is $v \in \mathrm{Poss}(\mathbf{V})$ such that $\Pr(\mathrm{Succ}_{\mathcal{A}} \mid V = v)$ is non–negligible. Using $\mathcal{A}$, we build an efficient FXCR forger $\mathcal{S}$ such that $\Pr(\mathrm{Succ}_{\mathcal{S}} \mid V = v)$ is non–negligible. The forger $\mathcal{S}$ works as described in Figure 10.

Under the random oracle model, the simulation in Figure 10 is perfect, except with negligible probability; a deviation occurs when the same message–challenge pair (m_{2_j}, Y_j) is chosen twice in two signature queries on the same pair (m_{1_j}, X_j). As the simulator chooses its challenges uniformly at random in $\mathcal{G}^*$, this occurs with probability $L/(p-1)$ which is negligible. Also, the probability the attacker provides a valid forgery without issuing $\bar{H}(X_0, Y_0, m_{1_0}, m_{2_0})$ and $\bar{H}(Y_0, X_0, m_{1_0}, m_{2_0})$ is smaller than 2^{-l}, which is negligible. Hence, if $\mathcal{A}$ succeeds with non–negligible probability in a real environment, it succeeds also with non–negligible probability under this simulation. Furthermore $\mathcal{S}$ succeeds with probability

$$\Pr(\mathrm{Succ}_{\mathcal{S}} \mid V = v) \geqslant \Pr(\mathrm{Succ}_{\mathcal{A}} \mid V = v) - \frac{L}{(p-1)} - 2^{-l},$$

which is non–negligible if $\Pr(\mathrm{Succ}_{\mathcal{A}} \mid V = v)$ is non–negligible. As already shown in Theorem 1, this is impossible under the RO model and the CDH assumption. Hence, for all $v \in \mathrm{Poss}(\mathbf{V})$, $\Pr(\mathrm{Succ}_{\mathcal{A}} \mid V = v)$ is negligible, and then $\Pr(\mathrm{Succ}_{\mathcal{A}})$ is negligible. □

Figure 10. A FXCR Forger $\mathcal{S}$ from $\mathcal{A}$

Run of $\mathcal{A}$:

1) When $\mathcal{S}$ is activated with (m_{1_j}, A), it does the following:
   a) Choose $s_{j,B} \in_R \{1, \cdots, p-1\}$, $e_j \in_R \{0,1\}^l$, $m_{2_j} \in \cdot \{0,1\}^{F(|p|)}$ for some positive polynomial F, set $Y_j = G^{s_{j,B}} B^{e_j^{-1}}$.
   b) Create an empty list $\mathcal{L}_{e_j, Y_j, s_{j,B}, m_{1_j}, m_{2_j}}$.
   c) Provides the attacker with $(m_{1_j}, m_{2_j}, Y_j, B)$.
2) At $\mathcal{A}$'s digest query on a message which does not have format (Y, Z, m_1, m_2), the simulator $\mathcal{S}$ responds with $e \in_R \{0,1\}^l$.
3) At digest query on messages with format (Y, Z, m_1, m_2), $\mathcal{S}$ does as follows:
   a) If it provided the attacker with $(m_{1_j}, m_{2_j}, Y_j, A, B)$ such that $m_{1_j} \| m_{2_j} = m_1 \| m_2$, $Y = Y_j$ and if $|\mathcal{L}_{e_j, Y_j, s_{j,B}, m_{1_j}, m_{2_j}}| = v_j - 1$, it provides the attacker with e_j and appends Z to $\mathcal{L}_{e_j, Y_j, s_{j,B}, m_{1_j}, m_{2_j}}$.
   b) Otherwise, it responds with $e \in_R \{0,1\}^l$, and if $m_{1_j} \| m_{2_j} = m_1 \| m_2$ and $Y_j = Y$ then it appends Z to $\mathcal{L}_{e_j, Y_j, s_{j,B}, m_{1_j}, m_{2_j}}$.
4) When $\mathcal{A}$ provides $(m_{1_j}, m_{2_j}, X_j, Y_j, A, B)$, if no value is already assigned to $d = \hat{H}(X_j, Y_j, m_{1_j}, m_{1_j})$ $\mathcal{S}$ chooses $d \in_R \{0,1\}^l$, and responds with $(m_{1_j}, m_{2_j}, X_j, Y_j, A, B, (X_j A^d)^{s_{j,B}})$.
5) At $\mathcal{A}$'s halt with a non–null output $(m_{1_0}, m_{2_0}, X_0, Y_0, A, B, \sigma_0)$ $\mathcal{S}$ verifies that the following conditions are satisfied; if not it aborts.
   – $Y_0 \in \mathcal{G}^*$ and $d_0 = \bar{H}(Y_0, X_0, m_{1_0}, m_{2_0})$ and $e_0 = \bar{H}(X_0, Y_0, m_{1_0}, m_{2_0})$ were issued from the hashing oracle.
   – $\mathcal{S}$ never issued a signature $(m_1', m_2', X_0, Y_0, A, B, \sigma_0)$ such that $m_1' \| m_2' = m_{1_0} \| m_{2_0}$.

Output: If all the conditions at step 5 are satisfied, $\mathcal{S}$ outputs $\sigma_0 (Y_0 B^{e_0})^{-d_0 a} = (Y_0 B^{e_0})^{x_0}$ as a FXCR forgery on $m_{1_0} \| m_{2_0}$.

This shows that all the FDCR security attributes remain intact in the interaction order considered in [21].

5   Separation Between FHMQV and HMQV

Security Separation

The sensitivity of the HMQV protocol to partial leakages on intermediate exponents s_A and s_B [26], exploited again with KCI attack in Sect. 2, motivated the FHMQV design which is resilient to such leakages. FHMQV was shown secure in a mixture of two security definitions (termed ck and eck in [26]), which was latter refined into the seCK model [28]. In the CK_{FHMQV} model (ck model in [26]), it is assumed that at all parties the ephemeral keys are as protected as the static ones. This assumption matches some common implementations; such as (EC)DSA signature generation (where a leakage of an ephemeral private key leads to a disclosure of the signer's static private key). However, it does not seem reasonable to assume implementations performed in the same way at all parties, and then consider the same leakages at all parties. We point out that

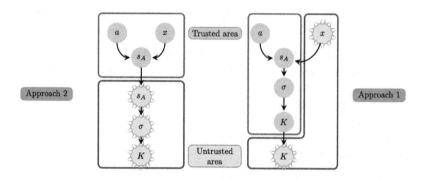

Fig. 1. (F)HMQV implementation approaches in the seCK model

the CK, CK_{HMQV}, eCK and CK_{FHMQV} models define the information that can be leaked in the same way for all parties. While in real word settings, implementations may be different, depending on environments specificities (presence of a desktop computer (DC) only, of a power limited smart–card and a DC, etc.). This observation is one of the motivations of seCK model and corresponds to real–world vulnerabilities [15, 31–34].

Broadly, in the seCK model, it is assumed for DH protocols that at each party, implementation is performed using one of the two following approaches[5] (Fig. 1).

Approach 1. It is assumed that the ephemeral keys are generated in an untrusted area, and the session keys are used also in this area. All the other intermediate results are computed in the trusted area. At a party using this approach, the attacker is allowed the following queries:
  - EphemeralKeyReveal(session) to learn a session's ephemeral private key;
  - SessionKeyReveal(session) to learn a session key;
  - Corrupt$_{SC}$(party) to model an attacker which bypasses the tamper protection mechanisms and learns the party's static key;
  - EstablishParty(party, key) to register a static key on behalf of the party; a party against which this query is not issued is said to be *honest*.

Approach 2. In this approach, it is supposed that both the static and ephemeral keys are computed and used in the trusted area, and all other computations are performed in the untrusted area. So, the attacker is provided with
  - SessionKeyReveal(session) and EstablishParty(party, key) queries, and
  - a reveal query to learn any intermediate result that is computed or used in the untrusted area.

Matching sessions. A session at a party $\hat{P}_i$ is identified with a quintuple $(\hat{P}_i, \hat{P}_j, \text{out}, \text{in}, \text{role})$ wherein $\hat{P}_j$ is the peer, out is the list of the messages sent to the peer, in is the list of the messages received, and role is $\hat{P}_i$'s role, initiator

---

[5] These implementation approaches are not the only possible, however they seem to be common enough in real word to be considered in the model.

$\mathcal{I}$ or responder $\mathcal{R}$. Two sessions $(\hat{P}_i, \hat{P}_j, \text{out}, \text{in}, \text{role})$ and $(\hat{P}'_i, \hat{P}'_j, \text{out'}, \text{in'}, \text{role'})$ are said to be matching if $\hat{P}_i = \hat{P}'_j$, $\hat{P}_j = \hat{P}'_i$, out = in', in = out', and role $\neq$ role'.

Session freshness. A session at an honest party following the implementation approach 1 is said to be *locally exposed* if it were issued a SessionKeyReveal query, or if it were issued an EphemeralKeyReveal query and its owner were issued a Corrupt$_{SC}$ query. At an honest party following the second approach, a session is said to be locally exposed, if it were issued a SessionKeyReveal query or an intermediate result query. A session is said to be *exposed* if it is locally exposed or if its matching session (if any) is locally exposed. A non–exposed session is said to be *seCK–fresh*.

seCK Security. A protocol is said to be secure if (i) when two honest parties complete matching session, then they both derive the same session key; and (ii) an efficient attacker in total control of communication links cannot distinguish a fresh session key from a random value chosen uniformly from the distribution of session keys with probability significantly greater than 1/2.

As already reported in [28], seCK security implies eCK security[6]; seCK security is also strictly stronger than CK$_{\text{FHMQV}}$ security. The seCK model and the CK$_{\text{HMQV}}$ security models are formally incomparable, as the seCK model considers only role–asymmetric protocols while the CK$_{\text{HMQV}}$ model considers only role–symmetric protocols [8][7]. Nevertheless, as shown in [28], there are attacks which are captured in the seCK model but not captured in the eCK and CK$_{\text{HMQV}}$ models. While any real word attack that is captured in the CK$_{\text{HMQV}}$ and eCK models is also captured in the seCK model.

We stress that even when $\mathcal{G}$–tests are performed, HMQV is insecure in the seCK model, for two reasons. First, HMQV is known to be vulnerable to a KCI impersonation attack when leakages on the shared secret σ are considered [16, pp. 17–18]. Second, in the case of a ("sufficient" partial) leakage on ephemeral secret exponents s_A or s_B in a session, an attacker can indefinitely impersonate the session owner; the HMQV protocol cannot then, meet a security definition which allows total leakages on both the shared secret σ and the ephemeral secret exponents s_A and s_B.

Theorem 3. *Under the RO model and the Gap Diffie–Hellman Assumption in $\mathcal{G}$, the FHMQV protocol is seCK–secure.*

Although we already analyzed the main ingredients of the proof of Theorem 3 (the FXCR and FDCR schemes), for lack of space, we do not provide the proof here. We defer the security reduction to the extended version of this paper.

---

[6] There is no dynamic key registration query in the eCK model [19]; the adversary is only allowed to select dishonest parties before starting its game. Dynamic key registration permits the adversary to select the parties it sets as dishonest *after* having seen their behaviour; this is an advantage for the adversary, and does not affect the comparability between the seCK and the eCK models.

[7] Given the work [8], the *Claim 1* from [21] about the formal incomparability between CK$_{\text{FHMQV}}$ and the CK$_{\text{HMQV}}$ models is trivial.

Protocol 11. The FHMQV Key Exchange

I) The initiator $\hat{A}$ does the following:
    a) Choose $x \in_R \{1, \cdots p - 1\}$ and compute $X = G^x$.
    b) Send $(\hat{A}, \hat{B}, X)$ to $\hat{B}$.

II) At receipt of $(\hat{A}, \hat{B}, X)$, $\hat{B}$ does the following:
    a) Verify that $X \in \mathcal{G}^*$.
    b) Choose $y \in_R \{1, \cdots p - 1\}$ and compute $Y = G^y$.
    c) Send $(\hat{B}, \hat{A}, Y)$ to $\hat{A}$.
    d) Compute $d = \bar{H}(X, Y, \hat{A}, \hat{B}), e = \bar{H}(Y, X, \hat{A}, \hat{B})$ and $s_B = y + eb \bmod p$.
    e) Compute $\sigma_B = (XA^d)^{s_B}$ and $K = H(\sigma_B, \hat{A}, \hat{B}, X, Y)$.

III) At receipt of $(\hat{B}, \hat{A}, Y)$, $\hat{A}$ does the following:
    a) Verify that $Y \in \mathcal{G}^*$.
    b) Compute $d = \bar{H}(X, Y, \hat{A}, \hat{B}), e = \bar{H}(Y, X, \hat{A}, \hat{B})$ and $s_A = x + da \bmod p$.
    c) Compute $\sigma_A = (YB^e)^{s_A}$ and $K = H(\sigma_A, \hat{A}, \hat{B}, X, Y)$.

IV) The shared session key is K.

Efficiency Separation. Without a proper validation of ephemeral keys, the HMQV protocol cannot achieve its security goals. When ephemeral keys are validated in HMQV, the FHMQV protocol is as efficient as HMQV in the implementation approach 1. Moreover, for FHMQV, in approach 2, if ephemeral keys are computed in idle–time, only one digest computation, one modular integer addition and one modular integer multiplication has to be performed in the trusted area in non–idle–time; *no exponentiation is performed in the trusted area (usually a smart–card or a hardware security module) in non–idle time.* As neither HMQV, nor MQV can confine the effects of a secret exponent (s_A or s_B) leakage to the leaked session, none of these protocols can achieve such a performance.

6   Concluding Remarks

We revisited the FXCR and the FDCR signature schemes which are the building blocks of the FHMQV protocol, clarifying their strengths, independence to interaction order, and security advantages compared to the XCR and DCR schemes. We clarified also both the security and efficiency separation between HMQV and FHMQV, showing that even if ephemeral keys are validated in HMQV, the FHMQV protocol is strictly stronger than HMQV both in security and efficiency. In settings wherein a trusted device is used to store static and ephemeral keys, a FHMQV implementation can achieve performances which cannot be achieved by MQV or HMQV.

We pointed out a Key Compromise Impersonation attack against HMQV. Namely we showed that omitting ephemeral key validation *only once* is sufficient for a Key Compromise Impersonation. Besides, we revisited the motivations of the seCK model, showing that it is formally stronger than the eCK model, and from a real word perspective, stronger than the CK_{HMQV} model.

In a future work we will be interested in generalizing the compiler from [9] to security models allowing dynamic key registration and intermediate results leakage in the multiple CAs setting [4,5].

References

1. Barker, E., Barker, W., Burr, W., Polk, W., Smid, M.: NIST Special Publication 800–57 Recommendation for Key Management - Part 1: General (Revision 3), (see also the draft of Revision 4 at http://tinyurl.com/qdluuqj)
2. Bellare, M., Neven, G.: Multi-signatures in the plain public-key model and a general forking lemma. In: Proceedings of the 13th ACM Conference on Computer and Communications Security, pp. 390–399. ACM (2006)
3. Bellare, M., Rogaway, P.: Entity authentication and key distribution. In: Stinson, D.R. (ed.) CRYPTO 1993. LNCS, vol. 773, pp. 232–249. Springer, Heidelberg (1994)
4. Boyd, C., Cremers, C., Feltz, M., Paterson, K.G., Poettering, B., Stebila, D.: ASICS: authenticated key exchange security incorporating certification systems. In: Crampton, J., Jajodia, S., Mayes, K. (eds.) ESORICS 2013. LNCS, vol. 8134, pp. 381–399. Springer, Heidelberg (2013)
5. Boyd, C., Cremers, C., Feltz, M., Paterson, K.G., Poettering, B., Stebila, D.: ASICS: authenticated key exchange security incorporating certification systems. Cryptology ePrint Archive: Report 2013/398 (2013)
6. Canetti, R., Krawczyk, H.: Analysis of key-exchange protocols and their use for building secure channels. In: Pfitzmann, B. (ed.) EUROCRYPT 2001. LNCS, vol. 2045, pp. 453–474. Springer, Heidelberg (2001)
7. Chalkias, K., Baldimtsi, F., Hristu-Varsakelis, D., Stephanides, G.: Two types of key-compromise impersonation attacks against one-pass key establishment protocols. In: Filipe, J., Obaidat, M.S. (eds.) E-business and Telecommunications. Communications in Computer and Information Science, vol. 23, pp. 227–238. Springer, Heidelberg (2009)
8. Cremers, C.: Examining indistinguishability-based security models for key exchange protocols: the case of CK, CK-HMQV, and eCK. In: Proceedings of the 6th ACM Symposium on Information, Computer and Communications Security, pp. 80–91. ACM (2011)
9. Cremers, C., Feltz, M.: Beyond eCK: perfect forward secrecy under actor compromise and ephemeral-key reveal. Des. Codes Crypt. **74**(1), 183–218 (2013). Springer
10. Cullinan, J., Hajir, F.: Primes of prescribed congruence class in short intervals. Integers **12**, A56 (2012). De Gruyter
11. Ellison, W., Ellison, F.: Prime Numbers. Wiley and Hermann Editions, New York (1985)
12. Gopalakrishnan, K., Thériault, N., Yao, C.Z.: Solving discrete logarithms from partial knowledge of the key. In: Srinathan, K., Rangan, C.P., Yung, M. (eds.) INDOCRYPT 2007. LNCS, vol. 4859, pp. 224–237. Springer, Heidelberg (2007)
13. Gordon, D.M.: Discrete logarithms in GF(P) using the number field sieve. SIAM J. Discrete Math. **6**(1), 124–138 (1993). SIAM
14. Güneysu T., Pfeiffer G., Paar C., Schimmler M.: Three years of evolution: cryptanalysis with COPACOBANA. In: Workshop Record of Special-Purpose Hardware for Attacking Cryptographic Systems–SHARCS 2009 (2009)

15. Huq, N.: PoS RAM Scraper Malware: Past, Present, and Future. A Trend Micro Research Paper (2014). http://tinyurl.com/jcwc8wz
16. Krawczyk, H.: HMQV: a hight performance secure diffie-hellman protocol. Cryptology ePrint Archive, Report 2005/176 (2005)
17. Krawczyk, H.: HMQV in IEEE P1363. Submission to the IEEE P1363 working group. http://tinyurl.com/opjqknd
18. Kumar, S., Paar, C., Pelzl, J., Pfeiffer, G., Rupp, A., Schimmler, M.: How to break DES for € 8,980. In: International Workshop on Special-Purpose Hardware for Attacking Cryptographic Systems – SHARCS 2006, Cologne, April 2006
19. LaMacchia, B.A., Lauter, K., Mityagin, A.: Stronger security of authenticated key exchange. In: Susilo, W., Liu, J.K., Mu, Y. (eds.) ProvSec 2007. LNCS, vol. 4784, pp. 1–16. Springer, Heidelberg (2007)
20. Law, L., Menezes, A., Qu, M., Solinas, J., Vanstone, S.: An efficient protocol for authenticated key agreement. Des. Codes Crypt. **28**(2), 119–134 (2003). Springer
21. Liu, S., Sakurai, K., Weng, J., Zhang, F., Zhao, Y.: Security model and analysis of FHMQV, revisited. In: Lin, D., Xu, S., Yung, M. (eds.) Inscrypt 2013. LNCS, vol. 8567, pp. 255–269. Springer, Heidelberg (2014)
22. Menezes, A.: Another look at HMQV. J. Math. Cryptology **1**(1), 47–64 (2007). De Gruyter
23. Menezes, A.: Another Look at HMQV. Cryptology ePrint Archive: Report 2005/205 (2005)
24. Menezes, A., Ustaoglu, B.: On the importance of public-key validation in the MQV and HMQV key agreement protocols. In: Barua, R., Lange, T. (eds.) INDOCRYPT 2006. LNCS, vol. 4329, pp. 133–147. Springer, Heidelberg (2006)
25. Odlyzko, A.M.: Discrete logarithms in finite fields and their cryptographic significance. In: Beth, T., Cot, N., Ingemarsson, I. (eds.) EUROCRYPT 1984. LNCS, vol. 209, pp. 224–314. Springer, Heidelberg (1985)
26. Sarr, A.P., Elbaz-Vincent, P., Bajard, J.-C.: A secure and efficient authenticated diffie–hellman protocol. In: Martinelli, F., Preneel, B. (eds.) EuroPKI 2009. LNCS, vol. 6391, pp. 83–98. Springer, Heidelberg (2010)
27. Sarr, A.P., Elbaz-Vincent, P., Bajard, J.C.: A Secure and Efficient Authenticated Diffie-Hellman Protocol. Cryptology ePrint Archive: Report 2009/408 (2009)
28. Sarr, A.P., Elbaz-Vincent, P., Bajard, J.-C.: A new security model for authenticated key agreement. In: Garay, J.A., De Prisco, R. (eds.) SCN 2010. LNCS, vol. 6280, pp. 219–234. Springer, Heidelberg (2010)
29. Schirokauer, O.: Using number fields to compute logarithms in finite fields. Math. Comput. **69**(231), 1267–1283 (2000). AMS
30. Thomé, E.: Théorie algorithmique des nombres et applications à la cryptanalyse de primitives cryptographiques. Habilitation to conduct research. Université de Lorraine, p. 218 (2012). https://hal.inria.fr/tel-00765982
31. Trend Labs Security Intelligence Blog: RawPOS Technical Brief, April 2015. http://tinyurl.com/joyazja
32. VISA Data Security Alert: Debugging Software Memory–Parsing Vulnerability (2008). http://tinyurl.com/joyazja
33. VISA Data Security Alert: Targeted Hospitality Sector Vulnerabilities (2009). http://tinyurl.com/nnpsl3a
34. VISA Data Security Alert: Retail Merchants Targeted by Memory-Parsing Malware (2013). http://tinyurl.com/j3duvlg

Non-Interactive Verifiable Secret Sharing
for Monotone Circuits

Ge Bai[1], Ivan Damgård[2], Claudio Orlandi[2(✉)], and Yu Xia[1]

[1] IIIS, Tsinghua University, Beijing, China
[2] Aarhus University, Aarhus, Denmark
orlandi@cs.au.dk

Abstract. We propose a computationally secure and *non-interactive* verifiable secret sharing scheme that can be *efficiently* constructed from any monotone Boolean circuit. By non-interactive we mean that the dealer needs to be active only once, where he posts a public message as well as a private message to each shareholder. In the random oracle model, we can even avoid interaction between shareholders. By efficient, we mean that we avoid generic zero-knowledge techniques. Such efficient constructions were previously only known from linear secret sharing schemes (LSSS). It is believed that the class of access structures that can be handled with polynomial size LSSS is incomparable to the class that can be recognized by polynomial size monotone circuits, so in this sense we extend the class of access structures with efficient and non-interactive VSS.

1 Introduction

Secret-Sharing. Secret sharing schemes are fundamentally important tools in many areas of cryptography, because they allow us to strike a balance between confidentiality and security against loss of data, by storing shares of the data in separate locations. This is useful, e.g., when storing cryptographic keys.

A secret sharing scheme is defined by two algorithms: a probabilistic *sharing* algorithm which takes a secret message m as input and produces n shares $s_1, ..., s_n$; and a *reconstruction* algorithm that takes a subset of shares $\{s_i | i \in I\}$ as input and outputs m, provided I is a *qualified* set. The family of qualified sets is called the *access* structure of the scheme. An access structure $\mathcal{A}$ is always monotone: if $I \in \mathcal{A}$ and $I \subset J$ then $J \in \mathcal{A}$. We also require that if a set J is not qualified, then the subset of shares $\{s_i | i \in J\}$ gives no information on m. More precisely, in a perfect scheme, such a subset has distribution independent of m. Or in case of *computational* secret sharing, an unqualified set of shares has a distribution that can be simulated with computationally indistinguishable distribution without knowing m. In a perfect scheme, shares must be at least as large as the secret, while in computational secret sharing they can be much smaller, which is a main motivation for considering computational schemes.

Work done while visiting Aarhus University.

D. Pointcheval et al. (Eds.): AFRICACRYPT 2016, LNCS 9646, pp. 225–244, 2016.
DOI: 10.1007/978-3-319-31517-1_12

Naturally, we want the total size of the shares to be minimal, and certainly polynomial in n, and also we would like the sharing algorithm to run in time polynomial in n.[1]

Access Structures. In the following, we will speak of the access structure *characterised* by (for instance) a monotone Boolean circuit with n input bits: we think of the input bits as being in 1-1 correspondence with n players, and a player subset I can then be translated to a bit string by setting the bits corresponding to members of I to 1 and the rest to 0. The access structure characterised by the formula now consists of those subsets whose corresponding bit string causes the formula to output 1. This notion generalises naturally to any other computational model that computes monotone functions of bit strings.

Verifiable Secret-Sharing. One well-known and natural extension of secret sharing is *verifiable* secret sharing (VSS), where the party generating the shares (called the *dealer* in the following), and in addition some unqualified subset of the players (or shareholders) may be corrupted by a malicious adversary. We now execute a protocol in which the dealer sends shares to the players and some verification is performed. If the sharing phase is successful, all honest players must output a share, and otherwise they all reject. Later, any qualified player subset can go together and reconstruct the secret.

We now want the following properties: if the dealer is corrupt and the sharing phase is successful, the secret is well defined in the sense that any qualified subset will later reconstruct the same secret. If the dealer is honest, the sharing is always successful, and the secret is what the dealer intended; furthermore no unqualified player subset has any information on the secret (information theoretically or computationally). A VSS can be seen as a distributed commitment scheme that allows to open the committed information even if the dealer is not present.

This is an important property and can prevent different attacks depending on the scenario: one can imagine using a VSS as a distributed commitment scheme where we want the commitment to be binding. Think of a powerful cheating dealer who can "shut down" players arbitrarily; VSS guarantees that such a dealer cannot change its mind and control the output of the reconstruction by shutting down selected players. A perhaps more practically oriented application is in multiparty computation: consider a client with some secret input that he wants to supply to a multiparty computation run by a set of servers. A natural solution that will handle malicious attacks is to VSS the input among the servers. However, if this takes place in an asynchronous environment like the Internet, we clearly need a scheme with as little interaction as possible, ideally a client should be able to just post some information and then leave.

Non-Interactive VSS. The amount of interaction needed for the sharing phase of a VSS depends on the model for communication that is assumed. In this paper, we focus on the model where players have public encryption keys and

---

[1] However, since there are doubly exponentially many (families of) access structures, an easy counting argument shows that we cannot hope to handle all access structures with polynomial time sharing algorithms.

hold corresponding secret keys, and where no secure channels are assumed to be given "for free". But we assume that the dealer can publish information that all honest players will agree on (using, e.g., a bulletin-board or a broadcast channel). In this model, we clearly cannot get security without making computational assumptions. Moreover, the best we can hope for in terms of interaction is that the dealer publishes a single message, and each player then computes his output from this message and his secret key. We call such a scheme *non-interactive*. A slightly stronger property that many *non-interactive* schemes are born to satisfy is *public verifiability*. Namely, anyone, even outside the scheme, but with access to the public information, can perform the verification. One may also aim for a weaker property called *non-interactive with complaints*, where the scheme is only non-interactive if corrupt players behave honestly – but (motivated by the Internet scenario explained above) we require that even if interaction is needed, the dealer does not need to take part in this.

Our Contributions. In this paper, we present two VSS protocols with computational security, the first is non-interactive with complaints, while the second is non-interactive and *publicly verifiable* in the random oracle model (used for the Fiat-Shamir heuristic). Both schemes are built on top of the same *locally verifiable* scheme, which is our main technical contribution and is based on the standard RSA assumption.

The complexity of the scheme is polynomial in the size of a monotone Boolean circuit characterising the access structure. It is the first scheme with this property and it allows us to efficiently handle access structures that cannot be done efficiently with linear schemes. We emphasise that although we assume that some set-up information is available, we do not use generic non-interactive zero-knowledge (which could be used to solve the problem in a rather trivial way). In particular the communication complexity of our scheme does not depend on the circuit complexity of the dealer's computation when he generates shares, only on the security parameter and the number of fan-outs in the circuit characterising the access structure.

Related Work. The notion of information theoretic secret sharing was independently discovered by Blakley [Bla79] and by Shamir [Sha79], who constructed efficient schemes for simple threshold access structures. The introduction of general secret sharing is due to Ito, Saito and Nishiziki [ISN89]. Later, Benaloh and Leichter [BL90] constructed schemes that are efficient in the size of a Boolean formula characterising the target access structure. Karchmer and Wigderson [KW93] introduced the notion of a monotone linear span program (MSP) and showed that any MSP induces a linear secret sharing scheme (with complexity polynomial in the size of the MSP) for the access structure characterised by the MSP. In fact, any linear secret sharing scheme can be seen as being derived from an MSP, including the schemes by Shamir and Benaloh-Leichter.

This gives us efficient secret sharing schemes for any access structure that can be characterized by an MSP of size polynomial in the number of players. If one now wants to extend the class of access structures we can handle efficiently, it is natural to consider those characterised by polyomial size Boolean *circuits* (rather than formulas as considered in [BL90]). The reason for this is that the classes of access structures characterised by polynomial size MSPs, respectively monotone Boolean circuits are incomparable as far as we know, and hence being able to construct secret sharing schemes efficiently from monotone circuits would indeed give us something new. However, we know no such construction resulting in a perfect scheme. On the other hand, Vinod et al. [VNS+03] proposed a construction yielding computational security (based on an unpublished idea by Yao [Yao89]).

As for construction of verifiable secret sharing schemes, all constructions we know of are schemes that start from a regular secret sharing scheme and add verifiability on top. In a model where there are secure point-to-point channels between all pairs of players, this can be done with perfect security for any linear scheme, under certain conditions on the access structure [CDM00]. One can even convert any secret sharing in a black-box fashion to VSS with statistical security [CDD00] but the construction uses a lot of interaction and generic zero-knowledge techniques to some extent. In the communication model we consider in this paper, any linear scheme cane made verifiable under computational assumptions. The basic idea was first proposed by Pedersen [Ped92] based on Shamir's scheme, but the principle easily extends to any linear scheme. The resulting VSS is "almost" non-interactive with complaints, i.e., the dealer needs to help resolve conflicts if there are complaints. This problem was resolved in [DT07]. The notion of publicly verifiable secret sharing is introduced by Stadler in [Sta96].

Technical Overview. Our main result is to extend the class of access structures we can handle efficiently and verifiably, in the same way that [VNS+03] extended what we can do with regular secret sharing. We do so by constructing a verifiable version of the scheme from [VNS+03][2]. Prior to our work, no such construction was known.

The idea is that the dealer runs (a version of) the sharing scheme from [VNS+03], resulting in each player receiving his share. But in addition he also makes public a "tag" for each share (which is constructed in such a way as to preserve computational privacy). Every shareholder can check the share they received against the public tag. Furthermore, it is now possible (very simplistically speaking) to run the reconstruction algorithm on the set of all tags and make checks under way such that if the "homomorphic reconstruction" does not abort, then the set of actual shares is indeed a consistent sharing of a well defined secret.

We obtain this by constructing a symmetric encryption scheme (as required in the secret sharing scheme from [VNS+03]) and a tagging scheme that "lives" in the same RSA group and has certain convenient homomorphic properties.

This does not yet ensure that everything will always be fine: the dealer could make a good set of tags, but send an incorrect share to one or more of the players. A player P can detect that he has been sent a bad share, but this will

---

[2] On the way to our result, as a secondary contribution, we also prove that the construction of [VNS+03] satisfies a strong, simulation based notion of privacy, while the original paper only argues for a weaker, "one-way" definition of privacy.

not be clear to the other players: they have only seen a ciphertext meant for P and cannot tell if the dealer is corrupt or P is complaining for no good reason.

We can resolve this in two different ways, in both cases by using a specific encryption scheme for sending shares to players. The first approach is to use an encryption scheme that has so-called verifiable opening, allowing P to reveal his share along with a proof that this was indeed what the dealer sent him in encrypted form. This technique was introduced in [DT07] and formalized in [DHKT08]. This gives a scheme that is non-interactive with complaints. The second approach is to use an encryption scheme we suggest that is designed to allow the dealer to give an efficient proof (in the random oracle model) that the shares he encrypts indeed correspond to the tags he publishes. This gives a non-interactive scheme with only a small computational overhead compared to the work needed to compute shares in a non-verified way.

Roadmap. Section 2 introduces the notation and some preliminaries which will be used in the rest of the paper. Section 3 briefly reviews the computational secret sharing scheme of Vinod et al. Section 4 introduces the notion of locally verifiable secret sharing and presents our novel construction. Finally in Sect. 5 we discuss how to combine the locally verifiable scheme with different kind of encryption schemes to achieve a *non-interactive publicly verifiable* scheme and a *non-interactive scheme with complaints*.

2   Notation and Preliminaries

Basic Notation. We use the shorthand $[m, n]$ for $\{m, m + 1, \ldots, n - 1, n\}$ and $[n]$ for $[1, n]$. If S is a set $x \leftarrow S$ is a random element from S, if A is an algorithm $y \leftarrow A(x)$ is the output of A run on input x on an uniformly random tape. If $S \subseteq [n]$, then $s = \mathsf{set2bits}(S)$ is an n-bit string where the i-th bit of s is 1 iff $i \in S$. We call a function $f : \mathbb{N} \to \mathbb{R}$ *negligible* if for all c, for all big enough κ, $f(\kappa) < \kappa^{-c}$. We use $\mathsf{negl}(\kappa)$ for a generic negligible function. We say two families of distributions $U_0 = \{U_{0,\kappa}\}_{\kappa \in \mathbb{N}}, U_1 = \{U_{1,\kappa}\}_{\kappa \in \mathbb{N}}$ are *computationally indistinguishable* if for all probabilistic polynomial time (PPT) distinguisher D, $|\Pr[D(U_{b,\kappa}) = b] - \frac{1}{2}| = \mathsf{negl}(\kappa)$ and we write $U_0 \approx_c U_1$ for short.

Access Structure. An access structure $\mathcal{A}$ of $[n]$ is a monotone subset $\mathcal{A} \subseteq 2^{[n]}$. Given a set $I \subseteq [n]$ we say that I is a *qualified set* if $I \in \mathcal{A}$ or that I is an *unqualified set* if $I \notin \mathcal{A}$. We say that a Boolean circuit $C : \{0,1\}^n \to \{0,1\}$ *describes* $\mathcal{A}$ if $C(\mathsf{set2bits}(I)) = 1 \Leftrightarrow I \in \mathcal{A}$.

Circuits. We use the following notation for circuits: a *circuit* $C : \{0,1\}^n \to \{0,1\}$ is described by $\ell > n$ wires $\{w_1, \ldots, w_\ell\}_{i \in [\ell]}$ and λ gates $\{g_i\}_{i \in [\lambda]}$. We call $\{w_i\}_{[1,n]}$ the *input wires*, $\{w_i\}_{[n+1,\ell-1]}$ the *internal wires* and we call w_ℓ the *output wire*. Every internal wire connects exactly two gates, while the input wires and the output wire are only connected to one gate (therefore, there are exactly $\lambda = (2\ell + n + 1)/3$ gates). Wires carry values: at the beginning all wires are initialized to $\perp$, and we say a wire is *assigned* if $w_i \neq \perp$.

Each *gate* g_i is described by a tuple from $[\ell]^3$ (representing pointers to their input/output wires) and a type $\mathsf{type}_i \in \{\mathsf{and}, \mathsf{or}, \mathsf{fanout}\}$, which determines the semantic of the wires: if g_i has type $\mathsf{type}_i \in \{\mathsf{and}, \mathsf{or}\}$, then g_i has two input wires $(w_{i_{LI}}, w_{i_{RI}})$ (for left input and right input) and one output wire w_{i_O}. Finally, if g_i has type $\mathsf{type}_i = \mathsf{fanout}$, then g_i has one input wire w_{i_I} and two output wires $(w_{i_{LO}}, w_{i_{RO}})$. We assume (without loss of generality) that the output wire w_ℓ is not the output of a fan-out gate.

We say that a gate is *ready to be evaluated* if: both $w_{i_{LI}} \neq \bot$ and $w_{i_{RI}} \neq \bot$ when $\mathsf{type}_i = \mathsf{and}$; either $w_{i_{LI}} \neq \bot$ or $w_{i_{RI}} \neq \bot$ when $\mathsf{type}_i = \mathsf{or}$; $w_{i_I} \neq \bot$ when $\mathsf{type}_i = \mathsf{fanout}$. Finally, we say that a gate is *assigned* if all its output wires have been assigned.

The type of a gate determines how a circuit is evaluated. To exemplify our notation, we describe how to evaluate a simple Boolean circuit $C : \{0,1\}^n \to \{0,1\}$ on input $x \in \{0,1\}^n$.

1. Parse the input $x = (x_1, \ldots, x_n) \in \{0,1\}^n$ and assign $w_i = x_i$ for all $i \in [n]$;
2. While $w_\ell = \bot$, find the first gate g_i which is *ready to be evaluated*:
   - If $\mathsf{type}_i = \mathsf{and}$: Assign $w_{i_O} = w_{i_{LI}} \wedge w_{i_{RI}}$;
   - If $\mathsf{type}_i = \mathsf{or}$: Assign $w_{i_O} = w_{i_{LI}} \vee w_{i_{RI}}$;
   - If $\mathsf{type}_i = \mathsf{fanout}$: Assign $w_{i_{LO}} = w_I$ and $w_{i_{RO}} = w_I$;
3. Output w_ℓ;

Finally, we call $(\ell_{\mathsf{and}}, \ell_{\mathsf{or}}, \ell_{\mathsf{fanout}})$ respectively the number of $(\mathsf{and}, \mathsf{or}, \mathsf{fanout})$ gates in the circuit.

3   Computational Secret Sharing

In this section we review the basic definitions of a computational secret-sharing scheme and present the construction of Vinod et al. [VNS+03] in our notation. A *computational secret-sharing scheme* (CS3) is a tuple of algorithms $\pi = (\mathsf{Setup}, \mathsf{Share}, \mathsf{Rec})$ which are defined and used as follows:

Setup: The randomized setup algorithm $pp \leftarrow \mathsf{Setup}(1^\kappa)$ (run once and for all) outputs some public system parameters pp for the secret sharing scheme (which contains, among other things, some message space $\mathcal{M}$ from which the secret can be chosen)[3].

Share: A dealer can share a secret message $m \in \mathcal{M}$ with n parties $P_1, \ldots, P_n$ according to an access structure described by a circuit $C : \{0,1\}^n \to \{0,1\}$ by running the randomized algorithm $(s_0, s_1, \ldots, s_n) \leftarrow \mathsf{Share}_{pp}(C, m)$ which outputs $n + 1$ shares $s_0, s_1, \ldots, s_n$. The dealer sends to each party P_i the shares (s_0, s_i). Sometimes we refer to s_0 as the *public share* and to the s_i's, $i \in [n]$ as the *private shares*.

---

[3] In case where no trusted party exists to run this setup, a secure computation protocol can be used instead. We note that our setup algorithm will output an RSA modulus, and that several efficient protocols for this task exist, depending on the desired security guarantees and threshold.

Reconstruct: A set of parties $\{P_i\}_{i \in I}$ such that $C(\mathsf{set2bits}(I)) = 1$ can reconstruct the secret message m by running $m \leftarrow \mathsf{Rec}_{pp}(s_0, \{s_i\}_{i \in I})$.

Intuitively, we want such a scheme to be *correct* (any qualified set of parties can reconstruct the secret m) and *private* (any unqualified set of parties does not learn any information about m). This can be formalized as follows:

Definition 1 (Correctness). *A CS3 π is* correct *if for all $m \in \mathcal{M}$, for all circuits C describing an access structure $\mathcal{A}$, and for all $I \in \mathcal{A}$,*

$$\Pr[m \neq \mathsf{Rec}_{pp}(s_0, \{s_i\}_{i \in I})] \leq \mathsf{negl}(\kappa)$$

where $(s_0, s_1, \ldots, s_n) \leftarrow \mathsf{Share}_{pp}(C, m), pp \leftarrow \mathsf{Setup}(1^\kappa)$ and the probabilities are taken over the choices of all algorithms.

In the following definition we ask for privacy in a strong, simulation based sense, while in the original work [VNS+03] only a weaker "one-way" version of privacy is considered.

Definition 2 (Privacy). *A CS3 π is* private *if for all circuits C describing an access structure $\mathcal{A}$, and for all $I \notin \mathcal{A}$, there exist a PPT simulator Sim such that for all $m \in \mathcal{M}$:*

$$\mathsf{Sim}(pp, C, I) \approx_c (s_0, \{s_i\}_{i \in I})$$

where $pp \leftarrow \mathsf{Setup}(1^\kappa)$, $(s_0, s_1, \ldots, s_n) \leftarrow \mathsf{Share}_{pp}(C, m)$

Remarks on the Model. Note that at this point we make the assumption that there are secure point-to-point channels between the dealer and the parties. This assumption will be removed in Sect. 5 where we will show two techniques for distributing the shares which make the overall scheme verifiable against a malicious dealer.

Constructing CS3s. Vinod et al. [VNS+03] proposed a CS3 with the following communication complexity: $|s_0| = O(|C| + \kappa \cdot \ell_{\mathsf{fanout}})$ and $s_i = O(\kappa)$ for all $i \in [n]$. Their scheme uses a symmetric encryption scheme $(\mathsf{G}, \mathsf{E}, \mathsf{D})$, where the key space and message space of the encryption scheme are the group used by the secret sharing scheme (for instance, the group of κ-bit strings with bitwise XOR of strings as the group operation).

Setup: The setup algorithm outputs a cyclic group $\mathbb{G}$ (which we write here in multiplicative notation) which is used both as the message space $\mathcal{M}$ and as the working group for the scheme, as well as an IND-CPA secure symmetric encryption scheme $(\mathsf{G}, \mathsf{E}, \mathsf{D})$ where the key space and the message space is $\mathbb{G}$.

Share: To share a secret $m \in \mathbb{G}$ with an access structure described by a circuit C among n parties, the dealer runs the following algorithm:
1. Assign $w_\ell = m$ and let $s_0 = C$;
2. While $w_j = \bot$ for some $j \in [n]$ (the input wires), find the first *assigned* gate g_i and:

- If type$_i$ = and: Secret share the value of the output wire between the two input wires i.e., pick a random $w_{i_{LI}} \leftarrow \mathbb{G}$ and assign $w_{i_{RI}} = w_{i_O} \cdot (w_{i_{LI}})^{-1}$;
- If type$_i$ = or: Copy the value of the output wire to both input wires i.e., assign $w_{i_{LI}} = w_{i_O}$ and $w_{i_{RI}} = w_{i_O}$;
- If type$_i$ = fanout: Assign a fresh key to the input wire, and append encryptions of the output wires to the public share i.e., compute $k \leftarrow G(1^\kappa)$, $c_{i_{LO}} = E(k, w_{i_{LO}})$ and $c_{i_{RO}} = E(k, w_{i_{RO}})$; finally, assign $w_{i_I} = k$ and let $s_0 = s_0 || (i, c_{i_{LO}}, c_{i_{RO}})$;

3. Let $s_i = w_i$ for all $i \in [n]$, and output $(s_0, s_1, \ldots, s_n)$;

Reconstruct: To reconstruct a secret m given a set of shares $(s_0, \{s_i\}_{i \in I})$ from some qualified set I run the following algorithm:

1. Assign $w_i = s_i$ for all $i \in I$ and recover C from s_0.
2. While $w_\ell = \bot$, find the first gate g_i which is *ready to be evaluated* and:
   - If type$_i$ = and: Assign $w_{i_O} = w_{i_{LI}} \cdot w_{i_{RI}}$;
   - If type$_i$ = or: Assign $w_{i_O} = w_{i_{LI}}$ if $w_{i_{LI}} \neq \bot$ or $w_{i_O} = w_{i_{RI}}$ otherwise;
   - If type$_i$ = fanout: Recover $(i, c_{i_{LO}}, c_{i_{RO}})$ from s_0 and assign

$$w_{i_{LO}} \leftarrow D(w_{i_I}, c_{i_{LO}}) \text{ and } w_{i_{RO}} \leftarrow D(w_{i_I}, c_{i_{RO}})$$

3. Output w_ℓ;

The scheme is correct by inspection. Privacy can be proven by constructing a simulator Sim who runs the sharing scheme for a random secret m', and then arguing that any distinguisher can be used to break the IND-CPA security of the underlying encryption scheme. We will prove this in detail for the locally verifiable variant of this scheme construction (see proof of Theorem 1).

4   Locally Verifiable Secret-Sharing Scheme

In this section we show how to make the CS3 of Vinod et al. [VNS+03] *verifiable* i.e., even if the dealer is corrupt, we want to make sure that the secret message is well defined. In particular, we need to make sure that the output of the reconstruction phase does not depend on which of the possibly many set of qualified parties reconstructs the secret. Thus we define a *locally verifiable* computational secret-sharing scheme (VCS3) by adding an algorithm as follows:

Verification: The randomized algorithm $f \leftarrow \text{Ver}_{pp}(s_0, s_i)$ outputs a flag bit $f \in \{\text{true}, \text{false}\}$ which indicates whether (s_0, s_i) is a valid share for party P_i.

We now ask the following property:

Definition 3 (Local Verifiability). *We say a CS3 scheme is locally verifiable if for all $n \in \mathbb{Z}$, for all circuits C describing an access structure, and for all PPT algorithms D^* the following holds: Let $pp \leftarrow \text{Setup}(1^\kappa)$ and $(s_0, s_1, \ldots, s_n) \leftarrow D^*(pp)$. If $\forall i \in [n], \text{Ver}_{pp}(s_0, s_i) = \text{true}$, then with a overwhelming probability there exists a value $m \in \mathcal{M}$ such that $\text{Rec}_{pp}(s_0, \{s_i\}_{i \in I}) = m$ for all qualified sets $I \in \mathcal{A}$ (where $\mathcal{A}$ is defined in s_0).*

Remarks on the Model. It is clear that without some degree of interaction between the parties, it is impossible to achieve even a locally verifiable scheme. Think of a setting with two parties P_1, P_2 and a simple access structure $\mathcal{A} = \{\{P_1\}, \{P_2\}\}$. Now the dealer can simply send $s_1 \neq s_2$ to the two parties, which will therefore reconstruct to two different secrets. Therefore in this section we make the assumption that there is a broadcast channel from the dealer to the parties, which ensures that all parties P_i receive the same public share s_0. The private shares s_i's are still sent over private channels from the dealer to the parties. Note that local verifiability does not say anything about what to do when one of the parties rejects her share. We deal with complaints later in Sect. 5.

Feasibility of Locally Verifiable CS3. Note that it would be possible to enhance any CS3 scheme (such as the one presented in Sect. 3) with the local verifiability property described above by sending to each party, together with s_i, a non-interactive zero-knowledge proof (NIZK) that s_i was computed correctly (this could be achieved by generating the crs for the NIZK during the setup phase, and letting the dealer append a commitment to m and the randomness used in Share, and then the Ver algorithm simply checks the NIZK). However the communication of the resulting verifiable scheme would depend on the dealer's local computation (i.e., the circuit complexity of the original Share algorithm) and thus add a very significant overhead. In the following, we look for a solution which avoids this problem and essentially preserves both the communication and computational complexity of the original scheme by Vinod et al.

Locally Verifiable CS3. We present here our locally verifiable CS3. The scheme is based on the standard RSA assumption. Intuitively, we make the scheme locally verifiable by having the dealer publish some "tags" of all the private shares in the public share, using some function $t_i = \mathsf{Tag}(s_i)$. Now, since Tag is deterministic, every party can check that their private share is consistent with the public tag. In addition, the Tag function and the Rec_{pp} function are designed so that they "commute", meaning that (from a very high level point of view) it is possible to compute the reconstruction function on the tags (instead of the actual values) and verify if the obtained tag is equal to the published one i.e.,

$$\mathsf{Tag}(m) = \mathsf{Rec}_{pp}(s_0, \mathsf{Tag}(s_1), \ldots, \mathsf{Tag}(s_n))$$

4.1 Building Blocks

The Group. The scheme works in an RSA group $\mathbb{Z}_N^*$ where the RSA modulus is generated during the setup phase (hence its factorization is unknown to both the dealer and the parties) (see footnote 3). All operations are carried out in the group $\mathbb{Z}_N^*$, hence if $x \in \mathbb{Z}_N^*$ and $e \in \mathbb{Z}$ we write "$y = x^e$" instead of "$y = x^e$ mod N".

The Tags. The scheme uses a "tag" function $\mathsf{Tag}(x) = x^\tau$ where τ is a prime number larger than N which is generated by the dealer – it is easy to see that Tag is multiplicatively homomorphic i.e., $\mathsf{Tag}(x) \cdot \mathsf{Tag}(y) = \mathsf{Tag}(x \cdot y)$.

The Encryption Scheme. We also use a symmetric encryption scheme $(\mathsf{G}, \mathsf{E}, \mathsf{D})$ where G outputs a random $k \leftarrow \mathbb{Z}_N^*$; The encryption function $c \leftarrow \mathsf{E}(k, m)$ chooses a random prime $\rho > n$ and outputs it together with $\sigma = k^\rho \cdot m$; The decryption function $m \leftarrow \mathsf{D}(k, c)$ outputs $m = \sigma \cdot k^{-\rho}$. We note a useful property of our encryption scheme and the Tag function, namely that they commute nicely: if $\mathsf{D}(k, (\rho, \sigma)) = m$ then $\mathsf{D}(\mathsf{Tag}(k), \rho, \mathsf{Tag}(\sigma)) = \mathsf{Tag}(m)$ since

$$\mathsf{D}(\mathsf{Tag}(k), \rho, \mathsf{Tag}(\sigma)) = \mathsf{Tag}(\sigma) \cdot \mathsf{Tag}(k)^{-\rho} = \sigma^\tau \cdot (k^\tau)^{-\rho} = (\sigma \cdot k^{-\rho})^\tau$$
$$= \mathsf{Tag}(m) \tag{1}$$

Later in the proof we will need the following property from this scheme (intuitively, the Lemma says that the scheme is *one-way secure under single-query chosen plaintext attack*):

Lemma 1. *Consider the following game: (1) a challenger runs $k \leftarrow \mathsf{G}(1^\kappa)$ (2) the adversary picks a value $m \in \mathbb{Z}_N^*$; (3) the challenger picks a random $r \in \mathbb{Z}_N^*$ and sends $(\mathsf{E}(k, m), \mathsf{E}(k, r))$ to the adversary; (4) the adversary outputs r'; For all PPT adversary, $r' \neq r$ except with negligible probability if the RSA problem is hard.*

Proof. In step (3) the adversary receives a 4-tuple from $\mathbb{Z}_N^*$ composed of

$$(\rho_0, \sigma_0 = k^{\rho_0} \cdot m, \rho_1, \sigma_1 = k^{\rho_1} \cdot r)$$

We first claim that an adversary who computes $r' = r$ can be used to compute k efficiently in the following way: let a, b be the values such that $a \cdot \rho_0 + b \cdot \rho_1 = 1$ (which are guaranteed to exist since ρ_0, ρ_1 are different primes). Then

$$(\sigma_0 \cdot m^{-1})^a \cdot (\sigma_1 \cdot r^{-1})^b = k^{a \cdot \rho_0 + b \cdot \rho_1} = k$$

Now the reduction solves the RSA instance $(e, y = x^e)$ by setting $\rho_0 = e, \sigma_0 = m \cdot y$, sampling a random prime ρ_1 and a random element σ_1 from $\mathbb{Z}_N^*$. Note that this is the exact distribution that the adversary was expecting since this is equivalent to the choice of a random $r = y^{-1} \cdot \sigma_1$ in the game, and r is uniformly distributed in $\mathbb{Z}_N^*$ (unless some of the random choices are not invertible mod N, but in that case the reduction can trivially factor N).[4]

The Extractor. Finally, $\mathsf{Ext} : \mathbb{Z}_N^* \to \{0, 1\}^\mu$ is an extractor which extracts the $\mu = \log(\kappa)$ hard-core bits of the RSA function from some value in $\mathbb{Z}_N^*$ (the least significant $\log(\kappa)$ bits will do [ACGS88]).

---

[4] (Note that the scheme would not be secure if the adversary could make 2 CPA queries, since in that case it could recover k in the same way as the reduction does.).

4.2   The Construction

We are now ready to give the details of our construction:

Setup: Generate an RSA modulus N and a random prime number $\tau > N$ which defines the function $\mathsf{Tag}(x) = x^\tau$ and output $pp = (N, \tau)$;

Share: To share a secret $m \in \{0,1\}^\mu$ with an access structure described by a circuit C among n parties, the dealer runs the following algorithm.

1. Assign $w_\ell \leftarrow \mathbb{Z}_N^*$
2. Compute $u = \mathsf{Ext}(w_\ell) \oplus m$;
3. Compute $t_\ell = \mathsf{Tag}(w_\ell)$;
4. Let $s_0 = (C, u, \tau, t_\ell)$;
5. While $w_j = \bot$ for some $j \in [n]$ (the input wires), find the first *assigned* gate g_i and:
   - If $\mathsf{type}_i = $ and: Share the value of the output wire between the two input wires i.e., pick a random $w_{i_{LI}} \leftarrow \mathbb{Z}_N^*$ and assign $w_{i_{RI}} = w_{i_O} \cdot (w_{i_{LI}})^{-1}$; In addition, compute the tags for two input wires $t_{i_{LI}} = \mathsf{Tag}(w_{i_{LI}})$ and $t_{i_{RI}} = \mathsf{Tag}(w_{i_{RI}})$;
   - If $\mathsf{type}_i = $ or: Copy the value of the output wire to both input wires i.e., assign $w_{i_{LI}} = w_{i_O}$ and $w_{i_{RI}} = w_{i_O}$; In addition, copy $t_{i_{LI}} = t_{i_O}$ and $t_{i_{RI}} = t_{i_O}$;
   - If $\mathsf{type}_i = $ fanout: Assign a fresh key to the input wire, and append encryptions of the output wires to the public share i.e., compute $k \leftarrow \mathsf{G}(1^\kappa)$, $c_{i_{LO}} \leftarrow \mathsf{E}(k, w_{i_{LO}})$ and $c_{i_{RO}} \leftarrow \mathsf{E}(k, w_{i_{RO}})$; finally, assign $w_{i_I} = k$, compute $t_{i_I} = \mathsf{Tag}(w_{i_I})$ and let $s_0 = s_0 || (c_{i_{LO}}, c_{i_{RO}}, t_{i_{LO}}, t_{i_{RO}})$;
6. Let $s_i = w_i$ for all $i \in [n]$, append $s_0 = s_0 || (t_1, \ldots, t_n)$;

Verification: Upon receiving (s_0, s_i) party P_i runs the algorithm $\mathsf{Ver}_{pp}(s_0, s_i)$ described here:

1. From s_0, recover (C, u, τ, t_ℓ), t_i for every $i \in [n]$, and $c_{i_{LO}}, c_{i_{RO}}, t_{i_{LO}}, t_{i_{RO}}$ for every fanout gate g_i;
2. If $t_i \neq \mathsf{Tag}(s_i)$ stop and output `false`; else:
3. Assign $w_i = t_i$ for all $i \in [n]$.
4. While $w_\ell = \bot$, find the first gate g_i whose input wires are all assigned and:
   - If $\mathsf{type}_i = $ and: Assign $w_{i_O} = w_{i_{LI}} \cdot w_{i_{RI}}$;
   - If $\mathsf{type}_i = $ or: If $w_{i_{LI}} \neq w_{i_{RI}}$ stop and output `false`; else assign $w_{i_O} = w_{i_{LI}}$;
   - If $\mathsf{type}_i = $ fanout: Parse $(c_{i_{LO}}, c_{i_{RO}}) = (\rho_{i_{LO}}, \sigma_{i_{LO}}, \rho_{i_{RO}}, \sigma_{i_{RO}})$. If $\mathsf{D}(w_{i_I}, (\rho_{i_{LO}}, \mathsf{Tag}(\sigma_{i_{LO}}))) \neq t_{i_{LO}}$ or $\mathsf{D}(w_{i_I}, (\rho_{i_{RO}}, \mathsf{Tag}(\sigma_{i_{RO}}))) \neq t_{i_{RO}}$ stop and output `false`; else assign $w_{i_{LO}} = t_{i_{LO}}$ and $w_{i_{RO}} = t_{i_{RO}}$.
5. Stop and output `false` if $w_\ell \neq t_\ell$;
6. Else output `true`;

Reconstruct: To reconstruct a secret m given a set of shares $(s_0, \{s_i\}_{i \in I})$ from some qualified set I run the following algorithm.

1. Assign $w_i = s_i$ for all $i \in I$ and recover C from s_0.
2. While $w_\ell = \bot$, find the first gate g_i which is *ready to be evaluated* and:

- If $\mathsf{type}_i = \text{and}$: Assign $w_{i_O} = w_{i_{LI}} \cdot w_{i_{RI}}$;
- If $\mathsf{type}_i = \text{or}$: If $w_{i_{LI}} \neq \perp$, assign $w_{i_O} = w_{i_{LI}}$, else $w_{i_O} = w_{i_{RI}}$.
- If $\mathsf{type}_i = \text{fanout}$: Recover $(c_{i_{LO}}, c_{i_{RO}})$ from s_0 and assign

$$(w_{i_{LO}}, w_{i_{RO}}) \leftarrow (\mathsf{D}(w_{i_O}, c_{i_{LO}}), \mathsf{D}(w_{i_I}, c_{i_{RO}}))$$

3. Output $m = u \oplus \mathsf{Ext}(w_\ell)$;

The scheme satisfies *correctness* by inspection. We here state the theorems about the *privacy* and *local verifiability*, as well as giving a brief high-level overview of the proofs. The full proofs are presented in the following two subsections.

Theorem 1. *This construction is* private *according to Definition 2 under the assumption that the RSA problem is hard.*

The proof of this theorem proceeds in the following steps: it can be seen that the scheme is secure if the circuit contains no fanout gates, since in this case (roughly speaking) the adversary is given $(x^\tau, \mathsf{Ext}(x) \oplus m)$ and is asked to output some information about m. Any such adversary can be used to break the RSA function x^τ since Ext extracts the hard-core bits. To deal with the fanout gates we construct a series of hybrids where at each step we decompose the circuit C into a circuit C^* which does not contain any fanout gate and a circuit C' which takes two extra inputs (both set to be the output of C^* and one fanout gate less than the original circuit C. It is possible to argue that an adversary which succeeds in breaking the security for the original circuit C can be used to break the security of the scheme run on the circuit C' with one less fanout gate, roughly thanks to the security of the encryption scheme used in the construction of the fanout gates and the fact that the rest of the circuit C^* does not contain any fanout gate.

Theorem 2. *This construction is* locally verifiable *according to Definition 3.*

The proof of this theorem proceeds by first noting that the tag function is a permutation and hence the set of tags uniquely defines a set of numbers mod N that are supposed to act as the "wire values" w_i in the circuit C. One then checks that the verification ensures that the values on wires going into a gate correctly correspond to the value on the output wire. For fanout gates this check crucially relies on the observation above (1), that the tag function commutes with the encryption scheme. Hence, if everything checks out, the public data must correspond to a correct execution of the sharing algorithm, and therefore by the correctness property all qualified sets will reconstruct the same secret.

4.3   Proof of Privacy (Theorem 1)

Proof. We construct the simulator Sim as following:

1. Sim samples $r \leftarrow \{0, 1\}^\mu$;

2. Sim runs $(s_0, s_1, \ldots, s_n) \leftarrow \mathsf{Share}_{pp}(C, r)$
3. Sim outputs $(s_0, \{s_i\}_{i \in I})$.

We now prove that the output of Sim is computationally indistinguishable from the distribution of the real output.

From Decision to Search. Since τ is a prime number larger than N the function Tag is a one-way permutation. The only difference between a the real view and the simulated view is that in the real view $u = \mathsf{Ext}(w_\ell) \oplus m$ whereas in the simulated view u is a uniformly random value. Since Ext extracts the hard core bits of the RSA function Tag, any distinguisher $\mathcal{D}$ that distinguishes between the real view and the simulated view can be turned into an algorithm $\mathcal{D}'$ that outputs w_ℓ. We now prove that computing w_ℓ is impossible without breaking the RSA assumption.

Without Fanout Gates. We start by noting that if the circuit C had no fanout gates, then it is computationally hard to find the value w_ℓ: Without fanout gates the value corresponding to the output wire w_ℓ is simply the product of a subset of the input wires w_i i.e., $w_\ell = \prod_{i \in S} s_i$ for some set S (note the set may not be unique, in which case we consider the first such set in lexicographical order). In the privacy game the adversary only sees shares for an unqualified set i.e., the adversary receives $\{s_i\}_{i \in I}$ for some I such that $C(\mathsf{set2bits}(I)) = 0$, meaning that $T = S \setminus I \neq \emptyset$. Now an adversary who computes w_ℓ can be turned into an adversary who computes $y = \prod_{i \in T} s_i$, which is equivalent to breaking the one-way property of the permutation Tag. In particular, since Tag is homomorphic, given such an adversary a reduction can solve an RSA instance $(\tau, y = x^\tau)$ by choosing $|T| - 1$ random s_i's, defining the last s_i such that $y = \prod_{i \in T} s_i$ and computing the respective tags. Note that the same argument can be made for any internal wire: let $C_i(x)$ be the circuit which outputs the same value as gate g_i in C, then if $C_i(\mathsf{set2bits}(I)) = 0$ then no adversary can output the value w_{i_O}.

Removing Fanout Gates. The core of the proof is to show how we can "get rid of" the fanout gates by decomposing C into circuits without any fanout gates. We proceed as follows: $C^*_{\ell_{\mathsf{fanout}}}$ is a circuit which takes an input of length $n^*_{\ell_{\mathsf{fanout}}} = n + 2\ell_{\mathsf{fanout}}$ wires (the input wires plus all the output wires of the fanout gates) and contains all gates that can be reached from the output wire without traversing any fanout gate. As the next step, we take the input wire of the first fanout gate encountered in the previous process, and we define $C^*_{\ell_{\mathsf{fanout}}-1}$ as the circuit which takes an input of length $n^*_{\ell_{\mathsf{fanout}}-1} = n + 2(\ell_{\mathsf{fanout}} - 1)$ and contains all gates that can be reached from this wire without traversing any fanout gate. The process stop with C^*_0 which is guaranteed to take an input of length at most $n^*_0 = n$, that is its input is the same as the original circuit C and in particular none of its inputs come from the output of any fanout gates.

We now define the values $x_{n+1} = x_{n+2} = C^*_0(x_1, \ldots, x_n)$, $x_{n+3} = x_{n+4} = C^*_1(x_1, \ldots, x_{n^*_1})$ up to $x_{n+2\ell_{\mathsf{fanout}}-1} = x_{n+2\ell_{\mathsf{fanout}}} = C^*_{\ell_{\mathsf{fanout}}-1}(x_1, \ldots, x_{n^*_{\ell_{\mathsf{fanout}}-1}})$. It is convenient to define, given a set I such that $x = \mathsf{set2bits}(I)$, a set I^* such that $\mathsf{set2bits}(I^*) = (x_1, \ldots, x_{n+2\ell_{\mathsf{fanout}}})$. It is now clear by inspection that

$$C(x) = C^*_{\ell_{\mathsf{fanout}}}(x_1, \ldots, x_{n^*_{\ell_{\mathsf{fanout}}}}) \quad \forall x$$

Crucially all the circuits $(C_0^*, \ldots, C_{\ell_{\text{fanout}}}^*)$ do not contain any fanout gates. We construct also circuits $(C_0', \ldots, C_{\ell_{\text{fanout}}}')$, where C_j' takes an input of size at most $n_j' = n + 2j$ and has exactly $\ell_{\text{fanout}} - j$ fanout gates. We define $C_{\ell_{\text{fanout}}}' = C_{\ell_{\text{fanout}}}^*$ and C_{j-1}' to be equal to

$$C_j'(x_1, \ldots, x_{n_{j-1}'}, C_j^*(x_1, \ldots, x_{n_{j-1}'}), C_j^*(x_1, \ldots, x_{n_{j-1}'}))$$

It is clear by inspection that $C = C_0'$.

After the heavy but necessary notation, we are ready for showing our reduction. We construct a series of adversaries $\mathcal{D}_j$ who get as input

$$(s_0, s_1, \ldots, s_{n+2j}) \leftarrow \mathsf{Share}_{pp}(C_j', r)$$

We have already argued that no adversary can output the value corresponding to the output wire (which is necessary to distinguish between the real and simulated execution) if the circuit does not contain any fanout gates, which implies that $\mathcal{D}_{\ell_{\text{fanout}}}$ cannot either without breaking the RSA assumption. We then show that if $\mathcal{D}_{j-1}$ succeeds in outputting the value of the output wire for C_{j-1}' with noticeable probability then we can construct $\mathcal{D}_j$ who outputs the value of the output wire for C_j' as well. Using standard hybrid arguments we can therefore conclude that the adversary $\mathcal{D} = \mathcal{D}_0$ can only succeed in breaking privacy by breaking the RSA assumption.

The reduction goes as follows: $\mathcal{D}_j$ gets as input all the shares s_i' for all $i \in I^* \cap [n + 2j]$, where $(s_0', s_1', \ldots, s_{n+2 \cdot j}') \leftarrow \mathsf{Share}_{pp}(C_j', r)$. Remember that $C(\mathsf{set2bits}(I)) = 0$.

Now the $\mathcal{D}_j$ needs to construct a sharing for the circuit C_{j-1}' which is of the format expected by $\mathcal{D}_{j-1}$. Intuitively this is done by adding a single fanout gate to the circuit and running the share procedure for the circuit C_j^*. The complication here is that the shares $(s_0^*, s_1^*, \ldots, s_{n+2j-2}^*)$ for the circuit C_j^* must be consistent with the existing shares for all known shares i.e., it must be that $s_i^* = s_i'$ $i \in I^* \cap [n + 2j]$.

Case 1: *(the output of C_j^* is 1)*

The easier case is when $C_j^*(\mathsf{set2bits}(I^* \cap [n + 2j - 2])) = 1$ since in this case no values associated with the fanout gate we are introducing are supposed to stay hidden from $\mathcal{D}_{j-1}$ – in this case $\mathcal{D}_j$ runs the reconstruction procedure for C_j^* using the known shares and, since C^* does not contain any fanout gate, the reconstruction boils down to multiplying the shares for any qualified set S_j for C_j^* i.e., $k_j = \prod_{i \in S_j} s_i'$ and compute encryptions $c_{i_{LO}} \leftarrow \mathsf{E}(k_j, s_{n+2j-1}'), c_{i_{RO}} \leftarrow \mathsf{E}(k_j, s_{n+2j}')$. Finally $\mathcal{D}_j$ sets $s_0^* = s_0' || (c_{i_{LO}}, c_{i_{RO}})$ and $s_i^* = s_i'$ and gives these values as input to $\mathcal{D}_{j-1}$.

Now $\mathcal{D}_j$ simply outputs whatever $\mathcal{D}_{j-1}$ outputs and wins the game: In this case the values received by $\mathcal{D}_{j-1}$ are distributed exactly as in the real experiment, and therefore the probability with which $\mathcal{D}_{j-1}$ will output the value for the output wire is exactly the same.

Case 2: *(the output of C_j^* is 0, but s'_{n+2j-1} and s'_{n+2j} are known anyway)*
This case happens if, for example, both outputs of the fanout gate are input to OR gates which evaluate to 1. Since in the construction we set both input values of an OR gate to be equal to its output, this means that both $\mathcal{D}_j$ and $\mathcal{D}_{j-1}$ know these values. In this case $\mathcal{D}_j$ will simply choose a random k and encrypt those shares as above. Since $C_j^*(\text{set2bits}(I^* \cap [n + 2j - 2])) = 0$ and C_j^* does not contain any fanout gate, this implies that $\mathcal{D}_j$ cannot tell the difference without breaking the RSA assumption. Note that in this case we are *not* relying on the security of the encryption scheme – having access to both encrypted output $\mathcal{D}_j$ can actually recover k. The point here is that since $k = \prod_{i \in S_j} s'_i$ and the adversary does not know at least one of these s'_i's, we can conclude that an adversary which distinguishes successfully can be used to break the one-wayness of of Tag.

Case 3: *(the output of C_j^* is 0 and at least one between s'_{n+2j-1} and s'_{n+2j} is not known)*
The case where $C_j^*(\text{set2bits}(I^* \cap [n + 2j - 2])) = 1$ and at least one between s'_{n+2j-1} and s'_{n+2j} is not known is the most interesting, since in this case we rely on the security of the encryption scheme. Wlog say that $\mathcal{D}_{j-1}$ knows s'_{n+2j-1} but not s'_{n+2j} (the case where both are unknown follow in a straight-forward way): now $\mathcal{D}_{j-1}$ invokes the *one-query CPA oracle* for the encryption scheme with s'_{n+2j-1} and receives $c_{i_{LO}} \leftarrow \mathsf{E}(k, s'_{n+2j-1})$ and $c_{i_{RO}} \leftarrow \mathsf{E}(k, r)$, and if $\mathcal{D}_j$ outputs the value of the output wire $\mathcal{D}_{j-1}$ can reconstruct r and therefore break the security of the encryption scheme.

4.4   Proof of Verifiability (Theorem 2)

Proof. To prove local verifiability we first observe that since Tag is a permutation, given any y there exist a single x such that $\mathsf{Tag}(y) = x$.

Given any qualified set $I \subseteq [n]$ such that $C(\text{set2bits}(I)) = 1$ we define a set $I' \subseteq [\ell]$ such that $i \in I'$ if $w_i = 1$ during the (plain, Boolean) evaluation of $C(\text{set2bits}(I))$. Let w_i^V the values assigned to the wires during the verification phase, and w_i^R the values assigned to the wires during the reconstruction phase using the (qualified) set I. We prove that if $\mathsf{Ver}_{pp}(s_0, s_i) = \mathtt{true}$ for all $i \in [n]$, then for all $i \in I'$ it holds that $\mathsf{Tag}(w_i^R) = w_i^V$.

We prove this by induction. Thanks to Step 2 in Ver, it holds that

$$\mathsf{Tag}(w_i^R) = w_i^V \text{ for all } i \in I \subset I'$$

Now take the next wire $i \in I' \setminus I$ and wlog assume that i is the output of a gate g_j that only takes inputs from wires with index in $[n + i] \cap I'$ (one can always reorder the wires to make sure that this happens). We can now argue that:

- If g_j is an AND gate, then the value on wire $i = j_O$ is a function of the values on wires j_{LI} and j_{RI}, and since $i \in I'$ it must also be the case that j_{LI} and j_{RI} are in $I' \cap [n + i]$ (the output of the AND gate is set only if both input wires are set), which allows to use the induction hypothesis. By induction it holds that $\mathsf{Tag}(w_{j_{LI}}^R) = w_{j_{LI}}^V$ and $\mathsf{Tag}(w_{j_{RI}}^R) = w_{j_{RI}}^V$. Now since $w_{j_O}^R = w_{j_{LI}}^R \cdot w_{j_{RI}}^R$ and also $w_{j_O}^V = w_{j_{LI}}^V \cdot w_{j_{RI}}^V$ and, since the Tag function is homomorphic this implies that:

$$\mathsf{Tag}(w_{j_O}^R) = \mathsf{Tag}(w_{j_{LI}}^R \cdot w_{j_{RI}}^R) = w_{j_{LI}}^V \cdot w_{j_{RI}}^V = w_{j_O}^V$$

- If g_j is an OR gate, then the value on wire $i = j_O$ is a function of the values on wires j_{LI} and j_{RI}, and since $i \in I'$ it must also be the case that at least one between j_{LI} and j_{RI} are in $I' \cap [n+i]$ (the output of the OR gate is set only if at least one of the input wires are set), which allows to use the induction hypothesis: by induction it holds that at least one between $\mathsf{Tag}(w_{j_{LI}}^R) = w_{j_{LI}}^V$ and $\mathsf{Tag}(w_{j_{RI}}^R) = w_{j_{RI}}^V$ holds. During the verification phase $w_{j_O}^V = w_{j_{LI}}^V$ only if $w_{j_{LI}}^V = w_{j_{RI}}^V$, instead during the construction phase $w_{j_O}^R$ might be set to $w_{j_{LI}}^R$ or $w_{j_{LI}}^R$ depending on which qualified set is being used. But since Tag is a permutation, $\mathsf{Tag}(a) = \mathsf{Tag}(b)$ implies that $a = b$, and therefore during the reconstruction we have that if $w_{j_{LI}}^R \neq \bot$ and $w_{j_{RI}}^R \neq \bot$ then $w_{j_{LI}}^R = w_{j_{RI}}^R$ and therefore

$$\mathsf{Tag}(w_{j_O}^R) = \mathsf{Tag}(w_{j_{LI}}^R) = w_{j_{LI}}^V = w_{j_O}^V$$

- If g_j is a fanout gate, then the value on the wire $i = j_{LO}$ (the case where $i = j_{RO}$ can be argued exactly in the same way) is a function of the value on the wire j_I and in the public share s_0. During the verification phase $w_{j_{LO}}^V = t_{j_{LO}}$ only if $t_{j_{LO}} = \mathsf{D}(w_{j_I}^V, \rho_{j_{LO}}, \mathsf{Tag}(\sigma_{j_{LO}}))$ while in the reconstruction phase $w_{j_{LO}}^R = \mathsf{D}(w_{j_I}^R, c_{j_{LO}})$. By induction it holds that $\mathsf{Tag}(w_{j_I}^R) = w_{j_I}^V$. Using the fact that the encryption scheme and the Tag function commute we show that:

$$\mathsf{Tag}(w_{j_{LO}}^R) = \mathsf{D}(w_{j_I}^R, c_{j_{LO}})^\tau = \mathsf{D}((w_{j_I}^R)^\tau, \rho_{j_{LO}}, \sigma_{j_{LO}}^\tau)$$
$$= \mathsf{D}(w_{j_I}^V, \rho_{j_{LO}}, \mathsf{Tag}(\sigma_{j_L O})) = w_{j_{LO}}^V$$

Since $\ell \in I'$ (the set is qualified), this finally implies that $\mathsf{Tag}(w_\ell^R) = w_\ell^V$, and thanks to Step 5 in the Ver algorithm we can conclude that $w_\ell^V = t_\ell$ and therefore the value w_ℓ^R is the same for all qualified sets. Finally, since $m = u \oplus \mathsf{Ext}(w_\ell^R)$ is a deterministic function of u and w_ℓ^R, we can conclude that all qualified sets reconstruct the same secret m.

5   Globally Verifiable Secret Sharing Schemes

In the previous section we have presented a scheme where each player can check whether the private share received from the dealer is consistent with the public share. In this section we present two possible ways of implementing private channels from the dealer to the players, which also allow to deal with the case where the dealer is cheating and the honest players need to reach an agreement.

In both extensions, we let the setup algorithm output some additional parameters and decryption keys for all players. Now the dealer, instead of sending shares privately to each player, appends encryptions of the shares to the tags of the shares which are sent over the broadcast channel, and each player can recover her own share.

In the first proposal, we use an encryption scheme with the property that it is possible to verify, given an encryption and a tag, whether they contain the same value. This is done using efficient non-interactive zero-knowledge proofs by compiling efficient sigma-protocols for the statement using the Fiat-Shamir heuristic. Doing so makes the scheme *non-interactive* and *publicly-verifiable*, since everyone can check that the dealer sent correct shares to all players. In the second proposal, we let the dealer encrypt the shares using an encryption scheme which has a special property, namely it allows the receiver to prove to a third party what has been received: this scheme gets rid of the random oracle model, but requires each (complaining) party to send a single message to all other parties using a broadcast channel.

Notation. We need to redefine the syntax and the functioning of the setup phase and the sharing phase (the reconstruction phase is unchanged and the syntax of the verification phase differs for the two schemes). The scheme uses a public key encryption scheme (Gen, Enc, Dec).

Setup: The setup algorithm for the *globally-verifiable* CS3 outputs

$$(pp, \{d_i\}_{i \in [n]}) \leftarrow \mathsf{gvSetup}(1^\kappa)$$

where d_i is the decryption key for P_i (the corresponding encryption key for P_i can be derived from (pp, i)). The decryption keys are sent to the owner using private channels, where the public parameters are made public; The algorithm $\mathsf{gvSetup}(1^\kappa)$ simply runs $pp' \leftarrow \mathsf{Setup}(1^\kappa)$ to generate the public parameters for our underlying locally verifiable scheme and n copies of $\mathsf{Gen}_{pp}(1^\kappa)$ to generate n encryption/decryption key pairs (e_i, d_i), and finally outputs $pp = (pp', e_1, \ldots, e_n)$.

Share: The share algorithm for the *globally-verifiable* CS3 outputs

$$s \leftarrow \mathsf{gvShare}_{pp}(C, m)$$

and the dealer broadcasts s; The algorithm $\mathsf{gvShare}_{pp}(C, m)$ simply runs $\mathsf{Share}_{pp}(C, m)$ to generate $(s_0, s_1, \ldots, s_n)$, generates $z_i \leftarrow \mathsf{Enc}_{pp}(e_i, s_i)$ for all $i \in [n]$ and outputs $s = (s_0, z_1, \ldots, z_n)$.

We need also to redefine correctness and privacy as follows:

Definition 4 (Correctness). *A globally verifiable CS3 π is correct if for all $m \in \mathcal{M}$, for all circuits C describing an access structure $\mathcal{A}$, and for all $I \in \mathcal{A}$,*

$$\Pr[m \neq \mathsf{Rec}_{pp}(s, \{d_i\}_{i \in I})] \leq \mathsf{negl}(\kappa)$$

where $s \leftarrow \mathsf{gvShare}_{pp}(C, m)$, $(pp, \{d_i\}_{i \in [n]}) \leftarrow \mathsf{gvSetup}(1^\kappa)$ and the probabilities are taken over the choices of all algorithms.

It is trivial to see that combining a correct locally verifiable CS3 with an encryption scheme with a correct decryption leads to a correct globally verifiable CS3.

Definition 5 (Privacy). *A globally verifiable CS3 π is private if for all circuits C describing an access structure $\mathcal{A}$, and for all $I \notin \mathcal{A}$, there exist a PPT simulator Sim such that for all $m \in \mathcal{M}$,*

$$\text{Sim}(pp, C, I) \approx_c (s, \{d_i\}_{i \in I})$$

where $(pp, \{d_i\}_{i \in [n]}) \leftarrow \text{gvSetup}(1^\kappa)$, $s \leftarrow \text{gvShare}_{pp}(C, m)$.

It is trivial to see that combining a private locally verifiable CS3 with a semantically secure encryption scheme leads to a private globally verifiable CS3.

5.1   Non-Interactive and Publicly-Verifiable Scheme

Our first proposal is a non-interactive and publicly-verifiable VCS3. The syntax of the verification scheme here is:

Verification: The verification algorithm $f \leftarrow \text{niVer}_{pp}(s)$ outputs a bit $f \in \{\texttt{true}, \texttt{false}\}$ which indicates whether s is a valid sharing or not. Note that anyone can run the verification phase i.e., one does not need to know any of the decryption keys d_i to run this algorithm.

The scheme should satisfy the following property:

Definition 6 (Public-Verifiability). *We say a CS3 scheme is publicly verifiable if for all $n \in \mathbb{Z}$, for all circuits C describing an access structure, and for all PPT algorithms D^* the following holds: Let $(pp, d_1, \ldots, d_n) \leftarrow \text{Setup}(1^\kappa)$ and $s \leftarrow D^*(pp)$. If $\text{niVer}_{pp}(s) = \texttt{true}$, then with a overwhelming probability there exists a value $m \in \mathcal{M}$ such that $\text{Rec}_{pp}(s, \{d_i\}_{i \in I}) = m$ for all qualified sets $I \in \mathcal{A}$ (where $\mathcal{A}$ is defined in s).*

We construct the verification algorithm for this scheme, niVer, by replacing Step 2 in Ver (defined in Sect. 4.2) with the following:

3. If $\exists i \in [n] s.t., \text{Tag}(\text{Dec}_{pp}(d_i, e_i)) \neq t_i$ stop and output $\texttt{false}$; else:

This condition can be checked efficiently using the NIZKs. Soundness of the NIZKs together with the *local verifiability* of the underlying scheme implies *public verifiability*.

We finally describe how we construct the encryption scheme (Gen, Enc, Dec) and the NIZKs π_i. The scheme is essentially ElGamal encryption in the RSA group. We choose $N = pq$ where $p = 2p' + 1, q = 2q' + 1$ for primes p', q'. This ensures that the subgroup G of numbers with Jacobi symbol 1 mod N is cyclic of order $2p'q'$, and we let g be a generator of G. The encryption scheme we will construct is secure when applied to messages in G. We therefore need to slightly change the VSS constructed above so that wire values and tags are chosen from G and not from all of $\mathbb{Z}_N^*$. Since Jacobi symbols can be computed efficiently, one can always check that the dealer chooses his values correctly. The encryption scheme now works as follows:

Generation: In the set-up we sample a random decryption key $d_i \in \mathbb{Z}_N$ and output the corresponding encryption key $e_i = g^{d_i}$;

Encryption: sample a random $r \in \mathbb{Z}_N$ and output $z_i = (\alpha_i, \beta_i) = (g^r, e_i^r \cdot s_i)$;

Decryption: output $s_i = \beta_i \cdot \alpha_i^{-d_i}$;

We now need to construct a NIZK that allows a prover with witness d_i to persuade a verifier who knows $(\tau, e_i, t_i, \alpha_i, \beta_i)$ that

$$(g, e_i, \alpha_i^\tau, \beta_i^\tau \cdot t_i^{-1})$$

is a DDH tuple: note that when the dealer is honest this is indeed the case, i.e., the tuple in question is:

$$(g, g^{d_i}, g^{r \cdot \tau}, g^{d_i \cdot r \cdot \tau} \cdot (s_i^\tau / s_i^\tau)) = (g, g^{d_i}, g^{r \cdot \tau}, g^{(r \cdot \tau) \cdot d_i})$$

Very efficient sigma-protocols for this language are well-known (see e.g. [HL10]), which can be made non-interactive in the random oracle model using the Fiat-Shamir Heuristic.

5.2   Non-Interactive Scheme (with Complaints)

The main disadvantage of the previous solution is that it requires the random oracle model for the NIZKs. Our second proposal instead uses one-round complaints to ensure verifiability. The idea here is that every player retrieves her share from the public encryptions, and if the share does not match the public tag, the player can complain by broadcasting some information that allows all other parties to check that the dealer cheated. In particular we do not wish to allow a corrupt player to unfairly accuse an honest dealer, and the "proof of cheating" should not disclose any other information. Both these properties can be achieved using a technique introduced in [DT07] and formalized in [DHKT08]. We refer to [DHKT08] for the details of this method, and we only sketch the high-level idea here: The idea is to let the dealer encrypt the shares using an *identity-based encryption scheme with verifiable secret keys (IBE-VSK)*: in this setting the decryption key of each player is the master secret key for the IBE scheme (and the encryption key is the corresponding public key). When the dealer encrypts the shares for all the parties, he does so using a unique *id* as the identity in the IBE scheme. Each player can decrypt by generating the secret key sk_{id} corresponding to this *id* and then perform the IBE decryption. To complain, the player can broadcast the secret key sk_{id}, and all other parties are now able to retrieve the share and check whether it is consistent with the tag. The security of the IBE scheme implies that revealing sk_{id} does not disclose any information about the encrypted shares in other sessions. In addition, the VSK property allows all other players to verify that sk_{id} is indeed the secret key corresponding to the *id* for the public key of that player, and was not maliciously generated to accuse an honest dealer. We note that VSK is a mild assumption, and every proposed efficient IBE satisfies the VSK property [DT07, DHKT08].

References

[ACGS88] Alexi, W., Chor, B., Goldreich, O., Schnorr, C.P.: RSA and rabin functions: Certain parts are as hard as the whole. SIAM J. Comput. **17**(2), 194–209 (1988)

[BL90] Benaloh, J.C., Leichter, J.: Generalized secret sharing and monotone functions. In: Goldwasser, S. (ed.) CRYPTO 1988. LNCS, vol. 403, pp. 27–35. Springer, Heidelberg (1990)

[Bla79] Blakley, G.R.: Safeguarding cryptographic keys. In: National Computer Conference, pp. 313–317. IEEE Computer Society (1979)

[CDD00] Cramer, R., Damgård, I., Dziembowski, S.: On the complexity of verifiable secret sharing and multiparty computation. In: Proceedings of the Thirty-Second Annual ACM Symposium on Theory of Computing, pp. 325–334. ACM (2000)

[CDM00] Cramer, R., Damgård, I.B., Maurer, U.M.: General secure multi-party computation from any linear secret-sharing scheme. In: Preneel, B. (ed.) EURO-CRYPT 2000. LNCS, vol. 1807, pp. 316–334. Springer, Heidelberg (2000)

[DHKT08] Damgård, I., Hofheinz, D., Kiltz, E., Thorbek, R.: Public-key encryption with non-interactive opening. In: Malkin, T. (ed.) CT-RSA 2008. LNCS, vol. 4964, pp. 239–255. Springer, Heidelberg (2008)

[DT07] Damgård, I.B., Thorbek, R.: Non-interactive proofs for integer multiplication. In: Naor, M. (ed.) EUROCRYPT 2007. LNCS, vol. 4515, pp. 412–429. Springer, Heidelberg (2007)

[HL10] Hazay, C., Lindell, Y.: Efficient Secure Two-Party Protocols. Springer, Heidelberg (2010)

[ISN89] Ito, M., Saito, A., Nishizeki, T.: Secret sharing scheme realizing general access structure. Electron. Commun. Jpn. (Part III: Fundam. Electron. Sci.) **72**(9), 56–64 (1989)

[KW93] Karchmer, M., Wigderson, A.: On span programs. In: Structure in Complexity Theory Conference, pp. 102–111 (1993)

[Ped92] Pedersen, T.P.: Non-interactive and information-theoretic secure verifiable secret sharing. In: Feigenbaum, J. (ed.) CRYPTO 1991. LNCS, vol. 576, pp. 129–140. Springer, Heidelberg (1992)

[Sha79] Shamir, A.: How to share a secret. Commun. ACM **22**(11), 612–613 (1979)

[Sta96] Stadler, M.A.: Publicly verifiable secret sharing. In: Maurer, U.M. (ed.) EUROCRYPT 1996. LNCS, vol. 1070, pp. 190–199. Springer, Heidelberg (1996)

[VNS+03] Vinod, V., Narayanan, A., Srinathan, K., Pandu Rangan, C., Kim, K.: On the power of computational secret sharing. In: Johansson, T., Maitra, S. (eds.) INDOCRYPT 2003. LNCS, vol. 2904, pp. 162–176. Springer, Heidelberg (2003)

[Yao89] Yao, A.C.: Unpublished manuscript. Presented at Oberwolfach and DIMACS workshops (1989)

Fast Oblivious AES A Dedicated Application of the MiniMac Protocol

Ivan Damgård and Rasmus Zakarias[(✉)]

Department of Computer Science, Aarhus University, Aarhus, Denmark
{ivan,rwl}@cs.au.dk

Abstract. We present actively secure multi-party computation of the Advanced Encryption Standard (AES). To the best of our knowledge it is the fastest of its kind to date. We start from an efficient actively secure evaluation of general binary circuits that was implemented by the authors of [DLT14]. They presented an optimized implementation of the so-called MiniMac protocol [DZ13] that runs in the pre-processing model, and applied this to a binary AES circuit. In this paper we describe how to dedicate the pre-processing to the structure of AES, which improves significantly the throughput and latency of previous actively secure implementations. We get a latency of about 6 ms and amortised time about 0.4 ms per AES block, which seems completely adequate for practical applications such as verification of 1-time passwords.

Keywords: Multiparty computation · Arithmetic black box · Arithmetic circuit · Binary circuit · AES

1 Introduction

Secure Multi-party computation (MPC) allows a set of players (or computers) with *private inputs* to evaluate a function on these inputs. Security means that all players learns the output of the function and essentially nothing else. More precisely the problem of MPC for n players is to compute a function $f(x_1, ..., x_n) = (y_1, ..., y_n)$ such that Player, P_i, learns only y_i after evaluating f and x_i is the private input held by P_i. This problem was first proposed by Yao in [Yao82, Yao86] and has been an active area of research since.

The description of the function f can take different forms. In this work we consider descriptions of f as a circuit over the (AND, XOR) or (MUL, ADD) basis for binary and arithmetic circuits respectively. Protocols for evaluating such function typically implement an ideal functionality sometimes called an Arithmetic black-box [DN03]. The Arithmetic black-box is depicted in Fig. 1.

R. Zakarias—The authors acknowledge support from the Danish National Research Foundation and The National Science Foundation of China (under the grant 61061130540) for the Sino-Danish Center for the Theory of Interactive Computation, within part of this work was performed; and from the CFEM research center, supported by the Danish Strategic Research Council.

© Springer International Publishing Switzerland 2016
D. Pointcheval et al. (Eds.): AFRICACRYPT 2016, LNCS 9646, pp. 245–264, 2016.
DOI: 10.1007/978-3-319-31517-1_13

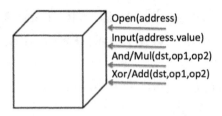

Fig. 1. The Arithmetic Black Box.

Intuitively, the players agree on a circuit over the actions *Open, Input, Xor/Add* and *And/Mul*. Players provide input values using the *Input* command, and then work their way through the circuit invoking the appropriate command for each gate. Finally the *Open* command is used to obtain the result.

MPC for the case where a majority of the players are corrupt requires public-key machinery and was therefore for a long time thought to be impractical, especially for the case of active security. To resolve this, the so-called pre-processing-model was proposed, where the heavy computations are pushed to a pre-processing phase. Using precomputed material one can evaluate the function securely much faster in the on-line phase. With recent result [NNOB12, DKL+13, DLT14, FJN14] in particular, practicality is within our reach, see [IKM+13] for an in depth discussion on the power of correlated randomness. We consider here the particular protocol nick-named *MiniMac* from [DZ13]. This is an arithmetic black-box protocol in the pre-processing-model, which was designed to handle arithmetic circuits over small fields efficiently. To do this, it computes on vectors of field elements instead of single values.

Benchmarking. Oblivious AES is a much used example of how practical MPC is becoming, see [PSSW09] and [NNOB12, DKL+12, GHS12, DLT14, HKS+10, HEKM11]. Oblivious AES distinguishes itself from the classical AES encryption by being distributed between two or more parties. All players know the plain-text and everybody learns the cipher-text. The key, however, is additively shared meaning that the key k is not known to any player.

Performance wise, previous state of the art for Oblivious AES with malicious security is the implementation using a binary circuit in [DLT14] where they report on amortized running times of less than 4 milliseconds per AES-block and 3–4 seconds of latency on ordinary consumer hardware[1]. In [KSS13] a different implementation was reported that uses the algebraic description of AES over $\mathbb{F}_{2^8}$. They achieve about 1 ms amortized time per AES block and a latency of 100 ms.

Our contributions. We show that both amortized time and latency can be improved significantly: in the fastest configuration, we obtain about 0.4 ms amortized time and a latency of about 6 ms. We present three constructions which are variations on the idea that if we exploit the special structure of AES, rather that seeing it as a general binary or arithmetic circuit, we can tailor the

---

[1] The concrete specifications of our experimental setup can be found in Appendix A.

No AES blocks	Time/AES ms	Latency ms	Prep. size $MB/player$
MiniMac [DLT14]			
680	4	9962	130, 0.2/AES
Protocol 1 incl. Key Expansion			
5	3	15	270, 54/AES
Protocol 1 without Key Expansion			
5	1.2	6	220, 44/AES
Estimated time, Protocol 2 (no on-line multiplications)			
15	0.4	6	650, 44/AES
Estimated time, Protocol 3 (minimized pre-processing)			
15	0.8	12	10.5, 0.7/AES

Fig. 2. Execution times of our AES protocol.

pre-processing such that we save on the number of rounds and also on local computation. We try out these ideas in practice using the implementation of MiniMac from [DLT14][2] as our starting point.

The basic approach in our first protocol is to pre-process a number of tables, each of which implement an AES S-box followed by the Mix-Column and Shift-Row operations applied to the bits output from the S-box in question. We first describe what the correlated randomness should look like to implement the tables, and then we present a slightly modified version of the MiniMac protocol using this material to perform AES. This solution computes 5 simultaneous AES blocks in only 10 rounds of online communication. In comparison [DLT14] required at least 6800 rounds.

For second protocol, we observe that after we introduce the tables, we no longer need to do secure multiplication in the on-line phase. This allows us to change the internal representation used in MiniMac to allow more parallelism at no extra cost. This immediately saves us a factor of about 3 in amortized time per AES instance.

For the third protocol we give a new construction that shows how to obtain much smaller pre-processing material. We save a factor of at least 60 in the size of pre-processed data at the cost of doing 1 extra round of communication and more local computation in the final protocol. Some explanation of the idea behind this optimization is in order as the idea may be of interest beyond oblivious AES: the efficiency of MiniMac is based on the idea of computing on vectors of values in a SIMD fashion, i.e. we add and multiply vectors coordinate-wise. Concretely, the implementation of MiniMac we started from uses vectors containing 85 data bytes. Now, the reason why it makes sense to compute the AES S-Box by table look-up is that the input is only 1 byte, so we need only 256 entries in the table. However, the result we get will be just one byte, and this result needs to go to the right place in the vector representing the state after the table look-up, of course without revealing what was output from the table. The simplest solution

---

[2] Available at http://tinyurl.com/q2dmcuw.

is to make the table entry be an entire MiniMac word containing data that only depends on the single byte that is output from that table entry. This will work, as we explain in more detail later, but of course means that tables get very large. What we do for protocol 3 is to show that with an appropriate combination of masking by random values and unconditional MACs, we can have table entries that only consist of a single data byte and some authentication information. This idea can be applied to computation of any function with small input and output, possibly followed by some linear function.

We implemented the Protocol 1 and based on this we calculated the size of pre-processed data for the other results and conservatively estimated their running times, as detailed in the following sections. A summary of this can be seen in Fig. 2.

The demands we have to the pre-processed data are quite specialized and it may not be clear how we can do the required pre-processing reasonably efficiently. In particular, the pre-processing assumed by the original MiniMac protocol does not produce data of the form we need. In principle, the problem can be solved by writing down an arithmetic circuit that takes some random bits as input and outputs the data players need; and then evaluate that circuit using the original MiniMac protocol. This would be an extremely large circuit, and therefore, in the final section, we give a recipe for how pre-processed material may be constructed more efficiently from a generic MiniMac instance.

A main take-home message from our paper is that the only structure we need from AES to speed up the computation is that its non-linear parts consist only of S-Boxes with small inputs, this is what allows us to use table look-up with tables of feasible size. In future work, it will therefore be interesting to investigate if secure computation of other ciphers or hash functions can be made practical using a similar approach.

The MiniMac protocol

This protocol is in a nutshell a SIMD Arithmetic black-box. That is, the protocol operates on a so called representation consisting of a vector of field elements. The actions of the Arithmetic black-box operates in parallel on all elements of the vector simultaneously. The details and proof of security can be found in [DZ13] while the extension for operating efficiently on binary fields was discovered in [DLT14]. For our purposes here we will think of MiniMac simply as an implementation of a SIMD Arithmetic Black-box and hence describe MiniMac's representation of data as containing an l-vector over a finite field[3].

$$[\![\mathbf{a}]\!] = [\![(a_1, ..., a_l)]\!]$$

The operations Input, Add, Mul and Open for the Arithmetic black-box operate on such vectors. E.g. adding two elements in the box with MiniMac yields the computation:

$$[\![\mathbf{a}]\!] + [\![\mathbf{b}]\!] = [\![(a_1 + b_1, ..., a_l + b_l)]\!] = [\![\mathbf{a} + \mathbf{b}]\!]$$

---

[3] Actually, the players in MiniMac have additive shares of the vectors and a special type of MACs are used to prevent cheating, but these details are not important here.

In a similar way Input requires the secret values to be loaded into the box to be l-vectors and Open gives the parties an l-vector back.

Advanced Encryption Standard

AES is described in [DR00]. Here we consider only 128-bit 10 round encryption with AES. We give a different description of AES in terms of matrix operations rather then the algorithmic approach in [DR00]. We do this for two reasons. Firstly, our implementation framework is geared towards matrix operations thus should the reader be interested in looking at our C code this section is extremely useful for understanding our code design choices. Secondly, this interpretation makes it completely natural and straight-forward to describe AES concisely as a series of matrix products interleaved with table lookups. In particular this make the description of our results in Sects. 2, 3, 4 and 5 easier to describe. The algorithm can be considered to have two distinct phases: The *Key Expansion* and the 10 *Rounds*. The Key Expansion operates on a 16 byte state of key material and the 10 rounds operate on a 16 byte state of plain/cipher-text.

The key expansion in more details operates on a 16 byte state of key material initially containing the encryption key. This state is updated in each round to contain the corresponding round-key.

In [DR00] the key expansion algorithm is explained in an algorithmic way over bytes. Our framework of implementation is geared for matrix operations thus we here give an alternative characterization of the key schedule in terms of matrices. Here we abstract the S-Box operations to merely a table lookup and explain later how such lookups can be done securely with MiniMac. Let $K_0 = (k_0, ..., k_{15})$ be the initial 16 bytes of encryption key. The key is divided in to 4 so called words w_0 through w_3 in the natural increasing order of indices.

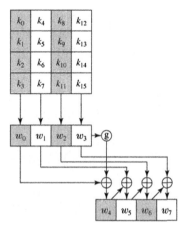

Fig. 3. The AES key schedule

Word $w_3 = (k_{12}, ..., k_{15})$ is passed through the *key schedule core*, denoted g in Fig. 3, which rotates the word one position left $rot(w_3) = (k_{13}, k_{14}, k_{15}, k_{12})$ and then forms the result $T_3 = (sb[k_{13}], sb[k_{14}], sb[k_{15}], sb[k_{12}])$. Here $sb[k]$ refers to the S-Box at index $k, k \in \mathbb{F}_{2^8}$. We have the four word state $T = (T_0, T_1, T_2, T_3) = (k_0, ..., k_{11}, sb[k_{13}], sb[k_{14}], sb[k_{15}], sb[k_{12}])$. From T the remaining of the key schedule is (almost) a linear transformation, KS, over $\mathbb{F}_{2^8}^{16}$ depicted below:

$$KS \times T \oplus w_3 = \begin{bmatrix} 1\,0\,0\,0 & 0\,0\,0\,0 & 0\,0\,0\,0 & 1\,0\,0\,0 \\ 0\,1\,0\,0 & 0\,0\,0\,0 & 0\,0\,0\,0 & 0\,1\,0\,0 \\ 0\,0\,1\,0 & 0\,0\,0\,0 & 0\,0\,0\,0 & 0\,0\,1\,0 \\ 0\,0\,0\,1 & 0\,0\,0\,0 & 0\,0\,0\,0 & 0\,0\,0\,1 \\ 1\,0\,0\,0 & 1\,0\,0\,0 & 0\,0\,0\,0 & 1\,0\,0\,0 \\ 0\,1\,0\,0 & 0\,1\,0\,0 & 0\,0\,0\,0 & 0\,1\,0\,0 \\ 0\,0\,1\,0 & 0\,0\,1\,0 & 0\,0\,0\,0 & 0\,0\,1\,0 \\ 0\,0\,0\,1 & 0\,0\,0\,1 & 0\,0\,0\,0 & 0\,0\,0\,1 \\ 1\,0\,0\,0 & 1\,0\,0\,0 & 1\,0\,0\,0 & 1\,0\,0\,0 \\ 0\,1\,0\,0 & 0\,1\,0\,0 & 0\,1\,0\,0 & 0\,1\,0\,0 \\ 0\,0\,1\,0 & 0\,0\,1\,0 & 0\,0\,1\,0 & 0\,0\,1\,0 \\ 0\,0\,0\,1 & 0\,0\,0\,1 & 0\,0\,0\,1 & 0\,0\,0\,1 \\ 1\,0\,0\,0 & 1\,0\,0\,0 & 1\,0\,0\,0 & 1\,0\,0\,0 \\ 0\,1\,0\,0 & 0\,1\,0\,0 & 0\,1\,0\,0 & 0\,1\,0\,0 \\ 0\,0\,1\,0 & 0\,0\,1\,0 & 0\,0\,1\,0 & 0\,0\,1\,0 \\ 0\,0\,0\,1 & 0\,0\,0\,1 & 0\,0\,0\,1 & 0\,0\,0\,1 \end{bmatrix} \times \begin{bmatrix} t_0 \\ t_1 \\ t_2 \\ t_3 \\ t_4 \\ t_5 \\ t_6 \\ t_7 \\ t_8 \\ t_9 \\ t_{10} \\ t_{11} \\ t_{12} \\ t_{13} \\ t_{14} \\ t_{15} \end{bmatrix} + \begin{bmatrix} 0 \\ 0 \\ 0 \\ 0 \\ 0 \\ 0 \\ 0 \\ 0 \\ 0 \\ 0 \\ 0 \\ 0 \\ k_{12} \\ k_{13} \\ k_{14} \\ k_{15} \end{bmatrix} \quad (1)$$

When expanding the key from round i to round $i + 1$ we denote $w_0, ..., w_3$ the four words of the current key and denote the new key $w_4, ..., w_7$.

To see how the computation above resembles the operations depicted in Fig. 3 we consider each word of the result in turn. The first word in the new key is w_4. We see from the figure that it should be $w_0 \oplus T_3 = (k_0 \oplus t_{12}, k_1 \oplus t_{13}, k_2 \oplus t_{14}, k_3 \oplus t_{15})$. Doing the inner product with the first four rows of KS and T yields this. Similar observations can be made progressing to rows 4 though 7 and rows 8 through 11 for words w_5 and w_6 respectively. For w_7 the operation is slightly different as it according to Fig. 3 should be $w_3 \oplus w_4 \oplus w_5 \oplus w_6$ where only w_4, w_5 and w_6 are present in our T vector. To complete the operation depicted in Fig. 3 we additionally XOR w_3 onto $KS \times T$ obtaining the same operation as in Fig. 3. The reason for laying out the computation as above will become clear later. Finally we add the round constant to obtain the new key.

The 10 rounds are the main encryption loop of AES. Algorithm 1 describes the algorithm. *Add Round Key* covers the operation of XORing the current round key with the current AES state obtaining a new AES state. *The KeyExpansion* step updates the current round key into the one needed for the following round. If the key expansion is computed beforehand as suggested above the KeyExpansion step can be ignored.

Algorithm 1. AES Encryption

Data: $S = (s_0, ..., s_{15})$ - /* the AES state   */
Data: $K = (k_0, ..., k_{15})$ - /* the AES key   */
/* Prepare the 11 round keys   */
/* Xor the 0th round key to the state   */
1 AddRoundKey(K,S,0);
2 **for** $round \in \{1, 2, 3, 4, 5, 6, 7, 8, 9\}$ **do**
3 | KeyExpansion(K,$round$);
4 | SubByte(S);
5 | ShiftRows(S);
   | /* Considered as polynomials over $\mathbb{F}_{2^8}$ the columns of the state
   | are multiplied with the fixed polynomial $3x^3 + x^2 + x + 2$
   | mod $x^4 + 1$.   */
6 | MixColumns(S);
   | /* Xor the [$round$]th key to the state   */
7 | AddRoundKey(K,S,$round$);
8 KeyExpansion(K,10);
9 SubByte(S);
10 ShiftRows(S);
11 AddRoundKey(K,S,10);

After the final step the 16 bytes AES state in S contains the cipher-text. Our results rely on a mathematical interpretation of the steps in AES which we will give in the following. We consider each step in Algorithm 1.

The *Sub-Byte* step is the operation of replacing each byte in the AES state with the S-Box lookup for that byte. More precisely, if the AES state is $\mathbf{S} = (s_0, ..., s_{15})$ after applying Sub-bytes the AES state is $\mathbf{S'} = (sb[s_0], ..., sb[s_{15}])$. Another interpretation of the S-Box can be found in [DK10]. Here the S-Box is considered a degree 254 polynomial over $\mathbb{F}_{2^8}$. In our case, we will use a lookup table however a low-depth binary circuit for the S-Box can be found in [BP11].

In the *Shift-rows* step we consider the 16 bytes AES state as a 4×4 matrix as in Fig. 4a. Then the shiftrows cycles the second row one element, the third row two elements and the fourth row three elements as depicted in Fig. 4b. This operation corresponds to the 16×16 linear transformation performed by the

$$S = \begin{bmatrix} s_0 & s_4 & s_8 & s_{12} \\ s_1 & s_5 & s_9 & s_{13} \\ s_2 & s_6 & s_{10} & s_{14} \\ s_3 & s_7 & s_{11} & s_{15} \end{bmatrix}$$

(a) 16 bytes of AES state laid out in a 4×4-matrix.

$$S = \begin{bmatrix} s_0 & s_4 & s_8 & s_{12} \\ s_5 & s_9 & s_{13} & s_1 \\ s_{10} & s_{14} & s_2 & s_6 \\ s_{15} & s_3 & s_7 & s_{11} \end{bmatrix}$$

(b) 16 bytes of AES state with *Shift-rows* applied.

Fig. 4. Illustration of *Shift-rows*.

matrix on the 16-vector holding the AES state, $S = (s_0, ..., s_{15})$

$$
\begin{bmatrix}
1\,0\,0\,0 & 0\,0\,0\,0 & 0\,0\,0\,0 & 0\,0\,0\,0 \\
0\,0\,0\,0 & 0\,1\,0\,0 & 0\,0\,0\,0 & 0\,0\,0\,0 \\
0\,0\,0\,0 & 0\,0\,0\,0 & 0\,0\,1\,0 & 0\,0\,0\,0 \\
0\,0\,0\,0 & 0\,0\,0\,0 & 0\,0\,0\,0 & 0\,0\,0\,1 \\
0\,0\,0\,0 & 1\,0\,0\,0 & 0\,0\,0\,0 & 0\,0\,0\,0 \\
0\,0\,0\,0 & 0\,0\,0\,0 & 0\,1\,0\,0 & 0\,0\,0\,0 \\
0\,0\,0\,0 & 0\,0\,0\,0 & 0\,0\,0\,0 & 0\,0\,1\,0 \\
0\,0\,0\,1 & 0\,0\,0\,0 & 0\,0\,0\,0 & 0\,0\,0\,0 \\
0\,0\,0\,0 & 0\,0\,0\,1 & 1\,0\,0\,0 & 0\,0\,0\,0 \\
0\,0\,0\,0 & 0\,0\,0\,0 & 0\,0\,0\,0 & 0\,1\,0\,0 \\
0\,0\,1\,0 & 0\,0\,0\,0 & 0\,0\,0\,0 & 0\,0\,0\,0 \\
0\,0\,0\,0 & 0\,0\,1\,0 & 0\,0\,0\,0 & 0\,0\,0\,0 \\
0\,0\,0\,0 & 0\,0\,0\,0 & 0\,0\,0\,1 & 1\,0\,0\,0 \\
0\,1\,0\,0 & 0\,0\,0\,0 & 0\,0\,0\,0 & 0\,0\,0\,0 \\
0\,0\,0\,0 & 0\,0\,1\,0 & 0\,0\,0\,0 & 0\,0\,0\,0 \\
0\,0\,0\,0 & 0\,0\,0\,0 & 0\,0\,0\,1 & 0\,0\,0\,0
\end{bmatrix}
\times
\begin{bmatrix}
s_0 \\ s_1 \\ s_2 \\ s_3 \\ s_4 \\ s_5 \\ s_6 \\ s_7 \\ s_8 \\ s_9 \\ s_{10} \\ s_{11} \\ s_{12} \\ s_{13} \\ s_{14} \\ s_{15}
\end{bmatrix}
=
\begin{bmatrix}
s_0 \\ s_5 \\ s_{10} \\ s_{15} \\ s_4 \\ s_9 \\ s_{14} \\ s_3 \\ s_8 \\ s_{13} \\ s_2 \\ s_7 \\ s_{12} \\ s_1 \\ s_6 \\ s_{11}
\end{bmatrix}
\tag{2}
$$

The *Mix-columns* step can also be described as a linear transformation, see Eq. 3. We are going to apply these matrices in the pre-processing phase and in the on-line phases of our protocol using a trick which will be explained later. In all of our application of *Shift-rows* and *Mix-columns* we compute on many AES blocks in parallel. For this we introduce one additional bit of notation. Let SR and MC be the *Shift-rows* and *Mix-columns* matrices as above respectively. To apply e.g. SR to a vector holding n AES states we write $SR_n \times S$ where SR_n is the $16n \times 16n$-matrix having n SR on the diagonal and zero everywhere else. We denote $SRMC$ the matrix that applies *Shift-rows* followed by *Mix-columns* to S. That is $SRMC \times S = MC \times SR \times S$. Likewise we also denote $SRMC_n$ the $16n \times 16n$ matrix having n $SRMC$ on its diagonal.

$$
\begin{bmatrix}
2\,3\,1\,1 & 0\,0\,0\,0 & 0\,0\,0\,0 & 0\,0\,0\,0 \\
1\,2\,3\,1 & 0\,0\,0\,0 & 0\,0\,0\,0 & 0\,0\,0\,0 \\
1\,1\,2\,3 & 0\,0\,0\,0 & 0\,0\,0\,0 & 0\,0\,0\,0 \\
3\,1\,1\,2 & 0\,0\,0\,0 & 0\,0\,0\,0 & 0\,0\,0\,0 \\
0\,0\,0\,0 & 2\,3\,1\,1 & 0\,0\,0\,0 & 0\,0\,0\,0 \\
0\,0\,0\,0 & 1\,2\,3\,1 & 0\,0\,0\,0 & 0\,0\,0\,0 \\
0\,0\,0\,0 & 1\,1\,2\,3 & 0\,0\,0\,0 & 0\,0\,0\,0 \\
0\,0\,0\,0 & 3\,1\,1\,2 & 0\,0\,0\,0 & 0\,0\,0\,0 \\
0\,0\,0\,0 & 0\,0\,0\,0 & 2\,3\,1\,1 & 0\,0\,0\,0 \\
0\,0\,0\,0 & 0\,0\,0\,0 & 1\,2\,3\,1 & 0\,0\,0\,0 \\
0\,0\,0\,0 & 0\,0\,0\,0 & 1\,1\,2\,3 & 0\,0\,0\,0 \\
0\,0\,0\,0 & 0\,0\,0\,0 & 3\,1\,1\,2 & 0\,0\,0\,0 \\
0\,0\,0\,0 & 0\,0\,0\,0 & 0\,0\,0\,0 & 2\,3\,1\,1 \\
0\,0\,0\,0 & 0\,0\,0\,0 & 0\,0\,0\,0 & 1\,2\,3\,1 \\
0\,0\,0\,0 & 0\,0\,0\,0 & 0\,0\,0\,0 & 1\,1\,2\,3 \\
0\,0\,0\,0 & 0\,0\,0\,0 & 0\,0\,0\,0 & 3\,1\,1\,2
\end{bmatrix}
\times
\begin{bmatrix}
s_0 \\ s_1 \\ s_2 \\ s_3 \\ s_4 \\ s_5 \\ s_6 \\ s_7 \\ s_8 \\ s_9 \\ s_{10} \\ s_{11} \\ s_{12} \\ s_{13} \\ s_{14} \\ s_{15}
\end{bmatrix}
=
\begin{bmatrix}
2s_0 + 3s_1 + s_2 + s_3 \\
s_0 + 2s_1 + 3s_2 + s_3 \\
s_0 + s_1 + 2s_2 + 3s_3 \\
3s_0 + s_1 + s_2 + 2s_3 \\
2s_4 + 3s_5 + s_6 + s_7 \\
s_4 + 2s_5 + 3s_6 + s_7 \\
s_4 + s_5 + 2s_6 + 3s_7 \\
3s_5 + s_5 + s_6 + 2s_7 \\
2s_8 + 3s_9 + s_{10} + s_{11} \\
s_8 + 2s_9 + 3s_{10} + s_{11} \\
s_8 + s_9 + 2s_{10} + 3s_{11} \\
3s_8 + s_9 + s_{10} + 2s_{11} \\
2s_{12} + 3s_{13} + s_{14} + s_{15} \\
s_{12} + 2s_{13} + 3s_{14} + s_{15} \\
s_{12} + s_{13} + 2s_{14} + 3s_{15} \\
3s_{12} + s_{13} + s_{14} + 2s_{15}
\end{bmatrix}
\tag{3}
$$

The organization of this paper is as follows. In Sect. 2 we present how to compute Oblivious AES as a multi-party computation with dedicated pre-processing.

We actually implemented this work as code and report on running times as low as 5 ms for 5 simultaneous AES instances in Fig. 2. Our work can be reproduced following our instructions in Appendix A. Then in Sect. 3 we discuss an optimization of the protocol from the fact that with dedicated pre-processing the online computation is all linear. In Sect. 4 we discuss another improvement reducing the size of our pre-processing material required. Finally in Sect. 5 we show how to efficiently get dedicated pre-processing from a general MiniMac instance evaluating arithmetic circuits over $\mathbb{F}_{2^8}$.

2 Fast AES Using Dedicated Pre-processing

In this Section we show how dedicated pre-processing can be used to efficiently compute Oblivious AES. We list demands for the pre-processing material needed and describe an online protocol using the material to compute Oblivious AES. Also we present an implementation and performance numbers on consumer grade hardware.

We employ the fastest version of MiniMac implementation from [DLT14] which allows us to compute on vectors containing 85 bytes. In such a vector we can pack 5 full AES states taking up 80 bytes. In this way we run a small number of AES circuits in parallel. However notice here that we are running "different" operations on individual bytes in the representation as we are not performing the same operations to all bytes in the AES state. This is not supported directly by MiniMac and hence we need help from the pre-processing.

The pre-processing will generate tables of correlated randomness corresponding to handling the entire AES round (except add round key). Loosely put, we pre-process random values with the AES round operation applied to them. Then, we correct these at runtime to yield the AES round operation on the actual input values.

We start by describing how we pre-process the *S-Box*. The S-Box can be thought of as a table with 256 entries. To apply the S-Box to a single byte in the AES state we look up the entry corresponding to that byte (e.i. the state byte has a numeric base 10 value between 0 and 255 which we use as index into the S-box). Lets consider how to do this for a single s_j in our representation with 5 AES states consisting of 80 bytes $\mathbf{S} = (s_0, ..., s_{79})$. To ensure s_j remains secret inside the box we disguise s_j with a uniformly random value R_j and open $R_j + s_j$ to all the parties. Now we will construct a pre-processed table $sb_{+R_j}[\cdot]$ that contains 256 MiniMac representations of S-Box values. However, the indexing into the table is permuted by adding the random R_j. More precisely, we want that

$$\forall s \in \mathbb{F}_{2^8}, j = 0 \ldots 79 : sb_{+R_j}[s + R_j] = [\![(0, \ldots, sb[s], 0, \ldots, 0)]\!]$$

where the value $sb[s]$ is placed in position j^4.

---

[4] Note that when we say an entry in the table is a MiniMac representation of some vector this actually means that players have additive shares of that vector as well as some MACs and corresponding keys, however, the details of this are not important here.

We will require the pre-processing to also output $[\![(R_0, \ldots, R_{79})]\!]$ and when the time comes to do the S-boxes, we add this to the current state and open $(s_0 + R_0, \ldots, s_{79} + R_{79})$. Then we do the 80 table look-ups and add all the outputs effective applying the S-Box to all 80 bytes.

This trick can be extended so that we can make the table look-up implicitly compute also the linear transformation constituted by *Shift-rows* and *Mix-columns*. Let the matrix $SRMC$ denote the Shift-rows matrix multiplied with the Mix-columns matrix from Sect. 2. Note that if taken directly from Sect. 2 this matrix would only operate on one AES state, but it can be extended in a natural way to operate on a vector containing 5 states.

Now, using the same random values R_j, we replace the S-box tables defined above by 80 tables denoted $AESBox_j, j = 0, \ldots, 79$, such that

$$\forall s \in \mathbb{F}_{2^8}, j = 0 \ldots 79 : AESBox_j[s + R_j] = [\![SRMC \times (0, .., 0, sb[s], 0, ..., 0)]\!]$$

where again the non-zero value is placed at position j. Because the multiplication by $SRMC$ is a linear operation, it follows that if we do the 80 table look-ups using $(s_0 + R_0, \ldots, s_{79} + R_{79})$ as indices and add the results, this time we will obtain

$$\sum_{i=0}^{79} [\![SRMC \times (0, .., 0, sb[s_j], 0, ..., 0)]\!] = [\![SRMC \times (sb[s_0], ..., sb[s_{79}])]\!]$$

and these are exactly the 5 new AES states we wanted. The protocol depicted in Fig. 5 describes how one AES round is handled using this approach.

AES Round: The AES round proceeds as follows:

1. Take a fresh $AESBox = \{AESBox_j\}_{j=0,\ldots,79}$ from the available ones and the corresponding $[\![R]\!] = [\![(R_0, ..., R_{79})]\!]$.
2. Let the current state be $[\![S]\!] = [\![(s_0, ..., s_{79})]\!]$. The parties compute $[\![\delta]\!] = [\![R + S]\!]$.
3. $\delta = (\delta_0, ..., \delta_{79})$ is opened
4. As δ is known in plain by all parties, they can look up $S'_j = AESBox[\delta_j], j = 0, ..., 79$ [a].
5. The parties form the state S' after *SubBytes, ShiftRows* and *MixColumns* by computing $S' = \sum_{i=0}^{79} S'_i$.
6. Finally (a MiniMac representation of) the round key is added to the state and the next round follows.

---

[a] For the 10th round we have a pre-processed AESBox which is the same except we only apply the Shift-Row matrix to the values in the tables.

Fig. 5. Online phase, $\Pi_{AES-round}$

2.1 Experiments with the Implementation

We have implemented the MiniMac on-line phase and a program for creating the pre-processing material for all parties running on one machine. In Fig. 2 we list timings of our experiments.

Execution time of out experiments are recorded as follows:

We have three test machines, two peers who will carry out the MPC and a third monitor who will record execution time. When the Peer processes have loaded pre-processing material from Disk and otherwise ready to commence computation they report "Ready" to the monitor. When both have done so, the monitor will record a time stamp and send "Start" to the Peers. Each Peer report to the monitor "Done" when it has reached completion of the MPC circuit. When all Peers have reported "Done" the monitor records the time and execution time is taken to the difference between our two time stamps. More precisely, for *Peer 0* and *Peer 1* the following happens:

- *Peer 0* starts, connects to the Monitor and listens for *Peer 1* to connect.
- *Peer 1* starts, connects to the Monitor and connects to *Peer 0*
- Then both peers loads pre-processing material and perform input-gates obtaining the initial shared AES state and reports "Ready" to the Monitor. Then they wait for the Monitor to signal start.
- When *All* peers has arrived at an initial AES-state the Monitor signals "Start" and the MPC begins.
- Upon completing the AES circuit each peer reports "Done" to the monitor.
- The Monitor records the time before the first "Start"-signal is issued until the last Peer reports back its computation has completed. The difference between these two time stamps is the computation time we report.

When to include the KeyExpansion requires a bit of discussion. When encrypting many blocks of data the *key expansion* can be computed once and reused. This requires that the round-keys are computed beforehand and stored. Thus using 11 representations, one for each round key, we can compute the key expansion once and reuse it for encrypting any number of blocks afterwards. Therefore, a good approximation of amortized execution time per block of encryption with large bodies of plain-text can be achieved with one round of encryption omitting the key expansion entirely and multiplying up. However, when measuring the latency (with a fresh key not pre-loaded) from when starting the encryption until the first block of cipher-text is ready, the key expansion does count and as we will see, it plays a significant role. In summary we care about two types of measurements: Latency from scratch and amortized execution time per block over many blocks . See the result in Fig. 2. Here, xxx/AES is the number of pre-processed Megabytes required per computed AES block.

3   Exploiting the Absence of On-Line Multiplications

The representation of data used in MiniMac is carefully designed to support secure coordinate-wise multiplication of vectors. However, using the techniques we have seen in the previous section, we do not need such multiplication operations.

In this section we describe how we can exploit this fact to change the data representation so that we can compute on more data at smaller cost.

One important step in representing a vector in MiniMac format (and the only one we need to worry about here) is to encode it in a linear code. In order to support multiplications, one needs two properties from this code: first, the encoding must be in systematic form, that is, the encoded vector appears in the first positions of the resulting codeword. Second, the so-called Schur transform of the code must have large enough minimum distance. To obtain the Schur transform of a code C is the linear span of all vectors in $\{a * b | \ a, b \in C\}$, where $a * b$ is the coordinate-wise (or Schur) product of a and b.

The implementation from [DLT14] obtains these properties by encoding vectors of length 85 into a Reed-Solomon code of length 255. Actually, one could use a larger value than 85 and still satisfy the two properties, but since the underlying field contains a root of unity of order 255 and 85 divides 255, these choices allow us to use the FFT algorithm to encode and decode and this speeds up the computations we need quite dramatically.

However, if we do not need to do multiplications, it turns out that the only demand we need to satisfy is that the code itself has large enough minimum distance, more precisely, it just has to be at least the security parameter divided by 8 (since each field element is 8 bits long). Furthermore we no longer need the code to be in systematic form.

With these relaxations, we can choose a Reed-Solomon code of length 255 and encode vectors of length 239. Because the codeword length is still 255, we can use FFT to encode and decode (the requirement for the data length to divide 255 was only necessary to have systematic encoding and still be able to use FFT).

This change will speed up our AES implementation in two ways: first we can pack 14 AES states into one vector instead of 5, immediately yielding a factor of 3 in computational capacity. Secondly, the encoding is faster than before because we no longer need systematic encoding. The reason for this is as follows: a Reed-Solomon codeword is computed by taking a polynomial of at most a certain degree and evaluating it in a set of fixed input points (255 in our case). For systematic encoding the polynomial must take the values specified by our input in the first points, so to encode one must first interpolate to get the right polynomial and then evaluate it in the other points to get the rest of the codeword. For non-systematic encoding one just thinks of the input as coefficients of a polynomial and then we just evaluate.

We did no do the resulting AES implementation, but since the number of rounds will be the same and local computation is simpler, we can safely assume that the total time for the protocol will not be larger than before. But we now compute 14 AES instances instead of 5, and in fact we can put 15 AES instances if we settle 120 bits of security in the authentication of data values which is more than enough in practice. So we can expect an amortized time of $0.4ms$ per AES block and the same latency of $6ms$.

The size of the pre-processing grows significantly as we now have 240 working bytes we need 240 S-Boxes and we store a 256 entry table for each. The estimated size of the pre-processing material is ≈ 650 Mb per player per AES block we will try to improve on this in the following.

4   Minimizing Size of the Pre-processing Material

The ideas we described so far require a rather large amount of pre-processed material. Each of the S-Box tables we have been using so far has 256 entries where each entry is an entire MiniMac codeword which requires 1056 bytes of storage for each player. This translates to approximately $21MB$ of pre-processed data per player per AES round using the first method we presented. We suggest in the following a different way to represent the tables that saves a factor of about 60 in the pre-processing size. The price we pay for this optimization is one extra round of communication per AES round and some extra local computation.

The Idea. We first describe our idea for organizing tables in a generic fashion because we believe it can be interesting in other contexts than secure AES. So assume that we are working with an arithmetic black-box, we have computed $[\![x]\!]$ and would like to compute $[\![f(x)]\!]$. We assume for concreteness that $x \in \mathbb{F}_{2^8}$, but this is not necessary in general. If f is rather complicated to compute via a circuit, as is the case for the AES S-Box, we can do better using a precomputed table. The first step towards this is similar to what we already did above: we will pre-process a random value $[\![R]\!]$ and also pre-process a table f_{+R} defined as

$$f_{+R}[z + R] = f(z), \text{ for } z = 0, \ldots 255.$$

Now we can compute and open $[\![x + R]\!] = [\![x]\!] + [\![R]\!]$ and look up in the table. This will hide x because we add R but is of course insecure because $f(x)$ will become public.

A slightly better idea is to pre-process a random $[\![v]\!]$ and re-define the table as

$$f_{+R}[z + R] = f(z) + v, \text{ for } z = 0, \ldots 255.$$

Now the table look-up will produce $f(x) + v$ and we can add this to $[\![v]\!]$ to get $[\![f(x)]\!]$. This is still not secure, however: we use the same mask v for all entries and so different table entries are not independent and we may reveal information on how the table was permuted and hence indirectly information on x.

So the final idea is to not store the table in the clear but instead secret share the entries additively between the players. We will only open the entry we actually look up, and now it is secure to use the same mask v for all entries. To prevent players from lying about their shares, we add standard message authentication codes (MACs) to the shares.

More concretely, this means that for each table entry $w = f(z) + v$, we choose in the pre-processing random r_1, r_2 such that $r_1 + r_2 = w$, and in addition we choose random vectors a_1, b_1, a_2, b_2, these will serve as MAC keys. Then we compute MACs, $m_1 = a_2 * (r_1, \ldots, r_1) + b_2$ and $m_2 = a_1 * (r_2, \ldots, r_2) + b_1$ and give r_1, m_1, a_1, b_1 to the first player and r_2, m_2, a_2, b_2 to the other. Here, $*$ denotes the coordinate-wise (or Schur) product of vectors.

We will use $(\!|w|\!)$ to denote all this data in the following. The reader should think of this as a randomized representation of w that can be reliably opened: the players would exchange shares and MACs and then use their keys to check the MACs. It is well known and easy to prove that having, say, the first player accept an incorrect value requires that you guess a_1. So if we choose the length of a_i and b_i to be 8 bytes, for instance, we get 64 bits of (unconditional) security which should be more than enough in most cases.

So the final table is of form

$$f_{+R}[z + R] = (\!|f(z) + v|\!), \text{ for } z = 0, \ldots 255.$$

We will need one representation $(\!|\cdot|\!)$ for each table entry, but the values a_1 and a_2 can be the same for all entries without affecting security. So in this case, a table entry requires essentially $1 + 8 + 8 = 17$ bytes for each player. This is a factor more than 60 less than the 1056 bytes we needed before.

A final observation is that if what we really want to compute is not $[\![f(x)]\!]$ but $[\![L(f(x))]\!]$ where L is a linear function, then we can precompute $[\![L(v)]\!]$. When players have computed $f(x) + v$ they can locally compute $L(f(x) + v) = L(f(x)) + L(v)$ and add this into $[\![L(v)]\!]$ to get $[\![L(f(x))]\!]$.

Using the Idea for AES. In the following we describe our observation above using plain codewords with 240 working bytes in each representation. We start by designing the content of the pre-processed S-Boxes differently as follows. We have a random $[\![R]\!] = [\![R_0, \ldots, R_{239}]\!]$ and 240 tables in mind namely one table for each byte in our 15 AES states. Let $[\![S]\!] = [\![S_0, \ldots, S_{239}]\!]$ denote the MiniMac representation holding our 15 AES states. For S_j the jth state byte we consider the table with S-Box values rotated by R_j and masked with a single v_j from a random $v \in \mathbb{F}_{2^8}^{240}$, see Fig. 6a. Now our idea is to additively share this table between the players with MACs. Thus each player m gets a table $r^{m,j}$ such that the entries $r_k^{m,j}$ in $r^{m,j}$ add up to $Sb[(R_j + k) \mod 256] + v_j = \sum_{m=0}^{n-1} r_k^{m,j}$ for fixed k, j summing over m adding each of the shares held by the players. This is illustrated in Fig. 6b for two players, e.g. $m \in \{0, 1\}$. Each table $r^{m,j}$ is MACed towards the other player(s). Thus in addition player m has a table $M(r^{l,j})$ for $l \in \{i | 0 \le i < n \land i \ne m\}$. For two players the situation is depicted in Fig. 7. Thus we define an *AESBox* in this new set up as a triple of three things: A random representation $[\![R]\!]$, a random representation $\mathbf{v}$ with the *SRMC*-linear transformation applied to it $[\![SRMC \times v]\!]$ and a set of 240 tables with 256 entries constructed as described above.

$$AESBox = \{[\![R]\!], [\![SRMC_{15} \times \mathbf{v}]\!], \{(\!|r_k^j|\!)\}_{j \in [239], k \in [255]}\}$$

Here we used $(\!|r_k^j|\!)$ to denote the set of values shared with MACs as described in Fig. 7 for all the players constituting table j.

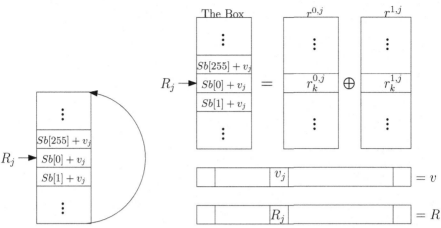

(a) Table with 256 S-Box entries rotated by a random value R_j and masked by v_j. Each entry is one byte.

(b) The Box intended for state position j now additively shared in $r^{0,j}$ and $r^{1,j}$. Player i gets table $r^{i,j} = (r_0^{i,j}, ..., r_{255}^{i,j})$. Each players also gets his shared state for $[\![v]\!]$ and $[\![R]\!]$.

Fig. 6. The new layout

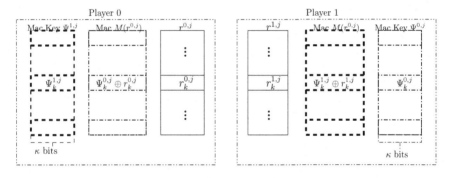

Fig. 7. Above we depict the $(\![x]\!) = \{\{x_0 \in \mathbb{F}_{2^8}, M(x_0) \in \mathbb{F}_{2^8}^{\kappa/8}, \Psi_1 \in \mathbb{F}_{2^8}^{\kappa/8}\}, \{x_1 \in \mathbb{F}_{2^8}, M(x_1) \in \mathbb{F}_{2^8}^{\kappa/8}, \Psi_0 \in \mathbb{F}_{2^8}^{\kappa/8}\}\}$ representation. Player 0 hold from left to right a MAC key table, a Mac table and a table of values. The Key table allows Player 0 to check the table of values held by Player 1. The MAC table allows player 0 to convince Player 1 his table of values is authentic.

The on-line phase is summarized in Fig. 5. Similar to our previous solution we start by taking an $AESBox = ([\![R]\!], [\![SRMC \times v]\!], \{(\![r_k^j]\!)\}_{j \in [239], k \in [255]})$ and "blind" the the current states in S by adding our random R to it obtaining $[\![\Delta]\!] = [\![S \oplus R]\!]$. Then we open $[\![\Delta]\!]$ to everyone. Now since our $r^{m,j}$ tables are shares of the S-Box masked by v_j and rotated by R_j we can lookup $(\![sb[S_j] \oplus v_j]\!)$ by letting each player m take $r_{\Delta_j}^{m,j}$ to be his share of $(\![T_j]\!) = (\![sb[S_j] \oplus v_j]\!)$. As $\mathbf{v}$ is randomly chosen it blinds the actual value looked up in the S-Box thus we can safely open $(\![T_j]\!)$. To open $(\![T_j]\!)$ the parties exchange the Values and MACs describe above in

---

$$\Phi_{AESRound}$$

1 Take an available $AESBox = (\llbracket R \rrbracket, \llbracket SRMC \times v \rrbracket, \{(\!|r_k^j|\!)\}_{j \in [239], k \in [255]})$
2 All parties compute $\llbracket \Delta \rrbracket = \llbracket S \rrbracket \oplus \llbracket R \rrbracket = \llbracket S \oplus R \rrbracket$
3 $\llbracket \Delta \rrbracket$ is opened to each player.
4 Each player uses Δ_j to lookup $r_{\Delta_j}^{m,j}$, obtaining $(\!|T_j|\!) = (\!|sb[S_j] \oplus v_j|\!)^a$ as a $(\!|\cdot|\!)$-representation between them.
5 T_j is opened to everyone for $j \in [239]$.
6 The players compute $T = \sum_{j=0}^{239} T_j = sb[S] \oplus v$
7 The players computes $SRMC \times T = SRMC \times sb[S] \oplus SRMC \times v$
8 The players take $S' = SRMC \times T \oplus \llbracket SRMC \times v \rrbracket = \llbracket SRMC \times sb[S] \rrbracket$ as the AES state after $SubBytes$, $Shift$-$rows$ and Mix-$columns$.
9 Finally the AES round key $\llbracket K_{round} \rrbracket$ is added obtaining the next state $S_{round+1} = S' \oplus K_{round}$.

---

^{a} $r_{\Delta_j}^{m,j}$ is the particular share each player can lookup. Here we use $(\!|T_j|\!)$ when referring to these shares as a combined shared value T_j mutually authenticated with MACs.

Fig. 8. $\Phi_{AESRound}$

Fig. 7 and the receiving parties checks that the MACs are correct. If no one aborts everybody know $T_j = sb[S_j] \oplus v_j$. Knowing all $T_j, \forall j \in [239]$ the players compute $T = \bigoplus_{j \in [239]} T_j$. Now the parties take the linear transformation $SRMC_{15}$ and apply it to T obtaining $SRMC_{15} \times T = SRMC_{15} \times S \oplus SRMC_{15} \times v$. Then the new state S' after $SubBytes$, $Shift$-$rows$ and Mix-$columns$ is computed as $SRMC_{15} \times T \oplus \llbracket SRMC_{15} \times v \rrbracket = \llbracket SRMC_{15} \times sb[S] \rrbracket$. Finally the round key is added to the S' and the following AES round follows.

For the 10th round $SRMC_{15}$ is replaced by the linear transformation SR_{15} which is the 240×240 matrix having 15 SR matrices on its diagonal. Note this influences $\llbracket SR_{15} \times v \rrbracket$ requiring a bit of book keeping taking a special $AESBox$ for the last round.

This protocol requires two rounds of communication instead of one for the first two protocols we presented. Also, it requires players to compute the linear mapping $SRMC$ locally. This, however, can be done in a simple way by a table look-up for each byte position in the input. Therefore we conservatively estimate that this protocol will require twice the time needed for protocol 1 (Fig. 8).

5   Pre-processing from the Original MiniMac Protocol

Our solutions above put some quite specialized requirements on the pre-processing material. In this section we show how one may generate such data by first running the pre-processing phase of the original MiniMac protocol and then using this to run the original MiniMac online phase. We set this up such that the function we compute will output the pre-processing material we need for

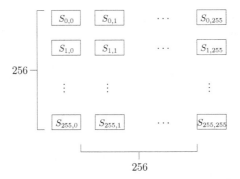

Fig. 9. 256×256 table for entry j in $\llbracket R \rrbracket$ with entry $S_{i,s} = SRMC_5 \times \Psi$ for Ψ having $sb[i+s]$ in position j.

our construction. We will describe how to generate the $AESBox$es as required by our protocol in Sect. 2. Generating pre-processing material for the protocol in Sects. 3 and 4 is a matter of applying appropriate linear transformations to the result presented here. Our goal is to generate $AESBox$ tables from a random representation $\llbracket (R_1, ..., R_{85}) \rrbracket$: Getting such a random value is directly supported by MiniMac. Now for each position j in $\llbracket (R_1, ..., R_{85}) \rrbracket$ we take 256^2 as depicted in Fig. 9. Each entry $S_{i,s}$ is a public MiniMac representation with value $SRMC_5 \times (0, ..., sb[i+s], ..., 0), i, s \in \mathbb{F}_{2^8}$.

Recall that our protocol in Sect. 2 requires an $AESBox$ to have the form:

$$s \in \mathbb{F}_{2^8}, AESBox_j[R_j + s] = \llbracket SRMC_5 \times (0, ..., sb[s], ..., 0) \rrbracket$$

This is exactly the values stored in row R_j of our 256^2 table above. The challenge is to lookup this row. To this end we start by computing the vector $\llbracket (0, ..., R_j, ..., 0) \rrbracket = (0, ..., 1, ..., 0) \times \llbracket R \rrbracket$. Recall that the original MiniMac protocol in [DZ13] allows its pre-processing to generate values of the form:

$$\llbracket R \rrbracket, \llbracket L \times R \rrbracket$$

for linear transformations L. The particular transformation we are after here is the one replicating R_j onto every position obtaining $\llbracket (R_j, ..., R_j) \rrbracket$. Then we compute $\Phi_i = (1 - \llbracket (R_j, ..., R_j) - \mathbf{i} \rrbracket^{255})$ for $i \in \mathbb{F}_{2^8}$ where $\mathbf{i}$ is the 85 vector with i in all entries. The resulting table for all $s \in \mathbb{F}_{2^8}$ is $AESBox_j[R_j + s] = \sum_{i=0}^{255} \Phi_i \times S_{i,s}$.

To see why this is actually what we wanted consider Φ_i. Because the subgroup of units in $\mathbb{F}_{2^8}$ has order 255 $\llbracket (R_j, ..., R_j) - \mathbf{i} \rrbracket^{255}$ is all ones when $R_j \neq i$ and zero only when $R_j = i$. As we want ones when they are equal we compute $(1 - (\llbracket (R_j, ..., R_j) - \mathbf{i} \rrbracket^{255})) = \Phi_i$ which is all ones only when $R_j = i$ and all zero otherwise. In this way Φ_i selects the row of $S_{i,s}$ where $i = R_j$ forming our $AESBox_j$ for each possible value of s. Now the steps above are repeated for all entries in $\llbracket R \rrbracket$ forming the full $AESBox = \{AESBox_j\}_{j=0,...,84}$.

We note that the $S_{i,j}$ tables do not all have to exist in memory at the same time; it is enough to generate the columns as needed on the fly.

6   Conclusion

We have seen that for dishonest majority protocols in the preprocessing model, the efficiency and in particular the latency of oblivious AES can be dramatically improved by tailoring the preprocessed data to the structure of AES. And that in particular that the only structure that matters is the fact that AES makes use of Sboxes with small input, so that we can use table look-up to circumvent the use of circuits to compute the non-linear parts.

Our study shows that we need only about 0.4 ms amortised time and 6 ms latency to do AES, which seems completely adequate for real life applications such as verifying 1-time passwords.

In future work, it would be interesting to see if other block ciphers or hash functions can be done securely and practically with a similar approach.

A   Reproducing our results with the implementation

Getting the code

The implementation of our work can be found on GitHub at http://tinyurl.com/qbx99jv

Requirements

- AutoMake 1.15
- Bash 3.2 or later
- Reasonable GCC compiler supporting C99 (or Windows SDK Visual Studio 2013 or later).

Building on Windows IA64

Install Visual Studio 2013 and open the solution file in `miniapps/dedicatedaes/winx64/daestest.sln`. Press `F7` in the x64-release build configuration to build the code. We have experienced problems with many small allocations on Windows making the `malloc` and `free` implementation on this system degenerate in performance.

Building on Linux and OSX

To build the code type `./build.sh release` or `./build.sh debug` depending on which configuration you want. To reproduce the performance numbers reported in the paper please build in the `release` configuration.

Generating pre-processing material for testing

Running the program with command line arguments `-prep` will generate the default set of preprocessing material needed to compute one block of cipher-text. `./miniapps/dedicatedaes/linux/src/cheetah -prep` or on windows setting the command-line arguments and pressing **F5**. Alternatively the windows .exe file can be located in `miniapps/dedicatedaes/winx64/daestest/Debug/daestest.exe`.

Running the protocol

Running the program with `-mpc -prepfile <filename>` will make the process given aes preprocessing material file for player zero listen and wait for the other players to connect.

E.g. for two players

`cheetah -mpc -prepfile ./aes_prep_4_player_0.rep` will start the listening peer listening on all interfaces port 2020. While `cheetah -mpc -prepfile ./aes_prep_4_player_1.rep -ip xxx.yyy.zzz.www -port 2020` will connect to a peer at ip-address xxx.yyy.zzz.www on port 2020.

Our experimental setup

The lab computers used for our experiments are interconnected by a 1 Gigabit LAN with the specifications below.

```
CPU: i7-3770K CPU @ 3.50GHz with 8 cores
Mem: 8Gb of Ram
Net: Gigabit LAN
OS:  3.13.0-59-generic #98-Ubuntu SMP Fri Jul 24
     21:05:26 UTC 2015 x86_64 x86_64 x86_64 GNU/Linux
```

We emphasize that our implementation carries out the computational steps of the protocol single threaded.

The implementation does use additional threads for receiving and sending network messages. In this way the computational thread is as independent of network delays as the protocol allows. More precisely, the concrete interleaving of sending and receiving messages does not effect computational progress unless data from other parties are strictly required for the protocol to continue.

References

[BP11] Boyar, J., Peralta, R.: A depth-16 circuit for the AES S-box. Cryptology ePrint Archive, Report 2011/332 (2011). http://eprint.iacr.org/

[DK10] Damgård, I., Keller, M.: Secure multiparty AES. In: Financial Cryptography and Data Security, 14th International Conference, FC , Tenerife, Canary Islands, 25–28 January 2010, Revised Selected Papers, pp. 367–374 (2010)

[DKL+12] Damgård, I., Keller, M., Larraia, E., Miles, C., Smart, N.P.: Implementing AES via an actively/covertly secure dishonest-majority MPC protocol. In: Visconti, I., De Prisco, R. (eds.) SCN 2012. LNCS, vol. 7485, pp. 241–263. Springer, Heidelberg (2012)

[DKL+13] Damgård, I., Keller, M., Larraia, E., Pastro, V., Scholl, P., Smart, N.P.: Practical covertly secure MPC for dishonest majority – or: breaking the SPDZ limits. In: Crampton, J., Jajodia, S., Mayes, K. (eds.) ESORICS 2013. LNCS, vol. 8134, pp. 1–18. Springer, Heidelberg (2013)

[DLT14] Damgård, I., Lauritsen, R., Toft, T.: An empirical study and some improvements of the minimac protocol for secure computation. IACR Cryptology ePrint Archive 2014:289 (2014)

[DN03] Damgård, I.B., Nielsen, J.B.: Universally composable efficient multiparty computation from threshold homomorphic encryption. In: Boneh, D. (ed.) CRYPTO 2003. LNCS, vol. 2729, pp. 247–264. Springer, Heidelberg (2003)

[DR00] Daemen, J., Rijmen, V.: Rijndael for AES. In: AES Candidate Conference, pp. 343–348 (2000)

[DZ13] Damgård, I., Zakarias, S.: Constant-overhead secure computation of boolean circuits using preprocessing. In: Sahai, A. (ed.) TCC 2013. LNCS, vol. 7785, pp. 621–641. Springer, Heidelberg (2013)

[FJN14] Frederiksen, T.K., Jakobsen, T.P., Nielsen, J.B.: Faster maliciously secure two-party computation using the GPU. In: Abdalla, M., De Prisco, R. (eds.) SCN 2014. LNCS, vol. 8642, pp. 358–379. Springer, Heidelberg (2014)

[GHS12] Gentry, C., Halevi, S., Smart, N.P.: Homomorphic evaluation of the AES circuit. IACR Cryptology ePrint Archive 2012:99 (2012)

[HEKM11] Huang, Y., Evans, D., Katz, J., Malka, L.: Faster secure two-party computation using garbled circuits. In: 20th USENIX Security Symposium, San Francisco, CA, USA, 8–12 August 2011

[HKS+10] Henecka, W., Kögl, S., Sadeghi, A.-R., Schneider, T., Wehrenberg, I.: TASTY: tool for automating secure two-party computations. In: Proceedings of the 17th ACM Conference on Computer and Communications Security, CCS, Chicago, Illinois, USA, 4–8 October 2010, pp. 451–462 (2010)

[IKM+13] Ishai, Y., Kushilevitz, E., Meldgaard, S., Orlandi, C., Paskin-Cherniavsky, A.: On the power of correlated randomness in secure computation. In: Sahai, A. (ed.) TCC 2013. LNCS, vol. 7785, pp. 600–620. Springer, Heidelberg (2013)

[KSS13] Keller, M., Scholl, P., Smart, N.P.: An architecture for practical actively secure MPC with dishonest majority. In: ACM SIGSAC Conference on Computer and Communications Security, CCS 2013, Berlin, Germany, 4–8 November 2013, pp. 549–560 (2013)

[NNOB12] Nielsen, J.B., Nordholt, P.S., Orlandi, C., Burra, S.S.: A new approach to practical active-secure two-party computation. In: Safavi-Naini, R., Canetti, R. (eds.) CRYPTO 2012. LNCS, vol. 7417, pp. 681–700. Springer, Heidelberg (2012)

[PSSW09] Pinkas, B., Schneider, T., Smart, N.P., Williams, S.C.: Secure two-party computation is practical. In: Matsui, M. (ed.) ASIACRYPT 2009. LNCS, vol. 5912, pp. 250–267. Springer, Heidelberg (2009)

[Yao82] Andrew Chi-Chih Yao: Protocols for secure computations (extended abstract). In: 27th Annual Symposium on Foundations of Computer Science, vol. 1982, pp. 160–164 (1986)

[Yao86] Yao, A.C.-C.: How to generate and exchange secrets. In: Foundations of Computer Science, vol. 1986, pp. 162–167. IEEE (1986)

Certificate Validation in Secure Computation and Its Use in Verifiable Linear Programming

Sebastiaan de Hoogh[1], Berry Schoenmakers[2], and Meilof Veeningen[1(✉)]

[1] Philips Research, Eindhoven, The Netherlands
`meilof.veeningen@philips.com`
[2] Eindhoven University of Technology, Eindhoven, The Netherlands

Abstract. For many applications of secure multiparty computation it is natural to demand that the output of the protocol is verifiable. Verifiability should ensure that incorrect outputs are always rejected, even if all parties executing the secure computation collude. Since the inputs to a secure computation are private, and potentially the outputs are private as well, adding verifiability is in general hard and costly.

In this paper we focus on privacy-preserving linear programming as a typical and practically relevant case for verifiable secure multiparty computation. We introduce certificate validation as an effective technique for achieving verifiable linear programming. Rather than verifying the computation proper, which involves many iterations of the simplex algorithm, we extend the output of the secure computation with a certificate. The certificate allows for efficient and direct validation of the correctness of the output. The overhead incurred by the computation of the certificate is marginal. For the validation of a certificate we design particularly efficient distributed-prover zero-knowledge proofs, fully exploiting the fact that we can use ElGamal encryption for this purpose, hence avoiding the use of more elaborate cryptosystems such as Paillier encryption.

We also formulate appropriate security definitions for our approach, and prove security for our protocols in this model, paying special attention to ensuring properties such as input independence. By means of several experiments performed in a real multi-cloud-provider environment, we show that the overall performance for verifiable linear programming is very competitive, incurring minimal overhead compared to protocols providing no correctness guarantees at all.

1 Introduction

When outsourcing a computation to the cloud, we want to be sure that the result is correct. But if the computation involves confidential inputs, e.g., of multiple mutually distrusting inputters, we also want to guarantee the privacy of the inputs. For instance, solving linear programs is useful for optimising global profits in supply chains [CdH10] or financial benchmarking [DDN+15]; confidentiality is important because the inputs are sensitive information from multiple companies but correctness is important because the outcome represents financial value.

Separately, privacy and correctness can each be achieved. Correctness can be achieved by replicating a computation and comparing the results (but this only

© Springer International Publishing Switzerland 2016
D. Pointcheval et al. (Eds.): AFRICACRYPT 2016, LNCS 9646, pp. 265–284, 2016.
DOI: 10.1007/978-3-319-31517-1_14

protects against uncorrelated failure); or by relying on the use of trusted hardware by the worker. Alternatively, correctness can be achieved without assuming uncorrelated failure or trusted hardware, by instead producing cryptographic proofs of correctness (e.g., [PHGR13]).

Achieving privacy is hard when outsourcing to a single cloud worker, but feasible if the computation is distributed between several workers. Indeed, having a single worker perform arbitrary computations on encryptions requires fully homomorphic encryption, a cryptographic primitive that is still impractical for realistic applications. But distributing computations between multiple workers in a privacy-preserving way is possible, and getting more and more practical, using multiparty computation protocols (e.g., [BD09, DKL+13]). Such protocols guarantee privacy and correctness up to a certain threshold of corrupted workers. The inputters can pick workers run at different cloud providers, thereby reducing the risk that too many of them collude or are compromised.

Unfortunately, using such techniques has a major drawback: apart from inputters having to trust the choice of workers for privacy, also recipients have to trust the choice of workers for correctness of their result. However, requiring this trust by the recipients is undesirable: it means recipients (potentially anybody, if the computation result is public) need to be involved in assessing the trustworthiness of workers; and the result may simply have too much value to allow the possibility of incorrectness.

In theory, privacy and correctness can be achieved by producing cryptographic proofs of correctness in a multi-party way. Indeed, this is the basic idea behind recent *universally verifiable* [SV15] (or *publicly auditable* [BDO14]) multiparty computation protocols. (Correctness holds regardless of the workers, but privacy only holds up to a certain maximum of corruptions: we cannot hope to circumvent this in outsourcing scenarios without resorting to fully homomorphic encryption.) However, the fact that cryptographic work (e.g., Paillier [SV15] or somewhat homomorphic [BDO14] encryption) needs to be performed for each gate in the computation, makes this expensive. Moreover, secure distributed set-up of these threshold cryptosystem is needed, which is hard in practice.

1.1   Our Contribution

In this paper, we present certificate validation as a general technique for achieving verifiable secure computation, and we demonstrate this in detail for verifiable linear programming. While solving a linear program, e.g., by means of the simplex algorithm, is complex and time-consuming, we make the critical observation that the so-called "dual solution" of a linear program allows one to efficiently verify the optimality of the result without redoing the full computation. Thus, we show how to use fast multiparty computation techniques for the computation itself, while limiting the use of slower verifiable techniques to prove the optimality of the result. We achieve further speedup by enabling the use of the ElGamal cryptosystem (implemented using elliptic curves) instead of the more expensive Paillier cryptosystem by combining the computation stage and the validation stage in a new way. We show how to enforce inputters to choose their inputs independently, and we prove security in a rigorous security model.

Concretely, our instantiation is with $n = 3$ workers (but can be easily generalised to $n = 2t + 1$ workers). We distribute the computation between all three workers using protocols that guarantee privacy if they do not collude or act maliciously (i.e., deviate from the protocol). Then, two (in general, $t + 1$) of the workers perform certificate validation to guarantee to anybody that the found solution is correct. Hence, we reach a compromise between passive and active security. One the one hand, we provide more security than passively secure multiparty computation because we guarantee correctness (in the sense that the solution is valid with respect to the certificate) regardless of corruptions. On the other hand, we provide less security than actively secure multiparty computation because we do not guarantee privacy if workers collude or act maliciously.

With our new protocol we demonstrate, for the first time, that certificate validation enables practical privacy-preserving outsourcing with correctness guarantees. We have implemented our protocol for linear programming and tested its performance in a real multi-cloud-provider environment. As mentioned, our security is in between passive and active; our experiments show that our performance is in fact much closer to passive, adding only little overhead in cases where using active security would be much slower.

1.2   Related Work

Verifiable computation, i.e., the question of how to verify correctness of computations performed by untrusted parties (without privacy) has a long history in the literature (e.g., [DFK+92, AS98, GKR08]). Recently, major practical improvements in efficiency (e.g., [PHGR13]) have allowed, in some cases, for computations to be verified faster than computed in practice.

Combining verifiability with privacy has traditionally only been considered for particular applications such as e-voting [CF85, SK95], but recent works [dH12, BDO14, SV15] have also started studying the problem of verifiability for general multiparty computation. In essence (like in our work) the correctness proofs of these works rely on zero-knowledge proofs of correct multiplication and decryption: of Paillier encryptions in [dH12, SV15], and of somewhat homomorphic encryptions in [BDO14]. Compared to these works, this work proposes a private multiplier approach that enables the use of the much more efficient ElGamal encryption scheme (besides introducing the approach of certificate validation).

Another recent line of work has combined verifiability and privacy when outsourcing computations to a single worker. However, known constructions in this line of work are unfortunately inpractical due to their use of costly primitives, e.g., fully homomorphic encryption and verifiable computation [LTV12, FGP14]; or functional encryption and garbled circuits [GKP+13]. Indeed, because such constructions require a single party to compute on encrypted data, even without offering verifiability they are inherently much heavier than our approach.

A final line of work on outsourcing computation combines privacy with "partial" verifiability in the sense that also correctness is only guaranteed if not all

$\mathcal{I}/\mathcal{P}/\mathcal{R}$	inputters/workers/recipients
party/ies P **do** S	party/ies P concurrently perform S
$\mathsf{Enc}_{\mathsf{pk}}(x;r)$	ElGamal encryption of x with public key pk, randomness r
$\mathsf{Dec}_{\mathsf{sk}}(x)$	ElGamal decryption with key (share) sk
p	prime order for ElGamal
$\oplus, \otimes$	ElGamal homomorphic addition/scalar multiplication
$\mathsf{send}(v;\mathcal{P}); \mathsf{recv}(\mathcal{P})$	send/receive v over secure, private channel (no $\mathcal{P}$ means $\mathcal{P}_1/\mathcal{P}_2$)
$\mathsf{bcast}(v)$	share v on bulletin board
$\mathrm{ZKVER}(\Sigma; v; \pi; a)$	Fiat-Shamir proof verification (p. 6)
$a \in_R S$	sample a uniformly random from S
$[x], [x']$	own/other party's additive share of x (for two workers)
$\mathcal{H}$	cryptographic hash function

Fig. 1. Notation in algorithms and protocols

workers collude. This is the case for normal multiparty computation protocols applied in an outsourcing setting [JNO14, DDN+15] as well as specialised outsourcing protocols [KMR11, ACG+14]. Compared to these works, we *do* offer correctness if all workers collude.

Finally, using short certificates to prove correctness of a larger operation has been proposed before. For instance, this idea was used to prove the correct execution of graph algorithms [ZPK14]; cf. [TT10] and references. As far as we know, we are the first to propose the use of certificates for verifiability of multiparty computation. Although we focus on linear programming, using our approach with the certificates from these works should be possible.

Outline. We first present a protocol for proving that encryptions satisfy certain polynomial relations (Sect. 2). We then combine this protocol with fast, non-verifiable multiparty computation (Sect. 3). We show with experiments that this gives practical verifiable secure linear programming (Sect. 4). We finally discuss related and future work (Sect. 5). Figure 1 shows our notation.

2   Proving Relations on ElGamal Encryptions

The main idea of our approach is to compute a function using multiparty computation, and then prove correctness of the result by proving that the input and result satisfy a number of polynomial relations. For this, we use a private multiplier-based protocol on ElGamal encryptions. Suppose that in the computation, ElGamal encryptions $X_1, \ldots, X_n$ have been produced representing the inputs and outputs of the computation, whose correctness we now want to prove. Say these ElGamal encryptions are encrypted under a private key s that is additively shared with threshold t (usually, $t = 1$), i.e., $t + 1$ workers have shares $[s]_1, \ldots, [s]_{t+1}$ such that $s = [s]_1 + \ldots + [s]_{t+1}$. Suppose the $t + 1$ workers have also additively shared the plain texts and randomness of these encryptions. Then the workers will together prove in zero knowledge that the encryptions satisfy certain relations, without learning any information about the encrypted values.

The overall approach for producing this proof is the following. For each polynomial relation $r(x_1, \ldots, x_n) = 0$ in values $x_1, \ldots, x_n$, the workers produce an encryption R of the left-hand side value. This requires additions, multiplications by a constant, and multiplications of two encryptions. The first two can be computed locally using homomorphic properties of ElGamal. Multiplications of an encryption Y by an encryption X_i of a shared plaintext x_i can be performed verifiably by letting the workers verifiably multiply their shares, and combining correctness proofs on the shares into an overall correctness proof using the multiparty Fiat-Shamir transform [SV15]. Finally, a proof that R decrypts to zero is made by homomorphically combining decryption proofs using shares $[s]_i$.

In the remainder of this section, we review the threshold homomorphic ElGamal cryptosystem and associated proofs of correct multiplication and decryption, and the multiparty variants from [SV15]. We then discuss how to use these multiplication and decryption proofs to obtain an overall proof that the polynomial relations hold. In [dHSV15], we give an explicit two-party protocol.

2.1   Threshold ElGamal and Zero-Knowledge Proofs

Recall the additively homomorphic ElGamal cryptosystem [El85]. Consider a discrete logarithm group of prime order p with generator g (e.g., points on an elliptic curve [Nat99]). Public keys are group elements h such that $s = \log_g h$ is unknown; the private key is s; encryption of $m \in \mathbb{Z}_p$ with randomness $r \in \mathbb{Z}_p$ is $(g^r, g^m h^r)$; and decryption of (a, b) is $g^m = ba^{-s}$. This cryptosystem is indeed additively homomorphic: if (a, b) encrypts m and (a', b') encrypts m', then $(a \cdot a', b \cdot b')$, denoted $(a, b) \oplus (a', b')$, encrypts $m + m'$. Moreover, if (a, b) encrypts m, then (a^α, b^α), denoted $(a, b) \otimes \alpha$, encrypts $m\alpha$; and $(a^\alpha g^r, b^\alpha h^r)$ is a random encryption of $m\alpha$. Because ElGamal decrypts to g^m and not to m, only small values can be decrypted for which the discrete logarithm problem with respect to g is feasible. In the threshold variant of ElGamal [Ped91], two parties together can perform decryption by sharing the private key s as $s = s_1 + s_2$: they publish decryptions $D_1 = ba^{-s_1}$, $D_2 = ba^{-s_2}$ from which the overall decryption is computed as $g^m = b(b^{-1} D_1)(b^{-1} D_2)$.[1] Public key shares $h_i = g^{s_i}$ are published that are used to prove correctness of decryption shares.

Correctness of decryption shares and multiplications can be proven using Σ-protocols [CDS94]. A Σ-protocol for a binary relation R is a three-move protocol in which a potentially malicious prover convinces a honest verifier that he knows a *witness* w for *statement* v such that $(v, w) \in R$. The prover sends an *announcement* to the verifier; the verifier responds with a uniformly random *challenge*; the prover sends his *response*, which the verifier verifies. We use three standard proofs: plaintext knowledge Σ_{PK}, correct multiplication Σ_{CM}, and correct decryption Σ_{CD}. Σ_{PK} proves knowledge of plaintext y and randomness r used in the statement $(a, b) = (g^r, h^r g^y)$. Σ_{CM} proves the following: given a statement consisting of encryptions (a_1, b_1), (a_2, b_2), and (a_3, b_3), the prover

---

[1] Of course, parties can alternatively share a^{-s_1}, a^{-s_2}; we prefer our description because it treats decryption and decryption shares uniformly.

knows (y, r, s) such that $a_2 = g^r$ and $b_2 = h^r g^y$ (i.e., (a_2, b_2) encrypts plaintext y with randomness r); and $a_3 = a_1^y g^s$ and $b_3 = b_1^y h^s$ (i.e., (a_3, b_3) encrypts the product encryption, randomised with s). For Σ_{CD}, recall that the decryption of plaintext (a, b) with private key (share) s is $D = ba^{-s}$. Correctness of D with respect to public key (share) h is proven by proving knowledge of the value s such that $h = g^s$ and $D^{-1}b = a^s$ using a standard equality proof.

Σ-protocols are turned into non-interactive zero-knowledge proofs with the *Fiat-Shamir heuristic* [FS86]. Namely, a party proves knowledge of a witness for statement v by generating announcement a; setting challenge $c = \mathcal{H}(v\|a\|aux)$ with some auxiliary information aux; and using this to computing response r. The proof (a, c, r) can be verified by checking that (a, c, r) is a valid Σ-protocl transcript and $\mathcal{H}(v\|a\|aux) = c$. If a can be computed from c and r, then the proof can be shortened to (c, r) and verification consists of computing a and checking $\mathcal{H}(v\|a\|aux) = c$, denoted $\text{ZKVER}(\Sigma; v; c, r; aux)$. Security is in the random oracle model, an idealised model of hash functions. Multiple proofs can use a combined challenge $c = \mathcal{H}(v_1\|a_1\|v_2\|a_2\|\dots\|aux)$.

For our Σ-protocols Σ_{PK}, Σ_{CD} and Σ_{CM}, *homomorphisms* [SV15] exist that allow provers to combine proofs for different statements into one single proof. Suppose we have an encryption X and a series of encryptions Y_i, Z_i such that Z_i is an encryption of the product of the plaintexts of X and Y_i. Then separate instances of Σ_{CM} can be used to prove that the Z_i are indeed product encryptions. If the transcripts (a_i, c, r_i) of these proofs all share the same challenge, then these transcript can be "homomorphically combined" into one transcript that proves that $\oplus Z_i$ is the product encryption of X and $\oplus Y_i$. (This combination is the pointwise product of the announcements and the pointwise sum of the responses.) If a verifier is just interested in X, $\oplus Y_i$ and $\oplus Z_i$, he can verify this combined proof instead of the individual proofs. Similarly, a homomorphism for Σ_{PK} combines proofs of plaintext knowledge for (a_i, b_i) into a proof of knowledge for $(\prod a_i, \prod b_i)$. That is, it combines proofs of knowledge of the plaintexts of X_i into one proof of (collective) knowledge of the plaintext of $\oplus X_i$. A homomorphism for Σ_{CD} combines proofs of correctness of decryption shares D_i for ciphertext (a, b) and public keys h_i into a proof of correctness of decryption $b \prod_i (b^{-1} d_i)$ for ciphertext (a, b) and public key $h = \prod_i h_i$. As above, these homomorphisms are by pointwise products and sums.

The above homomorphisms also give non-interactive zero-knowledge proofs of combined statements. Suppose parties with statements $v_1, \dots, v_t$ want to produce a proof for combined statement v. They exchange announcements $a_1, \dots, a_t$ for their shares $v_1, \dots, v_t$; compute combined announcement a; take challenge $h = \mathcal{H}(v\|a\|aux)$; and exchange responses $r_1, \dots, r_t$. The combined r with the challenge h proves collective knowledge of the witness corresponding to statement v. For security reasons, no party should be able to choose its a_i based those of others. To ensure this, before exchanging a_i the parties should first exchange commitments to these values. As above, it is possible to use the same challenge for multiple combined proofs. [SV15] proves the desirable notions of soundness and zero knowledge.

2.2   Proving and Verifying Polynomial Relations

We now present an overview of our POLYPROVE protocol that produces a proof that ElGamal encryptions $X_1, \ldots, X_n$ satisfy a given set of polynomial relations. POLYPROVE$^{\mathcal{E},\mathcal{G}}$(pk; [pk]; [sk]; $X_1, \ldots, X_n$; $[x_1], \ldots, [x_n]$; $[r_1], \ldots, [r_n]$) has two sets of inputs. First, the ElGamal public key pk and secret-shares [pk], [sk] of this key and the corresponding private key. Second, encryptions $X_1, \ldots, X_n$, and secret-shares of the respective plaintexts $[x_i]$ and randomness $[r_i]$. The set of relations to be proven is formalised by structures $\mathcal{E}$ and $\mathcal{G}$. $\mathcal{E}$ is a set of equations $x_j = 0$ ($1 \le j \le N$ for some $N \ge n$). $\mathcal{G}$ is an arithmetic circuit to compute values x_j for $j > n$. Specifically, $\mathcal{G}$ consists of gates $x_k = v$, $x_k = x_i + x_j$, $x_k = x_i \cdot v$, and $x_k = x_i \cdot x_j$ (v any constant). For multiplication $x_k = x_i \cdot x_j$, we require $1 \le j \le n$: then the workers have shared the plaintext and randomness for X_j, which we need to produce the proof. (Clearly, any set of polynomial relations can be described by such $\mathcal{E}$ and $\mathcal{G}$.) The protocol proceeds in the following steps:

- The first step of the protocol is to evaluate the circuit to obtain encryptions $X_{n+1}, \ldots, X_N$. All gates except $x_k = x_i \cdot x_j$ can be evaluated locally; for $x_k = x_i \cdot x_j$, the parties use their additive shares of the plaintext of X_j to obtain freshly randomized shares of X_k. The parties exchange these shares so that, at the end of this step, they know all encryptions $X_1, \ldots, X_N$.
- Then, the parties use the multiparty Fiat-Shamir transform to produce combined proofs of correctness of the multiplications in the arithmetic circuit $\mathcal{G}$ for $X_{n+1}, \ldots, X_N$.
- After verifying the correctness of all multiplication proofs, the parties can now safely decrypt encryptions X_j for all equations $x_j = 0$: first, they produce decryption shares with associated proofs of correctness, and then they use the multiparty Fiat-Shamir transform to produce a proof that the combination of the decryption shares produces zero. (Note that it is not necessary to exchange the decryption shares since the result is zero by assumption.)

The proof consists of the product encryptions X_k and the proofs of correct multiplication and decryption.

We remark that in the case of two parties, a slight optimization to the multiparty Fiat-Shamir transform is possible. Namely, instead of each party having to commit to each announcement before opening it, it is sufficient for the first worker to commit to its announcement; the second party to provide its announcement; and the first party to open its commitment. In [dHSV15] we explicitly give the above POLYPROVE algorithm which includes this optimization.

The corresponding algorithm POLYVER$^{\mathcal{E},\mathcal{G}}$(pk; $X_1, \ldots, X_n$; π) checks if the proof π produced by POLYPROVE is correct. The algorithm takes as arguments the public key pk, encryptions $X_1, \ldots, X_n$, and proof π as above. First, it computes missing encryptions in $\{X_{n+1}, \ldots, X_N\}$, i.e., of gates that are not inputs or multiplication results, using the homomorphic properties of ElGamal. Then, it verifies all multiplication and decryption proofs (as described in Sect. 2.1: by recomputing and hashing the announcements of the Σ-protocols). Details appear in [dHSV15].

3   Combining Computation and Validation

We now present our main protocol for privacy-preserving outsourcing with correctness guarantees. We compute a solution and a so-called "certificate" using normal multiparty computation, and then produce a proof that the solution is valid with respect to the certificate using the above ElGamal-based proofs.

3.1   Certificates and Validating Functions

To efficiently validate a computation result, we use certificates. In complexity theory, a certificate is a proof that a value lies in a certain set, that can be verified in polynomial time (see [Hro01]):

Definition 1. *Let S_1, S_2 be sets and $X \subseteq S_1$. A polynomial time computable predicate $\phi \subseteq S_1 \times S_2$ is called a* validating function *for X if $X = \{w \in S_1 \mid \exists c \in S_2 : \phi(w, c)\}$. If $\phi(w, c)$, then c is a* certificate *of the fact that $w \in X$.*

E.g., let $X = \{x \in \mathbb{N} \mid \exists y \in \mathbb{Z} : y^2 = x\}$ be the squares, then $\phi(x, y) := x \overset{?}{=} y^2$ is a validating function, and, ± 2 are certificates of the fact that $4 \in X$.

In our case, a computation is given by a computation function $(a, y) = f(x)$ and a validating function $\phi(x, a, y)$. Here, on input x, function f computes function output y and certificate a; validating function ϕ checks that y is a valid output with respect to x and a. We require that if $(a, y) = f(x)$, then $\phi(x, a, y)$, but we do not demand the converse: the outcome of the computation might not be unique, and ϕ might merely check that *some* correct solution was found, not that it was produced according to algorithm f. (For instance, a square root finder may return the positive square root while negative square root is also valid.) In our case study, we use a certificate to prove correctness of the solution to a linear program, but certificates have other applications as well, e.g., see [TT10, ZPK14] and references therein.

3.2   Security Properties

We consider the following setting. Say m inputters $\mathcal{I}_1, \ldots, \mathcal{I}_m$ want to perform a computation on their respective inputs $x = x_1, \ldots, x_m$. The computation is given by a function $(a, y) = f(x)$ and validating function $\phi(x, a, y)$, where y is the outcome of the computation and a is the certificate. The computation is distributed among n workers $\mathcal{P}_1, \ldots, \mathcal{P}_n$, and there is a *privacy threshold* $t < n$. One single recipient $\mathcal{R}$ obtains the result (we later discuss changes when multiple parties need to get the result).

We guarantee different security properties in different situations. We guarantee *correctness* of the computation result, in the sense that it satisfies ϕ, regardless of which parties are corrupted. *Privacy* means that nobody learns information about the honest parties' inputs (apart from the recipient learning the function result); we guarantee it if the workers are non-malicious (i.e., they do not deviate from the protocol) and at most t collude with each other

Table 1. Security properties and conditions on workers

Property	Satisfied if...
Correctness	Always
Input independence	Always
Privacy	No malicious and $\leq t$ colluding workers
Independence of robustness	No malicious and $\leq t$ colluding workers

(they may collude with inputters or the recipient). *Input independence* means that corrupted inputters cannot choose their input depending on honest inputs (note that this is not implied by privacy as we also want to prevent corrupted inputters from copying honest inputs); we guarantee this property regardless of which parties are corrupted. A final property often considered in this setting is *robustness*, i.e., the guarantee that parties cannot stop the computation from reaching a result; we do not aim for this property, and in fact, any inputter can make the computation break down by providing incorrect inputs. However, we do guarantee *independence of robustness* in the sense that parties cannot decide to make the computation break down depending on the inputs of honest inputters, if the workers are non-malicious or at most t collude.

Our security guarantees indeed (as discussed before) lie strictly between active and passive security for multiparty computation. Indeed, passively secure protocols do not guarantee correctness or input independence if there are malicious workers (which we do); but actively secure protocols guarantee correctness, privacy, and independence of robustness also with malicious workers (which we do not). We summarise our security properties, and the conditions on the workers under which they are satisfied, in Table 1. In Sect. 3.4, we will formalise these properties and state a security theorem for our protocol.

3.3   The VERMPC Protocol

We now present our VERMPC protocol providing the above security guarantees with privacy threshold $n = 2t + 1$. To compute $(\boldsymbol{a}, \boldsymbol{y}) = f(\boldsymbol{x})$, we use multiparty computation protocols based on (t, n)-Shamir sharing. In (t, n)-Shamir sharing, values are information-theoretically shared between the n workers such that $t + 1$ workers are needed to recover the value. In this setting, protocols exist for, e.g., multiplication, bit-decomposition, and comparison (see [dH12] for an overview); these protocols are secure against up to t passively corrupted workers. For POLYPROVE, we use additive sharing between $t + 1$ of the n workers, which also guarantees privacy as long as at most t workers collude. It is easy to switch between additive and Shamir sharing: $t + 1$ parties holding additive shares can Shamir-share them among all n; and $t + 1$ of the n parties holding Shamir shares can locally convert them to additive shares by Lagrange interpolation.

Given a multiparty computation protocol to compute $([a_1], \ldots, [y_1], \ldots, [y_l]) \leftarrow f([x_1], \ldots, [x_m])$ and the protocol POLYPROVE$^{\mathcal{E}_\phi, \mathcal{G}_\phi}(X_1, \ldots; [x_1], \ldots;$

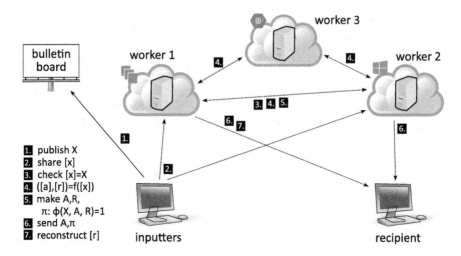

Fig. 2. VERMPC protocol with three workers (dotted lines are private, secure channels)

$[r_{x,1}], \ldots$) to prove that this result is correct, the question is how to combine them in a secure way. Figures 2, 3 show our VERMPC protocol, for concreteness instantiated with three workers ($t = 1$, $n = 3$). It consists of the following steps:

Step 1. First, the inputters announce their inputs. Each inputter encrypts its input (line 3), and makes a proof of knowledge $\pi_{x,i}$ of the corresponding plaintext (lines 4). (Here, Σ_{PK}.ann and Σ_{PK}.res denote the announcement and response function of the Σ-protocol for proving knowledge of the plaintext; see [dHSV15]; for here, it suffices that $\pi_{x,i}$ is the resulting proof of knowledge.)

These encryptions and proofs are posted on a bulletin board. To prevent corrupted inputters from adaptively choosing their input based on the inputs of others, this happens in two rounds: first, each inputter provides a hash as commitment to its input; having received the commitments of the other inputters, it then reveals the actual encrypted input and proof (line 6).

If anybody gives an incorrect input/proof, the protocol is terminated (line 7).

Step 2. Next, the inputters provide additive secret shares of the plaintext x_i and randomness $r_{x,i}$ of the encryption to the $t+1$ workers who will later perform the POLYPROVE protocol (line 8).

Step 3. The $t+1$ workers check if the provided sharing of the input is consistent with the encryptions that were posted in step 1. (Without this check, the recipient could learn information of the function output both on the encrypted and the secret-shared inputs, which should not be possible.) They do this by simply encrypting their shares of the inputs using their shares of the randomness; exchanging the result; and checking correctness using the homomorphic property of the cryptosystem (lines 11–12). (Note that this works because ElGamal is not only homomorphic in the plaintext but also in the randomness.)

Step 4. Then, the actual computation takes place (line 13). This is the only step that involves the additional workers. In this computation, the $t + 1$ workers holding additive shares of the input first Shamir-share them between all n

Require: pk/sk ElGamal public/private keys shared by $\mathcal{P}_1$, $\mathcal{P}_2$; $\boldsymbol{x} = x_1, \ldots, x_m$ inputs
Ensure: Recipient $\mathcal{R}$ returns either $\boldsymbol{y}$ with $\phi(\boldsymbol{x}, \boldsymbol{a}, \boldsymbol{y})$ for some $\boldsymbol{a}$, or $\bot$

```
 1:  protocol VERMPC^{f,φ}(pk; [pk]; [sk]; {xᵢ}ᵢ∈ᵢ)
```

$\quad$ 2: $\quad$ **parties** $\mathcal{I}_1, \ldots, \mathcal{I}_m$ **do** $\hfill \triangleright$ step 1

$\quad$ 3: $\qquad r_{\mathrm{x},i} \in_R \mathbb{Z}_p$; $X_i \leftarrow \mathsf{Enc}_{\mathsf{pk}}(x_i; r_{\mathrm{x},i})$

$\quad$ 4: $\qquad (u_i, v_i) \leftarrow \Sigma_{\mathrm{PK}}.\mathsf{ann}(X_i; x_i, r_{\mathrm{x},i})$; $c_i \leftarrow \mathcal{H}(X_i \| u_i \| i)$

$\quad$ 5: $\qquad w_i \leftarrow \Sigma_{\mathrm{PK}}.\mathsf{res}(X_i; x_i, r_{\mathrm{x},i}; u_i; v_i; c_i)$; $\pi_{\mathrm{x},i} \leftarrow (c_i, w_i)$

$\quad$ 6: $\qquad h_i \leftarrow \mathcal{H}(i \| X_i \| \pi_{\mathrm{x},i})$; $\mathsf{bcast}(h_i)$; $\mathsf{bcast}(X_i, \pi_{\mathrm{x},i})$

$\quad$ 7: $\qquad$ **if** $\exists j : h_j \neq \mathcal{H}(j \| X_j \| \pi_{\mathrm{x},j}) \vee \neg\mathrm{ZKVER}(\Sigma_{\mathrm{PK}}; X_j; \pi_{\mathrm{x},j}; j)$ **then return** $\bot$

$\quad$ 8: $\qquad x_i' \in_R \mathbb{Z}_p$; $r_{\mathrm{x},i}' \in_R \mathbb{Z}_p$; $\mathsf{send}(x_i', r_{\mathrm{x},i}'; \mathcal{P}_1)$; $\mathsf{send}(x_i - x_i', r_{\mathrm{x},i} - r_{\mathrm{x},i}'; \mathcal{P}_2)$ $\hfill \triangleright$ st 2

$\quad$ 9: $\quad$ **parties** $\{\mathcal{P}_1, \mathcal{P}_2\}$ **do**

10: $\qquad$ **for all** $1 \leq i \leq m$ **do**

11: $\qquad\quad [x_i], [r_{\mathrm{x},i}] \leftarrow \mathsf{recv}(\mathcal{I}_i)$; $[X_i] \leftarrow \mathsf{Enc}_{\mathsf{pk}}([x_i]; [r_{\mathrm{x},i}])$; $\mathsf{send}([X_i])$ $\hfill \triangleright$ step 3

12: $\qquad\quad [X_i'] \leftarrow \mathsf{recv}()$; **if** $X_i \neq [X_i] \oplus [X_i']$ **then return** $\bot$

13: $\quad$ **parties** $\{\mathcal{P}_1, \mathcal{P}_2, \mathcal{P}_3\}$ **do** $([a_1], \ldots, [a_k], [y_1], \ldots, [y_l]) \leftarrow f([x_1], \ldots, [x_m])$ $\hfill \triangleright$ st 4

14: $\quad$ **parties** $\{\mathcal{P}_1, \mathcal{P}_2\}$ **do** $\hfill \triangleright$ step 5

15: $\qquad$ **for all** $1 \leq i \leq k$ **do**

16: $\qquad\quad [r_{\mathrm{a},i}] \in_R \mathbb{Z}_p; [A_i] \leftarrow \mathsf{Enc}_{\mathsf{pk}}([a_i]; [r_{\mathrm{a},i}]); \mathsf{send}([A_i]); [A_i'] \leftarrow \mathsf{recv}(); A_i \leftarrow [A_i] \oplus [A_i']$

17: $\qquad$ **for all** $1 \leq i \leq l$ **do**

18: $\qquad\quad [r_{\mathrm{y},i}] \in_R \mathbb{Z}_p; [Y_i] \leftarrow \mathsf{Enc}_{\mathsf{pk}}([y_i]; [r_{\mathrm{y},i}]); \mathsf{send}([Y_i]); [Y_i'] \leftarrow \mathsf{recv}(); Y_i \leftarrow [Y_i] \oplus [Y_i']$

19: $\qquad \pi \leftarrow \mathrm{POLYPROVE}^{\mathcal{E}_\phi, \mathcal{G}_\phi}(\mathsf{pk}; [\mathsf{pk}]; [\mathsf{sk}]; X_1, \ldots, Y_l; [x_1]; \ldots; [r_{\mathrm{x},1}], \ldots)$

20: $\qquad \mathsf{send}(\{[y_i], [r_{\mathrm{y},i}]\}_{i=1,\ldots,l}; \mathcal{R})$ $\hfill \triangleright$ step 6

21: $\quad$ **party** $\mathcal{P}_1$ **do** $\mathsf{send}(A_1, \ldots, A_k, \pi; \mathcal{R})$ $\hfill \triangleright$ step 7

22: $\quad$ **party** $\mathcal{R}$ **do**

23: $\qquad \{[y_i]^{(1)}, [r_{\mathrm{y},i}]^{(1)}\}_{i=1,\ldots,l} \leftarrow \mathsf{recv}(\mathcal{P}_1)$; $\{[y_i]^{(2)}, [r_{\mathrm{y},i}]^{(2)}\}_{i=1,\ldots,l} \leftarrow \mathsf{recv}(\mathcal{P}_2)$

24: $\qquad (A_1, \ldots, A_k, \pi) \leftarrow \mathsf{recv}(\mathcal{P}_1)$

25: $\qquad$ **for all** $1 \leq i \leq m$ **do if** $\neg\mathrm{ZKVER}(\Sigma_{\mathrm{PK}}; X_i; \pi_{\mathrm{x},i}; j)$ **then return** $\bot$

26: $\qquad$ **for all** $1 \leq i \leq l$ **do** $Y_i \leftarrow \mathsf{Enc}_{\mathsf{pk}}([y_i]^{(1)} + [y_i]^{(2)}; [r_{\mathrm{y},i}]^{(1)} + [r_{\mathrm{y},i}]^{(2)})$

27: $\qquad$ **if** $\neg\mathrm{POLYVER}^{\mathcal{E}_\phi, \mathcal{G}_\phi}(\mathsf{pk}; X_1, \ldots, Y_l; \pi)$ **then ret** $(y_1, \ldots, y_l)$ **else ret** $\bot$

Fig. 3. VERMPC protocol with three workers

workers; then the computation is performed between the n workers; and finally, $\mathcal{P}_1, \ldots, \mathcal{P}_{t+1}$ locally convert their Shamir shares to additive shares $[a_i], [y_i]$.

Step 5. The $t+1$ workers produce the encrypted result and prove its correctness: They exchange encryptions of their respective additive shares of the certificate and result (lines 15–18). They run the POLYPROVE protocol from Sect. 2.2 to obtain a proof that $\phi(X, A, Y) = 1$ (line 19). The arithmetic circuit for ϕ should be such that each certificate value A_i and result value Y_i occurs at least once as right-hand side of a multiplication: because the workers prove knowledge of these right-hand sides, this guarantees that they know the corresponding plaintexts. Circuits usually satisfy this; otherwise dummy equations $1 \cdot Y_i = Y_i$ can be added.

Step 6. The workers send their additive shares of the result and the randomness of their encryption shares $[Y_i]$ to the recipient (line 20).

Step 7. One worker sends the encrypted and proof of correctness (line 21). The recipient checks the proofs of knowledge provided by the inputters (read from

1: **function** $\mathrm{IVERMPC}^{f,\phi}$
2:     **for all** honest inputters $\mathcal{I}_i$ **do** get x_i from party $\mathcal{I}_i$ ▷ input phase
3:     **for all** corrupted inputters $\mathcal{I}_i$ **do** get x_i from adversary $\mathcal{S}$
4:     **if** $\leq t$ passively corrupted workers **then**   ▷ computation phase
5:         compute certificate, result $\boldsymbol{a}; \boldsymbol{y} \leftarrow f(\boldsymbol{x})$
6:     **else if** $> t$ passively corrupted workers **then**
7:         send honest inputs $\{x_i\}_{i\in\mathcal{I}\setminus C}$ to adversary $\mathcal{S}$
8:         compute certificate, resut $\boldsymbol{a}; \boldsymbol{y} \leftarrow f(\boldsymbol{x})$
9:         **if** ≥ 1 actively corrupted inputter $\wedge$ $\mathcal{S}$ sends $\perp$ **then** $\boldsymbol{y} \leftarrow \perp, \ldots, \perp$
10:     **else** ▷ actively corrupted workers
11:         send honest inputs $\{x_i\}_{i\in\mathcal{I}\setminus C}$ to adversary $\mathcal{S}$
12:         get certificate, result $\boldsymbol{a}; \boldsymbol{y}$ from adversary $\mathcal{S}$
13:     **if** any x_i is $\perp$ or $\phi(\boldsymbol{x}; \boldsymbol{a}; \boldsymbol{r})$ does not hold **then** set result $\boldsymbol{r} \leftarrow \perp, \ldots, \perp$
14:     send result $\boldsymbol{r}$ to recipient $\mathcal{R}$ ▷ result phase

Fig. 4. Ideal-world trusted party capturing security guarantees for privacy threshold t

the bulletin board) (line 25); computes the encrypted result $Y_1, \ldots, Y_l$ from its shares (line 26); and calls POLYVER to verify correctness (line 27): if the proof verifies, plaintext $y_1, \ldots, y_l$ is the computation outcome.

3.4 Formal Security Model and Theorem

To formally state and prove the security of our protocol, we use the standard formalism used for multiparty computation: the ideal/real world paradigm [Can98]. We demand that the outputs of the recipient and the adversary in a protocol execution are distributed similarly to those outputs in an ideal world where the function is computed by an incorruptible trusted party. Because we provide different security guarantees under different conditions (Table 1), the trusted party gives the adversary the chance to learn inputs or manipulate outputs depending on the number and type of corruptions (cf. [SV15, BDO14]). In the ideal world, the adversary has no chances to break privacy or correctness apart from those explicitly given to it by the trusted party. If for every real-world adversary $\mathcal{A}$ there is an ideal-world adversary $\mathcal{S}_\mathcal{A}$ such that the real-world outputs are distributed the same as in the ideal world, then also real-world adversaries cannot learn or influence more than allowed by the ideal-world trusted party.

Figure 4 shows the algorithm $\mathrm{IVERMPC}^{f,\phi}$ of the ideal-world trusted party that captures the privacy and correctness guarantees discussed in Sect. 3.2. In the *input phase*, the trusted party obtains the inputs from the honest inputters (line 2) and then asks the adversary to provide the inputs on behalf of the corrupted inputters (line 3). In particular, regardless of corruptions, corrupted inputters cannot choose their inputs depending on those of honest inputters: this captures input independence. (However, they can provide $\perp$ in which case the whole computation will fail, capturing that we do not guarantee robustness.)

In the *computation phase*, we distinguish three different cases. The first, simplest case is when there are at most t passively corrupted workers: in this case, the trusted party simply evaluates the function f (line 5). In the second case, if there are more than t corrupted workers but they are all passive, we can no longer guarantee privacy. So the trusted party sends the inputs to the adversary (line 7), but still correctly computes f (line 8). If there are any actively corrupted inputters, then we do not guarantee independence of robustness. Namely, the $> t$ corrupted workers learn the honest inputs before the corrupted inputters provide their input shares, so the corrupted inputters can stop participating (but not change their inputs) depending on the honest inputs. We capture this by letting the trusted party ask $\mathcal{S}$ whether it wants to send $\perp$, in which case it sets all inputs to $\perp$ (line 9). In the third case, i.e., there are actively corrupted workers, then the passively secure protocols we use guarantee neither privacy nor correctness, so the trusted party provides the inputs to the adversary (line 11) and asks it to provide the computation result (line 12). Finally, in the *result phase*, the trusted party checks if the computation result satisfies ϕ, and otherwise sets the result to $\perp$, capturing correctness (line 13). The result is then sent to $\mathcal{R}$ (line 14).

We now precisely define the real-world and ideal-world execution models. Let C be a set of corrupted parties, of which A are actively corrupted. Let k be a security parameter. Let adversary $\mathcal{A}$ be a probabilistic polynomial time Turing machine. Define real-world execution

$$\mathrm{REAL}^{C,A}_{\mathrm{VerMPC}^{f,\phi},\mathcal{A}}(k; x_1, \ldots, x_m)$$

as the distribution consisting of the output of the recipient $\mathcal{R}$ and the adversary $\mathcal{A}$ in a protocol run. This run starts with a trusted set-up of the threshold ElGamal cryptosystem, i.e., secure paramters are chosen, everybody learns public key pk and its shares [pk]; and $\mathcal{P}_1, \mathcal{P}_2$ learn secret key shares [sk]. Next is an execution of the protocol $\mathrm{VerMPC}^{f,\phi}(\mathsf{pk}; [\mathsf{pk}]; [\mathsf{sk}]; \{x_i\}_{i\in\mathcal{I}})$ with adversary $\mathcal{A}$. We assume the communication model of [Can98], i.e., a fully connected, synchronous network with rushing; parties can use private channels and a bulletin board, and all communication is ideally authenticated (see [Can98, SV15] for details).

Similarly, the ideal-world execution given set C of corrupted parties of which A active, adversary $\mathcal{S}$, security parameter k, and inputs $x_1, \ldots, x_m$ is called

$$\mathrm{IDEAL}^{C,A}_{\mathrm{IVerMPC}^{f,\phi},\mathcal{S}}(k; x_1, \ldots, x_m);$$

it is defined as the distribution consisting of the outputs of the recipient $\mathcal{R}$ and the adversary $\mathcal{S}$ in an ideal-world protocol execution. In this execution, all parties communicate securely with an incorruptible trusted party $\mathcal{T}$ executing algorithm $\mathrm{IVerMPC}^{f,\phi}$ (Fig. 4). Honest inputters send their inputs to $\mathcal{T}$; a honest recipient gets its output from $\mathcal{T}$; and the adversary $\mathcal{S}$ can send arbitrary messages to $\mathcal{T}$ and return an arbitrary value.

Definition 2. *Protocol Π is a t-passively secure multiparty computation protocol with certificate validation if, for all probabilistic polynomial time adversaries corrupting set C of parties and actively corrupting $A \subseteq C$, there exists a probabilistic polynomial time adversary $\mathcal{S}$ such that for all possible inputs $\boldsymbol{x}$:*

$$\mathrm{REAL}^{C,A}_{\Pi,\mathcal{A}}(k; \boldsymbol{x}) \approx \mathrm{IDEAL}^{C,A}_{\mathrm{IVerMPC}^{f,\phi},\mathcal{S}}(k; \boldsymbol{x}),$$

where $\approx$ denotes computational indistinguishability in security parameter k.

Theorem 1. *Protocol* VERMPC *is a 1-passively secure multiparty computation protocol with certificate validation in the random oracle model assuming the decisional Diffie-Hellman problem in the ElGamal encryption group is hard.*

We prove this theorem in [dHSV15].

Because we use the Fiat-Shamir heuristic for non-interactive zero-knowledge proofs, our construction is only secure in the random oracle model. In this model, evaluations of the hash function $\mathcal{H}$ are modelled as queries to a "random oracle" $\mathcal{O}$ that evaluates a perfectly random function. Although security in the random oracle model does not generally imply security in the standard model, the model is commonly used to devise simple and efficient protocols, and no security problems due to its use are known. In particular, our variant of the model [SV15] assumes that the random oracle has not been used before the protocol starts: in practice, it should be instantiated with a keyed hash function, with every computation using a fresh random key.

3.5 Extensions

Input Range Checking. Some multiparty computation protocols for computing f only work if their inputs x are bounded, e.g., $-2^k \leq x \leq 2^k$. To guarantee this, input parties can use range proofs, this they can be avoided if inputs are smaller than 2^{k-1} by a statistical security parameter. In this case, input parties use statistically secure additive shares over the integers in line 8 of the protocol, i.e., they choose x'_i at random from $[-2^{k-1}, \ldots, 2^{k-1}]$. The workers check if the shares they receive in line 11 lie in this range.

Multiple Recipients and Universal Verifiability. In our model, only one party learns the result. If multiple parties need to learn the result, then the encrypted outputs $Y_1, \ldots, Y_l$ should be posted on a bulletin board to ensure consistency. Note that we cannot guarantee fairness as the workers can always choose to send their shares of the result to some recipients but not others.

At the end of the protocol, the recipient obtains not only the result; but also a non-interactive zero-knowledge proof that this result is correct. In particular, the recipient can also convince third parties that the encrypted outputs $Y_1, \ldots, Y_l$ are correct. In effect, this protocol achieves what is known as "universal" verifiability [dH12, SV15], although some small changes are needed to obtain security in the [SV15] model.

Basing it on Commitments. Our protocols could be based on Pedersen commitments instead of ElGamal encryptions. This requires a few changes; in particular, to prove that a commitments is zero, one needs to know the randomness, hence the randomness of product commitments needs to be computed in a multiparty way. Using Pedersen commitments likely leads to smaller proofs and quicker verification. Also, it is no longer needed to distribute decryption keys to the workers, hence a computation can be outsourced to anybody without preparation. On the other hand, when using Pedersen commitments, whoever knows the trapdoor $\log_g h$ used to set up the commitment scheme, can produce correctness proofs of incorrect computation results. (In the present construction, knowing the trapdoor breaks privacy but not correctness.)

Load Balancing of the 2PC. In the present protocol, two of the three workers produce the proof in line 19 while the third worker does nothing. If it is important to balance the computation load, then it is possible to let the three pairs of workers each produce one third of this proof.

Reducing Memory Load with Less Batching. In the present setup, all multiplication proofs and all decryption proofs share the same challenge. Although this gives the smallest proofs and fastest computation, it also means that the announcements for all those proofs need to be in memory at the same time. Memory usage can be reduced at the expense of a slight increase in proof and computation time by splitting the set of all equations into "blocks" and executing POLYPROVE and POLYVER for each block.

4 Secure and Verifiable Linear Programming

To demonstrate the feasibility of our approach, we apply it to linear programming. Linear programming is a broad class of optimisation problems occurring in many applications; for instance, it can be used for optimising global profits in supply chains [CdH10] or balancing risks in financial portfolios. A linear program (LP) it is given by a matrix $\mathbf{A}$ and vectors b and c. The problem is to minimize the linear function $c \cdot x = c_1 \cdot x_1 + \ldots + c_n \cdot x_n$ in variables $x = (x_1, \ldots, x_n)$, subject to constraints $\mathbf{A} \cdot x \leq b$ (where $\mathbf{A}$ is a m-by-n matrix). In addition to these constraints, we require $x_i \geq 0$. For instance, the LP

$$\mathbf{A} = \begin{pmatrix} 1 & 2 & 1 \\ 1 & -1 & 2 \end{pmatrix}, b = \begin{pmatrix} 2 \\ 1 \end{pmatrix}, c = \begin{pmatrix} -10 \\ 3 \\ -4 \end{pmatrix}$$

represents the problem to find x_1, x_2, x_3 satisfying $x_1 + 2x_2 + x_3 \leq 2$, $x_1 - x_2 + 2x_3 \leq 1$, and $x_1, x_2, x_3 \geq 0$, such that $-10x_1 + 3x_2 - 4x_3$ is minimal.

To find the optimal solution of a linear program, typically an iterative algorithm called the *simplex algorithm* is used. Each iteration involves several comparisons and a Gaussian elimination step, making it quite heavy for multiparty computation. For relatively small instances, passively secure linear programming is feasible [BD09,CdH10]; but actively secure MPC much less so when including preprocessing (as we discuss later). Fortunately, given a solution x to an LP, there is an easy way to prove that it is optimal using the optimal solution p of the so-called *dual LP* "maximise $b \cdot p$ such that $\mathbf{A} \cdot p \leq c, p \leq 0$". Namely, it is well known that solutions $(\frac{x_1}{q}, \ldots, \frac{x_n}{q})$ and $(\frac{p_1}{q}, \ldots, \frac{p_m}{q})$ $(x \in \mathbb{Z}^n, p \in \mathbb{Z}^m, q \in \mathbb{N}^+)$ are both optimal if the following conditions hold: (1) $q \geq 1$; (2) $p \cdot b = c \cdot x$; (3) $\mathbf{A} \cdot x \leq q \cdot b$; (4) $x \geq 0$; (5) $\mathbf{A}^T \cdot p \leq q \cdot c$; (6) $p \leq 0$. Also, the simplex algorithm for finding x turns out to also directly give p. To turn conditions

(1)–(6) into a set of polynomial equations, we use a certificate consisting of bit decompositions of $(q \cdot \boldsymbol{b} - \mathbf{A} \cdot \boldsymbol{x})_i$, $\boldsymbol{x}_i$, $(q \cdot \boldsymbol{c} - \mathbf{A}^T \cdot \boldsymbol{p})_i$, and $-\boldsymbol{p}_i$, and prove that each bit decomposition $b_0, b_1, \ldots$ sums up to the correct value v (with equation $v = b_0 + 2 \cdot b_1 + \ldots$) and contains only bits (with equations $b_i \cdot (1 - b_i) = 0$).

4.1 Cloud Experiments

To assess the performance of our solution, we have performed experiments in a realistic cloud outsourcing setting. Our experiments used a specially developed prototype implementation of our protocols. We took the simplex implementation from the TUeVIFF distribution of VIFF[2], and modified it to produce the certificate of correctness, i.e., the dual solution and required bit decompositions. We implemented the VERMPC protocol from Sect. 3.3 using SCAPI (http://crypto.biu.ac.il/about-scapi), a high-level cryptographic library that supports ElGamal encryption, Σ-protocols Σ_{PK} and Σ_{CD}, and the Fiat-Shamir heuristic. To implement VERMPC, we added threshold decryption, Σ_{CM}, and the POLYPROVE and POLYVER protocols from Sect. 2.2. For ElGamal we use the NIST P-224 elliptic curve using the MIRACL library.

To obtain a realistic outsourcing setting, we have deployed the three workers on three different cloud instances from different providers on different continents. See Table 2 for their specifications. In this setup, the inputters and recipient do not have to rely on any single cloud provider or jurisdiction for their privacy: they simply have to assume that the different instances do not collude. All machines ran Ubuntu 14.04.2 LTS. The recipient ran Windows 7 on an Intel i5-5300 (2.30 GHz). The VIFF part of the computation (i.e., step 4) requires authenticated and private channels; these were implemented using SSL. We did not implement steps 1–3 as they contribute minimally to the overall performance of the protocol.

We ran our experiments on several LPs: randomly-generated small LPs and larger LPs based on Netlib test programs[3]. We measured the time for VIFF to solve the LP and to compute the certificate (this depends on the LP size, number of iterations needed, and the bit length for internal computations); the time for POLYPROVE to produce a proof; and for POLYVER to verify it (this depends on the LP size and bit length for the proof).[4] Figure 5 shows the performance

Table 2. Specifications of the workers used in our cloud experiments

#	Location	Provider	Instance	Processor	Memory	€/hour
1	Ireland	Amazon EC2	m3.medium	Xeon @ 2.50 GHz	3.75 GB	€0.067
2	Virginia	MS Azure	Standard_D1	Xeon @ 2.20 GHz	3.5 GB	€0.070
3	Taiwan	Google GCE	n1-standard-1	Xeon @ 2.50 GHz	3.75 GB	€0.050

[2] Available at http://www.win.tue.nl/~berry/TUeVIFF/.

[3] http://www.netlib.org/lp/data/; coefficients rounded for performance.

[4] We took the minimal bitlengths needed for correctness. In practice, these are not known in advance: for VIFF, one takes a safe margin; for the proof, one can reveal and use the maximal bit length of all bit decompositions in the certificate.

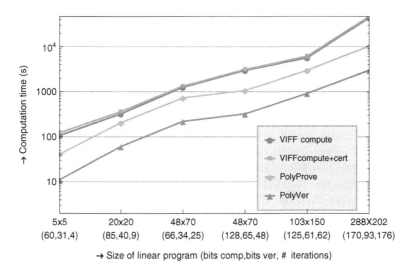

Fig. 5. Computation times of VERMPC on various LPs (the x-axis shows the LP size, bit length for VIFF, bit length for the certificate, and number of iterations)

numbers of our experiments. For the LPs in our experiments, we find that producing a proof adds little overhead to computing the solution, and that verifying the proof is much faster than participating in the computation. As a consequence, for the recipient, outsourcing both adds correctness and saves time compared to participating in the computation. Already in a setting with three inputters/recipients, privacy-preserving outsourcing makes sense; with more inputters/recipients, this performance effect is even bigger because computation scales linearly in the number of parties involved.

In general, one expects the difference between computing the solution and proving its correctness to be more pronounced for larger problems: indeed, both the computation and the correctness verification scale in the size of the LP; but computation additionally scales in the number of iterations needed to reach the optimal solution. This number of iterations typically grows with the LP size. However, we only found this for the biggest linear program, where proving is over four times faster than computing; for the other programs, this factor was around two. An explanation for this is that also the bitlength of solutions (which influences proving time) typically grows with the number of iterations.

4.2 Certificate Validation Versus Active Security

As discussed, the security guarantees of our model lie in between passive security (that does not guarantee correctness in case of active attacks) and active security (that guarantees privacy in this case). Above we showed that the overhead of our approach compared to passive security is small; we now compare our performance to that of active security. To get an idea of the performance difference between our approach and active security, we have solved several of our LP instances with an LP solver based on state-of-the-art protocols [DKL+13]. [DKL+13] combines a slow preprocessing phase, in which many random values are shared between workers, with a fast on-line phase with complexity comparable to passively secure protocols. Hence, after preprocessing has been performed, [DKL+13] can perform a computation with full privacy and correctness guarantees in about the same time as VIFF (in fact, the tested implementation is even a bit faster).

However, preprocessing is slow. No public implementation of the preprocessing phase is available, but it is possible to estimate the time it takes by measuring the amount of randomness needed for the on-line phase and combining this with available preprocessing performance figures [DKL+13]. Even with estimates that are very generous to [DKL+13], one finds that the preprocessing time is at least 15 times more than the VIFF computation time. For instance, for the first 48-by-70 linear program, we estimate that preprocessing for an actively secure computation takes at least 13 h; conversely, for out implementation, computation and proving time is close to 35 min and verification time is 3.7 min. Also, note that the timings from [DKL+13] were on a local network whereas our workers are spread over the world; and that the timings were for two parties while preprocessing scales linearly with the number of parties involved, including all inputters and recipients (whereas our performance does not depend on the number of inputters and recipients). These numbers suggest that outsourcing with certificate validation has favourable performance.

5   Concluding Remarks

In this paper, we have shown how to use certificate validation to obtain correctness guarantees for privacy-preserving outsourcing. In particular, we efficiently instantiate this idea by combining passively secure three-party computation with ElGamal-based proofs. For linear programming, verifying results takes much less time than participating in an actively secure computation; in fact, it even takes less time than participating in a passively secure computation without any correctness guarantees. Hence, for computations on inputs of mutually distrusting parties, privacy-preserving outsourcing with correctness guarantees provides a compelling combination of correctness (always) and privacy (against semi-honest, non-collaborating cloud workers).

We see several directions for improvement of our work. We have used passively secure protocols for computation; using protocols that guarantee privacy (but not correctness) also against active attacks would offer stronger protection, possibly at a low performance cost. Our implementation can be optimised, and our alternative construction using Pedersen commitment should have smaller proofs and faster verification. Much bigger speed-ups, however, (especially for linear programming) would come from using efficient zero-knowledge proofs for specific tasks, e.g., for showing that certain values are positive. In particular, range proofs are much faster to verify than our bit-wise proofs; the work of Keller et al. [KMR12] suggests ways of distributing these proofs that could be adapted to our setting. Alternatively, as we show in a recent pre-print [SVdV15], it may be possible to achieve even faster certificate validation by combining verifiable outsourcing techniques with the privacy guarantees of multiparty computation.

Acknowledgements. We thank Dan Bernstein, Thijs Laarhoven, Peter Nordholt, and Niels de Vreede for useful discussions, and the anonymous reviewers for their suggestions. This work was supported in part by the European Commission through the ICT program under contract INFSO-ICT-284833 (PUFFIN); through the FP7 programme under grant 609611 (PRACTICE); and through the H2020 programme under grant 643964 (SUPERCLOUD).

References

[ACG+14] Ananth, P., Chandran, N., Goyal, V., Kanukurthi, B., Ostrovsky, R.: Achieving privacy in verifiable computation with multiple servers – without FHE and without pre-processing. In: Krawczyk, H. (ed.) PKC 2014. LNCS, vol. 8383, pp. 149–166. Springer, Heidelberg (2014)

[AS98] Arora, S., Safra, S.: Probabilistic checking of proofs: a new characterization of NP. J. ACM **45**(1), 70–122 (1998)

[BD09] Bogetoft, P., et al.: Secure multiparty computation goes live. In: Dingledine, R., Golle, P. (eds.) FC 2009. LNCS, vol. 5628, pp. 325–343. Springer, Heidelberg (2009)

[BDO14] Baum, C., Damgård, I., Orlandi, C.: Publicly auditable secure multi-party computation. In: Abdalla, M., De Prisco, R. (eds.) SCN 2014. LNCS, vol. 8642, pp. 175–196. Springer, Heidelberg (2014)

[Can98] Canetti, R.: Security and composition of multi-party cryptographic protocols. J. Cryptology **13**, 2000 (1998)

[CdH10] Catrina, O., de Hoogh, S.: Secure multiparty linear programming using fixed-point arithmetic. In: Gritzalis, D., Preneel, B., Theoharidou, M. (eds.) ESORICS 2010. LNCS, vol. 6345, pp. 134–150. Springer, Heidelberg (2010)

[CDS94] Cramer, R., Damgård, I., Schoenmakers, B.: Proof of partial knowledge and simplified design of witness hiding protocols. In: Desmedt, Y.G. (ed.) CRYPTO 1994. LNCS, vol. 839, pp. 174–187. Springer, Heidelberg (1994)

[CF85] Cohen, J., Fischer, M.: A robust and verifiable cryptographically secure election scheme. In: Proceedings of FOCS 1985, pp. 372–382. IEEE (1985)

[DDN+15] Damgård, I., Damgård, K., Nielsen, K., Nordholt, P.S., Toft, T.: Confidential benchmarking based on multiparty computation. Cryptology eprint 2015/1006 (2015)

[DFK+92] Dwork, C., Feige, U., Kilian, J., Naor, M., Safra, M.: Low communication 2-prover zero-knowledge proofs for NP. In: Brickell, E.F. (ed.) CRYPTO 1992. LNCS, vol. 740, pp. 215–227. Springer, Heidelberg (1993)

[dH12] de Hoogh, S.: Design of large scale applications of secure multiparty computation: secure linear programming. Ph.D. thesis, Eindhoven University of Technology (2012)

[dHSV15] de Hoogh, S., Schoenmakers, B., Veeningen, M.: Certificate validation in secure computation and its use in verifiable linear programming. Cryptology eprint 2015/339 (full version of this paper) (2015)

[DKL+13] Damgård, I., Keller, M., Larraia, E., Pastro, V., Scholl, P., Smart, N.P.: Practical covertly secure MPC for dishonest majority – Or: breaking the SPDZ limits. In: Crampton, J., Jajodia, S., Mayes, K. (eds.) ESORICS 2013. LNCS, vol. 8134, pp. 1–18. Springer, Heidelberg (2013)

[El85] El Gamal, T.: A public key cryptosystem and a signature scheme based on discrete logarithms. IEEE Trans. Inf. Theory **31**(4), 469–472 (1985)

[FGP14] Fiore, D., Gennaro, R., Pastro, V.: Efficiently verifiable computation on encrypted data. In: Proceedings of CCS 2014 (2014)

[FS86] Fiat, A., Shamir, A.: How to prove yourself: practical solutions to identification and signature problems. In: Odlyzko, A.M. (ed.) CRYPTO 1986. LNCS, vol. 263, pp. 186–194. Springer, Heidelberg (1987)

[GKP+13] Goldwasser, S., Kalai, Y.T., Popa, R.A., Vaikuntanathan, V., Zeldovich, N.: Reusable garbled circuits and succinct functional encryption. In: Proceedings of STOC 2013 (2013)

[GKR08] Goldwasser, S., Kalai, Y.T., Rothblum, G.N.: Delegating computation: interactive proofs for muggles. In: Proceedings of STOC 2008, pp. 113–122 (2008)

[Hro01] Hromkovic, J.: Algorithmics for Hard Problems - Introduction to Combinatorial Optimization, Randomization, Approximation, and Heuristics. Springer, Heidelberg (2001)

[JNO14] Jakobsen, T.P., Nielsen, J.B., Orlandi, C.: A framework for outsourcing of secure computation. In: Proceedings of CCSW 2014, pp. 81–92 (2014)

[KMR11] Kamara, S., Mohassel, P., Raykova, M.: Outsourcing multi-party computation. Cryptology eprint 2011/272 (2011)

[KMR12] Keller, M., Mikkelsen, G.L., Rupp, A.: Efficient threshold zero-knowledge with applications to user-centric protocols. In: Smith, A. (ed.) ICITS 2012. LNCS, vol. 7412, pp. 147–166. Springer, Heidelberg (2012)

[LTV12] López-Alt, A., Tromer, E., Vaikuntanathan, V.: On-the-fly multiparty computation on the cloud via multikey fully homomorphic encryption. In: Proceedings of STOC 2012, pp. 1219–1234 (2012)

[Nat99] National Institute of Standards and Technology: Recommended elliptic curves for federal government use (1999). http://csrc.nist.gov/encryption

[Ped91] Pedersen, T.P.: A threshold cryptosystem without a trusted party (extended abstract). In: Davies, D.W. (ed.) EUROCRYPT 1991. LNCS, vol. 547, pp. 522–526. Springer, Heidelberg (1991)

[PHGR13] Parno, B., Howell, J., Gentry, C., Raykova, M.: Pinocchio: nearly practical verifiable computation. In: Proceedings of S&P 2013 (2013)

[SK95] Sako, K., Kilian, J.: Receipt-free mix-type voting scheme. In: Guillou, L.C., Quisquater, J.-J. (eds.) EUROCRYPT 1995. LNCS, vol. 921, pp. 393–403. Springer, Heidelberg (1995)

[SV15] Schoenmakers, B., Veeningen, M.: Universally verifiable multiparty computation from threshold homomorphic cryptosystems. In: Malkin, T., Kolesnikov, V., Lewko, A.B., Polychronakis, M. (eds.) ACNS 2015. LNCS, vol. 9092, pp. 3–22. Springer, Heidelberg (2015). doi:10.1007/978-3-319-28166-7_1

[SVdV15] Schoenmakers, B., Veeningen, M., de Vreede, N.: Trinocchio: privacy-friendly outsourcing by distributed verifiable computation. Cryptology eprint 2015/480 (2015)

[TT10] Tamassia, R., Triandopoulos, N.: Certification and authentication of data structures. In: Proceedings of AMW 2010 (2010)

[ZPK14] Zhang, Y., Papamanthou, C., Katz, J.: ALITHEIA: towards practical verifiable graph processing. In: Proceedings of CCS 2014, pp. 856–867 (2014)

Software-Only Two-Factor Authentication Secure Against Active Servers

Julien Bringer[1], Hervé Chabanne[1,2], and Roch Lescuyer[1(✉)]

[1] Morpho, Issy-les-moulineaux, France
roch.lescuyer@morpho.com
[2] Télécom ParisTech, Paris, France

Abstract. In most password-based authentication protocols, the server owns a value, the so-called verifier, that depends on the registered password. This verifier is often a one-way function of the password. Despite this protection, an unauthorized person who gets access to the verifier can mount a brute-force attack to recover the password. If the entropy of the password is low, which is often the case in practice, such an attack might be successful. Motivated by the growing need to face databases compromises, we propose a two-factor password-based authentication protocol where no information about the password leak from the server's side nor from the client's side, and where the password is not sent to the server when the user authenticates. During the registration, a user gets a value, called the token, while the server records the verifier. Our security model ensures that brute-force attacks are impossible if the server is compromised. Moreover, only on-line attempts are possible if a token is stolen. The solutions that we describe fit well into scenarios where the token is stored on a mobile phone. We provide constructions, proven secure in the random-oracle model, under standard assumptions.

1 Introduction

Password-based authentication (PA) is the most wide-spread way deployed to authenticate users. A lot of advanced forms of authentications have been developed by the research community. However, the simplest form of PA, consisting of a (login, password) pair, widely remains the method in use. Moreover, this form of authentication takes more and more place in citizen's life. Each entity and service, own their websites and information systems, and each of them asks the user for a password to get accessed.

The server managing the access rights stores the users' information. Fortunately, most of the time, the password is not directly stored. Instead, a one-way, hard to invert, function is applied to the password, and the output, aka the verifier, is stored on the server. This is a first step towards password protection. Given a verifier, a brute-force recovering of the password consists of computing the output of the function for all possible values of the password and comparing the results with the verifier, sometimes with the help of dictionaries. Such an attack is implicitly assumed to be impossible. However, the password is often

D. Pointcheval et al. (Eds.): AFRICACRYPT 2016, LNCS 9646, pp. 285–303, 2016.
DOI: 10.1007/978-3-319-31517-1_15

chosen within a limited set of passwords, which makes the brute-force recovering possible. A lot of protocols aims at being secure up to dictionary attacks. In other words, the protocol ensures that the best possible attack is the dictionary attack. In this paper, we ask whether it would be possible to go beyond this bound. Ideally, we would like the verifier to leak no information at all about the password. Thus, the consequences of server compromises would be mitigated.

OUR APPROACH. First of all, we would like to enhance the security of the most common password based authentication, so we do not want to rely on specific hardware. We rather use a two-factor software-only approach. During the registration process, a user gets a value, called the token, while the server records the verifier. The two factors needed for the authentication are the password and the token. We want the brute-force recovering of the password to be impossible given only the verifier or the token. If a token is stolen, we want that only online attempts are possible. As usual, a bound on the number of attempts will protect the password. Such an authentication is well-suited to a mobile scenario, where the token is stored on the phone. Moreover, we do not want to rely on advanced cryptographic mechanisms, such as bilinear pairings in group of prime orders. All operations in our constructions are based on operations in a group of prime order. We propose two solutions, called pw-com and pw-hom, we now briefly introduce in this introduction. In the main body of this paper, we prove them secure in the random oracle model.

HIGH-LEVEL VIEW OF OUR SOLUTION BASED ON COMMITMENTS. Our first solution, pw-com, uses the standard notion of commitments and zero-knowledge proofs. A commitment scheme enables a user to commit to a value without revealing it. The commitment binds the user to the committed value, but the user is ensured that the value is not disclosed. Then, with a zero-knowledge (ZK) proofs of knowledge (PK) of a committed value, the user proves that he knows a pair (m, r) such that $c = \texttt{Commit}(m; r)$ without revealing any information about (m, r). Let COM be a commitment scheme, $\mathcal{P}$ be a set of password and $\mathsf{H}_h : \mathcal{P} \to \mathcal{M}_C$ be an injective encoding from the set $\mathcal{P}$ to the message space $\mathcal{M}_C$ of the commitment scheme. The main idea is to store a statistically hiding commitment as verifier (on the server), and the random value used to commit to the password as token (on the client). The global parameters of the scheme are a commitment key ck. In the registration phase, the user draws a uniform value $t \leftarrow \mathcal{R}_C$ in the random space of the commitment scheme, stores t as token and sends $c := \texttt{Commit}(\mathsf{H}_h(pw); t)$. The server stores c as verifier. In the authentication phase, the user supplies a ZKPK of (h, t) such that $c = \texttt{Commit}(h; t)$. From a token t and a verifier c, a brute-force attack can be used to recover a password pw such that $c = \texttt{Commit}(\mathsf{H}_h(pw); t)$. However, the knowledge of t or c alone does not help to recover the password. On the one hand, the token t leaks no information about the password (in the information-theoretic sense), so if the token is disclosed, only guesses on-line may help to recover the password. On the other hand, the verifier c statistically hides the password, so brute-force attacks are impossible, even for an unlimited adversary. Last, but not least, an authentication session does not leak any information about the password, nor the token, thanks to the zero-knowledge property of the proof of knowledge.

HIGH-LEVEL VIEW OF OUR SOLUTION BASED ON A HOMOMORPHIC ENCRYPTION SCHEME. Updates of the password in the user's side are not possible in the pw-com solution. We now ask the question whether it is possible for a user to update its password without interaction with the server. We introduce a solution, denoted pw-hom, based on a homomorphic encryption scheme over a prime order group that achieves this property. The basic idea is the following. Let $\mathbb{G}$ be a group of prime order – in additive notation here by pure convention – and HOM-PKE be a public key homomorphic encryption scheme over $\mathbb{G}$ as message space. After an interactive registration, the client got a pair of ciphertexts encrypting two elements $(K, [h] \cdot K)$ where $K \in \mathbb{G}$ is a user-specific element and $h \in \mathbb{Z}_p^*$ an encoding of the password. The verifier contains the decryption keys (there is one key pair per user) and a hash value of K. No information about the password leak from the verifier. In the authentication step, the client computes a proof of knowledge of the password over the ciphertexts thanks to the homomorphic properties of the encryption scheme. In other words, he computes a Schnorr signature [35] over the ciphertexts. Thanks to the verifier, the server is able to decrypt the ciphertexts and to check the proof. However the server could retrieve the password from an authentication, since it could retrieve the pair $(K, [h] \cdot K)$, then the password by brute-force recovering. Therefore, the password is first masked with a fresh random value, then the proof of knowledge is performed. As a result, no information about the password leaks from the server point of view. The user-specific element K allows for authentication on behalf of the user. A crucial point is to use two independent keys to produce the pair of ciphertexts, to prevent the computation of a ciphertext of $[h] \cdot K$ from a ciphertext of K.

ON THE SALTING. In both solutions, H_h is just an encoding function without any security property. Common solutions include a salt in the password hashing, but there is no need here to include such a salt. A point to be noticed is that, without salt, the $\mathsf{H}_h(pw)$ values are not uniform, because of the distribution of pw. If H_h were a programmable random oracle, the defect would disappear. However, the hiding property of the commitment scheme and the semantic security of the encryption scheme are sufficient to hide the password and to avoid the random oracle assumption on H_h, a probably too strong and not realistic assumption. The only property we require is that H_h must be injective.

RELATED WORK. The literature about password based authentication is vast. A lot of protocols are designed to derive a session key, an issue we do not address here. The seminal work of [3] addresses authentication base on passwords only (without additional assumptions) and proposes the Encrypted Key-Exchange protocol (EKE). EKE was followed by several works [4,21,31,38,40]. The IEEE P1363.2 Password-Based Public-Key Cryptography Working Group [20] contains several of the proposals designed during the nineties. Formal models for Password-based Authenticated Key Exchange (PAKE) appeared in [2,7]. The GL's framework [16] is an abstraction of the construction of [26] (KOY), and was the first to propose a solution in the standard model. This framework underlies a lot of subsequent constructions [15,24,25,27,28]. The GK's framework [18] is an abstraction of the construction in [23], also achieves PAKE in the standard model, but without trusted setup. A third framework (KV) [29] achieves one-round PAKE in the standard model. Recent work explicitly includes the verifier in Authenticated Key Exchange [5,30]. However, they only consider security up to the brute-force attack, according to the min-entropy of the passwords.

Turning our attention to two-factor password authentication, the constructions of [36] achieve some of the properties we look for. However, they are based on pairings, a tool we want to avoid in this paper, and they lack a formal analysis. Several commercial solutions exist, such as Google Authenticator [13], Duo Security [10], HotPin [19], and PhoneFactor [33]. The work of [37] introduces a framework to analyse these two-factor authentication protocols, then proposes several efficient constructions, and apply them in different scenarios. The participants in the protocols above are, apart from the user, a client (say a web browser), a server, and a device (say a smartphone). When authenticating, the user submits a password and some additional information supplied by the device. Our model is not the one they follow. We only assume a device (say a smartphone) authenticating to a server; *i.e.*, we do not split between a client and a device. Nevertheless, our solutions can be adapted in some of the scenarios described in [37]. We elaborate on this point in the full version of our paper.

In most existing solutions, including [37], a hashed password is stored on the server and the password is sent during the authentication protocol. To the contrary, in our solutions, the password is never sent when the user authenticates.

In anonymous password authentication [39,41], several password-based sessions from the same user cannot be linked. Although, several constructions of anonymous password authentication use homomorphic encryption, our solutions do not address the same problem – we protect the passwords against the servers, without privacy of the identities –, and are more efficient.

Finally, let us mention recent concurrent and independent work which also aims at mitigating server breaches for diverse authentication tasks. [8] introduces Virtual Smart Card, a software-only solution for signature generation, in which a signature is jointly generated by a device and a server while the user owning the device authenticates to the server with a password. The signing key is distributed between the device and the server, however the server's data alone is not sufficient to produce a signature or to mount a brute-force password-recovery attack, and the same holds for the device's data. Another work [22] introduces Device-Enhanced PAKE, in which the presence of a device in the client's side is integrated into the notion of Password-based AKE. Their model is not exactly the same as ours: the value stored in the device in their model could be the token in our model, but they also assume computation abilities in the user's side, whereas we only consider the user as the owner of a device. Last, [6] also aims at mitigating server breaches in PAKE, but with a different approach: the password database is split among several servers.

ORGANISATION OF THE PAPER. Section 2 introduces some notations. Section 3 formally defines the two factor authentication we consider in this paper. Section 4 describes our solution based on commitments, and Sect. 5 our solution based on homomorphic encryption.

2   Notations

NOTATIONS. If x and y are strings over some alphabet, $x\|y$ denotes their concatenation. ε denotes the empty string. $\mathbb{R}$ denotes the set of real numbers, $\mathbb{R}^+$ the set of non-negative real numbers, $\mathbb{N}$ the set of non-negative integers, and $\mathbb{Z}_n$ the ring $\mathbb{Z}/n\mathbb{Z}$ of modular integers modulo the integer n. 1^n denotes the unary representation of the integer n. For an integer $n \geq 1$, $[1, n]$ denotes the set $\{1, \ldots, n\}$. If S is a set and D a distribution, $x \leftarrow_D S$ means that x is drawn from S according to D. $x \leftarrow S$ means that x is drawn according to the uniform distribution. $D \approx E$ denotes that two distributions D and E are indistinguishable (in a computational, statistical, or perfect sense depending on the context). If A is a (probabilistic) algorithm, $x \leftarrow \mathsf{A}(y)$ means that x is the result of the execution of A on input y, for some internal random coins. If these random coins used by A are made explicit, we note $x := \mathsf{A}(y; r)$, and A is deterministic. We note $x \in \mathsf{A}(y)$ to denote that x belongs to the support of A on input y, *i.e.*, that x might be an output of A on input y. The assignment phrase $a := E$ means that the value a receives the result of the evaluation of the (deterministic) expression E. A function $f : \mathbb{N} \to \mathbb{R}^+$ is said negligible if it decreases faster than any polynomial. $\mathsf{negl}(\cdot)$ denotes some unspecified negligible function.

3   Security Model

In this section, we formally defined the primitive we consider in this paper and the security properties it should satisfy. A *two-factor password-based authentication* TFPA scheme is given by a finite set $\mathcal{U} \subseteq \mathbb{N}$ of users and a set of functionalities $\{\mathsf{Setup}, \mathsf{Join}, \mathsf{Issue}, \mathsf{Prove}, \mathsf{Verify}\}$ described as follows.

Setup. This algorithm derives global parameters param together with a master key mk, according to a security parameter λ. We note: $\mathsf{Setup}(1^\lambda) \to (\mathsf{param}, \mathsf{mk})$.

Registration: Join↔Issue. During the registration step, the user supplies a password $pw \in \{0, 1\}^*$ (possibly with low entropy) and gets a token T. The token is recorded on the user's side and the password pw is discarded. The issuer owns the master key mk, and might additionally use some user-specific auxiliary information $info \in \{0, 1\}^*$ as input. The issuer outputs a user-specific value V, called the verifier, stored on a server. We note: $T \leftarrow \mathsf{Join}(\mathsf{param}, i, pw) \leftrightarrow \mathsf{Issue}(\mathsf{param}, \mathsf{mk}, i, info) \to V$.

Authentication: Prove↔Verify. On input a token T and a fresh password $\tilde{pw}$, a user authenticates to a server. The latter knows the verifier V recorded during the registration, and outputs a decision $dec \in \{\mathsf{accept}, \mathsf{reject}\}$. We note: $\mathsf{Prove}(\mathsf{param}, i, \tilde{pw}, T) \leftrightarrow \mathsf{Verify}(\mathsf{param}, i, V) \to dec$.

We assume that the registration protocol is carried out over a secure channel. We stress that the password is discarded after the registration. If the token is stolen (for instance on a mobile phone), the UF-pw security property below ensures that only guesses on-line are possible with the token. The password chosen by the user might depend on the information used to identify him. For instance, a 4 digits PIN on a mobile phone might be chosen by the user according to its mobile number (say the last 4 digits), the phone number being precisely the information used to enrol the user and to index the verifier. Our model takes into account such dependencies through the auxiliary information $info$.

PARAMETERS AND PASSWORD ENTROPY. A TFPA scheme is parametrized by three integers $\lambda, \beta, \tau \in \mathbb{N}$. λ manages the length of the keys, as in standard cryptographic primitives. β manages the min-entropy of the password. τ is a bound on the number of attempts to authenticate on behalf of a given user. An adversary could always try to guess the password on-line and brute-forcing the password takes 2^β attempts on average. In practice, τ is set according to β. One usually sets $\tau < \beta$ (such that $\tau \ll 2^\beta$). For instance, let us assume that a PIN number is composed of four uniform digits. The server usually aborts after $\tau = 3 < \beta \approx 13$ attempts that failed.

SECURITY PROPERTIES FOR TFPA. Intuitively, we address two kinds of problems. From the *authentication* point of view, we want that only a registered user can authenticate to the server, knowing a valid (token, password) pair. We handle this with two unforgeability games UF-token and UF-pw. In each game, the adversary tries to authenticate knowing a factor among (password, token) and ignoring the other. From the *password protection* point of view, we do not want the password to be guessable, neither from an adaptive external adversary, nor from corrupted authorities. In the security game, Password-Leakage, the adversary tries to guess the password, knowing the server's data (including the master key) but without knowing the client's data.

GAME-BASED DEFINITIONS. Properties are expressed by games played between an adversary A against a scheme Π and a challenger C. The adversary has access to a set of oracles, described below. The set $\mathcal{L}$ records what leaks to A. The tables pw, client and server record respectively the password, data on the client's side (aka the token) and data on the server's database (aka the verifier).

Password sampling. First of all, each security game is carried out according to a set $\mathcal{P}$ of passwords and a password distribution P with min-entropy β. The challenger uses them to sample the passwords (if needed).

AddUser. A supplies $(i, info, pw)$ where i is new. If $pw \neq \perp$, the challenger adds (i, pw) to $\mathcal{L}$. If $pw = \perp$, the challenger picks a random password in $pw \leftarrow_P \mathcal{P}$ according to the distribution P. In both cases the challenger computes the token T and the verifier V from pw, $info$, mk, records client$[i] := T$, server$[i] := V$ and pw$[i] := pw$.

SendToIssuer. A may interact with the issuer, and potentially deviate from the protocol during the registration. A supplies $(i, info)$. Values pw$[i]$ and client$[i]$ stay undefined. The challenger adds (i, pw), (i, client) to $\mathcal{L}$.

UserData. The adversary might ask for the i-th user's data: pw$[i]$, client$[i]$, or server$[i]$. The corresponding pair $(i, table)$ is added to $\mathcal{L}$, where $table \in \{\text{pw}, \text{client}, \text{server}\}$.

IssuerImpersonation: SendToUser$_{issuer}$. The adversary might impersonate the issuer in front of a new user i (and deviate from the protocol). The challenger plays the role of the user i, records the corresponding pw[i] and client[i], and sets server[i] := $\perp$. The pair (i, server) is added to $\mathcal{L}$.

SendToServer. An adversary tries to authenticate on behalf of a user i of her choice. According to the functionality, the challenger accepts up to τ attempts per registered user.

The verification procedure enables authentication of *registered* users only. As a consequence, the challenger responds only if i has already been enrolled.

ServerImpersonation: SendToUser$_{server}$. The adversary might impersonate the server in front of a registered user i. There is no restriction on the number of attempts for this oracle.

UF: UNFORGEABILITY. In the unforgeability games, the adversary tries to authenticate to the challenger on behalf of an existing user. The adversary may attempt an authentication without a token (she might know the password of the target user, and data from other users) or without knowing a password (she might know the token of the target user, and data from other users). The first property prevents an adversary to authenticate itself without being registered. The second property prevents an adversary to authenticate itself if it stole the token. We stress that the adversary knows whether an authentication attempt is successful or not (Fig. 1).

Property UF-token. Given a scheme Π, a probabilistic polynomial adversary A and security parameters $\lambda, \beta, \tau \in \mathbb{N}$, the probability of success in the Experiment$_{\Pi,\text{A}}^{\text{UF}-\text{token}}$ game is negligible as a function of λ:

$$\Pr\left[\text{Experiment}_{\Pi,\text{A}}^{\text{UF}-\text{token}}(\lambda, \beta, \tau) \Rightarrow 1\right] < \text{negl}(\lambda).$$

Property UF-pw. Given a scheme Π, a probabilistic polynomial adversary A and security parameters $\lambda, \beta, \tau \in \mathbb{N}$, the probability of success in the Experiment$_{\Pi,\text{A}}^{\text{UF}-\text{pw}}$ game is negligible as a function of λ, up to on-line guesses:

$$\Pr\left[\text{Experiment}_{\Pi,\text{A}}^{\text{UF}-\text{pw}}(\lambda, \beta, \tau) \Rightarrow 1\right] < \tau/2^{\beta} + \text{negl}(\lambda).$$

Remarks. In the UF-pw game, the possibility to query server[i^*] is disallowed, and it is inherent to the notion: from server[i^*] and client[i^*], an adversary could brute-force recover pin[i^*]. Moreover, we assume that the registration protocol is carried out over a secure channel; so the adversary has no access to a SendToUser$_{issuer}$ oracle, neither to transcripts of registration sessions.

PL: PASSWORD LEAKAGE. We do not want the issuer or the verifier to be able to recover the password. This point is not addressed by the UF games above. We formalize a game where the adversary knows the master key and the verifiers, but is not allowed to know the password nor the token for a target user i (it might know these data for other users). If it knew the token, it could brute-force recover the password (Fig. 2).

Experiment$_{\Pi,\mathsf{A}}^{\mathsf{UF}\text{-}\{\mathsf{token}|\mathsf{pw}\}}(\lambda,\beta,\tau)$:
- $\mathcal{L} := \emptyset$; $table[i] := \bot$, $\forall table \in \{\mathsf{pw},\mathsf{client},\mathsf{server}\}$
- $(\mathsf{param},\mathsf{mk}) \leftarrow \mathsf{Setup}(1^\lambda)$
- $(i^*, state) \leftarrow \mathsf{A}^{\mathcal{O}}(\mathsf{param})$ where $\mathcal{O} := \{\mathsf{AddUser}, \mathsf{UserData}, \mathsf{SendToIssuer},$
   $\mathsf{SendToServer}, \mathsf{SendToUser}_{server}\}$
- if ($\mathsf{server}[i^*] = \bot$): return 0
- C engages an authentication with A (acting as user i^*):
   $\mathsf{A}(state) \leftrightarrow \mathsf{Verify}(\mathsf{param}, i^*, \mathsf{server}[i^*]) \rightarrow dec$
- return 1 if ($dec = \mathsf{accept}$) and
   · in the weak UF-token case: $((i^*, \mathsf{server}) \notin \mathcal{L})$ and $((i^*, \mathsf{client}) \notin \mathcal{L})$
   · in the strong UF-token case: $((i^*, \mathsf{client}) \notin \mathcal{L})$
   · in the UF-pw case: $((i^*, \mathsf{server}) \notin \mathcal{L})$ and $((i^*, \mathsf{pw}) \notin \mathcal{L})$
   otherwise return 0

Fig. 1. The unforgeability experiment

Experiment$_{\Pi,\mathsf{A}}^{\mathsf{PL}}(\lambda,\beta,\tau)$:
- $\mathcal{L} \leftarrow \emptyset$; $table[i] := \bot$, $\forall table \in \{\mathsf{pw},\mathsf{client},\mathsf{server}\}$
- $(\mathsf{param},\mathsf{mk}) \leftarrow \mathsf{Setup}(1^\lambda)$
- $\mathcal{O} := \{\mathsf{UserData}, \mathsf{SendToUser}_{issuer}, \mathsf{SendToUser}_{server}\}$
- $(i^*, pw^*) \leftarrow \mathsf{A}^{\mathcal{O}}(\mathsf{param},\mathsf{mk},\mathsf{sk})$
- return 1 if $(\mathsf{pw}[i^*] = pw^*)$ and $(\{(i^*, \mathsf{pw}), (i^*, \mathsf{client})\} \cap \mathcal{L} = \emptyset)$,
   otherwise return 0

Fig. 2. The password leakage experiment

Property Password-Leakage. Given a scheme Π, a probabilistic polynomial adversary A and security parameters $\lambda, \beta, \tau \in \mathbb{N}$, the probability of success in the Experiment$_{\Pi,\mathsf{A}}^{\mathsf{PL}}$ game is negligible as a function of λ, up to a simple guess of the password:

$$\Pr\left[\mathsf{Experiment}_{\Pi,\mathsf{A}}^{\mathsf{PL}}(\lambda,\beta,\tau) \Rightarrow 1\right] < 1/2^\beta + \mathsf{negl}(\lambda).$$

THE NON-INTERACTIVE AUTHENTICATION SETTING. By non-interactive setting, we mean that the authentication procedure is the non-interactive signing of a random message. The message to be signed is fresh for each authentication. It might be a random challenge chosen by the verifier, or a hash value of a context-dependent message (time, verifier identity, *etc.*), determined by the protocol specification. In this setting, the $\mathsf{SendToUser}_{server}$ oracle becomes a signature oracle and the $\mathsf{SendToServer}$ oracle becomes a verification oracle. According to the standard existential unforgeability notion for signatures [17], we allow the adversary to choose the message to be signed in the security experiment.

$\mathsf{SendToUser}_{server}$. The adversary supplies (i, m), where i is a user, and m a message; and receives a signature σ on m on behalf of i.

SendToServer. The adversary supplies (i, m, σ); and receives the decision of the verifier about the validity of the signature.

At the end of the game, there is no interaction with the challenger. The adversary eventually outputs a tuple (i^*, m^*, σ^*) and wins if the non-triviality conditions hold (server$[i^*] \neq \bot$, etc.), if m^* has not been queried to the signature oracle, and if Verify(param, i^*, server$[i^*]$, m^*, σ^*) = accept.

LOCAL UPDATES. We now discuss the possibility of password updates on the client's side into our definition. May the user change its password and update its token accordingly without interaction with the server? In fact, this property is not contradictory with the notion. However, if such a procedure exists, the client cannot check if the token is correctly updated. By contradiction, he could carry out a verification protocol on its own, and therefore the adversary could try, given a token, to guess the password and check its guess. In practice, one can imagine that the old token is backed up and an authentication protocol is done with the new token. If successful, then the old token is removed. If not, the new token is removed and the old one is kept.

4   A Solution Based on Commitments

In this section, we give a simple solution that uses the standard notions of commitments and proofs of knowledge of committed values.

COMMITMENTS AND ZK PROOFS. A *commitment* scheme COM is composed of a message space $\mathcal{M}_C$, a random value space $\mathcal{R}_C$, a commitment space $\mathcal{C}_C$, and a set of functionalities {Setup, Commit, Open} as follows. On input a security parameter $\lambda \in \mathbb{N}$, the setup procedure Setup outputs a commitment key ck. The (deterministic) commitment procedure Commit takes as input a commitment key ck, a message $m \in \mathcal{M}_C$ and a random value $r \in \mathcal{R}_C$, and outputs a committed value $c \in \mathcal{C}_C$. Given a commitment key ck and a commitment c, the Open procedure simply reveals a pair (m, r): everyone can check whether $c =$ Commit(ck, $m; r$). A commitment scheme is *binding* if it is impossible to reveal a distinct pair $(m', r') \neq (m, r)$ such that $c =$ Commit$(m'; r')$. It is *hiding* if c does not leak any information about m. Both security notions can be defined in a computational, statistical or perfect (information-theoretic) sense. In our constructions, we use standard ZK proofs of knowledge, known as Σ-protocols, and their standard transformation into signatures of knowledge [9] through the Fiat-Shamir heuristic [12]. A Σ-protocol is proof of knowledge that consists of three messages: a commitment message R, a random uniform challenge $c \leftarrow \{0, 1\}^{\lambda_c}$ for some security parameter λ_c, and a response s. The Fiat-Shamir heuristic makes this ZKPK non-interactive by generating the random challenge with a hash function. The resulting signature – denoted SoK – is as secure as the underlying Σ-protocol in the programmable random oracle model.

user(ck, pw)	issuer($info$)
[user]: hash $h := \mathtt{H}_h(pw)$; pick $t \leftarrow \mathcal{R}_C$	
[user]: set $C := \mathrm{Com}.\mathtt{Commit}(\mathsf{ck}, h; t)$; store $T = t$	
send $\quad C$ $\longrightarrow$	
[issuer]:	store $V = C$

The pw-com registration protocol

user(ck, $T = t, \tilde{pw}, \mathsf{m}$)	server(ck, $V = C, \mathsf{m}$)
[user]: $\tilde{h} := \mathtt{H}_h(\tilde{pw})$; compute $\sigma \leftarrow \mathrm{SoK}.\mathtt{Sign}(\mathsf{ck}, (\tilde{h}, t), \mathsf{m})$	
send $\quad \sigma$ $\longrightarrow$	
[server]:	check $\mathrm{SoK}.\mathtt{Verify}(\mathsf{ck}, C, \sigma, \mathsf{m}) = \mathsf{accept}$

The two factor pw-com authentication protocol

Fig. 3. The two factor pw-com solution

DESCRIPTION OF THE SOLUTION. Let COM be a computationally binding, statistically hiding commitment scheme, and $\mathcal{P}$ be a set of passwords. Let $\mathtt{H}_h : \mathcal{P} \to \mathcal{M}_C$ be an injective encoding function from the passwords to the message space of the COM scheme. The Setup procedure picks a commitment key $\mathsf{ck} \leftarrow \mathrm{Com}.\mathtt{Setup}(1^\lambda)$ and returns ck as global parameter. The registration and authentication protocols are described Fig. 3.

SECURITY ANALYSIS. The pw-com solution is token-unforgeable in the programmable random oracle for $\mathtt{H}_c$ if the SoK scheme is sound, zero-knowledge, and if the COM scheme is computationally binding and statistically hiding. It is password-unforgeable in the programmable random oracle for $\mathtt{H}_c$ if the SoK scheme is sound, zero-knowledge, and if the COM scheme is computationally binding and statistically hiding. Finally, no information about the password is available from the issuer's and server's point of view, in the random oracle for $\mathtt{H}_c$ under the zero-knowledge property of the SoK scheme and the statistical hiding property of the COM scheme. For the sake of reading, we use the following notations in the proofs: S is the signature oracle (instead of $\mathsf{SendToUser}_{server}$), V the verification oracle (instead of $\mathsf{SendToServer}$). Moreover we analyse the security for a single user, the extension to several users being straightforward.

Theorem 1. *Let* A *be an adversary against the* UF*-token property of the* pw-com *scheme with a single user, running in time at most* t*, making at most* q_s *signature queries, and at most* q_v *verification queries. Then:*

$$\mathbf{Adv}^{\mathsf{UF}\text{-}\mathsf{token}}_{\mathsf{pw}\text{-}\mathsf{com},\mathsf{ck}}(\mathsf{A}) \leq \frac{q_s \cdot (q_s + q_h)}{|\mathcal{C}_C|} + \frac{q_h}{2^{\lambda_c}} + \sqrt{q_h \cdot \mathbf{Adv}^{\mathsf{bind}}_{\mathsf{ck}}(2 \cdot t)}.$$

Proof. Let $T = t$ be the token of the user, pw its password, and $V = C$ its verifier. Game G_0 is the token-unforgeability security experiment. The adversary gets pw and V but does not have access to T. In game G_1, the S oracle simulates the signature with the simulator of signature of knowledge, in the random oracle, without knowing the password, nor the token t, but knowing the verifier C. Games G_0 and G_1 are identical up to the simulation failure. Game $G_2(\text{ck}, pw, C)$ takes as input a commitment key ck, a random password $pw \leftarrow \mathcal{P}$, and a random commitment $C \leftarrow \mathcal{C}_C$. During the registration step, G_2 sets $T = \perp$ and $V = C$. The token is not available to the simulation, but is not needed to simulate the signatures. From the statistical property of the commitment scheme, we know that for all pw and commitment C, there exists $t \in \mathcal{R}_C$ such that $C = \text{Commit}(\text{H}_h(pw); t)$ with overwhelming probability. It remains to show that the probability of success of A in game G_2 is negligible. Let B be the following reduction. B receives a challenge ck for the computational binding property of the commitment scheme, picks $pw \leftarrow_D \mathcal{P}$, $C \leftarrow \mathcal{C}_C$, and runs A simulating game $G_2(\text{ck}, pw, C)$. A eventually outputs a valid signature σ for some message m. If A is successful, then the soundness of the SoK scheme is broken. $\square$

Theorem 2. *Let A be an adversary against the UF-pw property of the pw-com scheme with a single user, running in time at most t, making at most q_h random oracle queries, q_s signature queries, and $q_v < \tau$ verification queries. Then:*

$$\mathbf{Adv}^{\text{UF-pw}}_{\text{pw-com,ck}}(A) \leq \frac{\tau}{2^\beta} + \frac{(q_s \cdot (q_s + q_h) - q_v)}{|\mathcal{C}_C|} + \frac{q_h}{2^{\lambda_c}} + \sqrt{q_h \cdot \mathbf{Adv}^{\text{bind}}_{\text{ck}}(2 \cdot t)}.$$

The proof is very similar to the proof of Theorem 1. However, we must take care of the password distribution, which is not assumed to induce a uniform distribution over the message space of the commitment scheme.

Proof. Let $T = t$ be the token of the user, pw its password, and $V = C$ its verifier. Game G_0 is the password-unforgeability security experiment. The adversary gets t but does not have access to pw nor to C. The only difference between G_0 and G_1 lies in the responses from the verification oracle for 'fresh' queries, *i.e.*, message/signature pairs that do not correspond to a previous query/response pair from the signature oracle. When a valid fresh verification query is supplied, G_1 stops and returns 1. G_0 and G_1 returns 1 with the same probability, but by doing this we ensure that all fresh verification queries returns reject before the forgery. In game G_2, the signatures are simulated in the random oracle for H_c with the simulator of the signature of knowledge. Game $G_3(\text{ck}, t, C)$ is a transition step. It takes as input a commitment key ck, a random value $t \leftarrow \mathcal{R}_C$, a commitment $C := \text{Commit}(\text{H}_h(pw); t)$ for some password $pw \leftarrow_D \mathcal{P}$, and sets $T = t$ and $V = C$. The password pw is not given, but is not needed for the simulation. Game $G_4(\text{ck}, t, C)$ is as game G_3, except that $C \leftarrow \mathcal{C}_C$. If $2^\beta \ll |\mathcal{C}_C|$, it is unlikely that (event E): there exists pw such that $C = \text{Commit}(\text{H}_h(pw); t)$. The probability of E is at most $2^\beta / |\mathcal{C}_C|$, given that the encoding H_h is injective. However, the commitment key, the token and the simulated signatures are identical in both games. The S oracle can still simulate signatures since the SoK simulator can simulate signatures even for false statements. A gets information about C only through the verification oracle. If E does not happen, a potential bias of $1/2^\beta$ is introduced per verification query from A's point of view. This is because a negative response from the verification oracle reveals at most one bit of information about the password (recall that all responses from V are negative). So we bound the difference between G_3 and G_4 as follows: $\Pr\left[A^{G_3} \Rightarrow 1\right] - \Pr\left[A^{G_4} \Rightarrow 1\right] \leq q_v/2^\beta \cdot \left(1 - 2^\beta/|\mathcal{C}_C|\right)$. It remains to show that the probability of success of A in game G_4 is negligible. This is shown by reduction to the soundness of the SoK signature, as in Theorem 1. □

Theorem 3. *Let* A *be an adversary against the* PL *property of the* pw-com *scheme making at most* q_h *queries to the random oracle and* q_s *queries to the signature oracle. Then:*

$$\mathbf{Adv}^{\mathsf{PL}}_{\mathsf{pw\text{-}com},N}(A) \leq \frac{1}{2^\beta} + \frac{q_s \cdot (q_s + q_h)}{|\mathcal{C}_C|}.$$

Proof. Game G_0 is the PL security game. A eventually outputs (i^*, pw^*) and wins if it correctly guesses the password. In game G_1, signatures are simulated as in Theorem 1. In game G_2, in the registration protocol, a random commitment $C \leftarrow \mathcal{C}_C$ is drawn, instead of computing C according to the user's password. Thanks to the mask value t, the simulated C is statistically close from a real one, under the hiding property of the COM scheme. Finally, in game G_2, no password is used. The probability to win is then bound by the probability to guess a password. □

A DL INSTANTIATION. Let $(p, \mathbb{G}, G)$ be a group of prime order. During the registration step, the client takes a password hash $h := \mathsf{H}_h(pw)$, picks $t \leftarrow \mathbb{Z}_p$, and sets $C := [h] \cdot G + [t] \cdot H$. It stores t as token and sends C to the server which stores it as verifier. During the authentication step, the client takes a password hash $\tilde{h} := \mathsf{H}_h(\tilde{pw})$, picks $r_h, r_t \leftarrow \mathbb{Z}_p$, sets $R := [r_h] \cdot G + [r_t] \cdot H$;, $c := \mathsf{H}_c(R, \mathsf{m})$, $s_h := r_h + c \cdot \tilde{h} \mod p$, and $s_t := r_t + c \cdot t \mod p$. It sends (c, s_h, s_t) to the server, which computes $\tilde{R} := [s_h] \cdot G + [s_t] \cdot H - [c] \cdot C$, $\tilde{c} := \mathsf{H}_c(\tilde{R}, \mathsf{m})$;, and checks $c = \tilde{c}$. For 128 bits of security, the user's response (c, s_h, s_t) takes 640 bits.

5   A Solution Based on Homomorphic Encryption

In this section, we give a construction based on a homomorphic encryption scheme over group of prime order, which includes local updates of the password.

HASH FUNCTIONS. A *hash function* H is given by two procedures {KeyGen, Eval} as follows. On input a security parameter λ, the KeyGen procedure outputs some public parameters. In case of *keyed* hash function, the procedure also outputs a random uniform evaluation key $\mathsf{ek} \leftarrow \{0,1\}^\lambda$. On input a message $m \in \{0,1\}^{\ell(\lambda)}$ for some polynomial ℓ, the evaluation function Eval outputs a hash value in some space $\mathcal{H}$. We need *collision-resistant* (unkeyed) hash functions and *pseudo-random* keyed hash functions. The collision-resistance requires that it should impossible to exhibit two distinct messages that share the same hash value. A keyed hash function is pseudo-random if a polynomial adversary cannot distinguish whether it interacts with a pseudo-random function or a truly random one.

HOMOMORPHIC ENCRYPTION. Our constructions make use of an encryption scheme that supports some homomorphic operation and its efficient iteration. By pure convention we use here the additive symbol. A *homomorphic public key encryption* scheme HOM-PKE is composed of a message space $\mathcal{M}_E$ supporting some operation $+$, a ciphertext space $\mathcal{C}_E$, and a set of functionalities {KeyGen, Enc, Dec, Add, SMul} as follows.

On input a security parameter $\lambda \in \mathbb{N}$ and the particular message space $\mathcal{M}_E$, the key generation procedure KeyGen outputs a pair of public/private keys $(\mathsf{pk}, \mathsf{sk})$. The (probabilistic) encryption procedure Enc takes as input a public key pk and a message $m \in \mathcal{M}_E$, and outputs a ciphertext c. The (deterministic) decryption procedure takes as input a secret key sk and a ciphertext c, and outputs a message m. The homomorphic procedure takes as input a public key pk and two ciphertexts c_1, c_2, and outputs a ciphertext c'. The homomorphic operation Add is extended to an efficient scalar multiplication SMul which takes as input a public key pk, a ciphertext c and a scalar $n \in \mathbb{N}$, $n > 1$, and outputs a ciphertext c' such that for all $m \in \mathcal{M}_E$ we have: $n \times m \leftarrow \mathsf{Dec}(\mathsf{sk}, \mathsf{SMul}(\mathsf{pk}, \mathsf{Enc}(\mathsf{pk}, m), n))$. Sometimes we note $c_1 \oplus c_2$ as a shortcut for $\mathsf{Add}(\mathsf{pk}, c_1, c_2)$, and $[n] \cdot c$ for $\mathsf{SMul}(\mathsf{pk}, c, n)$, the public key being clear from the context.

The one-wayness OW property states that it is impossible given a ciphertext to recover the underlying plaintext. The semantic security, or indistinguishability against chosen plaintext attacks IND-CPA, state that no information about the plaintext leak from the ciphertext. We also assume that ciphertexts produced by the homomorphic procedures are indistinguishable from those directly produced by the encryption procedure.

DESCRIPTION OF THE SOLUTION. The Setup procedure picks a prime order group $(p, \mathbb{G}, G)$ in additive notation, with null element $\mathbf{0}$, and a master key $\mathsf{mk} \leftarrow \{0,1\}^\lambda$. Let HOM-PKE $:=$ {KeyGen, Enc, Dec, Add, SMul} be an additively homomorphic encryption scheme over $\mathcal{M}_E := (p, \mathbb{G}, G)$. Let $\mathsf{H}_u : \{0,1\}^\lambda \times \{0,1\}^* \to \mathbb{Z}_p$ be a pseudo-random hash function, $\mathsf{H}_c : \{0,1\}^\lambda \times \mathcal{C}_E^2 \times \{0,1\}^\ell \to \{0,1\}^\lambda$ a hash function (for a message length ℓ), $\mathsf{H}_v : \mathcal{C}_E \to \{0,1\}^\lambda$ another hash function, and $\mathsf{H}_h : \mathcal{P} \to \mathbb{Z}_p^*$ an injective encoding function of the passwords.

The Setup procedure returns the global parameters param $:= (p, \mathbb{G}, G)$. The registration and authentication protocols are described Fig. 4. In the registration step, the user has got a password pw, the server knows the master key mk and some user information $info$, and both know the parameters $(p, \mathbb{G}, G)$. In the authentication step, the user supplies a fresh $\tilde{pw}$ value and owns a token $(\mathsf{pk}_1, \mathsf{pk}_2, B, C)$, the server knows the verifier $(H, \mathsf{sk}_1, \mathsf{sk}_2)$, and both formerly agreed on a message m to be signed.

LOCAL UPDATES. Local updates on the client's side are possible in the pw-hom construction: (i) ask the user for pw_{old}, pw_{new}; set $h_{\mathsf{old}} := \mathsf{H}_h(pw_{\mathsf{old}})$, $h_{\mathsf{new}} := \mathsf{H}_h(pw_{\mathsf{new}})$; (ii) given $T := (\mathsf{pk}_1, \mathsf{pk}_2, B, C)$, update $C \leftarrow \mathsf{SMul}(\mathsf{pk}_2, C, h_{\mathsf{new}} \cdot (h_{\mathsf{old}})^{-1} \bmod p)$.

user(pw)	parameters $\lambda, (p, \mathbb{G}, G)$	issuer($\mathsf{mk}, info$)
[user]: compute $h := \mathtt{H}_h(pw)$; pick $a \leftarrow \mathbb{Z}_p^*$; set $Y := [a \cdot h] \cdot G$		
	send Y $\longrightarrow$	
[issuer]:		compute $k := \mathtt{H}_u(\mathsf{mk}, info)$
[issuer]:		pick $(\mathsf{pk}_1, \mathsf{sk}_1), (\mathsf{pk}_2, \mathsf{sk}_2) \leftarrow \mathtt{KeyGen}(1^\lambda, (p, \mathbb{G}, G))$
[issuer]:		set $\tilde{C} \leftarrow \mathtt{Enc}(\mathsf{pk}_2, [k] \cdot Y)$ $B \leftarrow \mathtt{Enc}(\mathsf{pk}_1, [k] \cdot G)$; $H := \mathtt{H}_v([k] \cdot G)$
[issuer]:		store $(H, \mathsf{sk}_1, \mathsf{sk}_2)$
	$\longleftarrow$ send $\mathsf{pk}_1, \mathsf{pk}_2, B, \tilde{C}$	
[user]: set $C \leftarrow \mathtt{SMul}(\mathsf{pk}_2, \tilde{C}, a^{-1} \bmod p))$		
[user]: store $(\mathsf{pk}_1, \mathsf{pk}_2, B, C)$		

The pw-hom registration protocol

user($T = (\mathsf{pk}_1, \mathsf{pk}_2, B, C), \tilde{pw}, \mathsf{m}$)	server($V = (H, \mathsf{sk}_1, \mathsf{sk}_2), \mathsf{m}$)
[user]: set $\tilde{h} := \mathtt{H}_h(\tilde{pw})$; pick $a, r \leftarrow \mathbb{Z}_p^*$	
[user]: $E \leftarrow \mathtt{SMul}(\mathsf{pk}_1, B, r)$; $F \leftarrow \mathtt{SMul}(\mathsf{pk}_2, C, a)$	
[user]: $c := \mathtt{H}_c(E, F, \mathsf{m})$; $s := r + c \cdot a \cdot \tilde{h} \bmod p$	
	send E, F, s $\longrightarrow$
[server]:	$\tilde{c} := \mathtt{H}_c(E, F, \mathsf{m})$; $R := \mathtt{Dec}(\mathsf{sk}_1, E)$; $A := \mathtt{Dec}(\mathsf{sk}_2, F)$
[server]:	check $A \neq \mathbf{0}$; $K := [s^{-1} \bmod p] \cdot (R + [\tilde{c}] \cdot A)$; check $\mathtt{H}_v(K) = H$

The pw-hom two factor authentication protocol

Fig. 4. The pw-hom two factor solution

SECURITY ANALYSIS. The pw-hom scheme is UF-pw secure under the semantic security of HOM-PKE, the pseudo-randomness of $\mathtt{H}_u$ and the collision-resistance of $\mathtt{H}_v$ in the random oracle for $\mathtt{H}_c$. It is (weakly) UF-token secure under the one-wayness of HOM-PKE, the pseudo-randomness of $\mathtt{H}_u$ and the collision-resistance of $\mathtt{H}_v$ in the random oracle for $\mathtt{H}_c$. It is PL resistant in the random oracle for $\mathtt{H}_c$.

Theorem 4. *Let* A *be an adversary against the* UF-pw *property of the* pw-hom *scheme with a single user, running in time at most t, making at most q_h random oracle queries, q_s signature queries and $q_v < \tau$ verification queries. Then:*

$$\mathbf{Adv}_{\mathsf{pw\text{-}hom}, \mathbb{G}}^{\mathsf{UF\text{-}pw}}(\mathsf{A}) \leq \frac{\tau^2}{2^\beta} + \frac{(q_s(q_s + q_h) - \tau^2)}{p} + \mathbf{Adv}_{\mathtt{H}_u}^{\mathsf{PRF}}(t) +$$
$$\tau \cdot \left(\mathbf{Adv}_{\mathsf{HOM\text{-}PKE}, \mathbb{G}}^{\mathsf{IND\text{-}CPA}}(t) + \frac{q_h}{2^{\lambda_c}} + \sqrt{q_h \cdot \left(\mathbf{Adv}_{\mathsf{HOM\text{-}PKE}, \mathbb{G}}^{\mathsf{OW}}(2 \cdot t) + \mathbf{Adv}_{\mathtt{H}_v}^{\mathsf{CR}}(2 \cdot t)\right)}\right).$$

Note on the bound. We are not able to prove that the bound is negligible beyond $\tau/2^\beta$, but only beyond $\tau^2/2^\beta$. This means that one should set $\tau^2 < \beta$ rather than $\tau < \beta$ in practice. If the encryption scheme is one-way under a computational problem, this bound could be taken back to $\tau/2^\beta$ at the cost of an interactive

Gap assumption [32], according to which the adversary has access to an oracle for the corresponding decision problem.

Proof. Let $T = (\mathsf{pk}_1, \mathsf{pk}_2, B, C)$ be the token of the user, pw its password, and $V = (H, \mathsf{sk}_1, \mathsf{sk}_2)$ its verifier. Game G_0 is the UF-pw security experiment. The adversary gets T but does not have access to pw nor to V. In game G_1, the S oracle simulates the signatures, in the random oracle, without knowing the password, nor the ciphertext C, but knowing the ciphertext B and the public encryption keys. During the n-th query, on input a message m_n, game G_1 generates the signature as follows: pick $s_n \leftarrow \mathbb{Z}_p$; $c_n \leftarrow \{0,1\}^\lambda$; $L_n \leftarrow \mathbb{G}$, set $F_n \leftarrow \mathtt{Enc}(\mathsf{pk}_2, L_n)$; $E_n \leftarrow \mathtt{Enc}(\mathsf{pk}_1, [-c_n] \cdot L_n) \oplus \mathtt{SMul}(\mathsf{pk}_1, B, s_n)$, if $\mathsf{H}_c(E_n, F_n, m_n) \neq \bot$, abort, otherwise program $\mathsf{H}_c(E_n, F_n, m_n) := c_n$, return $\sigma := (E_n, F_n, s_n)$. Game G_2 is game G_1, except that the master key mk is dropped and G_2 has an oracle access to $\mathsf{H}_u(\mathsf{mk}, \cdot)$. In game G_3, the oracle access to H_u is replaced by an oracle which draws random values in $\mathbb{Z}_p^*$. The success probability of a distinguisher between G_2 and G_3 is bound by the advantage to break the PRF property of H_u within the same time. In game G_3, A supplies $q_v + 1$ message/signature pairs, either to the verification oracle, or at the end of the game. Game G_4 guesses the first query $\mathsf{j} \in [1, q_v + 1]$ where A gives a valid 'fresh' message/signature pair. For all $j < \mathsf{j}$, the verification oracle returns 0. At the j-th query (or at the end of the game if $\mathsf{j} = q_v + 1$), game G_4 returns 1 and stops. The oracle V no longer needs the secret keys, at the price of a security loss linear in the number of verification queries. Game $\mathsf{G}_5(\mathsf{pk}_1, B, \mathsf{pk}_2, C)$ is a transition step. It takes as input two public keys pk_1, pk_2, a ciphertext B of a group element $[k] \cdot G$ under pk_1, and a ciphertext C of a group element $[\mathsf{H}_h(pw) \cdot k] \cdot G$ under pk_2 for some $pw \leftarrow_P \mathcal{P}$ and $k \leftarrow \mathbb{Z}_p$. The password pw is not given, but the simulation of signatures is not affected. Likewise, the secret keys are not given, but they are not needed in the simulation of the verification oracle. In game G_6, a random exponent $h \leftarrow \mathbb{Z}_p^*$ is taken instead of computing h from the password. With probability at most $2^\beta/p$, there exists $pw \in \mathcal{P}$ such that $h = \mathsf{H}_h(pw)$. Otherwise, the distribution of C is not as in the real protocol. Under the semantic security of HOM-PKE, C does not leak information about the password. However the verification queries might leak information about the password. We bound the difference between G_5 and G_6 as follows: $\Pr\left[\mathsf{A}^{\mathsf{G}_5} \Rightarrow 1\right] - \Pr\left[\mathsf{A}^{\mathsf{G}_6} \Rightarrow 1\right] \leq \left(q_v/2^\beta + \mathbf{Adv}_{\mathrm{HOM\text{-}PKE},\mathbb{G}}^{\mathrm{IND\text{-}CPA}}(t)\right) \cdot \left(1 - 2^\beta/p\right).$ It remains to bound the probability of success of A is game G_6. We adapt the standard forking lemma techniques [1,34] to our case, which do not involve difficulties. For the sake of place, we postpone it to the full version of our paper. □

Theorem 5. *Let* A *be an adversary against the* UF-token *property of the* pw-hom *scheme with a single user, running in time at most* t, *making at most* q_h *random oracle queries,* q_s *signature queries and* q_v *verification queries. Then:*

$$\mathbf{Adv}^{\mathsf{UF\text{-}token}}_{\mathsf{pw\text{-}hom},\mathbb{G}}(A) \leq \frac{(q_s \cdot (q_s + q_h) + q_v \cdot q_h)}{p} + \mathbf{Adv}^{\mathsf{PRF}}_{\mathsf{H}_u}(t) +$$

$$(q_v + 1) \cdot \sqrt{q_h \cdot \left(\mathbf{Adv}^{\mathsf{OW}}_{\mathsf{HOM\text{-}PKE},\mathbb{G}}(2 \cdot t) + \mathbf{Adv}^{\mathsf{CR}}_{\mathsf{H}_v}(2 \cdot t)\right)}.$$

Proof. Let $T = (\mathsf{pk}_1, \mathsf{pk}_2, B, C)$ be the token of the user, pw its password, and $V = (H, \mathsf{sk}_1, \mathsf{sk}_2)$ its verifier. Game G_0 is the token-unforgeability security experiment. The adversary gets pw but does not have access to T nor V. Games G_1 to G_4 are as in Theorem 4. Game $\mathsf{G}_5(\mathsf{pk}, \mathsf{B})$ takes as input a public key pk, and the ciphertext B of $[k] \cdot G$ for some uniform $k \leftarrow \mathbb{Z}_p$. G_4 sets $\mathsf{pk}_1 := \mathsf{pk}$ and $B := \mathsf{B}$. The distributions of B in G_4 and G_5 are identical. The ciphertext C is not computed and is never used. It remains to show that the probability of success of A in game G_5 is negligible. This is done as in Theorem 4, under the one-wayness of HOM-PKE and the collision-resistance of H_v. $\qquad \square$

Theorem 6. *Given an adversary* A *making at most* q_h *requests to the random oracle and* q_s *requests to the signature oracle, we have:*

$$\mathbf{Adv}^{\mathsf{PL}}_{\mathsf{pw\text{-}hom},\mathbb{G}}(A) \leq \frac{1}{2^\beta} + \frac{q_s \cdot (q_s + q_h)}{p}.$$

Proof. The proof is fairly straightforward, as in the pw-com scheme. $\qquad \square$

SECURITY IN PRESENCE OF LOCAL UPDATES. We add an oracle to the security game, to catch the possibility of local updates. The adversary supplies (i, pw'), for i such that $\mathsf{pw}[i]$ and $\mathsf{client}[i]$ are well-defined. If $pw' = \bot$, C picks a new password $pw' \leftarrow_P \mathcal{P}$ at random, according to the password distribution P. In both cases, C updates the tables. If UserData has already been called on $\mathsf{pw}[i]$ (or $\mathsf{client}[i]$), C sends the corresponding updated value to A. This oracle has an effect during the game G_6 in the UF-pw proof. The ciphertext C is replaced by another ciphertext $C = \mathsf{SMul}(\mathsf{pk}_2, C, t)$ for some uniform $t \leftarrow \mathbb{Z}_p^*$. The value t is not distributed as in the real protocol, but the simulation remain correct under the semantic security of the encryption scheme.

A CONCRETE INSTANTIATION WITH ELGAMAL. The black-box construction above might be instantiated with the ElGamal encryption scheme [14]. Let ELG be the following scheme. The key generation procedure takes as input a prime order group $(p, \mathbb{G}, G)$, picks a secret key $\mathsf{sk} \leftarrow \mathbb{Z}_p^*$, computes the public key $\mathsf{pk} := [\mathsf{sk}] \cdot G$, and returns the key pair $(\mathsf{pk}, \mathsf{sk}) \in \mathbb{G} \times \mathbb{Z}_p^*$. Then encryption procedure takes as input an element $M \in \mathbb{G}$ and a public key pk, picks $r \leftarrow \mathbb{Z}_p$ and outputs a ciphertext $C([r] \cdot G, M + [r] \cdot \mathsf{pk}) \in \mathbb{G}^2$. The decryption step takes as input a ciphertext (C_1, C_2) and a secret key sk, and outputs $M = C_2 - [\mathsf{sk}] \cdot C_1$. The one-wayness of the ELG scheme is equivalent to the CDH problem and its semantic security is equivalent to the DDH problem. When instantiated with the ELG scheme, a token is given by 6 group elements, a verifier by two scalars and a hash, and a signature by 4 elements plus a scalar. According to the state of the art (see for instance [11]), for 100 bits of security, the ELG scheme might be instantiated with an elliptic curve over a prime field $\mathbb{F}_q$ with a 200-bits prime q.

As a result, for 128 bits of security, a token needs $\approx 3k$ bits to be stored without compression (and half this value with compression), a verifier takes ≈ 640 bits, and a signature ≈ 2300 bits without compression (≈ 1300 with compression).

Acknowledgements. This work has been partially funded by the European FP7 EKSISTENZ (SEC-2013-607049) project. The opinions expressed in this document only represent the authors' view. They reflect neither the view of the European Commission nor the view of their employer. The authors would like to thanks Rodolphe Hugel, Olivier Cipière and Victor Servant for useful discussions, and the anonymous reviewers for their valuable comments and suggestions.

References

1. Bellare, M., Neven, G.: Multi-signatures in the plain public-key model and a general forking lemma. In: CCS 2006, pp. 390–399. ACM (2006)
2. Bellare, M., Pointcheval, D., Rogaway, P.: Authenticated key exchange secure against dictionary attacks. In: Preneel, B. (ed.) EUROCRYPT 2000. LNCS, vol. 1807, pp. 139–155. Springer, Heidelberg (2000)
3. Bellovin, S.M., Merritt, M.: Encrypted key exchange: password-based protocols secure against dictionary attacks. In: SP 1992, pp. 72–84. IEEE (1992)
4. Bellovin, S.M., Merritt, M.: Augmented encrypted key exchange: a password-based protocol secure against dictionary attacks and password file compromise. In: Computer and Communications Security (CCS 1993), pp. 244–250. ACM (1993)
5. Benhamouda, F., Pointcheval, D.: Verifier-based password-authenticated key exchange: new models and constructions. IACR ePrint Archive, 2013/833 (2013)
6. Blazy, O., Chevalier, C., Vergnaud, D.: Mitigating server breaches in password-based authentication: secure and efficient solutions. In: CT-RSA 2016 (2016). to appear
7. Boyko, V., MacKenzie, P.D., Patel, S.: Provably Secure password-authenticated key exchange using Diffie-Hellman. In: Preneel, B. (ed.) EUROCRYPT 2000. LNCS, vol. 1807, pp. 156–171. Springer, Heidelberg (2000)
8. Camenisch, J., Lehmann, A., Neven, G., Samelin, K.: Virtual smart cards: how to sign with a password and a server. IACR ePrint Archive, 2015/1101 (2015)
9. Chase, M., Lysyanskaya, A.: On signatures of knowledge. In: Dwork, C. (ed.) CRYPTO 2006. LNCS, vol. 4117, pp. 78–96. Springer, Heidelberg (2006)
10. Duo Security two-factor authentication. https://www.duosecurity.com/
11. ECRYPT II NoE. Yearly report on algorithms and keysizes. D.SPA.20 Rev. 1.0, ICT-2007-216676 ECRYPT II, 09/2012
12. Fiat, A., Shamir, A.: How to prove yourself: practical solutions to identification and signature problems. In: Odlyzko, A.M. (ed.) CRYPTO 1986. LNCS, vol. 263, pp. 186–194. Springer, Heidelberg (1987)
13. Google Authenticator. http://code.google.com/p/google-authenticator/
14. El Gamal, T.: A public key cryptosystem and a signature scheme based on discrete logarithms. In: Blakely, G.R., Chaum, D. (eds.) CRYPTO 1984. LNCS, vol. 196, pp. 10–18. Springer, Heidelberg (1985)
15. Gennaro, R.: Faster and shorter password-authenticated key exchange. In: Canetti, R. (ed.) TCC 2008. LNCS, vol. 4948, pp. 589–606. Springer, Heidelberg (2008)
16. Gennaro, R., Lindell, Y.: A framework for password-based authenticated key exchange. ACM Trans. Inf. Syst. Secur. **9**(2), 181–234 (2006)

17. Goldwasser, S., Micali, S., Rivest, R.L.: A digital signature scheme secure against adaptive chosen-message attacks. SIAM J. Comput. **17**(2), 281–308 (1988)
18. Groce, A., Katz, J.: A new framework for efficient password-based authenticated key exchange. In: CCS 2010, pp. 516–525. ACM Press (2010)
19. Celestix HotPin. http://www.celestixworks.com/HOTPin.asp
20. IEEE P1363.2. Password-based public-key cryptography working group
21. Jablon, D.P.: Extended password key exchange protocols immune to dictionary attacks. In: WET-ICE 1997, pp. 248–255. IEEE Computer Society (1997)
22. Jarecki, S., Krawczyk, H., Shirvanian, M.: Saxena device-enhanced password protocols with optimal online-offline protection. IACR Archive, 2015/1099 (2015)
23. Jiang, S., Gong, G.: Password based key exchange with mutual authentication. In: Handschuh, H., Hasan, M.A. (eds.) SAC 2004. LNCS, vol. 3357, pp. 267–279. Springer, Heidelberg (2004)
24. Katz, J., MacKenzie, P.D., Taban, G., Gligor, V.D.: Two-server password-only authenticated key exchange. In: Ioannidis, J., Keromytis, A.D., Yung, M. (eds.) ACNS 2005. LNCS, vol. 3531, pp. 1–16. Springer, Heidelberg (2005)
25. Katz, J., MacKenzie, P.D., Taban, G., Gligor, V.D.: Two-server password-only authenticated key exchange. J. Comput. Syst. Sci. **78**(2), 651–669 (2012)
26. Katz, J., Ostrovsky, R., Yung, M.: Efficient password-authenticated key exchange using human-memorable passwords. In: Pfitzmann, B. (ed.) EUROCRYPT 2001. LNCS, vol. 2045, pp. 475–494. Springer, Heidelberg (2001)
27. Katz, J., Ostrovsky, R., Yung, M.: Forward secrecy in password-only key exchange protocols. In: Cimato, S., Galdi, C., Persiano, G. (eds.) SCN 2002. LNCS, vol. 2576, pp. 29–44. Springer, Heidelberg (2003)
28. Katz, J., Ostrovsky, R., Yung, M.: Efficient and secure authenticated key exchange using weak passwords. J. ACM **57**(1), 78–116 (2009)
29. Katz, J., Vaikuntanathan, V.: Round-optimal password-based authenticated key exchange. J. Cryptol. **26**(4), 714–743 (2013)
30. Kiefer, F., Manulis, M.: Zero-knowledge password policy checks and verifier-based PAKE. In: Kutyłowski, M., Vaidya, J. (eds.) ICAIS 2014, Part II. LNCS, vol. 8713, pp. 295–312. Springer, Heidelberg (2014)
31. Lucks, S.: Open key exchange: how to defeat dictionary attacks without encrypting public keys. In: Christianson, B., Crispo, B., Lomas, M., Roe, M. (eds.) Security Protocols 1997. LNCS, vol. 1361. Springer, Heidelberg (1998)
32. Okamoto, T., Pointcheval, D.: The Gap-problems: a new class of problems for the security of cryptographic schemes. In: Kim, K. (ed.) PKC 2001. LNCS, vol. 1992. Springer, Heidelberg (2001)
33. Microsoft PhoneFactor. https://www.phonefactor.com/
34. Pointcheval, D., Stern, J.: Security arguments for digital signatures and blind signatures. J. Cryptol. **13**(3), 361–396 (2000)
35. Schnorr, C.: Efficient signature generation by smart cards. J. Cryptol. **4**(3), 161–174 (1991)
36. Scott, M.: Replacing username/password with software-only two-factor authentication. IACR IACR ePrint Archive, 2012/148 (2012)
37. Shirvanian, M., Jarecki, S., Saxena, N., Nathan, N.: Two-factor authentication resilient to server compromise using mix-bandwidth devices. In: Network and Distributed System Security - NDSS 2014. The Internet Society (2014)
38. Steiner, M., Tsudik, G., Waidner, M.: Refinement and extension of encrypted key exchange. Oper. Syst. Rev. **29**(3), 22–30 (1995)

39. Viet, D.Q., Yamamura, A., Tanaka, H.: Anonymous password-based authenticated key exchange. In: Maitra, S., Veni Madhavan, C.E., Venkatesan, R. (eds.) INDOCRYPT 2005. LNCS, vol. 3797, pp. 244–257. Springer, Heidelberg (2005)
40. Wu, T.D.: The secure remote password protocol. In: Network and Distributed System Security - NDSS 1998. The Internet Society (1998)
41. Yang, Y., Zhou, J., Weng, J., Bao, F.: A new approach for anonymous password authentication. In: ACSAC 2009, pp. 199–208. IEEE Computer Society (2009)

Public-Key Cryptography

Attribute-Based Fully Homomorphic Encryption with a Bounded Number of Inputs

Michael Clear$^{(\boxtimes)}$ and Ciarán McGoldrick

School of Computer Science and Statistics, Trinity College Dublin, Dublin, Ireland
{clearm,Ciaran.McGoldrick}@scss.tcd.ie

Abstract. The only known way to achieve Attribute-based Fully Homomorphic Encryption (ABFHE) is through indistinguishability obfuscation. The best we can do at the moment without obfuscation is Attribute-Based Leveled FHE which allows circuits of an a priori bounded depth to be evaluated. This has been achieved from the Learning with Errors (LWE) assumption. However we know of no other way without obfuscation of constructing a scheme that can evaluate circuits of unbounded depth. In this paper, we present an ABFHE scheme that can evaluate circuits of unbounded depth but with one limitation: there is a bound N on the number of inputs that can be used in a circuit evaluation. The bound N could be thought of as a bound on the number of independent senders. Our scheme allows N to be exponentially large so we can set the parameters so that there is no limitation on the number of inputs in practice. Our construction relies on multi-key FHE and leveled ABFHE, both of which have been realized from LWE, and therefore we obtain a concrete scheme that is secure under LWE.

1 Introduction

Attribute Based Encryption (ABE) is a cryptographic primitive that realizes the notion of cryptographic access control. ABE owes its roots to a simpler primitive called Identity Based Encryption (IBE), proposed in 1985 by Shamir [1] and first realized in 2001 by Boneh and Franklin [2] and Cocks [3]. IBE is centered around the notion that a user's public key can be efficiently derived from an identity string and a system-wide *public parameters*.

The identity string may be a person's email address, IP address or staff number, depending on the application. The public parameters along with a secret trapdoor (master secret key) are generated by a trusted third party referred to as the Trusted Authority (TA). The primary purpose of the TA is to issue a secret key to a user that corresponds to her identity string (we abbreviate this to *identity*) over a secure channel. The means by which the users authenticate to the TA or establish a secure channel are outside the scope of IBE. The TA uses the master secret key to derive the secret keys for identities. It is assumed

M. Clear—The author's work is funded by the Irish Research Council EMBARK Initiative.

D. Pointcheval et al. (Eds.): AFRICACRYPT 2016, LNCS 9646, pp. 307–324, 2016.
DOI: 10.1007/978-3-319-31517-1_16

that all parties have a priori access to the public parameters. For instance, the public parameters may be hard-coded in the software used by the participants, or made available on a public website.

ABE was proposed in 2005 by Sahai and Waters [4]. ABE can be viewed as a generalization of IBE. In ABE, the TA generates secret keys instead for *access policies* (an access policy prescribes the types of data a user is authorized to access). An encryptor Alice can use the public parameters to encrypt data, and embed within the ciphertext a *descriptor* of her choice that suitably describes her data. The descriptor is referred to as an *attribute*. We caution the reader that although the term *attribute* is used here in its singular form, it may in fact incorporate a *collection* of descriptive elements (which we call "subattributes"). To illustrate this, an example of an attribute is { "CS", "CRYPTO"}; it consists of the subattributes "CS" and "CRYPTO". Let us assume that this is the attribute chosen by Alice. Suppose the TA has issued a user Bob a secret key for his access policy. Keeping with the above example, suppose his access policy "accepts" an attribute if it contains **both** the subattributes "CS" and "CRYPTO". It follows that Alice's chosen attribute satisfies Bob's access policy. As such, Bob can use his secret key to decrypt Alice's ciphertext. Notice that IBE is a special case of ABE. One way of looking at an IBE scheme is that each attribute corresponds to a unique identity string such as an email address or phone number. In IBE, there is a one-to-one mapping between attributes and access policies, so Alice is given a secret key for a policy that is singularly satisfied by her identity string.

We will return to identity/attribute-based encryption momentarily. First we need to introduce the notion of fully homomorphic encryption (FHE). An FHE scheme can evaluate all polynomial-time computable functions. Strikingly, it achieves this without expanding the ciphertext size. For many applications, we need only the capability to evaluate circuits of some limited depth. Leveled FHE is a relaxation of FHE that can evaluate circuits of depth at most some positive integer d.

FHE was first constructed in 2009 in a breakthrough work by Gentry [5]. Most work on FHE has focused on the public-key setting but there has been some work in recent years in achieving FHE in the identity/attribute-based setting. Gentry et al. [6] constructed the first leveled Identity-Based Fully Homomorphic Encryption (IBFHE) scheme and the first leveled Attribute-Based Fully Homomorphic Encryption (ABFHE) scheme from the Learning with Errors (LWE) problem. Clear and McGoldrick [7] extended the former to achieve "multi-identity" leveled IBFHE where evaluation can be performed on ciphertexts associated with different identities. These schemes are leveled; that is, they are not "pure" FHE schemes insofar as all circuits cannot be evaluated, only those of limited depth.

The only known way to achieve "pure" ABFHE (i.e. where all circuits can be evaluated) is through indistinguishability obfuscation [8], namely the construction in [9]. The best we can do at the moment without obfuscation is Attribute-Based Leveled FHE which allows circuits of an a priori bounded depth to be evaluated. However we know of no other way in the identity/attribute-based setting (without obfuscation) of constructing a scheme that can evaluate circuits of unbounded depth. This has particular significance in the attribute-based setting because the public parameters are generated once and the chosen bound on the circuit may not cater for all applications where deeper circuits are needed, and it would be unwieldy to generate new public parameteres.

The technique of bootstrapping is currently the only known way to evaluate circuits of unbounded depth. Obtaining ABFHE for circuits of unbounded depth has been impeded by the fact that employing bootstrapping in the attribute-based setting (non-interactively) is particularly challenging since bootstrapping requires encryptions of the secret key bits to be available as part of the public key. Even in the identity-based setting this is a difficult challenge because one has to non-interactively derive encryptions of the secret key bits for any identity string from the public parameters alone. The only known way of doing bootstrapping is via indistinguishability obfuscation [9]. Without obfuscation, we have not been able to achieve "pure" ABFHE.

In this work we construct an almost "pure" ABFHE with one catch, namely, there is a pre-established bound N on the number of inputs to the circuits that can be evaluated where each input is a bitstring of arbitrary size. Another way of looking at it is that there is a limit on the number of independent senders who can contribute inputs to the circuit. Our construction allows N to be exponentially large because the parameter sizes grow logarithmically in N so it can be set large enough to accomodate most reasonable applications. For example by setting $N = 2^{32}$, the parameter sizes do not grow much and over 4 billion inputs can be accomodated, which is more than one would expect in reasonable applications, since each input (contributed by an independent sender) can be of arbitrary size.

1.1   Our Construction

Our construction relies on multi-key FHE and leveled ABFHE. Our use of multi-key FHE is similar to that of [10] which uses it to a achieve a non-compact form of ABFHE. If we have a leveled ABFHE with a class of access policies $\mathbb{F}$, then we get a ("pure") ABFHE for the class of policies $\mathbb{F}$ with a bound N on the number of inputs. The main idea behind our approach is that an encryptor generates a key-pair (pk, sk) for the multi-key FHE scheme and it encrypts the secret key sk with the leveled ABFHE scheme to obtain ciphertext ψ. Then the encryptor encrypts every bit of plaintext (say w bits) with the multi-key FHE scheme using pk to obtain ciphertext $c_1, \ldots, c_w$. It sends the ciphertext $\mathsf{CT} := (\psi, c_1, \ldots, c_w)$. The evaluator evaluates the circuit on the multi-key FHE ciphertexts and obtains an encrypted result c'. Then it evaluates with the leveled ABFHE scheme the decryption circuit of the multi-key FHE scheme on c' together with the encrypted secret keys (the ψ ciphertexts). We obtain a ciphertext in the leveled ABFHE scheme that encrypts the result of the computation (i.e. what c' encrypts). The size of the resulting ciphertext is independent of N and the size of the circuit. By using the multi-key FHE scheme of Clear and McGoldrick [7], we only need the leveled ABFHE scheme to have $L = O(\log N)$ levels where N is the bound on the number of inputs.

We say a scheme is *single-attribute* if it only allows homomorphic evaluation on ciphertexts with the same attribute. Otherwise, if it allows evaluation on ciphertexts with different attributes, we refer to the scheme as *multi-attribute*. Whether our construction is single-attribute or multi-attribute depends on the underlying leveled ABFHE scheme that is used. Single-attribute leveled ABFHE has been achieved from LWE as has multi-identity leveled IBFHE. However multi-attribute leveled ABFHE is an open problem. Hence we cannot obtain "pure" multi-attribute ABFHE with a bounded number of inputs because there are no multi-attribute leveled schemes. The closest we have is multi-identity leveled IBFHE. The only known way of achieving "pure" multi-attribute ABFHE is via indistinguishability obfuscation.

1.2   Organization

This paper is organized as follows. In Sect. 2, we introduce definitions that we use throughout the paper including a definition of Attribute-Based Homomorphic Encryption. In Sect. 3, we provide security definitions and introduce a new security notion which we call EVAL-SIM security. In Sect. 4, we present our construction of ABFHE with a bounded number of inputs. We prove security of the construction in Sect. 5. We review our main result and its corollaries in Sect. 6.

2   Definitions

Let us briefly recall the definition of key-policy attribute based encryption (KP-ABE). A trusted authority (TA) generates public parameters and a master secret key. It uses its master secret key to generate secret keys for *access policies*. Alice encrypts her data, using the public parameters, under an "attribute" of her choice in some designated set of "attributes". An "attribute" serves as a descriptor for the data she is encrypting. Suppose the TA issues a secret key for some *access policy* to Bob. This access policy essentially describes which attributes he is authorized to access. Bob can decrypt Alice's ciphertext if its associated "attribute" satisfies his *access policy*.

We refer to the result of an evaluation on a set of ciphertexts as an *evaluated ciphertext*.

2.1   Models of Access Control for Decryption

A model of access control for decryption specifies how decryption of an evaluated ciphertext is to be performed. Consider an evaluated ciphertext c' associated with d attributes $a_1, \ldots, a_d \in \mathbb{A}$. There are two primary models of decryption, each with different strengths and weaknesses. Both models will be considered in turn.

Atomic Access. The intended semantics of this model is that a user should only be able to decrypt an evaluated ciphertext c' if she has a secret key for a policy f that satisfies *all* d attributes $a_1, \ldots, a_d$. In other words, policies are enforced in an "all or nothing" manner. So in order to decrypt a ciphertext c', the decryptor needs a secret key for a policy f with $f(a_1) = \cdots = f(a_d) = 1$. Furthermore, it captures the natural requirement that a decryptor be authorized *completely* to access data associated with a particular attribute.

Non-atomic Access - Collaborative Decryption. The interpretation in this model is that a group of users can pool together their secret keys to decrypt a ciphertext c'. In other words, there may not be a single $f \in \mathbb{F}$ that satisfies all d attributes (or no user holds a secret key for such an f), but the users may share secret keys for a set of policies that "covers all" d attributes. In other words, suppose the group of users have (between them) secret keys for policies $f_1, \ldots, f_k \in \mathbb{F}$. In this model, they can decrypt c' if and only if for every $i \in [d]$, there exists a $j \in [k]$ such that $f_j(a_i) = 1$.

How is decryption performed? There are a few possible approaches:

1. Every user in the group shares their secret keys with each other, and all users can decrypt. However, this violates the *principle of least privilege* and gives users in the group access to data they might not have been explicitly authorized to access.
2. Perform decryption collaboratively using a multi-party computation (MPC) protocol. This approach has been suggested in other works including [11]. The advantage of this approach is that it does not leak any party's secret key to the other parties.
3. It is possible that a user has been issued secret keys for several policies. For example: ABE for disjunctive policies can be achieved with an IBE scheme where the TA issues secret keys for different identities (treated as "attributes") to the same user.
4. Collaborative decryption subsumes the *functionality* of the atomic model i.e. a user with a single policy f satisfying all d attributes can still decrypt on her own.

Our syntax for attribute based homomorphic encryption (ABHE) presented in the next section generalizes both models. We do this by parameterizing an ABHE scheme with an integer $\mathcal{K} \in [\mathcal{D}]$, which specifies the maximum number of keys that can be passed to the decryption algorithm. The setting $\mathcal{K} = 1$ specifies the atomic model whereas the setting $\mathcal{K} = \mathcal{D}$ specifies the collaborative model. Note that this is only a syntactic rule, it does not pertain to enforcing the security property of either model. Our "default" model, assumed implicitly without further qualification, is the collaborative model. This is for several reasons, which we will enumerate now:

- In the identity-based setting, collaborative decryption is necessary. In this context, a single f is satisfied by only one attribute (i.e. identity). Suppose an evaluation is performed on ciphertexts with *different* identities to yield an evaluated ciphertext c'. Clearly, there is no single secret key that is sufficient to decrypt c', since each secret key corresponds to exactly one identity.

Because IBE is a special case of ABE, and very important in its own right, we want to ensure we allow multi-identity evaluation.

- As noted above, the collaborative model subsumes the *functionality* of the atomic model. The greater flexibility of permitting multiple users to collaboratively decrypt (such as via MPC) invites more applications.

2.2   Definition of Attribute-Based Homomorphic Encryption

Recall the definition of ABE from the introduction. An ABE scheme with message space $\mathcal{M}$, attribute space $\mathbb{A}$ and class of supported access policies $\mathbb{F}$ is a tuple of probabilistic polynomial time (PPT) algorithms (Setup, KeyGen, Encrypt, Decrypt).

Definition 1 (Degree of Composition). *Let $c_1, \ldots, c_\ell$ be input ciphertexts to an evaluation. Each ciphertext c_i is associated with an attribute $a_i \in \mathbb{A}$. The* **degree of composition** *of the evaluation is the number of **distinct** attributes among the a_i; that is, the cardinality of the set $|\{a_1, \ldots, a_\ell\}|$.*

We use the symbol d to denote the degree of composition. When the context is unambiguous, the term is abbreviated to *degree*. We use the symbol $\mathcal{D}$ to denote the *maximum* degree of composition supported by a particular system.

Definition 2. *A (Key-Policy) Attribute-Based Homomorphic Encryption (ABHE) scheme $\mathcal{E}^{(\mathcal{D}, \mathcal{K})}$ for an integer $\mathcal{D} > 0$ and an integer $\mathcal{K} \in [\mathcal{D}]$ is defined with respect to a message space $\mathcal{M}$, an attribute space $\mathbb{A}$, a class of access policies $\mathbb{F} \subseteq \mathbb{A} \to \{0, 1\}$, and a class of circuits $\mathbb{C} \subseteq \mathcal{M}^* \to \mathcal{M}$. An ABHE scheme is a tuple of PPT algorithms (Setup, KeyGen, Encrypt, Decrypt, Eval) where Setup, KeyGen, Encrypt are defined equivalently to KP-ABE. We denote by $\mathcal{C}$ the ciphertext space. The decryption algorithm Decrypt and evaluation algorithm Eval are defined as follows:*

- Decrypt($\mathsf{sk}_{f_1}, \ldots, \mathsf{sk}_{f_\mathcal{K}}, c$): *On input a sequence of $\mathcal{K} \leq \mathcal{K}$ secret keys for policies $f_1, \ldots, f_\mathcal{K} \in \mathbb{F}$ and a ciphertext c, output a plaintext $\mu' \in \mathcal{M}$ iff every attribute associated with c is satisfied by at least one of the f_i; output $\bot$ otherwise.*
- Eval($\mathsf{PP}, C, c_1, \ldots, c_\ell$): *On input public parameters PP, a circuit $C \in \mathbb{C}$ and ciphertexts $c_1, \ldots, c_\ell \in \mathcal{C}$, output an evaluated ciphertext $c' \in \mathcal{C}$.*

More precisely, Eval is required to satisfy the following properties:

- *Over all choices of $(\mathsf{PP}, \mathsf{MSK}) \leftarrow \mathsf{Setup}(1^\lambda)$, $C : \mathcal{M}^\ell \to \mathcal{M} \in \mathbb{C}$, every $d \leq \mathcal{D}$, $a_1, \ldots, a_\ell \in \mathbb{A}$ s.t $|\{a_1, \ldots, a_\ell\}| = d$, $\mu_1, \ldots, \mu_\ell \in \mathcal{M}$, $c_i \leftarrow \mathsf{Encrypt}(\mathsf{PP}, a_i, \mu_i)$ for $i \in [\ell]$, and $c' \leftarrow \mathsf{Eval}(\mathsf{PP}, C, c_1, \ldots, c_\ell)$:*
  - **Correctness**

$$\mathsf{Decrypt}(\mathsf{sk}_{f_1}, \ldots, \mathsf{sk}_{f_\mathcal{K}}, c') = C(\mu_1, \ldots, \mu_\ell) \ \text{iff} \ \forall i \in [d] \ \exists j \in [\mathcal{K}] \quad f_j(a_i) = 1 \tag{2.1}$$

  for any $\mathcal{K} \in [\mathcal{K}]$, any $f_1, \ldots, f_\mathcal{K} \in \mathbb{F}$, and any $\mathsf{sk}_{f_j} \leftarrow \mathsf{KeyGen}(\mathsf{MSK}, f_j)$ for $j \in [\mathcal{K}]$.

- **Compactness** *There exists a fixed polynomial* $s(\cdot, \cdot)$ *for the scheme such that*

$$|c'| \leq s(\lambda, d). \tag{2.2}$$

The complexity of all algorithms may depend on $\mathcal{D}$. Furthermore, the size of freshly encrypted ciphertexts, the size of the public parameters and the size of secret keys may depend on $\mathcal{D}$. On the other hand, the size of the evaluated ciphertext c' must remain independent of $\mathcal{D}$ (along with the size of the circuit C), but it may depend on the *actual* number of distinct attributes, d, used in the evaluation. Note that single-attribute ABHE is the special case where $\mathcal{D} = 1$ i.e. evaluation is correct only for ciphertexts associated with the same attribute. As mentioned earlier, $\mathcal{K} = 1$ represents the atomic model of decryption whereas $\mathcal{K} = \mathcal{D}$ represents the collaborative model. When the parameter $\mathcal{K}$ is omitted, it can be assumed that $\mathcal{K} = \mathcal{D}$; that is, the notation $\mathcal{E}^{(\mathcal{D})}$ is shorthand for $\mathcal{E}^{(\mathcal{D},\mathcal{D})}$.

Definition 3. *Multi-Attribute ABHE (MA-ABHE) is a primitive with the same syntax as ABHE except that its* Setup *algorithm takes an additional input* $\mathcal{D} > 0$, *which is the maximum degree of composition to support. An instance of MA-ABHE can be viewed as a family of ABHE schemes* $\{\mathcal{E}^{(\mathcal{D})} = (\mathsf{Setup}, \mathsf{KeyGen}, \mathsf{Encrypt}, \mathsf{Decrypt}, \mathsf{Eval})\}_{\mathcal{D}>0}$.

Remark 1. In the constructions considered in this work, $\mathbb{A}$ consists of attributes of fixed length. However the above definition is easily generalized to capture variable-length attributes, by letting $|c'|$ grow with the total length of the d distinct attributes.

A concrete ABHE scheme is characterized by three facets: (1). its supported computations (i.e. the class of circuits $\mathbb{C}$); (2). its supported access policies (the class of access policies $\mathbb{F}$); and (3). its supported composition defined by its maximum degree of composition, $\mathcal{D}$.

3   Security Definitions

3.1   Semantic Security

The semantic security definition for ABHE is the same as that for ABE, except that the adversary has access to the Eval algorithm as well. There are two definitions of semantic security for ABE: selective and adaptive security. In the selective security game, the adversary chooses the attribute to attack before receiving the public parameters whereas in the adaptive game, the adversary chooses its target attribute after receiving the public parameters. We denote the selective definition by IND-sel-CPA and the adaptive definition by IND-AD-CPA.

3.2   Simulation Model of Evaluation

Let $\mathcal{D}$ and $\mathcal{K} \leq \mathcal{D}$ be fixed parameters denoting the maximum degree of composition and the maximum number of keys passed to the decryption algorithm respectively. Consider ciphertexts $c_1, \ldots, c_\ell$ encrypted under attributes $a_1, \ldots, a_\ell$ respectively. We expect that a ciphertext c' resulting from an evaluation on $c_1, \ldots, c_\ell$ be decryptable by a set of policies $\{f_i\}_{i \in [k]}$ with $k \in [\mathcal{K}]$ if the following two conditions are satisfied: (1). the degree of composition d is less than $\mathcal{D}$ (i.e. $d := |\{a_1, \ldots, a_\ell\}| \leq \mathcal{D}$) - for convenience we re-label the d *distinct* attributes as $a_1, \ldots, a_d$; and (2). for every $i \in [d]$, there exists a $j \in [k]$ with $f_j(a_i) = 1$.

Ideally a user who does not have keys for such a set of policies $\{f_i\}_{i \in [k]}$ should not learn anything about c' except that it is associated with the attributes $a_1, \ldots, a_d$. This implies that such a user should not be able to efficiently decide whether c' was produced from $c_1, \ldots, c_\ell$ or an alternative sequence of ciphertexts $d_1, \ldots, d_{\ell'}$ with the same collection of distinct attributes $a_1, \ldots, a_d$. We now give a definition of security that captures the fact that an adversary learns nothing from an evaluated ciphertext other than that it was generated from a particular circuit and is associated with the attributes $a_1, \ldots, a_d$.

Definition 4 (EVAL-SIM Security). *Let $F \subseteq \mathbb{F}$ be a set of policies, and let $A \subseteq \mathbb{A}$ be a set of attributes. For convenience, we define the predicate*

$$\mathsf{compat}(F, A) = \begin{cases} 1 & \text{if } \exists a \in A \; \forall f \in F \; f(a) = 0 \\ 0 & \text{otherwise.} \end{cases}$$

Let $\mathcal{E}$ be an ABHE scheme with parameters $\mathcal{D}$ and $\mathcal{K}$. We define the following experiments for a pair of PPT adversarial algorithms $\mathcal{A} = (\mathcal{A}_1, \mathcal{A}_2)$ and a PPT algorithm $\mathcal{S}$.

- **$\mathbf{Exp}_{\mathcal{E},\mathcal{A}}^{\mathsf{REAL}}(\lambda)$ *(Real World):***
  1. $(\mathsf{PP}, \mathsf{MSK}) \leftarrow \mathcal{E}.\mathsf{Setup}(1^\lambda)$.
  2. $(C, (a_1, \mu_1), \ldots, (a_\ell, \mu_\ell), \mathsf{st}) \leftarrow \mathcal{A}_1^{\mathcal{E}.\mathsf{KeyGen}(\mathsf{MSK}, \cdot)}(\mathsf{PP})$.
  3. *Let F be the set of policies queried by $\mathcal{A}_1$.*
  4. *Let $A := \{\mathfrak{a}_1, \ldots, \mathfrak{a}_d\}$ be the distinct attributes in the collection $a_1, \ldots, a_\ell$.*
  5. *Assert $d \leq \mathcal{D}$ and $\mathsf{compat}(F, A) = 1$; otherwise output a random bit and abort.*
  6. $c_j \leftarrow \mathcal{E}.\mathsf{Encrypt}(\mathsf{PP}, a_j, \mu_j)$ *for $j \in [\ell]$.*
  7. $c' \leftarrow \mathcal{E}.\mathsf{Eval}(\mathsf{PP}, C, c_1, \ldots, c_\ell)$.
  8. $b \leftarrow \mathcal{A}_2^{\mathcal{O}(\mathsf{MSK}, \cdot)}(\mathsf{st}, c', c_1, \ldots, c_\ell)$
  9. *Output b.*
- **$\mathbf{Exp}_{\mathcal{E},\mathcal{A},\mathcal{S}}^{\mathsf{IDEAL}}(\lambda)$ *(Ideal World):***
  1. $(\mathsf{PP}, \mathsf{MSK}) \leftarrow \mathcal{E}.\mathsf{Setup}(1^\lambda)$.
  2. $(C, (a_1, \mu_1), \ldots, (a_\ell, \mu_\ell), \mathsf{st}) \leftarrow \mathcal{A}_1^{\mathcal{E}.\mathsf{KeyGen}(\mathsf{MSK}, \cdot)}(\mathsf{PP})$.
  3. *Let F be the set of policies queried by $\mathcal{A}_1$.*
  4. *Let $A := \{\mathfrak{a}_1, \ldots, \mathfrak{a}_d\}$ be the distinct attributes in the collection $a_1, \ldots, a_\ell$.*
  5. *Assert $d \leq \mathcal{D}$ and $\mathsf{compat}(F, A) = 1$; otherwise output a random bit and abort.*
  6. $c_j \leftarrow \mathcal{E}.\mathsf{Encrypt}(\mathsf{PP}, a_j, \mu_j)$ *for $j \in [\ell]$.*
  7. $c' \leftarrow \mathcal{S}(\mathsf{PP}, C, A)$.

8. $b \leftarrow \mathcal{A}_2^{\mathcal{O}(\mathsf{MSK}, \cdot)}(\mathsf{st}, c', c_1, \ldots, c_\ell)$

9. *Output b.*

where $\mathcal{O}(\mathsf{MSK}, \cdot)$ *is defined as:*

- $\mathcal{O}(\mathsf{MSK}, f)$:
  1. *If* $\mathsf{compat}(F \cup \{f\}, A) = 1$: *set* $F \leftarrow F \cup \{f\}$ *and output* $\mathcal{E}.\mathsf{KeyGen}(\mathsf{MSK}, f)$.
  2. *Else output* $\perp$.

Then $\mathcal{E}$ *is said to be* EVAL-SIM-*secure if there exists a PPT simulator* $\mathcal{S}$ *such that for every pair of PPT algorithms* $\mathcal{A} := (\mathcal{A}_1, \mathcal{A}_2)$, *it holds that*

$$|\Pr[\mathbf{Exp}_{\mathcal{E}, \mathcal{A}}^{\mathsf{REAL}} \to 1] - \Pr[\mathbf{Exp}_{\mathcal{E}, \mathcal{A}, \mathcal{S}}^{\mathsf{IDEAL}} \to 1]| < \mathsf{negl}(\lambda).$$

Note that the above definition relates to adaptive security. For selective security, the adversary must choose the attributes before receiving the public parameters. As a result, in the modified definition, $\mathcal{A}$ consists of three PPT algorithms $(\mathcal{A}_1, \mathcal{A}_2, \mathcal{A}_3)$. Furthermore, $\mathcal{A}_1$ outputs a set of $d \leq \mathcal{D}$ attributes $A := \{\mathfrak{a}_1, \ldots, \mathfrak{a}_d\}$; $\mathcal{A}_2$ receives PP and outputs a circuit C along with a sequence of ℓ pairs (μ_i, a_i) for $i \in [\ell]$ where $\mu_i \in \mathcal{M}$ and $a_i \in A$. Finally, $\mathcal{A}_3$ is defined equivalently to $\mathcal{A}_2$ in the above definition. We denote the selective variant by sel-EVAL-SIM.

4 Construction

4.1 Building Blocks

Multi-key FHE. Multi-Key FHE allows multiple independently-generated keys to be used together in a homomorphic evaluation. The syntax of multi-key FHE imposes a limit N on the number of such keys that can be supported. Furthermore, the size of the evaluated ciphertext does not depend on the size of the circuit (or number of inputs), but instead on the number of independent keys N that is supported. In order to decrypt, the parties who have the corresponding secret keys must collaborate such as in an MPC protocol.

Definition 5 (Based on Definition 2.1 in [11]). *A multi-key* $\mathbb{C}$-*homomorphic scheme family for a class of circuits* $\mathbb{C}$ *and message space* $\mathcal{M}$ *is a family of PPT algorithms* $\{\mathcal{E}^{(N)} := (\mathsf{Gen}, \mathsf{Encrypt}, \mathsf{Decrypt}, \mathsf{Eval})\}_{N>0}$ *where* $\mathcal{E}^{(N)}$ *is defined as follows:*

- MKFHE.Gen *takes as input the security parameter* 1^λ *and outputs a tuple* $(\mathsf{pk}, \mathsf{sk}, \mathsf{vk})$ *where* pk *is a public key,* sk *is a secret key and* vk *is an evaluation key.*
- MKFHE.Encrypt *takes as input a public key* pk *and a message* $m \in \mathcal{M}$, *and outputs an encryption of* m *under* pk.
- MKFHE.Decrypt *takes as input* $1 \leq k \leq N$ *secret keys* $\mathsf{sk}_1, \ldots, \mathsf{sk}_k$ *and a ciphertext* c, *and outputs a message* $m' \in \mathcal{M}$.

- MKFHE.Eval *takes as input a circuit* $C \in \mathbb{C}$, *and* ℓ *pairs* $(c_1, \mathsf{vk}_1), \ldots, (c_\ell, \mathsf{vk}_\ell)$ *and outputs a ciphertext* c'.

Informally, evaluation is only required to be *correct* if at most N keys are used in MKFHE.Eval; that is, $|\{\mathsf{vk}_1, \ldots, \mathsf{vk}_\ell\}| \leq N$. Furthermore, the size of an evaluated ciphertext c' must only depend polynomially on the security parameter λ and the number of keys N, and not on the size of the circuit.

The IND-CPA security game for multi-key homomorphic encryption is the same as that for standard public-key encryption; note that the adversary is given the evaluation key vk.

There are two multi-key FHE schemes in the literature: the scheme of López-Alt et al. [11] based on NTRU and the scheme of Clear and McGoldrick [7] based on Learning with Errors (LWE). Although our construction can work with any multi-key FHE, we obtain better efficiency if we use the multi-key FHE scheme of Clear and McGoldrick, which we call CM. More precisely, the depth of the decryption circuit of CM is $O(\log N)$ (as opposed to $O(\log^2 N)$ in the case of the multi-key FHE from [11]) which results in fewer levels needed for the leveled ABFHE.

For the remainder of the paper, we will denote an instance of a multi-key FHE by $\mathcal{E}_{\mathsf{MKFHE}}$.

Leveled ABFHE. Our approach uses a leveled ABFHE scheme in an essential way. A leveled ABFHE scheme allows one to evaluate a circuit of bounded depth. The bound on the depth L is chosen in advance of generating the public parameters. Gentry et al. [6] presented the first leveled ABFHE where the class of access policies consists of bounded-depth circuits. They based security on LWE. A leveled Identity-Based FHE (IBFHE) scheme from LWE is also presented in [6]. Furthermore a leveled IBFHE that is multi-identity (supports evaluation on ciphertexts with different identities) was constructed in [7] from LWE.

Any of the above schemes can be used to instantiate our construction and its properties are inherited by our construction. Therefore if we use an identity-based scheme, our resulting construction is identity-based etc.

For the rest of the paper, we will denote a leveled ABFHE scheme by $\mathcal{E}_{\mathsf{IABFHE}}$ with message space $\mathcal{M}_{\mathcal{E}_{\mathsf{IABFHE}}}$, attribute space $\mathbb{A}_{\mathcal{E}_{\mathsf{IABFHE}}}$ and class of predicates $\mathbb{F}_{\mathcal{E}_{\mathsf{IABFHE}}}$.

4.2   Overview of Our Approach

The main idea behind our approach is to exploit multi-key FHE and leveled ABFHE to construct a new ABFHE scheme that can evaluate circuits with up to N inputs, where N is chosen before generating the public parameters. Let $\mathcal{E}_{\mathsf{MKFHE}}$ be a multi-key FHE scheme whose decryption circuit has depth $\delta(\lambda, N)$ where N is the number of independent keys tolerated and λ is the security parameter. Let $\mathcal{E}_{\mathsf{IABFHE}}$ be a leveled ABFHE scheme as described in Sect. 4.1 that can compactly evaluate circuits of depth $\delta(\lambda, N)$.

Let w be a positive integer. The supported message space of our scheme is $\mathcal{M} \triangleq \{0,1\}^w$. The supported attribute space is $\mathbb{A} \triangleq \mathbb{A}_{\mathcal{E}_{\mathsf{IABFHE}}}$ and the supported class of access policies is $\mathbb{F} \triangleq \mathbb{F}_{\mathcal{E}_{\mathsf{IABFHE}}}$. In other words, the attribute space and class of access policies is the same as the underlying leveled ABFHE scheme. Finally, the class of supported circuits is $\mathbb{C} \triangleq \mathcal{M}^N \to \mathcal{M}$.

Roughly speaking, to encrypt a message $\mu \in \mathcal{M}$ under attribute $a \in \mathbb{A}$ in our scheme, (1) a key triple $(\mathsf{pk}, \mathsf{sk}, \mathsf{vk})$ is generated for $\mathcal{E}_{\mathsf{MKFHE}}$; (2) μ is encrypted with $\mathcal{E}_{\mathsf{MKFHE}}$ under pk; (3) sk is encrypted with $\mathcal{E}_{\mathsf{IABFHE}}$ under attribute a; (4) the two previous ciphertexts along with vk constitute the ciphertext that is produced. Therefore, $\mathcal{E}_{\mathsf{MKFHE}}$ is used for hiding the message and for homomorphic computation, whereas $\mathcal{E}_{\mathsf{IABFHE}}$ enforces access control by appropriately hiding the secret keys for $\mathcal{E}_{\mathsf{MKFHE}}$.

The evaluator performs homomorphic evaluation on the multi-key FHE ciphertexts and obtains a result c'. It then homomorphically decrypts c' with the leveled ABFHE scheme using the encryptions of the secret keys for $\mathcal{E}_{\mathsf{MKFHE}}$. As a result we obtain a ciphertext whose length is independent of N and the circuit size, which satisfies our compactness condition.

In more concrete terms, we assume without loss of generality that the message space of $\mathcal{E}_{\mathsf{MKFHE}}$ is $\{0,1\}$, and we encrypt a w-bit message $\mu = (\mu_1, \ldots, \mu_w) \in \{0,1\}^w$ one bit at a time using $\mathcal{E}_{\mathsf{MKFHE}}$. Furthermore, let N be the maximum number of keys supported by $\mathcal{E}_{\mathsf{MKFHE}}$. Our construction can therefore support the class of circuits $\mathbb{C} = \{(\{0,1\}^w)^N \to \{0,1\}^w\}$. We remind the reader that w can be arbitrarily large, and in practice, the length of plaintexts may be shorter than w. In practice, each sender's input may be of arbitrary size. However, there is a limit, N, on the number of independent senders i.e. the number of inputs to the circuit where the inputs are taken from the domain $\{0,1\}^w$.

4.3   Construction

We now present our construction, which we call bABFHE.

Setup. On input a security parameter λ and a bound N on the number of inputs to support, the following steps are performed:

1. Choose integer w.
2. Generate $(\mathsf{PP}_{\mathcal{E}_{\mathsf{IABFHE}}}, \mathsf{MSK}_{\mathcal{E}_{\mathsf{IABFHE}}}) \leftarrow \mathcal{E}_{\mathsf{IABFHE}}.\mathsf{Setup}(1^\lambda, 1^L)$ where $L = O(\log \lambda \cdot N)$ is the depth of the decryption circuit of $\mathcal{E}_{\mathsf{IABFHE}}$ for parameters λ and N.
3. Output $(\mathsf{PP} := (\mathsf{PP}_{\mathcal{E}_{\mathsf{IABFHE}}}, \lambda, N, w), \mathsf{MSK} := (\mathsf{PP}, \mathsf{MSK}_{\mathcal{E}_{\mathsf{IABFHE}}}))$.

Secret Key Generation. Given the master secret key $\mathsf{MSK} := (\mathsf{PP}, \mathsf{MSK}_{\mathcal{E}_{\mathsf{IABFHE}}})$ and a policy $f \in \mathbb{F}$, a secret key sk_f for f is generated as $\mathsf{sk}_f \leftarrow \mathcal{E}_{\mathsf{IABFHE}}.\mathsf{KeyGen}(\mathsf{MSK}_{\mathcal{E}_{\mathsf{IABFHE}}}, f)$. The secret key $\mathsf{SK}_f := (\mathsf{PP}, \mathsf{sk}_f)$ is issued to the user.

Encryption. On input public parameters $\mathsf{PP} := (\mathsf{PP}_{\mathcal{E}_{\mathsf{IABFHE}}}, \lambda, N, w)$, a binary string $\mu = (\mu_1, \dots, \mu_w) \in \{0,1\}^w$ and an attribute $a \in \mathbb{A}$: the sender first generates a key triple for $\mathcal{E}_{\mathsf{MKFHE}}$; that is, she computes $(\mathsf{pk}, \mathsf{sk}, \mathsf{vk}) \leftarrow \mathcal{E}_{\mathsf{MKFHE}}.\mathsf{Gen}(1^\lambda, 1^N)$. Then she runs $\psi \leftarrow \mathcal{E}_{\mathsf{IABFHE}}.\mathsf{Encrypt}(\mathsf{PP}_{\mathcal{E}_{\mathsf{IABFHE}}}, a, \mathsf{sk})$. Subsequently she uses pk to encrypt each bit $\mu_i \in \{0,1\}$ in turn using $\mathcal{E}_{\mathsf{MKFHE}}$ for $i \in [w]$; that is, she computes $c_i \leftarrow \mathcal{E}_{\mathsf{MKFHE}}.\mathsf{Encrypt}(\mathsf{pk}, \mu_i)$. Finally she outputs the ciphertext $\mathsf{CT} := (\mathsf{type} := 0, \mathsf{enc} := (\psi, \mathsf{vk}, (c_1, \dots, c_w)))$.

Remark 2. A ciphertext CT in our scheme has two components: the first is labeled with type and the second is labeled with enc. The former has two valid values: 0 and 1; 0 indicates that the ciphertext is "fresh" while 1 indicates that the ciphertext is the result of an evaluation. The value of the type component specifies how the enc component is to be parsed.

Evaluation. On input public parameters $\mathsf{PP} := (\mathsf{PP}_{\mathcal{E}_{\mathsf{IABFHE}}}, \lambda, N, w)$, a circuit $C \in \mathbb{C}$, and ciphertexts $\mathsf{CT}_1, \dots, \mathsf{CT}_\ell$ with $\ell \leq N$, the evaluator performs the following steps. Firstly, the ciphertexts are assumed to be "fresh" ciphertexts generated with the encryption algorithm. In other words, their type components are all 0. Otherwise the evaluator outputs $\bot$. Consequently, the evaluator can parse CT_i as $(\mathsf{type} := 0, \mathsf{enc} := (\psi_i, \mathsf{vk}_i, (c_1^{(i)}, \dots, c_w^{(i)})))$ for every $i \in [\ell]$. We denote by a_i the attribute associated with the $\mathcal{E}_{\mathsf{IABFHE}}$ ciphertext ψ_i. The maximum degree of composition of our construction is inherited from that of the underlying leveled ABFHE scheme $\mathcal{E}_{\mathsf{IABFHE}}$. We denote this as usual by $\mathcal{D}$. The evaluator derives the degree of composition as $d \leftarrow |\{a_1, \dots, a_\ell\}|$, and outputs $\bot$ and aborts unless $d \leq \mathcal{D}$.

Next the evaluator computes

$$c' \leftarrow \mathcal{E}_{\mathsf{MKFHE}}.\mathsf{Eval}(C, (c_1^{(1)}, \mathsf{vk}_1), \dots, (c_w^{(1)}, \mathsf{vk}_1), \dots, (c_1^{(\ell)}, \mathsf{vk}_\ell), \dots, (c_w^{(\ell)}, \mathsf{vk}_\ell))$$

and encrypts this ciphertext with the leveled ABFHE scheme under any arbitrary a_i, say a_1; that is, the evaluator computes $\psi_{c'} \leftarrow \mathcal{E}_{\mathsf{IABFHE}}.\mathsf{Encrypt}(\mathsf{PP}_{\mathcal{E}_{\mathsf{IABFHE}}}, a_1, c')$. The final step is to evaluate using $\mathcal{E}_{\mathsf{IABFHE}}$ the decryption circuit $D_{\langle N, \lambda \rangle}$[1] of $\mathcal{E}_{\mathsf{MKFHE}}$:

$$\psi \leftarrow \mathcal{E}_{\mathsf{IABFHE}}.\mathsf{Eval}\big(\mathsf{PP}_{\mathcal{E}_{\mathsf{IABFHE}}}, D_{\langle N, \lambda \rangle}, \psi_{c'}, \psi_1, \dots, \psi_\ell\big).$$

The evaluator outputs the *evaluated ciphertext* $\mathsf{CT}' := (\mathsf{type} := 1, \mathsf{enc} := \psi)$.

Remark 3. Observe that a "fresh" ciphertext has a different form to an evaluated ciphertext. Further evaluation with evaluated ciphertexts is not guaranteed by our construction. Hence it is a 1-hop homomorphic scheme using the terminology of Gentry et al. [12].

Decryption. To decrypt a ciphertext $\mathsf{CT} := (\mathsf{type}, \mathsf{enc})$ with a sequence of secret keys $\mathsf{SK}_{f_1} := (\mathsf{PP}, \mathsf{sk}_{f_1}), \dots, \mathsf{SK}_{f_k} := (\mathsf{PP}, \mathsf{sk}_{f_k})$ for respective policies $f_1, \dots, f_k \in \mathbb{F}$, a decryptor performs the following steps.

---

[1] for the specific case of parameters N and λ.

If CT is a "fresh" ciphertext (i.e. type $= 0$), then enc is parsed as $(\psi, \mathsf{vk}, (c_1, \ldots, c_w))$ and the decryptor computes $\mathsf{sk} \leftarrow \mathcal{E}_{\mathsf{IABFHE}}.\mathsf{Decrypt}(\mathsf{sk}_1, \ldots, \mathsf{sk}_{\hat{k}}, \psi)$. If $\mathsf{sk} = \bot$, then the decryptor outputs $\bot$ and aborts. Otherwise, she computes

$$\mu_j \leftarrow \mathcal{E}_{\mathsf{MKFHE}}.\mathsf{Decrypt}(\mathsf{sk}, c_j) \text{ for every } j \in [w]$$

and outputs the plaintext $\mu := (\mu_1, \ldots, \mu_w) \in \{0,1\}^w$.

If CT is an evaluated ciphertext (i.e. type $= 1$), then the decryptor parses enc as ψ and computes $x \leftarrow \mathcal{E}_{\mathsf{IABFHE}}.\mathsf{Decrypt}(\mathsf{sk}_1, \ldots, \mathsf{sk}_{\hat{k}}, \psi)$. If $x = \bot$ the decryptor outputs $\bot$ and aborts; otherwise the plaintext $\mu := x \in \{0,1\}^w$ is outputted.

4.4 Formal Description

A formal description of the construction bABFHE is given in Fig. 1. As mentioned previously, the parameters $\mathcal{D}$ (maximum degree of composition) and $\mathcal{K}$ (maximum number of decryption keys passed to Decrypt) are inherited directly from the underlying leveled ABFHE scheme $\mathcal{E}_{\mathsf{IABFHE}}$. Although circuits in the supported class send a sequence of elements in the message space $\mathcal{M} := \{0,1\}^w$ to another element in the message space $\mathcal{M}$, we simplify the description here and assume that each circuit C outputs a single bit. A circuit $\hat{C}$ in our supported class can then be modelled as w such circuits.

4.5 Correctness

In the evaluation algorithm, the desired N-ary circuit C whose N inputs are over the domain $\{0,1\}^w$ is evaluated using the multi-key FHE scheme. Observe that C can be of arbitrary depth since the size of the resultant multi-key FHE ciphertext only depends on λ and N. We then encrypt this resulting ciphertext with $\mathcal{E}_{\mathsf{IABFHE}}$ in order to homomorphically evaluate the decryption circuit of $\mathcal{E}_{\mathsf{MKFHE}}$ using $\mathcal{E}_{\mathsf{IABFHE}}$. Consequently, we obtain a ciphertext whose size is independent of N as required by the compactness condition for ABHE.

5 Security

5.1 Semantic Security

Without loss of generality we assume that the message space $\mathcal{M}_{\mathcal{E}_{\mathsf{IABFHE}}}$ of $\mathcal{E}_{\mathsf{IABFHE}}$ is big enough to represent secret keys in $\mathcal{E}_{\mathsf{MKFHE}}$ and binary strings in $\mathcal{M}$.

Lemma 1. *If $\mathcal{E}_{\mathsf{IABFHE}}$ is an* IND-X-CPA-*secure leveled ABFHE scheme and $\mathcal{E}_{\mathsf{MKFHE}}$ is an* IND-CPA-*secure multi-key FHE scheme, then* bABFHE *is* IND-X-CPA *where* $X \in \{\mathsf{sel}, \mathsf{AD}\}$.

Proof. We prove the lemma by means of a hybrid argument.

Hybrid 0 IND-X-CPA game for bABFHE.

Setup$(1^\lambda, 1^N)$:

1. Choose integer w.
2. Let $g(\cdot, \cdot)$ be a polynomial associated with $\mathcal{E}_{\mathsf{MKFHE}}$ that gives the number of inputs to the decryption circuit for N keys and security parameter λ. Let $L = g(\lambda, N)$.
3. Generate $(\mathsf{PP}_{\mathcal{E}_{\mathsf{IABFHE}}}, \mathsf{MSK}_{\mathcal{E}_{\mathsf{IABFHE}}})$ $\leftarrow$ $\mathcal{E}_{\mathsf{IABFHE}}.\mathsf{Setup}(1^\lambda, 1^L)$.
4. Output $(\mathsf{PP}$ $:=$ $(\mathsf{PP}_{\mathcal{E}_{\mathsf{IABFHE}}}, \lambda, N, w), \mathsf{MSK}$ $:=$ $\mathsf{MSK}_{\mathcal{E}_{\mathsf{IABFHE}}})$.

Encrypt(PP, a, μ) :

1. Parse PP as $(\mathsf{PP}_{\mathcal{E}_{\mathsf{IABFHE}}}, \lambda, N, w)$.
2. Parse μ as $(\mu_1, \ldots, \mu_w) \in \{0, 1\}^w$.
3. $(\mathsf{pk}, \mathsf{sk}, \mathsf{vk}) \leftarrow \mathcal{E}_{\mathsf{MKFHE}}.\mathsf{Gen}(1^\lambda, 1^N)$
4. ψ $\leftarrow$ $\mathcal{E}_{\mathsf{IABFHE}}.\mathsf{Encrypt}(\mathsf{PP}_{\mathcal{E}_{\mathsf{IABFHE}}}, a, \mathsf{sk})$.
5. $c_i \leftarrow \mathcal{E}_{\mathsf{MKFHE}}.\mathsf{Encrypt}(\mathsf{pk}, \mu_i)$ for $i \in [w]$.
6. Output $\mathsf{CT} := (\mathsf{type} := 0, \mathsf{enc} := (\psi, \mathsf{vk}, (c_1, \ldots, c_w)))$.

KeyGen(MSK, f) :

1. Parse MSK as $(\mathsf{PP}, \mathsf{MSK}_{\mathcal{E}_{\mathsf{IABFHE}}})$.
2. sk_f $\leftarrow$ $\mathcal{E}_{\mathsf{IABFHE}}.\mathsf{KeyGen}(\mathsf{MSK}_{\mathcal{E}_{\mathsf{IABFHE}}}, f)$.
3. Output $\mathsf{SK}_f := (\mathsf{PP}, \mathsf{sk}_f)$.

Decrypt$(\mathsf{SK}_{f_1}, \ldots, \mathsf{SK}_{f_k}, \mathsf{CT})$:

1. If $k > \mathcal{K}$: output $\bot$ and abort.
2. Parse SK_{f_i} as $(\mathsf{PP}, \mathsf{sk}_{f_i})$ for $i \in [k]$.
3. Parse PP as $(\mathsf{PP}_{\mathcal{E}_{\mathsf{IABFHE}}}, \lambda, N, w)$.
4. Parse CT as $(\mathsf{type}, \mathsf{enc})$.
5. If $\mathsf{type} = 0$:
   (a) Parse enc as $(\psi, \mathsf{vk}, (c_1, \ldots, c_w))$
   (b) Compute sk $\leftarrow$ $\mathcal{E}_{\mathsf{IABFHE}}.\mathsf{Decrypt}(\mathsf{sk}_1, \ldots, \mathsf{sk}_k, \psi)$.
   (c) If $\mathsf{sk} = \bot$: output $\bot$ and abort.
   (d) $\mu_i \leftarrow \mathcal{E}_{\mathsf{MKFHE}}.\mathsf{Decrypt}(\mathsf{sk}, c_i)$ for $i \in [w]$.
   (e) Output $\mu := (\mu_1, \ldots, \mu_w) \in \{0, 1\}^w$.
6. Else If $\mathsf{type} = 1$:
   (a) Parse enc as ψ.
   (b) Compute x $\leftarrow$ $\mathcal{E}_{\mathsf{IABFHE}}.\mathsf{Decrypt}(\mathsf{sk}_1, \ldots, \mathsf{sk}_k, \psi)$.
   (c) If $x = \bot$: output $\bot$ and abort.
   (d) Output $\mu := x \in \{0, 1\}^w$.
7. Else output $\bot$.

Eval$(\mathsf{PP}, C, \mathsf{CT}_1, \ldots, \mathsf{CT}_\ell)$:

1. If $\ell > N$: output $\bot$ and abort.
2. Parse PP as $(\mathsf{PP}_{\mathcal{E}_{\mathsf{IABFHE}}}, \lambda, N, w)$.
3. For $i \in [\ell]$:
   (a) Parse CT_i as $(\mathsf{type} := 0, \mathsf{enc} := (\psi_i, \mathsf{vk}_i, (c_1^{(i)}, \ldots, c_w^{(i)})))$.
   (b) Set a_i as the attribute associated with ψ_i.
4. Set $d \leftarrow |\{a_1, \ldots, a_\ell\}|$ (degree of composition).
5. If $d > \mathcal{D}$: output $\bot$ and abort.
6. $c' \leftarrow \mathcal{E}_{\mathsf{MKFHE}}.\mathsf{Eval}(C, (c_1^{(1)}, \mathsf{vk}_1), \ldots, (c_w^{(1)}, \mathsf{vk}_1), \ldots, (c_1^{(\ell)}, \mathsf{vk}_\ell), \ldots, (c_w^{(\ell)}, \mathsf{vk}_\ell))$.
7. $\psi_{c'} \leftarrow \mathcal{E}_{\mathsf{IABFHE}}.\mathsf{Encrypt}(\mathsf{PP}_{\mathcal{E}_{\mathsf{IABFHE}}}, a_1, c')$.
8. Let $D_{\langle N, \lambda \rangle}$ be the decryption circuit of $\mathcal{E}_{\mathsf{MKFHE}}$ for parameters N and λ.
9. $\psi \leftarrow \mathcal{E}_{\mathsf{IABFHE}}.\mathsf{Eval}(\mathsf{PP}_{\mathcal{E}_{\mathsf{IABFHE}}}, D_{\langle N, \lambda \rangle}, \psi_{c'}, \psi_1, \ldots, \psi_\ell)$.
10. Output $\mathsf{CT}' := (\mathsf{type} := 1, \mathsf{enc} := \psi)$.

Fig. 1. Formal Description of scheme bABFHE.

Hybrid 1 Same as Hybrid 0 except with one difference. Let $a^\star \in \mathbb{A}$ be the target attribute chosen by the adversary $\mathcal{A}$. The challenger uses a modified Encrypt algorithm to compute the leveled ABFHE ciphertext corresponding to a^* by replacing Step 4 with $\psi \leftarrow \mathcal{E}_{\mathsf{IABFHE}}.\mathsf{Encrypt}(\mathsf{PP}_{\mathcal{E}_{\mathsf{IABFHE}}}, a^*, 0^{|\mathsf{sk}|})$ where $0^{|\mathsf{sk}|}$ is a string of zeros whose length is the same as the multi-key FHE secret key generated in Step 3 of Encrypt. The algorithm is otherwise unchanged.

We claim that any poly-time $\mathcal{A}$ that can distinguish between Hybrid 0 and Hybrid 1 with a non-negligible advantage can break the IND-X-CPA security of $\mathcal{E}_{\mathsf{IABFHE}}$. An adversary $\mathcal{B}$ that uses $\mathcal{A}$ proceeds as follows. When $\mathcal{A}$ chooses a target attribute a^*, $\mathcal{B}$ generates a key-triple for $\mathcal{E}_{\mathsf{MKFHE}}$ i.e. it computes

$$(\mathsf{pk}, \mathsf{sk}, \mathsf{vk}) \leftarrow \mathcal{E}_{\mathsf{MKFHE}}.\mathsf{Gen}(1^\lambda, 1^N).$$

Then it gives a^* to its challenger along with two messages $x_0 := \mathsf{sk}$ and $x_1 := 0^{|\mathsf{sk}|}$. Note that we assume for simplicity that both messages are in $\mathcal{M}_{\mathcal{E}_{\mathsf{IABFHE}}}$; if multiple messages (say k) are required then the usual hybrid argument can be applied which loses a factor of k. Subsequently, $\mathcal{B}$ embeds the challenge leveled ABFHE ciphertext as the ψ component of its own challenge ciphertext CT^*. It computes the remaining components of CT^* as in the Encrypt algorithm. If ψ encrypts x_0, then $\mathcal{B}$ perfectly simulates Hybrid 0. Otherwise, $\mathcal{B}$ perfectly simulates Hybrid 1. Note that secret key queries made by $\mathcal{A}$ can be perfectly simulated by $\mathcal{B}$. Thus, if $\mathcal{A}$ has a non-negligible advantage distinguishing between the hybrids, then $\mathcal{B}$ has a non-negligible advantage attacking the IND-X-CPA security of $\mathcal{E}_{\mathsf{IABFHE}}$.

For $i \in [w]$:

Hybrid 1 + i Same as Hybrid $1 + (i-1)$ with the exception that the challenger does not encrypt message bit $\mu_i^{(0)}$ or $\mu_i^{(1)}$ (using $\mathcal{E}_{\mathsf{MKFHE}}$) chosen by $\mathcal{A}$. Instead it encrypts some fixed message bit $\beta \in \{0, 1\}$.

We now show that if $\mathcal{A}$ can efficiently distinguish between Hybrid $1 + i$ and Hybrid $1 + (i - 1)$, then there is a PPT algorithm $\mathcal{G}$ that can use $\mathcal{A}$ to attack the IND-CPA security of $\mathcal{E}_{\mathsf{MKFHE}}$. Let pk and vk be the public key and evaluation key that $\mathcal{G}$ receives from its challenger. When $\mathcal{A}$ chooses $\mu^{(0)} \in \{0, 1\}^w$ and $\mu^{(1)} \in \{0, 1\}$, $\mathcal{G}$ simply gives $\mu_i^{(b)}$ and β to its IND-CPA challenger where b is the bit it uniformly samples in its simulation of the IND-X-CPA challenger. Let c^* be the challenge ciphertext it receives from the IND-CPA challenger. It sets $c_i \leftarrow c^*$ in the challenge ciphertext CT^*. If c^* encrypts $\mu_i^{(b)}$, then the view of $\mathcal{A}$ is identical to Hybrid $1 + (i - 1)$. Otherwise, the view of $\mathcal{A}$ is identical to Hybrid $1 + i$. Therefore, a non-negligible advantage obtained by $\mathcal{A}$ implies a non-negligible advantage for $\mathcal{G}$ in the IND-CPA game, and thus contradicts the IND-CPA security of $\mathcal{E}_{\mathsf{MKFHE}}$.

Finally observe that the adversary has a zero advantage in Hybrid $1 + w$ because the challenge ciphertext contains no information about the challenger's bit. □

5.2   EVAL-SIM Security

Recall the simulation-based security definition from Sect. 3.2, which we called EVAL-SIM security. In the following lemma, we show that bABFHE inherits EVAL-SIM security from $\mathcal{E}_{\mathsf{IABFHE}}$.

Lemma 2. *Let $\mathcal{E}_{\mathsf{MKFHE}}$ be an* IND-CPA *secure multi-key FHE scheme. Let $\mathcal{E}_{\mathsf{IABFHE}}$ be an X-EVAL-SIM secure ABHE scheme with $X \in \{\mathsf{sel}, \mathsf{AD}\}$. Then bABFHE is X-EVAL-SIM secure.*

Proof. By the hypothesized X-EVAL-SIM security of $\mathcal{E}_{\mathsf{IABFHE}}$, there exists a PPT simulator $\mathcal{S}_{\mathcal{E}_{\mathsf{IABFHE}}}$ such that for all PPT adversaries $\mathcal{A}_{\mathcal{E}_{\mathsf{IABFHE}}} := (\mathcal{A}_{\mathcal{E}_{\mathsf{IABFHE}},1}, \mathcal{A}_{\mathcal{E}_{\mathsf{IABFHE}},2})$ we have

$$|\Pr[\mathbf{Exp}^{\mathsf{REAL}}_{\mathcal{E}_{\mathsf{IABFHE}}, \mathcal{A}_{\mathcal{E}_{\mathsf{IABFHE}}}} \to 1] - \Pr[\mathbf{Exp}^{\mathsf{IDEAL}}_{\mathcal{E}_{\mathsf{IABFHE}}, \mathcal{A}_{\mathcal{E}_{\mathsf{IABFHE}}}, \mathcal{S}_{\mathcal{E}_{\mathsf{IABFHE}}}} \to 1]| < \mathsf{negl}(\lambda). \quad (5.1)$$

Remark 4. Note that in this proof we use the definition for adaptive EVAL-SIM security, which is slightly different to that for sel-EVAL-SIM security, but the argument holds analogously for the latter.

A simulator $\mathcal{S}$ can be constructed using $\mathcal{S}_{\mathcal{E}_{\mathsf{IABFHE}}}$ in order to achieve X-EVAL-SIM security for bABFHE. The simulator $\mathcal{S}$ runs as follows:

- $\mathcal{S}(\mathsf{PP}, C, \{a_1, \ldots, a_d\})$ with $d \leq \mathcal{D}$, $a_1, \ldots, a_d \in \mathbb{A}$ and $C \in \mathbb{C}$:
  1. Parse PP as $(\mathsf{PP}_{\mathcal{E}_{\mathsf{IABFHE}}}, \lambda, N, w)$.
  2. Let $D_{\langle N, \lambda \rangle}$ be the decryption circuit of $\mathcal{E}_{\mathsf{MKFHE}}$ for parameters N and λ.
  3. Output $\mathcal{S}_{\mathcal{E}_{\mathsf{IABFHE}}}(\mathsf{PP}_{\mathcal{E}_{\mathsf{IABFHE}}}, D_{\langle N, \lambda \rangle}, \{a_1, \ldots, a_d\})$.

We claim that if there exists a PPT adversary $\mathcal{A} := (\mathcal{A}_1, \mathcal{A}_2)$ with a non-negligible advantage distinguishing the real distribution and ideal distribution for bABFHE (with respect to $\mathcal{S}$), then there exists a PPT adversary $\mathcal{A}_{\mathcal{E}_{\mathsf{IABFHE}}} := (\mathcal{A}_{\mathcal{E}_{\mathsf{IABFHE}},1}, \mathcal{A}_{\mathcal{E}_{\mathsf{IABFHE}},2})$ with a non-negligible advantage distinguishing the real distribution and ideal distribution for $\mathcal{E}_{\mathsf{IABFHE}}$ (with respect to $\mathcal{S}_{\mathcal{E}_{\mathsf{IABFHE}}}$). If this claim were to hold it would contradict the hypothesized X-EVAL-SIM security of $\mathcal{E}_{\mathsf{IABFHE}}$, which seals the lemma. To prove the claim, we show how to construct $(\mathcal{A}_{\mathcal{E}_{\mathsf{IABFHE}},1}, \mathcal{A}_{\mathcal{E}_{\mathsf{IABFHE}},2})$ from $(\mathcal{A}_1, \mathcal{A}_2)$. The algorithm $\mathcal{A}_{\mathcal{E}_{\mathsf{IABFHE}},1}$ is given as input the public parameters $\mathsf{PP}_{\mathcal{E}_{\mathsf{IABFHE}}}$ for $\mathcal{E}_{\mathsf{IABFHE}}$. We denote its key generation oracle by $\mathcal{O}_1$. It runs as follows.

1. Set $\mathsf{PP} := (\mathsf{PP}_{\mathcal{E}_{\mathsf{IABFHE}}}, \lambda, N, w)$ (the parameters N and w are fixed elsewhere).
2. Run $(C, (a_1, \mu_1), \ldots, (a_\ell, \mu_\ell), \mathsf{st}) \leftarrow \mathcal{A}_1^{\mathcal{O}_1}(\mathsf{PP})$.
3. For $i \in [\ell]$:
   (a) Parse μ_i as $(\mu_1^{(i)}, \ldots, \mu_w^{(i)}) \in \{0,1\}^w$.
   (b) $(\mathsf{pk}_i, \mathsf{sk}_i, \mathsf{vk}_i) \leftarrow \mathcal{E}_{\mathsf{MKFHE}}.\mathsf{Gen}(1^\lambda, 1^N)$
   (c) $c_j^{(i)} \leftarrow \mathcal{E}_{\mathsf{MKFHE}}.\mathsf{Encrypt}(\mathsf{pk}, \mu_j^{(i)})$ for $j \in [w]$.
4. Set $d \leftarrow |\{a_1, \ldots, a_\ell\}|$ (degree of composition).
5. $c' \leftarrow \mathcal{E}_{\mathsf{MKFHE}}.\mathsf{Eval}(C, (c_1^{(1)}, \mathsf{vk}_1), \ldots, (c_w^{(1)}, \mathsf{vk}_1), \ldots, (c_1^{(\ell)}, \mathsf{vk}_\ell), \ldots, (c_w^{(\ell)}, \mathsf{vk}_\ell))$.
6. Let $D_{\langle N, \lambda \rangle}$ be the decryption circuit of $\mathcal{E}_{\mathsf{MKFHE}}$ for parameters N and λ.

7. Set state $\leftarrow (\mathsf{st}, \mathsf{PP}, (\mathsf{vk}_1, (c_1^{(1)}, \ldots, c_w^{(1)})), \ldots, (\mathsf{vk}_\ell, (c_1^{(\ell)}, \ldots, c_w^{(\ell)})))$.
8. Output $(D_{\langle N, \lambda \rangle}, (a_1, c'), (a_1, \mathsf{sk}_1), \ldots, (a_\ell, \mathsf{sk}_\ell), \mathsf{state})$.

The algorithm $\mathcal{A}_{\mathcal{E}_{\mathsf{IABFHE}}, 2}$ is given as input the state state (generated in $\mathcal{A}_{\mathcal{E}_{\mathsf{IABFHE}}, 1}$), the evaluated ciphertext ψ' along with the $\ell + 1$ "input ciphertexts" (which we denote by $\psi_{c'}, \psi_1, \ldots, \psi_\ell$) and attributes $\{a_1, \ldots, a_d\}$. We denote its key generation oracle by $\mathcal{O}_2$. It runs as follows.

1. Parse state as $(\mathsf{st}, \mathsf{PP}, (\mathsf{vk}_1, (c_1^{(1)}, \ldots, c_w^{(1)})), \ldots, (\mathsf{vk}_\ell, (c_1^{(\ell)}, \ldots, c_w^{(\ell)})))$.
2. Parse PP as $(\mathsf{PP}_{\mathcal{E}_{\mathsf{IABFHE}}}, \lambda, N, w)$.
3. Generate bABFHE input ciphertext $\mathsf{CT}_i \leftarrow (\mathsf{type} := 0, \mathsf{enc} := (\psi_i, \mathsf{vk}_i, (c_1^{(i)}, \ldots, c_w^{(i)})))$ for $i \in [\ell]$.
4. Generate bABFHE evaluated ciphertext $\mathsf{CT}' \leftarrow (\mathsf{type} := 1, \mathsf{enc} := \psi')$.
5. Run $b \leftarrow \mathcal{A}_2^{\mathcal{O}_2}(\mathsf{st}, \mathsf{CT}', \mathsf{CT}_1, \ldots, \mathsf{CT}_\ell)$.
6. Output b.

If ψ' is generated with $\mathcal{E}_{\mathsf{IABFHE}}.\mathsf{Eval}$ (i.e. the real distribution) then CT' is distributed identically to the output of $\mathsf{bABFHE}.\mathsf{Eval}$. On the other hand, if ψ' is generated with $\mathcal{S}_{\mathcal{E}_{\mathsf{IABFHE}}}$ (i.e. the ideal distribution), then CT' is distributed identically to $\mathcal{S}$. Therefore, a non-negligible advantage against bABFHE implies a non-negligible advantage against $\mathcal{E}_{\mathsf{IABFHE}}$. $\square$

6 Main Result

Theorem 1. *Let N be a positive integer. Let w be a positive integer. Let λ be a security parameter. Suppose there exists an* IND-CPA *secure multi-key FHE scheme $\mathcal{E}_{\mathsf{MKFHE}}$ whose decryption circuit has depth $\delta(N, \lambda)$. Suppose there exists a leveled ABFHE scheme $\mathcal{E}_{\mathsf{IABFHE}}$ that can compactly evaluate circuits of depth δ. Then there exists an ABHE scheme $\mathcal{E}$ (whose parameters $\mathcal{D}$ and $\mathcal{K}$ are the same as $\mathcal{E}_{\mathsf{IABFHE}}$) that can compactly evaluate all Boolean circuits in $\{(\{0,1\}^w)^N \to \{0,1\}^w\}$ i.e. the class of Boolean circuits of unbounded depth with N inputs over the domain $\{0,1\}^w$, such that*

1. *$\mathcal{E}$ is* IND-X-CPA *secure if $\mathcal{E}_{\mathsf{IABFHE}}$ is* IND-X-CPA *secure.*
2. *$\mathcal{E}$ is X-EVAL-SIM secure if $\mathcal{E}_{\mathsf{IABFHE}}$ is X-EVAL-SIM secure.*

for $X \in \{\mathsf{sel}, \mathsf{AD}\}$.

Proof. Instantiating our scheme bABFHE from Sect. 4.3 with the multi-key FHE scheme $\mathcal{E}_{\mathsf{MKFHE}}$ and the ABHE scheme $\mathcal{E}_{\mathsf{IABFHE}}$, the theorem follows by appealing to Lemmas 1 (IND-X-CPA security) and 2 (X-EVAL-SIM security). $\square$

Corollary 1. *Let N be a positive integer. Assuming the hardness of LWE, there exists a* IND-sel-CPA *secure ABFHE that can compactly evaluate circuits with N inputs.*

Proof. We can instantiate the multi-key FHE scheme in our construction with the CM multi-key FHE from [7], whose security is based on LWE. Furthermore we can instantiate the leveled ABFHE in our construction with the leveled ABFHE of Gentry et al. [6], which is shown to be selectively secure under LWE. $\square$

References

1. Shamir, A.: Identity-based cryptosystems and signature schemes. In: Blakely, G.R., Chaum, D. (eds.) CRYPTO 1984. LNCS, vol. 196, pp. 47–53. Springer, Heidelberg (1985)
2. Boneh, D., Franklin, M.: Identity-based encryption from the weil pairing. In: Kilian, J. (ed.) CRYPTO 2001. LNCS, vol. 2139, pp. 213–229. Springer, Heidelberg (2001)
3. Cocks, C.: An identity based encryption scheme based on quadratic residues. In: Honary, B. (ed.) Cryptography and Coding 2001. LNCS, vol. 2260, pp. 360–363. Springer, Heidelberg (2001)
4. Sahai, A., Waters, B.: Fuzzy identity-based encryption. In: Cramer, R. (ed.) EUROCRYPT 2005. LNCS, vol. 3494, pp. 457–473. Springer, Heidelberg (2005)
5. Gentry, C.: Fully homomorphic encryption using ideal lattices. In: Proceedings of the 41st Annual ACM Symposium on Theory of Computing STOC 2009, pp. 169 (2009)
6. Gentry, C., Sahai, A., Waters, B.: Homomorphic encryption from learning with errors: conceptually-simpler, asymptotically-faster, attribute-based. In: Canetti, R., Garay, J.A. (eds.) CRYPTO 2013, Part I. LNCS, vol. 8042, pp. 75–92. Springer, Heidelberg (2013)
7. Clear, M., McGoldrick, C.: Multi-identity and multi-key leveled FHE from learning with errors. In: Gennaro, R., Robshaw, M. (eds.) CRYPTO 2015. LNCS, vol. 9216, pp. 630–656. Springer, Heidelberg (2015)
8. Garg, S., Gentry, C., Halevi, S., Raykova, M., Sahai, A., Waters, B.: Candidate indistinguishability obfuscation and functional encryption for all circuits. In: FOCS, IEEE Computer Society, pp. 40–49 (2013)
9. Clear, M., McGoldrick, C.: Bootstrappable identity-based fully homomorphic encryption. In: Gritzalis, D., Kiayias, A., Askoxylakis, I. (eds.) CANS 2014. LNCS, vol. 8813, pp. 1–19. Springer, Heidelberg (2014)
10. Clear, M., McGoldrick, C.: Policy-based non-interactive outsourcing of computation using multikey FHE and CP-ABE. In: Proceedings of the 10th International Conference on Security and Cryptography, SECRYPT 2013 (2013)
11. López-Alt, A., Tromer, E., Vaikuntanathan, V.: On-the-fly multiparty computation on the cloud via multikey fully homomorphic encryption. In: Proceedings of the 44th Symposium on Theory of Computing (STOC 2012), pp. 1219–1234. ACM, New York (2012)
12. Gentry, C., Halevi, S., Vaikuntanathan, V.: i-hop homomorphic encryption and rerandomizable yao circuits. In: Rabin, T. (ed.) CRYPTO 2010. LNCS, vol. 6223, pp. 155–172. Springer, Heidelberg (2010)

Adaptively Secure Unrestricted Attribute-Based Encryption with Subset Difference Revocation in Bilinear Groups of Prime Order

Pratish Datta$^{(\boxtimes)}$, Ratna Dutta, and Sourav Mukhopadhyay

Department of Mathematics, Indian Institute of Technology Kharagpur,
Kharagpur 721302, India
{pratishdatta,ratna,sourav}@maths.iitkgp.ernet.in

Abstract. Providing an efficient revocation mechanism for attribute-based encryption (ABE) is of utmost importance since over time a user's credentials may be revealed or expired. All previously known revocable ABE (RABE) constructions (a) essentially utilize the complete subtree (CS) scheme for revocation purpose, (b) are restricted in the sense that the size of the public parameters depends linearly on the size of the attribute universe and logarithmically on the number of users in the system, and (c) are either selectively secure, which seems unrealistic in a dynamic system such as RABE, or fully secure but built in a composite order bilinear group setting, which results in high computational cost. This paper presents the *first adaptively secure unrestricted* RABE using *subset difference* (SD) mechanism for revocation which greatly improves the broadcast efficiency compared to the CS scheme. Our RABE scheme is built on a prime order bilinear group setting resulting in practical computation cost, and its security depends on the Decisional Linear assumption.

Keywords: Revocable attribute-based encryption · Subset difference method · Prime order bilinear groups · Dual pairing vector spaces

1 Introduction

In recent times, the cost effectiveness and greater flexibility of cloud technology has triggered an emerging trend among individuals and organizations to outsource potentially sensitive private data to the "cloud", an external large and powerful server. Attribute-based encryption (ABE), a noble paradigm for public key encryption in which ciphertexts are encrypted for entities possessing specific decryption credentials, has been extensively deployed to realize complex access control functionalities in cloud environment. Specifically, in a (key-policy) ABE system, an encrypter may specify a set of attributes directly while encrypting a certain plaintext. A user in the system possesses a key associated with an access policy, stating what kind of ciphertext it can decrypt. In such a system, a user can decrypt a ciphertext if the policy associated with its key satisfies the attribute set associated with the ciphertext.

© Springer International Publishing Switzerland 2016
D. Pointcheval et al. (Eds.): AFRICACRYPT 2016, LNCS 9646, pp. 325–345, 2016.
DOI: 10.1007/978-3-319-31517-1_17

A crucial requirement in the context of ABE is user *revocation*, a tool for *changing* the users' decryption rights. Over time many users' private keys might get compromised, users might leave or be dismissed due to the revealing of malicious activities. In the literature several revocation mechanisms have been proposed in ABE setting [1–3,10,13,15]. The *direct* revocation technique [1,2,13], that controls revocation by specifying a revocation list directly during encryption, does not involve any additional proxy server [15] or key update phase [1,3,10]. Consequently, the non-revoked users remain unaffected and revocation can take effect instantly without requiring to wait for the expiration of the current time period.

Further the main design principle of the existing *revocable* ABE (RABE) constructions [1–3,10,13] essentially follows that of Boldyreva et al. [3] and employs the complete subtree (CS) scheme of Naor et al. [11] for user revocation. Replacing the CS technique by the *subset difference method* (SD) [11] or the *layered subset difference method* (LSD) [7] can reduce the size of the ciphertext component meant for enforcing revocation from $O(\widehat{r} \log \frac{N_{\max}}{\widehat{r}})$ to $O(\widehat{r})$ where $N_{\max}$ and $\widehat{r}$ respectively denote the total number of users and number of revoked users. This can provide significant improvement in the broadcast efficiency particularly when the number of users present in the system is very large compared to the number of revoked users.

Recently Lee et al. [9], utilized the SD scheme to manage revocation for identity-based encryption (IBE) and pointed out that their technique for RIBE cannot be extended to realize RABE via SD scheme.

Another important feature of an RABE scheme is its *independence* from the size of the attribute universe and the total number of users supported by the system. In all previous RABE schemes [1–3,10,13], the public parameter size grows linearly with the number of attributes in the attribute universe and logarithmically with the number of users. Furthermore, all previous RABE schemes except [13] provide only *selective security* which seems unrealistic in a dynamic system such as RABE. Although [13] achieves *full security*, it is built on a composite order bilinear group setting under non-standard assumptions. Note that the bit length of group elements is very large, as well as, group-operations and pairing computations are prohibitively slow in composite order bilinear groups than a comparable prime order group. From the security view point as well, Prime order bilinear groups are desirable compared to composite order ones [6].

Our Contribution: Our goal in this work is to explore applicability of SD mechanism in ABE setting to design an *unrestricted* RABE with *reduced communication bandwidth* and simultaneously achieve *adaptive security*. Instead of designing any new ABE scheme, we take the ABE construction of Okamoto and Takashima [12] as the starting point of our work. However, integrating SD revocation with ABE or replacing CS scheme by SD technique to construct a broadcast efficient RABE seems to be a quite challenging task.

We built the *first adaptively secure unrestricted* (key-policy) RABE scheme supporting *direct* revocation employing *subset difference* (SD) mechanism. Towards this end, we develop a rather non-trivial technique in order to integrate the SD scheme with the ABE construction of [12]. On a more positive note, the adaptive security of our RABE scheme is based on standard assumption, namely, the *Decisional Linear* (DLIN) assumption. Most importantly, due to the use of *prime order bilinear group* and *subset difference* revocation scheme, the *broadcast efficiency* of our RABE is much *higher* compared to the existing constructions [1,2,10,13] with reasonable computation cost. Furthermore, the proposed RABE scheme is the *first to achieve constant size public parameters* and thus overcomes the bottleneck of accommodating large attribute universe and an unlimited number of users.

Note that, as opposed to the CS scheme, an assigned key for a subset in SD scheme depends on the keys of some other subsets. This interdependence of keys makes the use of SD method in attribute-based setting quite challenging. We assign uniformly and independently chosen secrets to each users and split that secret into two parts – one for the access structure and the other for revocation. This latter part is further subdivided into random secrets to solve the complex key assignment problem of the SD method. We integrate SD with ABE by enforcing the condition that a user can retrieve the first part of its secret if and only if its access structure is satisfied by the set of attributes specified in the ciphertext, and all the subdivided components of the other part of its secret can be extracted if and only if its subscription is valid according to the conditions of SD scheme.

For proving security of our RABE scheme, the main intricacy of this work, we utilize the *(extended) dual system encryption* methodology over *dual pairing vector spaces* introduced in [12]. However, in order to adopt the technique of [12], for our RABE scheme, we extend some of the problems and methodology employed in [12].

Although we use the monotone version of the ABE scheme of [12] to present our RABE construction for simplicity, we would like to mention that our technique can also be applied to combine the original non-monotone ABE scheme of [12] with the SD method.

2   Preliminaries

2.1   Notations

- $y \xleftarrow{\$} A$: y is randomly selected from A according to its distribution, when A is a random variable, and y is uniformly selected from A, when A is a set.
- $\mathcal{G} \to x$: x is the output of the algorithm or experiment $\mathcal{G}$.
- $\vec{x}$: a vector $(x_1, \ldots, x_n) \in \mathbb{F}_q^n$ of length n for some $n \in \mathbb{N}$.
- $\boldsymbol{x}$: an element of vector space $\mathbb{V} \neq \mathbb{F}_q^n$.
- $\mathsf{span}\langle \boldsymbol{b}_1, \ldots, \boldsymbol{b}_m \rangle \subseteq \mathbb{V}$: the subspace of $\mathbb{V}$ generated by $\{\boldsymbol{b}_1, \ldots, \boldsymbol{b}_m\} \subseteq \mathbb{V}$.
- $\mathsf{span}\langle \vec{x}_1, \ldots, \vec{x}_m \rangle \subseteq \mathbb{F}_q^n$: the subspace of $\mathbb{F}_q^n$ spanned by $\{\vec{x}_1, \ldots, \vec{x}_m\} \subseteq \mathbb{F}_q^n$.
- $(x_1, \ldots, x_m)_{\mathbb{B}}$: $\sum\limits_{i=1}^{m} x_i \boldsymbol{b}_i$ that is a linear combination of vectors in $\mathbb{B} = \{\boldsymbol{b}_1, \ldots, \boldsymbol{b}_m\} \subseteq \mathbb{V}$ with scalars $x_1, \ldots, x_m$.
- $\mathsf{GL}(m, \mathbb{F}_q)$: The general linear group of degree m over $\mathbb{F}_q$.

2.2 Dual Pairing Vector Spaces by Direct Product of Symmetric Pairing Groups

Definition 1 (Symmetric Bilinear Pairing Groups). *A symmetric bilinear pairing group* $(q, \mathbb{G}, \mathbb{G}_T, G, e)$ *is a tuple of a prime* q, *cyclic additive group* $\mathbb{G}$ *and multiplicative group* $\mathbb{G}_T$ *of order* q *each,* $G \neq 0 \in \mathbb{G}$, *and a polynomial time computable non-degenerate bilinear pairing* $e : \mathbb{G} \times \mathbb{G} \to \mathbb{G}_T$, *i.e.,* $e(sG, tG) = e(G, G)^{st}$ *for all* $s, t \in \mathbb{F}_q$ *(bilinearity) and* $e(G, G) \neq 1$ *(non-degeneracy). Let* $\mathcal{G}_{\mathsf{bpg}}$ *be an algorithm that takes input* 1^λ *and outputs a description of bilinear pairing group* $(q, \mathbb{G}, \mathbb{G}_T, G, e)$ *with security parameter* λ.

Definition 2 (Dual Pairing Vector Spaces (DPVS)). *As defined in* [12], *a dual pairing vector space (DPVS)* $(q, \mathbb{V}, \mathbb{G}_T, \mathbb{A}, E)$ *by a direct product of symmetric pairing groups* $(q, \mathbb{G}, \mathbb{G}_T, G, e)$ *is a tuple of prime* q, n *dimensional vector space* $\mathbb{V} = \mathbb{G}^n = \overbrace{\mathbb{G} \times \ldots \times \mathbb{G}}^{n}$ *over* $\mathbb{F}_q$, *cyclic group* $\mathbb{G}_T$ *of order* q, *canonical basis* $\mathbb{A} = \{ \boldsymbol{a}_1, \ldots, \boldsymbol{a}_n \}$ *of* $\mathbb{V}$, *where* $\boldsymbol{a}_i = (\overbrace{0, \ldots, 0}^{i-1}, G, \overbrace{0, \ldots, 0}^{n-i})$, *and pairing* $E : \mathbb{V} \times \mathbb{V} \to \mathbb{G}_T$. *The pairing* E *is defined by* $E(\boldsymbol{x}, \boldsymbol{y}) = \prod_{i=1}^{n} e(G_i, H_i) \in \mathbb{G}_T$ *where* $\boldsymbol{x} = (G_1, \ldots, G_n) \in \mathbb{V}$ *and* $\boldsymbol{y} = (H_1, \ldots, H_n) \in \mathbb{V}$. *The map* E *is non-degenerate bilinear, i.e.,* $E(s\boldsymbol{x}, t\boldsymbol{y}) = E(\boldsymbol{x}, \boldsymbol{y})^{st}$ *for* $s, t \in \mathbb{F}_q$, *and if* $E(\boldsymbol{x}, \boldsymbol{y}) = 1$ *for all* $\boldsymbol{y} \in \mathbb{V}$, *then* $\boldsymbol{x} = 0$. *For all* i *and* j, $E(\boldsymbol{a}_i, \boldsymbol{a}_j) = e(G, G)^{\delta_{i,j}}$ *where* $\delta_{i,j} = 1$ *if* $i = j$, *and 0 otherwise, and* $e(G, G) \neq 1 \in \mathbb{G}_T$. *DPVS generation algorithm* $\mathcal{G}_{\mathsf{dpvs}}$ *takes input* 1^λ $(\lambda \in \mathbb{N})$, $n \in \mathbb{N}$ *together with* $\mathsf{param}_{\mathbb{G}} = (q, \mathbb{G}, \mathbb{G}_T, G, e)$, *and outputs a description of* $\mathsf{param}_{\mathbb{V}} = (q, \mathbb{V}, \mathbb{G}_T, \mathbb{A}, E)$ *with security parameter* λ *and* n-*dimensional* $\mathbb{V}$. *It can be constructed by using* $\mathcal{G}_{\mathsf{bpg}}$ *as a subroutine.*

For a matrix $\boldsymbol{W} = (w_{i,j})_{i,j=1,\ldots,n} \in \mathbb{F}_q^{n \times n}$ *and element* $\boldsymbol{x} = (G_1, \ldots, G_n)$ *in* n-*dimensional* $\mathbb{V}$, $\boldsymbol{x}\boldsymbol{W}$ *denotes* $\left(\sum_{i=1}^{n} G_i w_{i,1}, \ldots, \sum_{i=1}^{n} G_i w_{i,n} \right) = \left(\sum_{i=1}^{n} w_{i,1} G_i, \ldots, \sum_{i=1}^{n} w_{i,n} G_i \right)$ *by a natural multiplication of an* n-*dimensional row vector and an* $n \times n$ *matrix. Thus it satisfies an associative law, i.e.,* $(\boldsymbol{x}\boldsymbol{W}_1)\boldsymbol{W}_2 = \boldsymbol{x}(\boldsymbol{W}_1\boldsymbol{W}_2)$.

Below we describe *random dual orthogonal basis generator* $\mathcal{G}_{\mathsf{ob}}$, which is used as a subroutine in our RABE scheme.

$\mathcal{G}_{\mathsf{ob}}(1^\lambda, (n_t)_{t=0,1})$: This algorithm performs the following operations:

– Generate $(\mathsf{param}_{\mathbb{G}} = (q, \mathbb{G}, \mathbb{G}_T, G, e) \xleftarrow{\$} \mathcal{G}_{\mathsf{bpg}}(1^\lambda)$, $\psi \xleftarrow{\$} \mathbb{F}_q^\times$, where $\mathbb{F}_q^\times = \mathbb{F}_q \backslash \{0\}$.
– For $t = 0, 1$ execute the following:
   • Obtain $\mathsf{param}_{\mathbb{V}_t} = (q, \mathbb{V}_t, \mathbb{G}_T, \mathbb{A}_t, E) \xleftarrow{\$} \mathcal{G}_{\mathsf{dpvs}}(1^\lambda, n_t, \mathsf{param}_{\mathbb{G}})$ such that $\mathbb{V}_t = \overbrace{\mathbb{G} \times \ldots \times \mathbb{G}}^{n_t}$ and $\mathbb{A}_t = \{ \boldsymbol{a}_{t,1}, \ldots, \boldsymbol{a}_{t,n_t} \}$ is the canonical basis of $\mathbb{V}_t$.
   • Choose $\boldsymbol{X}_t = (\chi_{t,i,j})_{i,j=1,\ldots,n_t} \xleftarrow{\$} \mathsf{GL}(n_t, \mathbb{F}_q)$.

- Compute $\boldsymbol{X}_t^* = (\vartheta_{t,i,j})_{i,j=1,\ldots,n_t} = \psi(\boldsymbol{X}_t^\mathsf{T})^{-1}$, where $\boldsymbol{Y}^\mathsf{T}$ denotes transpose of the matrix $\boldsymbol{Y}$. Hereafter, $\vec{\chi}_{t,i}$ and $\vec{\vartheta}_{t,i}$ represent the i-th rows of $\boldsymbol{X}_t$ and $\boldsymbol{X}_t^*$ respectively, for $i = 1,\ldots,n_t$. Note that, for $i, i' = 1,\ldots,n_t$,

$$\vec{\chi}_{t,i} \cdot \vec{\vartheta}_{t,i'} = \sum_{j=1}^{n_t} \chi_{t,i,j}\vartheta_{t,i',j} = \psi, \text{ if } i = i', \text{ and } 0, \text{ otherwise.}$$

- Set $\boldsymbol{b}_{t,i} = (\vec{\chi}_{t,i})_{\mathbb{A}_t} = \sum_{j=1}^{n_t} \chi_{t,i,j}\boldsymbol{a}_{t,j} = (\chi_{t,i,1}G,\ldots,\chi_{t,i,n_t}G), \ \boldsymbol{b}_{t,i}^* = (\vec{\vartheta}_{t,i})_{\mathbb{A}_t}$

$$= \sum_{j=1}^{n_t} \vartheta_{t,i,j}\boldsymbol{a}_{t,j} = (\vartheta_{t,i,1}G,\ldots,\vartheta_{t,i,n_t}G) \text{ for } i = 1,\ldots n_t, \text{ and define } \mathbb{B}_t =$$

$\{\boldsymbol{b}_{t,1},\ldots,\boldsymbol{b}_{t,n_t}\}, \ \mathbb{B}_t^* = \{\boldsymbol{b}_{t,1}^*,\ldots,\boldsymbol{b}_{t,n_t}^*\}.$
- Compute $g_T = e(G,G)^\psi$ and set $\mathsf{param} = (\{\mathsf{param}_{\mathbb{V}_t}\}_{t=0,1}, g_T)$.
- Return $(\mathsf{param}, \{\mathbb{B}_t, \mathbb{B}_t^*\}_{t=0,1})$.

Note that, $E(\boldsymbol{b}_{t,i}, \boldsymbol{b}_{t,i'}^*) = E((\vec{\chi}_{t,i})_{\mathbb{A}_t}, (\vec{\vartheta}_{t,i'})_{\mathbb{A}_t})$

$$= \prod_{j=1}^{n_t} e(G,G)^{\chi_{t,i,j}\vartheta_{t,i',j}} = e(G,G)^{\vec{\chi}_{t,i}\cdot\vec{\vartheta}_{t,i'}}$$

$$= g_T \text{ if } i = i', \text{ and } 0 \text{ otherwise for } t = 0,1; i,i' = 1,\ldots,n_t$$

Henceforth, for simplicity, we denote $n = n_1$, $\mathbb{V} = \mathbb{V}_1$, $\mathbb{A} = \mathbb{A}_1$, $\mathbb{B} = \mathbb{B}_1 = \{\boldsymbol{b}_1,\ldots,\boldsymbol{b}_{16}\}$ and $\mathbb{B}^* = \mathbb{B}_1^* = \{\boldsymbol{b}_1^*,\ldots,\boldsymbol{b}_{16}^*\}$ for variables with $t = 1$.

2.3 Complexity Assumptions Derived from the Decisional Linear (DLIN) Assumption

Definition 3 (DLIN: Decisional Linear Assumption). *To guess* $\beta \in \{0,1\}$ *given* $\varrho = (\mathsf{param}_{\mathbb{G}}, G, \xi G, \kappa G, \delta\xi G, \sigma\kappa G, Y_\beta) \xleftarrow{\$} \mathcal{G}_\beta^{\mathsf{DLIN}}(1^\lambda)$, *where* $\mathcal{G}_\beta^{\mathsf{DLIN}}(1^\lambda)$ *is defined in Fig. 1. For a probabilistic machine* $\mathcal{F}$, *we define advantage of* $\mathcal{F}$ *for* DLIN *problem as:*

$$\mathsf{Adv}_{\mathcal{F}}^{\mathsf{DLIN}}(\lambda) = \left| \Pr\left[\mathcal{F}(1^\lambda, \varrho) \to 1 | \varrho \xleftarrow{\$} \mathcal{G}_0^{\mathsf{DLIN}}(1^\lambda)\right] \right.$$

$$\left. - \Pr\left[\mathcal{F}(1^\lambda, \varrho) \to 1 | \varrho \xleftarrow{\$} \mathcal{G}_1^{\mathsf{DLIN}}(1^\lambda)\right] \right|.$$

The DLIN *assumption states that for any probabilistic polynomial-time adversary* $\mathcal{F}$, *the advantage* $\mathsf{Adv}_{\mathcal{F}}^{\mathsf{DLIN}}(\lambda)$ *is negligible in* λ.

We now introduce two new assumptions derived from the DLIN assumption that are used in proving the full security of our RABE scheme.

Definition 4 (Problem 1). *To guess* $\beta \in \{0,1\}$ *given* $\varrho = (\mathsf{param}, \widehat{\mathbb{B}}_0, \widehat{\mathbb{B}}_0^*, \widehat{\mathbb{B}}, \widehat{\mathbb{B}}^*,$ $\boldsymbol{e}_{\beta,0}, \{\boldsymbol{e}_{\beta,t,i}\}_{t=1,\ldots,d;i=1,2}, \{\boldsymbol{e}_{\beta,d+v,\varpi,i}\}_{v=1,2,\varpi=1,\ldots,\widehat{r}_{\max},i=1,2}) \xleftarrow{\$} \mathcal{G}_\beta^{\mathsf{P1}}(1^\lambda, d, \widehat{r}_{\max}),$

$$\mathcal{G}_\beta^{\mathsf{DLIN}}(1^\lambda): \quad \mathbf{param}_\mathbb{G} = (q, \mathbb{G}, \mathbb{G}_T, G, e) \xleftarrow{\$} \mathcal{G}_{\mathsf{bpg}}(1^\lambda),$$
$$\kappa, \delta, \xi, \sigma \xleftarrow{\$} \mathbb{F}_q, \ Y_0 = (\delta + \sigma)G, \ Y_1 \xleftarrow{\$} \mathbb{G},$$
$$\text{return } \varrho = (\mathbf{param}_\mathbb{G}, G, \xi G, \kappa G, \delta \xi G, \sigma \kappa G, Y_\beta).$$

Fig. 1. $\mathcal{G}_\beta^{\mathsf{DLIN}}(1^\lambda)$

$\mathcal{G}_\beta^{\mathsf{P1}}(1^\lambda, d, \widehat{r}_{\max}): \ (\mathsf{param}, \{\mathbb{B}_0, \mathbb{B}_0^*\}, \{\mathbb{B}, \mathbb{B}^*\}) \xleftarrow{\$} \mathcal{G}_{\mathsf{ob}}(1^\lambda, (n_0 = 5, n = 16)),$

$\varphi_0, \omega \xleftarrow{\$} \mathbb{F}_q, \ \tau \xleftarrow{\$} \mathbb{F}_q^\times,$
$\widehat{\mathbb{B}}_0 = \{b_{0,1}, b_{0,3}, b_{0,5}\}, \ \widehat{\mathbb{B}} = \{b_1, \ldots, b_4, b_{13}, b_{14}\},$
$\widehat{\mathbb{B}}_0^* = \{b_{0,1}^*, b_{0,3}^*, b_{0,4}^*\}, \ \widehat{\mathbb{B}}^* = \{b_1^*, \ldots, b_4^*, b_{11}^*, b_{12}^*\},$
$e_{0,0} = (\omega, 0, 0, 0, \varphi_0)_{\mathbb{B}_0}, \ e_{1,0} = (\omega, \tau, 0, 0, \varphi_0)_{\mathbb{B}_0},$
$\vec{e}_1 = (1, 0), \ \vec{e}_2 = (0, 1) \in \mathbb{F}_q^2,$

for $t = 1, \ldots, d,$

$\quad Z_t \xleftarrow{\$} \mathsf{GL}(2, \mathbb{F}_q),$
$\quad$ for $i = 1, 2,$
$\qquad \sigma_{t,i}, \varphi_{t,i,1}, \varphi_{t,i,2} \xleftarrow{\$} \mathbb{F}_q,$
$\qquad e_{0,t,i} = (\sigma_{t,i}(1, t), \omega \vec{e}_i, \qquad 0^6, \qquad 0^2, \varphi_{t,i,1}, \varphi_{t,i,2}, 0^2)_\mathbb{B},$
$\qquad e_{1,t,i} = (\sigma_{t,i}(1, t), \omega \vec{e}_i, \tau \vec{e}_i, 0^2, \tau \vec{e}_i Z_t, \ 0^2, \varphi_{t,i,1}, \varphi_{t,i,2}, 0^2)_\mathbb{B},$

for $v = 1, 2; \varpi = 1, \ldots, \widehat{r}_{\max},$

$\quad Z_{d+v, \varpi} \xleftarrow{\$} \mathsf{GL}(2, \mathbb{F}_q),$
$\quad$ for $i = 1, 2,$
$\qquad \sigma_{d+v, \varpi, i}, \varphi_{d+v, \varpi, i, 1}, \varphi_{d+v, \varpi, i, 2} \xleftarrow{\$} \mathbb{F}_q,$
$\qquad e_{0, d+v, \varpi, i} = (\sigma_{d+v, \varpi, i}(1, d+v), \omega \vec{e}_i, \qquad 0^6, \qquad\qquad 0^2, \varphi_{d+v, \varpi, i, 1}, \varphi_{d+v, \varpi, i, 2}, 0^2)_\mathbb{B},$
$\qquad e_{1, d+v, \varpi, i} = (\sigma_{d+v, \varpi, i}(1, d+v), \omega \vec{e}_i, \tau \vec{e}_i, 0^2, \tau \vec{e}_i Z_{d+v, \varpi}, \ 0^2, \varphi_{d+v, \varpi, i, 1}, \varphi_{d+v, \varpi, i, 2}, 0^2)_\mathbb{B},$
return $\varrho = (\mathsf{param}, \widehat{\mathbb{B}}_0, \widehat{\mathbb{B}}_0^*, \widehat{\mathbb{B}}, \widehat{\mathbb{B}}^*, e_{\beta, 0}, \{e_{\beta, t, i}\}_{t=1, \ldots, d; i=1, 2}, \{e_{\beta, d+v, \varpi, i}\}_{v=1, 2; \varpi=1, \ldots, \widehat{r}_{\max}; i=1, 2}).$

Fig. 2. $\mathcal{G}_\beta^{\mathsf{P1}}(1^\lambda, d, \widehat{r}_{\max})$

where $\mathcal{G}_\beta^{\mathsf{P1}}(1^\lambda, d, \widehat{r}_{\max})$ is defined in Fig. 2. For a probabilistic adversary $\mathcal{B}$ the advantage of $\mathcal{B}$ for Problem 1 is given by

$$\mathsf{Adv}_\mathcal{B}^{\mathsf{P1}}(\lambda) = \left| \Pr\left[\mathcal{B}(1^\lambda, \varrho) \to 1 | \varrho \xleftarrow{\$} \mathcal{G}_0^{\mathsf{P1}}(1^\lambda, d, \widehat{r}_{\max}) \right] \right.$$
$$\left. - \Pr\left[\mathcal{B}(1^\lambda, \varrho) \to 1 | \varrho \xleftarrow{\$} \mathcal{G}_1^{\mathsf{P1}}(1^\lambda, d, \widehat{r}_{\max}) \right] \right|.$$

Lemma 1. *Problem 1 is computationally intractable under the DLIN assumption. Formally, for any probabilistic polynomial-time adversary $\mathcal{B}$ there exist probabilistic machines $\mathcal{F}_1, \mathcal{F}_2, \mathcal{F}_3$, whose running times are essentially the same as that of $\mathcal{B}$, such that for any security parameter λ,*

$$\mathsf{Adv}_\mathcal{B}^{\mathsf{P1}}(\lambda) \leq \mathsf{Adv}_{\mathcal{F}_1}^{\mathsf{DLIN}}(\lambda) + \sum_{p=1}^d \sum_{j=1}^2 \mathsf{Adv}_{\mathcal{F}_{2\text{-}p\text{-}j}}^{\mathsf{DLIN}}(\lambda) + \sum_{v=1}^2 \sum_{\varpi=1}^{\widehat{r}_{\max}} \sum_{j=1}^2 \mathsf{Adv}_{\mathcal{F}_{3\text{-}(d+v)\text{-}\varpi\text{-}j}}^{\mathsf{DLIN}}(\lambda) + \epsilon,$$

where $\mathcal{F}_{2\text{-}p\text{-}j}(\cdot) = \mathcal{F}_2(p, j, \cdot), \mathcal{F}_{3\text{-}(d+v)\text{-}\varpi\text{-}j}(\cdot) = \mathcal{F}_3(d+v, \varpi, j, \cdot)$ and $\epsilon = [5 + 10d + 20\widehat{r}_{\max}]/q.$

Definition 5 (Problem 2). *To guess $\beta \in \{0, 1\}$ given $(\mathsf{param}, \widehat{\mathbb{B}}_0, \widehat{\mathbb{B}}_0^*, \widehat{\mathbb{B}}, \widehat{\mathbb{B}}^*,$*
$h_{\beta, 0}^*, e_0, \{h_{\beta, t, i}^*, e_{t,i}\}_{t=1, \ldots, d; i=1, 2}, \{h_{\beta, d+v, \varpi, i}^*, e_{d+v, \varpi, i}\}_{v=1, 2; \varpi=1, \ldots, \aleph; i=1, 2}) \xleftarrow{\$}$

$$
\begin{aligned}
&\mathcal{G}_\beta^{\mathsf{P2}}(1^\lambda, d, N_{\max}, \widehat{r}_{\max}): \quad (\mathsf{param}, \{\mathbb{B}_0, \mathbb{B}_0^*\}, \{\mathbb{B}, \mathbb{B}^*\}) \xleftarrow{\$} \mathcal{G}_{\mathsf{ob}}(1^\lambda, (n_0 = 5, n = 16)), \\
&\gamma, \eta_0, \varphi_0, \omega \xleftarrow{\$} \mathbb{F}_q, \ \tau, \delta \xleftarrow{\$} \mathbb{F}_q^\times, \\
&\widehat{\mathbb{B}}_0 = \{\boldsymbol{b}_{0,1}, \boldsymbol{b}_{0,3}, \boldsymbol{b}_{0,5}\}, \ \widehat{\mathbb{B}} = \{\boldsymbol{b}_1, \dots, \boldsymbol{b}_4, \boldsymbol{b}_{13}, \boldsymbol{b}_{14}\}, \\
&\widehat{\mathbb{B}}_0^* = \{\boldsymbol{b}_{0,1}^*, \dots, \boldsymbol{b}_{0,4}^*\}, \ \widehat{\mathbb{B}}^* = \{\boldsymbol{b}_1^*, \dots, \boldsymbol{b}_4^*, \boldsymbol{b}_{11}^*, \boldsymbol{b}_{12}^*\}, \\
&\boldsymbol{h}_{0,0}^* = (\gamma, 0, 0, \eta_0, 0)_{\mathbb{B}_0^*}, \ \boldsymbol{h}_{1,0}^* = (\gamma, \delta, 0, \eta_0, 0)_{\mathbb{B}_0^*}, \ \boldsymbol{e}_0 = (\omega, \tau, 0, 0, \varphi_0)_{\mathbb{B}_0}, \\
&\overrightarrow{\boldsymbol{e}}_1 = (1, 0), \ \overrightarrow{\boldsymbol{e}}_2 = (0, 1) \in \mathbb{F}_q^2, \\[4pt]
&\text{for } t = 1, \dots, d, \\
&\quad \boldsymbol{Z}_t \xleftarrow{\$} \mathsf{GL}(2, \mathbb{F}_q), \ \boldsymbol{U}_t = (\boldsymbol{Z}_t^{-1})^\mathsf{T}, \\
&\quad \text{for } i = 1, 2, \\
&\quad\quad \mu_{t,i}, \sigma_{t,i}, \eta_{t,i,1}, \eta_{t,i,2}, \varphi_{t,i,1}, \varphi_{t,i,2} \xleftarrow{\$} \mathbb{F}_q, \\
&\quad\quad \boldsymbol{h}_{0,t,i}^* = (\mu_{t,i}(t, -1), \gamma \overrightarrow{\boldsymbol{e}}_i, \quad 0^4, \quad\quad 0^2, \quad \eta_{t,i,1}, \eta_{t,i,2}, 0^2, 0^2)_{\mathbb{B}^*}, \\
&\quad\quad \boldsymbol{h}_{1,t,i}^* = (\mu_{t,i}(t, -1), \gamma \overrightarrow{\boldsymbol{e}}_i, \quad 0^4, \quad \delta \overrightarrow{\boldsymbol{e}}_i \boldsymbol{U}_t, \eta_{t,i,1}, \eta_{t,i,2}, 0^2, 0^2)_{\mathbb{B}^*}, \\
&\quad\quad \boldsymbol{e}_{t,i} = (\sigma_{t,i}(1, t), \omega \overrightarrow{\boldsymbol{e}}_i, \ \tau \overrightarrow{\boldsymbol{e}}_i, 0^2, \ \tau \overrightarrow{\boldsymbol{e}}_i \boldsymbol{Z}_t, \ 0^2, \varphi_{t,i,1}, \varphi_{t,i,2}, 0^2)_{\mathbb{B}}, \\[4pt]
&\text{for } \upsilon = 1, 2; \varpi = 1, \dots, \log^2 N_{\max} + \widehat{r}_{\max}, \\
&\quad \boldsymbol{Z}_{d+\upsilon,\varpi} \xleftarrow{\$} \mathsf{GL}(2, \mathbb{F}_q), \ \boldsymbol{U}_{d+\upsilon,\varpi} = (\boldsymbol{Z}_{d+\upsilon,\varpi}^{-1})^\mathsf{T}, \\
&\quad \text{for } i = 1, 2, \\
&\quad\quad \mu_{d+\upsilon,\varpi,i}, \sigma_{d+\upsilon,\varpi,i}, \eta_{d+\upsilon,\varpi,i,1}, \eta_{d+\upsilon,\varpi,i,2}, \varphi_{d+\upsilon,\varpi,i,1}, \varphi_{d+\upsilon,\varpi,i,2} \xleftarrow{\$} \mathbb{F}_q, \\
&\quad\quad \boldsymbol{h}_{0,d+\upsilon,\varpi,i}^* = (\mu_{d+\upsilon,\varpi,i}(d+\upsilon, -1), \gamma \overrightarrow{\boldsymbol{e}}_i, \quad 0^4, \quad\quad 0^2, \quad \eta_{d+\upsilon,\varpi,i,1}, \eta_{d+\upsilon,\varpi,i,2}, 0^2, 0^2)_{\mathbb{B}^*}, \\
&\quad\quad \boldsymbol{h}_{1,d+\upsilon,\varpi,i}^* = (\mu_{d+\upsilon,\varpi,i}(d+\upsilon, -1), \gamma \overrightarrow{\boldsymbol{e}}_i, \quad 0^4, \quad \delta \overrightarrow{\boldsymbol{e}}_i \boldsymbol{U}_{d+\upsilon,\varpi}, \eta_{d+\upsilon,\varpi,i,1}, \eta_{d+\upsilon,\varpi,i,2}, 0^2, 0^2)_{\mathbb{B}^*}, \\
&\quad\quad \boldsymbol{e}_{d+\upsilon,\varpi,i} = (\sigma_{d+\upsilon,\varpi,i}(1, d+\upsilon), \omega \overrightarrow{\boldsymbol{e}}_i, \ \tau \overrightarrow{\boldsymbol{e}}_i, 0^2, \ \tau \overrightarrow{\boldsymbol{e}}_i \boldsymbol{Z}_{d+\upsilon,\varpi}, \ 0^2, \varphi_{d+\upsilon,\varpi,i,1}, \varphi_{d+\upsilon,\varpi,i,2}, 0^2)_{\mathbb{B}}, \\
&\text{return } \varrho = (\mathsf{param}, \widehat{\mathbb{B}}_0, \widehat{\mathbb{B}}_0^*, \widehat{\mathbb{B}}, \widehat{\mathbb{B}}^*, \boldsymbol{h}_{\beta,0}^*, \boldsymbol{e}_0, \{\boldsymbol{h}_{\beta,t,i}^*, \boldsymbol{e}_{t,i}\}_{t=1,\dots,d;i=1,2}, \\
&\quad\quad\quad\quad \{\boldsymbol{h}_{\beta,d+\upsilon,\varpi,i}^*, \boldsymbol{e}_{d+\upsilon,\varpi,i}\}_{\upsilon=1,2;\varpi=1,\dots,\aleph;i=1,2}).
\end{aligned}
$$

Fig. 3. $\mathcal{G}_\beta^{\mathsf{P2}}(1^\lambda, d, N_{\max}, \widehat{r}_{\max})$

$\mathcal{G}_\beta^{\mathsf{P2}}(1^\lambda, d, N_{\max}, \widehat{r}_{\max})$, where $\mathcal{G}_\beta^{\mathsf{P2}}(1^\lambda, d, N_{\max}, \widehat{r}_{\max})$ is defined in Fig. 3. For a probabilistic adversary $\mathcal{B}$, the advantage of $\mathcal{B}$ for Problem 2 is given by

$$
\begin{aligned}
\mathsf{Adv}_{\mathcal{B}}^{\mathsf{P2}}(\lambda) = \Big| &\Pr\Big[\mathcal{B}(1^\lambda, \varrho) \to 1 \,|\, \varrho \xleftarrow{\$} \mathcal{G}_0^{\mathsf{P2}}(1^\lambda, d, N_{\max}, \widehat{r}_{\max})\Big] \\
&- \Pr\Big[\mathcal{B}(1^\lambda, \varrho) \to 1 \,|\, \varrho \xleftarrow{\$} \mathcal{G}_1^{\mathsf{P2}}(1^\lambda, d, N_{\max}, \widehat{r}_{\max})\Big] \Big|.
\end{aligned}
$$

Lemma 2. *Problem 2 is computationally intractable under the* DLIN *assumption. More formally, for any probabilistic polynomial-time adversary $\mathcal{B}$, there exist probabilistic machines $\mathcal{F}_1, \mathcal{F}_{2\text{-}1}, \dots, \mathcal{F}_{2\text{-}11}$, whose running times are essentially the same as that of $\mathcal{B}$, such that for any security parameter λ,*

$$
\begin{aligned}
\mathsf{Adv}_{\mathcal{B}}^{\mathsf{P2}}(\lambda) \leq\ &\mathsf{Adv}_{\mathcal{F}_1}^{\mathsf{DLIN}}(\lambda) + \sum_{j=1}^{2}\Bigg[\sum_{p=1}^{d}\Bigg\{\mathsf{Adv}_{\mathcal{F}_{2\text{-}p\text{-}1\text{-}j}}^{\mathsf{DLIN}}(\lambda) + \mathsf{Adv}_{\mathcal{F}_{2\text{-}p\text{-}2\text{-}j}}^{\mathsf{DLIN}}(\lambda) + \\
&\sum_{\substack{l=1 \\ l\neq p}}^{d+2}\Big(\mathsf{Adv}_{\mathcal{F}_{2\text{-}p\text{-}3\text{-}j\text{-}l}}^{\mathsf{DLIN}}(\lambda) + \mathsf{Adv}_{\mathcal{F}_{2\text{-}p\text{-}4\text{-}j\text{-}l}}^{\mathsf{DLIN}}(\lambda)\Big) + \mathsf{Adv}_{\mathcal{F}_{2\text{-}p\text{-}5\text{-}j}}^{\mathsf{DLIN}}(\lambda)\Bigg\} + \\
&\sum_{\upsilon=1}^{2}\sum_{\varpi=1}^{\aleph}\Bigg\{\mathsf{Adv}_{\mathcal{F}_{2\text{-}(d+\upsilon)\text{-}\varpi\text{-}6\text{-}j}}^{\mathsf{DLIN}}(\lambda) + \mathsf{Adv}_{\mathcal{F}_{2\text{-}(d+\upsilon)\text{-}\varpi\text{-}7\text{-}j}}^{\mathsf{DLIN}}(\lambda) + \sum_{\substack{l=1 \\ l\neq d+\upsilon}}^{d+2}\Big(\mathsf{Adv}_{\mathcal{F}_{2\text{-}(d+\upsilon)\text{-}\varpi\text{-}8\text{-}j\text{-}l}}^{\mathsf{DLIN}}(\lambda) \\
&+ \mathsf{Adv}_{\mathcal{F}_{2\text{-}(d+\upsilon)\text{-}\varpi\text{-}9\text{-}j\text{-}l}}^{\mathsf{DLIN}}(\lambda)\Big) + \sum_{\substack{\iota=1 \\ \iota\neq\varpi}}^{\aleph}\mathsf{Adv}_{\mathcal{F}_{2\text{-}(d+\upsilon)\text{-}\varpi\text{-}10\text{-}j\text{-}\iota}}^{\mathsf{DLIN}}(\lambda) + \mathsf{Adv}_{\mathcal{F}_{2\text{-}(d+\upsilon)\text{-}\varpi\text{-}11\text{-}j}}^{\mathsf{DLIN}}(\lambda)\Bigg\}\Bigg] + \epsilon,
\end{aligned}
$$

where $\mathcal{F}_{2\text{-}p\text{-}1\text{-}j}(\cdot) = \mathcal{F}_{2\text{-}1}(p, j, \cdot), \mathcal{F}_{2\text{-}p\text{-}2\text{-}j}(\cdot) = \mathcal{F}_{2\text{-}2}(p, j, \cdot),$

$\quad\quad \mathcal{F}_{2\text{-}p\text{-}3\text{-}j\text{-}l}(\cdot) = \mathcal{F}_{2\text{-}3}(p, j, l, \cdot), \mathcal{F}_{2\text{-}p\text{-}4\text{-}j\text{-}l}(\cdot) = \mathcal{F}_{2\text{-}4}(p, j, l, \cdot),$

$\quad\quad \mathcal{F}_{2\text{-}p\text{-}5\text{-}j}(\cdot) = \mathcal{F}_{2\text{-}5}(p, j, \cdot), \mathcal{F}_{2\text{-}(d+\upsilon)\text{-}\varpi\text{-}6\text{-}j}(\cdot) = \mathcal{F}_{2\text{-}6}(d + \upsilon, \varpi, j, \cdot)$ etc.,

$\aleph = \log^2 N_{\max} + \widehat{r}_{\max}$ and $\epsilon = \left[5 + 40d + 10d^2 + 2\aleph(30 + 10d + 10\aleph)\right]/q.$

Problems 1 and 2 are extended from Problems 1-ABE and 2-ABE in [12] respectively. The proofs of Lemmas 1 and 2 can be found in the full version.

2.4   The Notion of Revocable Attribute-Based Encryption

We assume familiarity with monotone access structures and secret-sharing schemes.

• **Syntax of Revocable Attribute-Based Encryption:** As described in [1, 13], a (key-policy) revocable attribute-based encryption (RABE) scheme that is associated with the attribute universe $\mathbb{U}$ of size d, each element of which is expressed by a pair of attribute id and value of attribute, i.e., $\mathbb{U} = \{(t, A_t)|t \in \{1,\ldots,d\} \wedge A_t \in \mathbb{F}_q\}$; a collection $\mathfrak{S}$ of admissible monotone access structures $\mathbb{S} = (\boldsymbol{M}, \rho)$, consisting of some matrix $\boldsymbol{M}$ over $\mathbb{F}_q$ together with a function ρ labeling the rows of $\boldsymbol{M}$ by attributes in $\mathbb{U}$; and message space $\mathbb{M}$, consists of the following algorithms:

RABE.Setup$(1^\lambda, N_{\max})$: Taking as input a security parameter 1^λ and the maximum number of users $N_{\max}$, the key generation center publishes public parameters PP and a state ST, while generates a master secret key MK for itself.

RABE.GenKey$(\text{PP}, \text{MK}, \text{ST}, ID, \mathbb{S})$: The key generation center takes as input the public parameters PP, the master secret key MK, the state ST, a user identity ID and the access structure $\mathbb{S} = (\boldsymbol{M}, \rho) \in \mathfrak{S}$ of that user. It provides a private key $\text{SK}_{\mathbb{S},ID}$ to that user and publishes an updated state ST.

RABE.Encrypt$(\text{PP}, \text{ST}, \varGamma, \text{RL}, M)$: On input the public parameters PP, the state ST, an attribute set $\varGamma \subseteq \mathbb{U}$, a set of revoked user identities RL and a message $M \in \mathbb{M}$, the encrypter outputs a ciphertext $\text{CT}_{\varGamma,\text{RL}}$.

RABE.Decrypt$(\text{CT}_{\varGamma,\text{RL}}, \text{SK}_{\mathbb{S},ID}, \mathbb{S}, ID, \text{PP}, \text{ST})$: A user takes as input a ciphertext $\text{CT}_{\varGamma,\text{RL}}$, its private key $\text{SK}_{\mathbb{S},ID}$, its access structure $\mathbb{S}$, user identity ID, the public parameters PP and the state ST. It obtains an encrypted message M or the distinguished symbol $\perp$.

• **Correctness:** The correctness of RABE is defined as follows: For all PP, ST, MK generated by RABE.Setup$(1^\lambda, N_{\max})$, $\text{SK}_{\mathbb{S},ID}$ generated by RABE.GenKey $(\text{PP}, \text{MK}, \text{ST}, ID, \mathbb{S})$ for any $\mathbb{S}, ID$, $\text{CT}_{\varGamma,\text{RL}}$ generated by RABE.Encrypt$(\text{PP}, \text{ST}, \varGamma,$ RL, $M)$ for any $\varGamma$, RL and M, it is required that (a) if $\mathbb{S}$ accepts $\varGamma$ and $ID \notin$ RL, then RABE.Decrypt$(\text{CT}_{\varGamma,\text{RL}}, \text{SK}_{\mathbb{S},ID}, \mathbb{S}, ID, \text{PP}, \text{ST}) = M$, and (b) if $\mathbb{S}$ does not accept $\varGamma$ or $ID \in$ RL, then RABE.Decrypt$(\text{CT}_{\varGamma,\text{RL}}, \text{SK}_{\mathbb{S},ID}, \mathbb{S}, ID, \text{PP}, \text{ST}) = \perp$ with all but negligible probability.

- **Security Model:** The adaptive security of RABE under chosen plaintext attacks (CPA) is defined in terms of the following experiment between a challenger $\mathcal{B}$ and a probabilistic polynomial-time adversary $\mathcal{A}$:

Setup: $\mathcal{B}$ obtains a master secret key MK, a state ST, together with public parameters PP by running RABE.Setup($1^\lambda, N_{\max}$); keeps MK to itself; and gives PP, ST to $\mathcal{A}$.

Phase 1: $\mathcal{A}$ adaptively requests a polynomial number of private keys for access structure-user identity pairs $(\mathbb{S}_1, ID_1), \ldots, (\mathbb{S}_{\widehat{q}_1}, ID_{\widehat{q}_1})$, and $\mathcal{B}$ gives the corresponding private keys $\mathsf{SK}_{\mathbb{S}_1, ID_1}, \ldots, \mathsf{SK}_{\mathbb{S}_{\widehat{q}_1}, ID_{\widehat{q}_1}}$ along with the updated state ST to $\mathcal{A}$ by executing RABE.Genkey(PP, MK, ST, $ID_\imath, \mathbb{S}_\imath$) for $\imath = 1, \ldots, \widehat{q}_1$.

Challenge: $\mathcal{A}$ submits a challenge attribute set Γ^*, a revocation list RL*, and two challenge messages M_0^*, M_1^* with equal length satisfying the following restriction: If a private key query for an access structure-user identity pair $(\mathbb{S}_\imath, ID_\imath)$ such that $\mathbb{S}_\imath$ accepts Γ^* was requested, then $ID_\imath$ must belong to RL*. $\mathcal{B}$ flips a random coin $b \in \{0, 1\}$ and gives the challenge ciphertext CT* to $\mathcal{A}$ by performing RABE.Encrypt(PP, ST, Γ^*, RL$^*, M_b^*$).

Phase 2: $\mathcal{A}$ may continue to make a polynomial number of additional private key queries for access structure-user identity pairs $(\mathbb{S}_{\widehat{q}_1+1}, ID_{\widehat{q}_1+1}), \ldots, (\mathbb{S}_{\widehat{q}}, ID_{\widehat{q}})$ subject to the same restriction as before, and $\mathcal{B}$ gives corresponding keys to $\mathcal{A}$.

Guess: Finally, $\mathcal{A}$ outputs a guess $b' \in \{0, 1\}$ and wins the game if $b = b'$.

The advantage of $\mathcal{A}$ is defined as $\mathsf{Adv}_{\mathcal{A}}^{\mathsf{RABE, IND\text{-}CPA}}(\lambda) = |\Pr[b = b'] - 1/2|$ where the probability is taken over all the randomness of the experiment. An RABE scheme is adaptively secure under chosen plaintext attacks if for all probabilistic polynomial-time adversary $\mathcal{A}$, the advantage of $\mathcal{A}$ in the above experiment is negligible in the security parameter λ.

2.5   The Subset Difference Revocation Scheme

- **Notations Related to Full Binary Tree:** A full binary tree $\mathcal{T}$ is a tree data structure where each node except the leaf nodes has two child nodes. We define some notations concerning a full binary tree used in subsequent discussions:

- $N_{\max}$: The number of leaf nodes in $\mathcal{T}$. The number of all nodes in $\mathcal{T}$ is $2N_{\max} - 1$.
- ν_i: A node in $\mathcal{T}$ for any i, $1 \leq i \leq 2N_{\max} - 1$.
- D_i: The depth of a node ν_i, i.e., the length of the path from the root node to the node ν_i. The root node is at depth zero. The depth of $\mathcal{T}$ is the length of the path from the root node to a leaf node.
- T_i: A subtree of $\mathcal{T}$ that is rooted at ν_i for any node ν_i in $\mathcal{T}$.
- $T_{i,j}$: The subtree $T_i \backslash T_j$ for any two nodes ν_i, ν_j in $\mathcal{T}$ such that ν_j is a descendant of ν_i, i.e., all nodes that are descendants of ν_i but not of ν_j.
- S_i: The set of leaf nodes in T_i.
- $S_{i,j}$: The set of leaf nodes in $T_{i,j}$, i.e., $S_{i,j} = S_i \backslash S_j$.
- L_i: An identifier for a node ν_i in $\mathcal{T}$, that is a fixed and unique string. The identifier of each node in $\mathcal{T}$ is assigned as follows: Each edge in $\mathcal{T}$ is assigned with 0 or 1 depending on whether the edge connects a node to its left or right

child node. The identifier L_i of a node ν_i is the bit string obtained by reading all the labels of edges in the path from the root to the node ν_i.

- $(L_i \| D_j)$: The integer representation of the string formed by concatenating the binary representation of D_j, depth of the node ν_j, with L_i, identifier of ν_i for any subset $S_{i,j}$ of leaf nodes defined by the nodes ν_i and ν_j.
- $(L_i \| L_j)$: The integer representation of the string obtained by concatenating L_j with L_i for any subset $S_{i,j}$ of leaf nodes defined by the nodes ν_i and ν_j.
- $ST(\mathcal{T}, R)$ (or simply $ST(R)$): The Steiner Tree induced by a subset R of leaf nodes and the root node of the full binary tree $\mathcal{T}$, i.e., the minimal subtree of $\mathcal{T}$ that connects all the leaf nodes in R and the root node.

• **Subset Difference Method:** The *subset difference* (SD) revocation method is a special instance of a general methodology for revocation schemes proposed by Naor et al. known as the subset cover (SC) framework [11]. The well-known complete subtree (CS) scheme used in all the previous RABE schemes [1,3,13] is another instance of the subset cover framework. The original subset cover framework consists of a subset assignment part and a key assignment part. As in [9], in this paper, we define the subset cover framework by using the subset assignment part only. The formal definition of subset cover framework is given as follows:

Definition 6 (Subset Cover Framework). *A subset cover (SC) scheme for the set* $\mathcal{N} = \{1, \ldots, N_{\max}\}$ *of users consists of following probabilistic polynomial-time algorithms:*

SC.Setup($N_{\max}$): *The trusted authority takes in the maximum number* $N_{\max}$ *of users and publishes a collection* $\mathcal{S}$ *of subsets* $S_1, \ldots, S_w$ *where* $S_i \subseteq \mathcal{N}$.

SC.Assign($\mathcal{S}, u$): *On input the collection* $\mathcal{S}$ *and a user serial number* $u \in \mathcal{N}$, *the trusted authority provides a private set* $\mathsf{PV}_u = \{S_{j_1}, \ldots, S_{j_v}\}$ *to the user with serial number* u.

SC.Cover($\mathcal{S}, R$): *Taking as input the collection* $\mathcal{S}$ *and a revoked set* $R \subset \mathcal{N}$ *of users, a cover generator outputs a covering set* $\mathsf{CV}_R = \{S_{i_1}, \ldots, S_{i_z}\}$, *that is a partition of the non-revoked users* $\mathcal{N} \backslash R$ *into disjoined subsets* $S_{i_1}, \ldots, S_{i_z}$ *such that* $\mathcal{N} \backslash R = \cup_{l=1}^{z} S_{i_l}$.

SC.Match($\mathsf{CV}_R, \mathsf{PV}_u$): *A user takes as input a covering set* $\mathsf{CV}_R = \{S_{i_1}, \ldots, S_{i_z}\}$ *together with its private set* $\mathsf{PV}_u = \{S_{j_1}, \ldots, S_{j_v}\}$ *and obtains* $(S_{i_l}, S_{j_{l'}})$ *such that* $S_{i_l} \in \mathsf{CV}_R$, $u \in S_{i_l}$, *and* $S_{j_{l'}} \in \mathsf{PV}_u$, *or obtains* $\bot$.

• *Correctness:* The correctness of subset cover framework is defined as follows: For all $\mathcal{S}$ generated by SC.Setup, all PV_u generated by SC.Assign, and any R, it is required that: (a) if $u \notin R$, then SC.Match($\mathsf{CV}_R, \mathsf{PV}_u$) $= (S_{i_l}, S_{j_{l'}})$ such that $S_{i_l} \in \mathsf{CV}_R$, $u \in S_{i_l}$ and $S_{j_{l'}} \in \mathsf{PV}_u$, and (b) if $u \in R$, then SC.Match($\mathsf{CV}_R, \mathsf{PV}_u$) $= \bot$.

As mentioned earlier, the SD scheme is a particular instance of the SC scheme and it was proposed by Naor et al. [11] as an improvement on the CS scheme. Below we describe the version of SD scheme almost verbatim from [9]:

SD.Setup(N_{max}): The trusted authority takes in the maximum number N_{max} of users. Let $N_{max} = 2^{n_{max}}$ for simplicity. It first sets a full binary tree $\mathcal{T}$ of depth n_{max}. Each user is assigned to a different leaf node in $\mathcal{T}$. The collection $\mathcal{S}$ of SD scheme is the set of all subsets $S_{j,k} = S_j \backslash S_k$ where ν_j, ν_k are nodes in $\mathcal{T}$, ν_k is a descendant of ν_j; where S_j (resp. S_k) is the set of leaf nodes of the subtree rooted at ν_j (resp. ν_k). It publishes the full binary tree $\mathcal{T}$.

SD.Assign($\mathcal{T}, u$): Taking as input the tree $\mathcal{T}$ and a user serial number $u \in \mathcal{N}$, the trusted authority computes the private set PV_u for the user u as follows: Let $\nu(u)$ be the leaf node of $\mathcal{T}$ that is assigned to the user u. Let $(\nu_{l_0}, \ldots, \nu_{l_{n_{max}}})$ be the path from the root node ν_{l_0} to the leaf node $\nu_{l_{n_{max}}} = \nu(u)$. It first sets $\mathsf{PV}_u = \varnothing$. For all $j, k \in \{l_0, \ldots, l_{n_{max}}\}$ such that ν_k is a descendant of ν_j, it adds into PV_u the subset $S_{j,k}$ defined by nodes ν_j and ν_k.

SD.Cover($\mathcal{T}, R$): On input the tree $\mathcal{T}$ and a revoked set R of users, a cover generator proceeds as follows: It first sets a subtree T as $ST(R)$, and then it outputs a covering set CV_R built iteratively by removing nodes from T until T consists of just a single node as described below:

1. It finds two leaf nodes ν_j and ν_k in T such that the least-common-ancestor ν of ν_j and ν_k does not contain any other leaf nodes of T in its subtree. Note that such a pair (ν_j, ν_k) can always be found. Let ν_l and ν_m be the two child nodes of ν such that ν_j is a descendant of ν_l and ν_k is a descendant of ν_m. If there is only one leaf node left, it makes $\nu_j = \nu_k$ to be that leaf node, ν to be the root of T and $\nu_l = \nu_m = \nu$.

2. If $\nu_l \neq \nu_j$, then it adds the subset $S_{l,j}$ to CV_R; likewise, if $\nu_m \neq \nu_k$, then it adds the subset $S_{m,k}$ to CV_R.

3. It removes from T all the descendants of ν and makes ν a leaf node.

SD.Match($\mathsf{CV}_R, \mathsf{PV}_u$): A user takes as input a covering set CV_R and its private set PV_u. If it finds two subsets $S_{j,k}$ and $S_{j',k'}$ such that $S_{j,k} \in \mathsf{CV}_R$, $S_{j',k'} \in \mathsf{PV}_u$, and $(j = j') \wedge (D_k = D_{k'}) \wedge (k \neq k')$, then it outputs $(S_{j,k}, S_{j',k'})$. Otherwise, it obtains $\perp$.

The correctness property of SD scheme is formally stated by the following lemma:

Lemma 3. *If $u \notin R$, then there exists a unique pair of subsets $(S_{j,k}, S_{j',k'})$ such that $S_{j,k} \in \mathsf{CV}_R$, $S_{j',k'} \in \mathsf{PV}_u$ and $(j = j') \wedge (D_k = D_{k'}) \wedge (k \neq k')$ holds. Otherwise, such a pair of subsets cannot be found.*

Observation: Note that for any fixed pair (ν_j, D) of node and depth value, there is at most one subset $S_{j,k} \in \mathsf{CV}_R$ and at most one subset $S_{j,k'} \in \mathsf{PV}_u$ such that $D_k = D_{k'} = D$. Moreover, the defining nodes ν_j, ν_k of the subsets $S_{j,k} \in \mathsf{CV}_R$ are all distinct. This observation is very important in our RABE construction.

Lemma 4 ([11]). *Let N_{max} be the number of leaf nodes in a full binary tree and $\widehat{r}$ be the size of a revoked set. In the SD scheme, the size of a private set is $O(\log^2 N_{max})$ and the size of a covering set is at most $2\widehat{r} - 1$.*

Note that the layered subset difference scheme (LSD) was proposed by Halevy and Shamir [7] to reduce the size of a private set in the SD scheme. The SD scheme in a cryptosystem generally can be replaced by the LSD scheme since the LSD scheme is a special case of the SD scheme.

Lemma 5 ([7]). *Let N_{max} be the number of leaf nodes in a full binary tree and $\widehat{r}$ be the size of a revoked set. In the* LSD *scheme, which is a variant of the* SD *mechanism, the size of a private set is $O(\log^{1.5} N_{max})$ and the size of a covering set is at most $4\widehat{r} - 2$.*

3   Our RABE Scheme

Let $\mathcal{N} = \{1, \ldots, N_{max} = 2^{n_{max}}\}$ be the universe of user key serial numbers. Let d be the total number of attributes in the attribute universe $\mathbb{U} = \{(t, A_t)|t \in \{1, \ldots, d\} \wedge A_t \in \mathbb{F}_q\}$. Further, assume that $\widehat{r}_{max}$ be the maximum of $\sharp\mathsf{CV}_{\mathsf{RI}}$, the cardinality of the covering set $\mathsf{CV}_{\mathsf{RI}}$, for all revoked set $\mathsf{RI} \subset \mathcal{N}$ used in the system. Our RABE scheme supports monotone access structures $\mathbb{S} = (M, \rho)$, the collection of which is denoted by $\mathfrak{S}$. In the proposed RABE scheme, we assume that ρ is injective for $\mathbb{S} = (M, \rho) \in \mathfrak{S}$. The message space in our RABE scheme is $\mathbb{M} = \mathbb{G}_T$. Our RABE scheme is described as follows:

RABE.Setup($1^\lambda, N_{max}$): The key generation center takes as input a security parameter 1^λ and the maximum number N_{max} of users and proceeds as follows:

1. It first runs $\mathcal{G}_{ob}(1^\lambda, (n_0 = 5, n = 16))$ to get (param $= (\mathsf{param}_{\mathbb{V}_0}, \mathsf{param}_{\mathbb{V}}, g_T)$, $\{\mathbb{B}_0, \mathbb{B}_0^*\}, \{\mathbb{B}, \mathbb{B}^*\}$). It sets $\widehat{\mathbb{B}}_0 = \{b_{0,1}, b_{0,3}, b_{0,5}\}$, $\widehat{\mathbb{B}} = \{b_1, \ldots, b_4, b_{13}, b_{14}\}$, $\widehat{\mathbb{B}}_0^* = \{b_{0,1}^*, b_{0,3}^*, b_{0,4}^*\}$, $\widehat{\mathbb{B}}^* = \{b_1^*, \ldots, b_4^*, b_{11}^*, b_{12}^*\}$.
2. It obtains $\mathcal{T}$ by running SD.Setup(N_{max}). Let $\mathcal{S}$ be the collection of all subsets $S_{j,k}$ of $\mathcal{T}$. It initializes the user list $\mathsf{UL} = \varnothing$.
3. It publishes the public parameters $\mathsf{PP} = (\mathsf{param}, \widehat{\mathbb{B}}_0, \widehat{\mathbb{B}})$, and a state $\mathsf{ST} = (\mathcal{T}, \mathsf{UL})$, while it keeps the master secret key $\mathsf{MK} = (\widehat{\mathbb{B}}_0^*, \widehat{\mathbb{B}}^*)$.

RABE.GenKey($\mathsf{PP}, \mathsf{MK}, \mathsf{ST}, \mathbb{S}, ID$): Taking as input the public parameters PP, the master secret key MK, the state $\mathsf{ST} = (\mathcal{T}, \mathsf{UL})$, an access structure $\mathbb{S} = (M, \rho) \in \mathfrak{S}$ such that M is an $\ell \times r$ matrix and ρ is a labeling of the rows of M by attributes in $\mathbb{U}$, and a user identity ID, the key generation center provides a private key to the corresponding user as follows:

1. It first chooses $\vec{f} \xleftarrow{\$} \mathbb{F}_q^r$, computes $\vec{s}^\mathsf{T} = (s_1, \ldots, s_\ell)^\mathsf{T} = M \cdot \vec{f}^\mathsf{T}$, $s_0' = \vec{1} \cdot \vec{f}^\mathsf{T}$, selects $\eta_0, s_0'' \xleftarrow{\$} \mathbb{F}_q$, and sets $s_0 = s_0' + s_0''$. Note that, $s_1, \ldots, s_\ell$ are shares of s_0'. Next, it computes

$$k_0^* = (-s_0, 1, \eta_0)_{\widehat{\mathbb{B}}_0^*} = -s_0 b_{0,1}^* + b_{0,3}^* + \eta_0 b_{0,4}^* = (-s_0, 0, 1, \eta_0, 0)_{\mathbb{B}_0^*}.$$

For $i = 1, \ldots, \ell$, if $\rho(i) = (t, A_t)$, it chooses $\mu_i, \theta_i, \eta_{i,1}, \eta_{i,2} \xleftarrow{\$} \mathbb{F}_q$ and computes

$$\begin{aligned} k_i^* &= (\mu_i(t, -1), s_i + \theta_i A_t, -\theta_i, \eta_{i,1}, \eta_{i,2})_{\widehat{\mathbb{B}}^*} \\ &= \mu_i t b_1^* - \mu_i b_2^* + (s_i + \theta_i A_t) b_3^* - \theta_i b_4^* + \eta_{i,1} b_{11}^* + \eta_{i,2} b_{12}^* \\ &= (\mu_i(t, -1), s_i + \theta_i A_t, -\theta_i, 0^6, \eta_{i,1}, \eta_{i,2}, 0^2, 0^2)_{\mathbb{B}^*}. \end{aligned}$$

We mention that, k_0^* and k_i^* are actually some linear combinations of vectors in $\widehat{\mathbb{B}}_0^*$ and $\widehat{\mathbb{B}}^*$ respectively, where $\widehat{\mathbb{B}}_0^*$ and $\widehat{\mathbb{B}}^*$ are extractable from MK. However, for ease of discussion, we have represented k_0^* and k_i^* as linear combinations of vectors in $\mathbb{B}_0^*$ and $\mathbb{B}^*$ respectively taking the coefficients of vectors in $\mathbb{B}_0^* \backslash \widehat{\mathbb{B}}_0^*$ and $\mathbb{B}^* \backslash \widehat{\mathbb{B}}^*$ as zeros. Hereafter, a similar notation will be followed for representing linear combinations in $\mathbb{V}_0$ and $\mathbb{V}$.

2. It assigns the user identity ID to a leaf node $\nu(u)$ in $\mathcal{T}$ that is not yet assigned, where $u \in \mathcal{N}$ is a serial number that is assigned to ID. It saves (ID, u) to UL. Next it obtains PV_u by running $\mathsf{SD.Assign}(\mathcal{T}, u)$.

3. For each $S_{j,k} \in \mathsf{PV}_u$, it performs the following operations: It first selects $s_{j,k,1}, s_{j,k,2} \xleftarrow{\$} \mathbb{F}_q$ such that $s_0'' = s_{j,k,1} + s_{j,k,2}$, i.e., it breaks s_0'' into two random parts. It further chooses $\mu_{j,k,1}, \mu_{j,k,2}, \theta_{j,k}, \eta_{j,k,1,1}, \eta_{j,k,1,2}, \eta_{j,k,2,1}, \eta_{j,k,2,2}, \xleftarrow{\$} \mathbb{F}_q$ and computes

$$k_{j,k,1}^* = (\mu_{j,k,1}(d+1,-1), s_{j,k,1} + \theta_{j,k}(L_j \| D_k), -\theta_{j,k}, 0^6,$$
$$\eta_{j,k,1,1}, \eta_{j,k,1,2}, 0^2, 0^2)_{\mathbb{B}^*}$$
$$k_{j,k,2}^* = (\mu_{j,k,2}(d+2,-1), s_{j,k,2}((L_j \| L_k), -1), 0^6, \eta_{j,k,2,1}, \eta_{j,k,2,2}, 0^2, 0^2)_{\mathbb{B}^*}$$

4. Finally, it publishes the updated state $\mathsf{ST} = (\mathcal{T}, \mathsf{UL})$ and provides a private key $\mathsf{SK}_{\mathbb{S},ID} = (\mathsf{PV}_u, k_0^*, \{k_i^*\}_{i=1,\dots,\ell}, \{k_{j,k,1}^*, k_{j,k,2}^*\}_{S_{j,k} \in \mathsf{PV}_u})$ to the user.

$\mathsf{RABE.Encrypt}(\mathsf{PP}, \mathsf{ST}, \Gamma, \mathsf{RL}, M)$: On input the public parameters PP, the state $\mathsf{ST} = (\mathcal{T}, \mathsf{UL})$, an attribute set $\Gamma \subseteq \mathbb{U}$, a revocation list RL of user identities and a message $M \in \mathbb{G}_T$, the encrypter executes the following steps:

1. It first extracts g_T from PP, chooses $\omega, \zeta, \varphi_0 \xleftarrow{\$} \mathbb{F}_q$ and computes

$$c_0 = (\omega, 0, \zeta, 0, \varphi_0)_{\mathbb{B}_0}, \text{ and } c = g_T^\zeta M.$$

2. For all $(t, A_t) \in \Gamma$, it selects $\sigma_t, \varphi_{t,1}, \varphi_{t,2} \xleftarrow{\$} \mathbb{F}_q$, and computes

$$c_t = (\sigma_t(1,t), \omega(1, A_t), 0^6, 0^2, \varphi_{t,1}, \varphi_{t,2}, 0^2)_{\mathbb{B}}.$$

3. Then it defines the revoked user serial number set $\mathsf{RI} \subseteq \mathcal{N}$ from RL by using UL. Next it obtains the covering set $\mathsf{CV}_{\mathsf{RI}}$ by executing $\mathsf{SD.Cover}(\mathcal{T}, \mathsf{RI})$.

4. For each $S_{j,k} \in \mathsf{CV}_{\mathsf{RI}}$, it performs the following steps: It chooses $\sigma_{j,k,1}, \sigma_{j,k,2}, \varphi_{j,k,1,1}, \varphi_{j,k,1,2}, \varphi_{j,k,2,1}, \varphi_{j,k,2,2} \xleftarrow{\$} \mathbb{F}_q$ and computes

$$c_{j,k,1} = (\sigma_{j,k,1}(1, d+1), \omega(1, (L_j \| D_k)), 0^6, 0^2, \varphi_{j,k,1,1}, \varphi_{j,k,1,2}, 0^2)_{\mathbb{B}},$$
$$c_{j,k,2} = (\sigma_{j,k,2}(1, d+2), \omega(1, (L_j \| L_k)), 0^6, 0^2, \varphi_{j,k,2,1}, \varphi_{j,k,2,2}, 0^2)_{\mathbb{B}}.$$

5. The encrypter outputs the ciphertext as $\mathsf{CT}_{\Gamma,\mathsf{RL}} = (\mathsf{CV}_{\mathsf{RI}}, c, c_0, \{c_t\}_{(t,A_t) \in \Gamma}, \{c_{j,k,1}, c_{j,k,2}\}_{S_{j,k} \in \mathsf{CV}_{\mathsf{RI}}})$.

$\mathsf{RABE.Decrypt}(\mathsf{CT}_{\Gamma,\mathsf{RL}}, \mathsf{SK}_{\mathbb{S},ID}, \mathbb{S}, ID, \mathsf{PP}, \mathsf{ST})$: A user takes as input a ciphertext $\mathsf{CT}_{\Gamma,\mathsf{RL}} = (\mathsf{CV}_{\mathsf{RI}}, c, c_0, \{c_t\}_{(t,A_t) \in \Gamma}, \{c_{j,k,1}, c_{j,k,2}\}_{S_{j,k} \in \mathsf{CV}_{\mathsf{RI}}})$, its private key $\mathsf{SK}_{\mathbb{S},ID} = (\mathsf{PV}_u, k_0^*, \{k_i^*\}_{i=1,\dots,\ell}, \{k_{j,k,1}^*, k_{j,k,2}^*\}_{S_{j,k} \in \mathsf{PV}_u})$, its access structure $\mathbb{S} = (M, \rho)$, identity ID, public parameters PP and state ST. It proceeds as follows:

1. If the access structure $\mathbb{S}$ accepts Γ, then it computes I and $\{\alpha_i\}_{i\in I}$ such that $\vec{1} = \sum_{i\in I} \alpha_i \boldsymbol{M}_i$ and hence $s'_0 = \sum_{i\in I} \alpha_i s_i$, where $\boldsymbol{M}_i$ is the i-th row of $\boldsymbol{M}$ and $I \subseteq \{i \in \{1,\dots,\ell\} | \rho(i) = (t, A_t) \in \Gamma\}$. Otherwise, it obtains $\perp$.

2. If $ID \notin \mathsf{RL}$ for $(ID, u) \in \mathsf{UL}$, then it obtains $(S_{j,k}, S_{j',k'})$ by running SD.Match$(\mathsf{CV}_{\mathsf{RI}}, \mathsf{PV}_u)$ such that $S_{j,k} \in \mathsf{CV}_{\mathsf{RI}}$, $S_{j',k'} \in \mathsf{PV}_u$ and $(j = j') \wedge (D_k = D_{k'}) \wedge (k \neq k')$. Otherwise, it outputs $\perp$.

3. It computes

$$\pi' = \prod_{\substack{i\in I \\ \rho(i)=(t,A_t)}} E(\boldsymbol{c}_t, \boldsymbol{k}_i^*)^{\alpha_i},$$

$$\pi'' = E(\boldsymbol{c}_{j,k,1}, \boldsymbol{k}_{j',k',1}^*) E(\boldsymbol{c}_{j,k,2}, \boldsymbol{k}_{j',k',2}^*)^{\overline{\frac{1}{(L_{j'}\|L_{k'})-(L_j\|L_k)}}} \text{ and } \pi = E(\boldsymbol{c}_0, \boldsymbol{k}_0^*)\pi'\pi''$$

4. It retrieves the message as $M = c/\pi$.

- **Correctness:** Let $\mathsf{SK}_{\mathbb{S},ID} = (\mathsf{PV}_u, \boldsymbol{k}_0^*, \{\boldsymbol{k}_i^*\}_{i=1,\dots,\ell}, \{\boldsymbol{k}_{j,k,1}^*, \boldsymbol{k}_{j,k,2}^*\}_{S_{j,k}\in\mathsf{PV}_u})$ be a private key for a user with identity ID together with an access structure $\mathbb{S} = (\boldsymbol{M}, \rho)$, and $\mathsf{CT}_{\Gamma,\mathsf{RL}} = (\mathsf{CV}_{\mathsf{RI}}, c, \boldsymbol{c}_0, \{\boldsymbol{c}_t\}_{(t,A_t)\in\Gamma}, \{\boldsymbol{c}_{j,k,1}, \boldsymbol{c}_{j,k,2}\}_{S_{j,k}\in\mathsf{CV}_{\mathsf{RI}}})$ be a ciphertext for an attribute set Γ together with a revocation list RL of user identities. If $ID \notin \mathsf{RL}$, then a pair of subsets $(S_{j,k}, S_{j',k'})$ such that $S_{j,k} \in \mathsf{CV}_{\mathsf{RI}}$, $S_{j',k'} \in \mathsf{PV}_u$ and $(j = j') \wedge (D_k = D_{k'}) \wedge (k \neq k')$, i.e., $(L_j\|D_k) = (L_{j'}\|D_{k'})$ and $(L_j\|L_k) \neq (L_{j'}\|L_{k'})$, can be found from correctness of SD (Lemma 3). Now,

$$\pi' = \prod_{\substack{i\in I \\ \rho(i)=(t,A_t)}} E(\boldsymbol{c}_t, \boldsymbol{k}_i^*)^{\alpha_i} = \prod_{\substack{i\in I \\ \rho(i)=(t,A_t)}} g_T^{\omega\alpha_i s_i} = g_T^{\omega s'_0}, \text{ as } \sum_{\substack{i\in I \\ \rho(i)=(t,A_t)}} \alpha_i s_i = s'_0,$$

since $\mathbb{S} = (\boldsymbol{M}, \rho)$ accepts Γ.

Also, $\pi'' = g_T^{\omega s_{j',k',1}} g_T^{\omega s_{j',k',2} \frac{(L_{j'}\|L_{k'})-(L_j\|L_k)}{(L_{j'}\|L_{k'})-(L_j\|L_k)}} = g_T^{\omega s''_0}$, as $s_{j',k',1} + s_{j',k',2} = s''_0$, since $S_{j',k'} \in \mathsf{PV}_u$.

Thus, $\pi = g_T^{[-\omega s_0 + \zeta + \omega s'_0 + \omega s''_0]} = g_T^{[\omega(-s_0 + s'_0 + s''_0)+\zeta]} = g_T^\zeta$.

So, $c/\pi = g_T^\zeta M/g^\zeta = M$.

4   Security Analysis

Theorem 1. *The* RABE *scheme, introduced in Sect. 3, is adaptively secure against chosen plaintext attacks* (CPA) *under the* DLIN *assumption, formally defined in Sect. 2.3. More precisely, for any probabilistic polynomial-time adversary* $\mathcal{A}$, *there exists probabilistic machines* $\mathcal{F}_{1\text{-}1},\dots,\mathcal{F}_{1\text{-}3}$, $\mathcal{F}_{2\text{-}1\text{-}1},\dots,\mathcal{F}_{2\text{-}1\text{-}12}$, $\mathcal{F}_{2\text{-}2\text{-}1},\dots,\mathcal{F}_{2\text{-}2\text{-}12}$ *whose running times are essentially the same as that of* $\mathcal{A}$, *such that for any security parameter* λ,

$$\mathsf{Adv}_{\mathcal{A}}^{\mathsf{RABE,IND\text{-}CPA}}(\lambda) \le \mathsf{Adv}_{\mathcal{F}_{1\text{-}1}}^{\mathsf{DLIN}}(\lambda) + \sum_{p=1}^{d}\sum_{j=1}^{2}\mathsf{Adv}_{\mathcal{F}_{1\text{-}2\text{-}p\text{-}j}}^{\mathsf{DLIN}}(\lambda) +$$

$$\sum_{v=1}^{2}\sum_{\varpi=1}^{\widehat{r}_{\max}}\sum_{j=1}^{2}\mathsf{Adv}_{\mathcal{F}_{1\text{-}3\text{-}(d+v)\text{-}\varpi\text{-}j}}^{\mathsf{DLIN}}(\lambda) + \sum_{h=1}^{\widehat{q}}\sum_{i=1}^{2}\left[\mathsf{Adv}_{\mathcal{F}_{2\text{-}h\text{-}i\text{-}1}}^{\mathsf{DLIN}}(\lambda) + \right.$$

$$\sum_{j=1}^{2}\left[\sum_{p=1}^{d}\left\{\mathsf{Adv}_{\mathcal{F}_{2\text{-}h\text{-}i\text{-}p\text{-}2\text{-}j}}^{\mathsf{DLIN}}(\lambda) + \mathsf{Adv}_{\mathcal{F}_{2\text{-}h\text{-}i\text{-}p\text{-}3\text{-}j}}^{\mathsf{DLIN}}(\lambda) + \sum_{\substack{l=1\\l\neq p}}^{d+2}\left(\mathsf{Adv}_{\mathcal{F}_{2\text{-}h\text{-}i\text{-}p\text{-}4\text{-}j\text{-}l}}^{\mathsf{DLIN}}(\lambda) + \right.\right.$$

$$\left.\left.\mathsf{Adv}_{\mathcal{F}_{2\text{-}h\text{-}i\text{-}p\text{-}5\text{-}j\text{-}l}}^{\mathsf{DLIN}}(\lambda)\right) + \mathsf{Adv}_{\mathcal{F}_{2\text{-}h\text{-}i\text{-}p\text{-}6\text{-}j}}^{\mathsf{DLIN}}(\lambda)\right\} + \sum_{v=1}^{2}\sum_{\varpi=1}^{\aleph}\left\{\mathsf{Adv}_{\mathcal{F}_{2\text{-}h\text{-}i\text{-}(d+v)\text{-}\varpi\text{-}7\text{-}j}}^{\mathsf{DLIN}}(\lambda) + \right.$$

$$\mathsf{Adv}_{\mathcal{F}_{2\text{-}h\text{-}i\text{-}(d+v)\text{-}\varpi\text{-}8\text{-}j}}^{\mathsf{DLIN}}(\lambda) + \sum_{\substack{l=1\\l\neq d+v}}^{d+2}\left(\mathsf{Adv}_{\mathcal{F}_{2\text{-}h\text{-}i\text{-}(d+v)\text{-}\varpi\text{-}9\text{-}j\text{-}l}}^{\mathsf{DLIN}}(\lambda) + \mathsf{Adv}_{\mathcal{F}_{2\text{-}h\text{-}i\text{-}(d+v)\text{-}\varpi\text{-}10\text{-}j\text{-}l}}^{\mathsf{DLIN}}(\lambda)\right)$$

$$\left.\left.+ \sum_{\substack{\iota=1\\\iota\neq\varpi}}^{\aleph}\mathsf{Adv}_{\mathcal{F}_{2\text{-}h\text{-}i\text{-}(d+v)\text{-}\varpi\text{-}11\text{-}j\text{-}\iota}}^{\mathsf{DLIN}}(\lambda) + \mathsf{Adv}_{\mathcal{F}_{2\text{-}h\text{-}i\text{-}(d+v)\text{-}\varpi\text{-}12\text{-}j}}^{\mathsf{DLIN}}(\lambda)\right\}\right]\right] + \epsilon,$$

where $\mathcal{F}_{1\text{-}2\text{-}p\text{-}j}(\cdot) = \mathcal{F}_{1\text{-}2}(p,j,\cdot)$, $\mathcal{F}_{1\text{-}3\text{-}(d+v)\text{-}\varpi\text{-}j}(\cdot) = \mathcal{F}_{1\text{-}3}(d+v,\varpi,j,\cdot)$, and, for $i = 1,2$,

$$\mathcal{F}_{2\text{-}h\text{-}i\text{-}1}(\cdot) = \mathcal{F}_{2\text{-}i\text{-}1}(h,\cdot), \quad \mathcal{F}_{2\text{-}h\text{-}i\text{-}p\text{-}2\text{-}j}(\cdot) = \mathcal{F}_{2\text{-}i\text{-}2}(h,p,j,\cdot),$$
$$\mathcal{F}_{2\text{-}h\text{-}i\text{-}p\text{-}3\text{-}j}(\cdot) = \mathcal{F}_{2\text{-}i\text{-}3}(h,p,j,\cdot), \quad \mathcal{F}_{2\text{-}h\text{-}i\text{-}p\text{-}4\text{-}j\text{-}l}(\cdot) = \mathcal{F}_{2\text{-}i\text{-}4}(h,p,j,l,\cdot),$$
$$\mathcal{F}_{2\text{-}h\text{-}i\text{-}p\text{-}5\text{-}j\text{-}l}(\cdot) = \mathcal{F}_{2\text{-}i\text{-}5}(h,p,j,l,\cdot), \quad \mathcal{F}_{2\text{-}h\text{-}i\text{-}p\text{-}6\text{-}j}(\cdot) = \mathcal{F}_{2\text{-}i\text{-}6}(h,p,j,\cdot),$$
$$\mathcal{F}_{2\text{-}h\text{-}i\text{-}(d+v)\text{-}\varpi\text{-}7\text{-}j}(\cdot) = \mathcal{F}_{2\text{-}i\text{-}7}(h,d+v,\varpi,j,\cdot) \text{ etc.},$$

$\widehat{q}$ is the maximum number of $\mathcal{A}$'s private key queries, d is the size of the attribute universe used in the system, $N_{\max}$ is the upper bound of user key serial numbers, $\widehat{r}_{\max}$ is the maximum size of a covering set of non-revoked users used in the system, $\aleph = \log^2 N_{\max} + \widehat{r}_{\max}$, and $\epsilon = \left[6 + 10d + 20\widehat{r}_{\max} + 14\widehat{q} + 80d\widehat{q} + 20d^2\widehat{q} + 4\widehat{q}\aleph(30 + 10d + 10\aleph)\right]/q$.

Proof. At the top level of strategy of the security proof, we follow the dual system encryption methodology over dual pairing vector space (DPVS) described in [12]. To prove the security of our RABE scheme, we use Problem 1 and 2 defined in Sect. 2.3.

To prove Theorem 1, we consider the following games. In Game 0, a part framed by a box indicates positions of coefficients to be changed in a subsequent game. In the other games, a part framed by a box indicates coefficients which were changed in a transition from the previous game. Games proceed as follows:

Game 0 $\implies$ Game 1 $\implies$

{Game 2-h-1 $\implies$ Game 2-h-2 $\implies$ Game 2-h-3}$_{h=1,\ldots,\widehat{q}}$ $\implies$ Game 3

Game 0: Game 0 is the original security game, i.e., the reply to a key query for an access structure-user identity pair $(\mathbb{S}_\iota = (\boldsymbol{M}_\iota, \rho_\iota), ID_\iota)$ is given by $\mathsf{SK}_{\mathbb{S}_\iota, ID_\iota} = (\mathsf{PV}_u, \boldsymbol{k}_0^*, \{\boldsymbol{k}_i^*\}_{i=1,\ldots,\ell}, \{\boldsymbol{k}_{j,k,1}^*, \boldsymbol{k}_{j,k,2}^*\}_{S_{j,k} \in \mathsf{PV}_u})$, where

$$\boldsymbol{k}_0^* = (-s_0, \boxed{0}, 1, \eta_0, 0)_{\mathbb{B}_0^*}, \tag{1}$$

for $i = 1, \ldots, \ell$ such that $\rho_\iota(i) = (t, A_t)$,

$$\boldsymbol{k}_i^* = (\mu_i(t, -1), s_i + \theta_i A_t, -\theta_i, 0^4, \boxed{0^2}, \eta_{i,1}, \eta_{i,2}, 0^2, 0^2)_{\mathbb{B}^*}, \tag{2}$$

in which $\nu_i, \theta_i, \eta_0, \eta_{i,1}, \eta_{i,2} \xleftarrow{\$} \mathbb{F}_q$, $s_0 = s_0' + s_0''$, $s_0' = \vec{1} \cdot \vec{f}^\mathsf{T}$, $\vec{s}^\mathsf{T} = (s_1, \ldots, s_\ell)^\mathsf{T} = \boldsymbol{M}_\iota \cdot \vec{f}^\mathsf{T}$, $s_0'' \xleftarrow{\$} \mathbb{F}_q$, $\vec{f} \xleftarrow{\$} \mathbb{F}_q^r$, $\boldsymbol{M}_\iota$ being an $\ell \times r$ matrix, and for all $S_{j,k} \in \mathsf{PV}_u$, such that $\nu(u)$ is the leaf node of $\mathcal{T}$ assigned to ID_ι,

$$\boldsymbol{k}_{j,k,1}^* = (\mu_{j,k,1}(d+1, -1), s_{j,k,1} + \theta_{j,k}(L_j \| D_k), -\theta_{j,k}, 0^4, \boxed{0^2},$$
$$\eta_{j,k,1,1}, \eta_{j,k,1,2}, 0^2, 0^2)_{\mathbb{B}^*}, \tag{3}$$

$$\boldsymbol{k}_{j,k,2}^* = (\mu_{j,k,2}(d+2, -1), s_{j,k,2}((L_j \| L_k), -1), 0^4, \boxed{0^2},$$
$$\eta_{j,k,2,1}, \eta_{j,k,2,2}, 0^2, 0^2)_{\mathbb{B}^*}, \tag{4}$$

such that $s_{j,k,1}, s_{j,k,2}, \mu_{j,k,1}, \mu_{j,k,2}, \theta_{j,k}, \eta_{j,k,1,1}, \eta_{j,k,1,2}, \eta_{j,k,2,1}, \eta_{j,k,2,2} \xleftarrow{\$} \mathbb{F}_q$ so that $s_0'' = s_{j,k,1} + s_{j,k,2}$.

The challenge ciphertext corresponding to challenge plaintext(M_0^*, M_1^*), attribute set $\Gamma^* = \{(t, A_t) | 1 \le t \le d\}$ and the revocation list RL^* is expressed as $\mathsf{CT}^* = (\mathsf{CV}_{\mathsf{RI}}, c, c_0, \{c_t\}_{(t,A_t) \in \Gamma^*}, \{c_{j,k,1}, c_{j,k,2}\}_{S_{j,k} \in \mathsf{CV}_{\mathsf{RI}}})$ where

$$c = g_T^\varsigma M_b^*, b \xleftarrow{\$} \{0, 1\}, \tag{5}$$

$$c_0 = (\omega, \boxed{0}, \boxed{\varsigma}, 0, \varphi_0)_{\mathbb{B}_0}, \tag{6}$$

for $(t, A_t) \in \Gamma^*$,

$$c_t = (\sigma_t(1, t), \omega(1, A_t), \boxed{0^2}, 0^2, \boxed{0^2}, 0^2, \varphi_{t,1}, \varphi_{t,2}, 0^2)_{\mathbb{B}}, \tag{7}$$

such that $\omega, \varsigma, \varphi_0, \sigma_t, \varphi_{t,1}, \varphi_{t,2} \xleftarrow{\$} \mathbb{F}_q$, and for all $S_{j,k} \in \mathsf{CV}_{\mathsf{RI}}$, RI being the set of revoked user serial numbers obtained from RL^*,

$$c_{j,k,1} = (\sigma_{j,k,1}(1, d+1), \omega(1, (L_j \| D_k)), \boxed{0^2}, 0^2, \boxed{0^2}, 0^2, \varphi_{j,k,1,1}, \varphi_{j,k,1,2}, 0^2)_{\mathbb{B}}, \tag{8}$$

$$c_{j,k,2} = (\sigma_{j,k,2}(1, d+2), \omega(1, (L_j \| L_k)), \boxed{0^2}, 0^2, \boxed{0^2}, 0^2, \varphi_{j,k,2,1}, \varphi_{j,k,2,2}, 0^2)_{\mathbb{B}}, \tag{9}$$

such that $\sigma_{j,k,1}, \sigma_{j,k,2}, \varphi_{j,k,1,1}, \varphi_{j,k,1,2}, \varphi_{j,k,2,1}, \varphi_{j,k,2,2} \xleftarrow{\$} \mathbb{F}_q$. Note that there is at most one $S_{j,k}$ in $\mathsf{CV}_{\mathsf{RI}}$ for any j and any k, as mentioned in Sect. 2.5.

Game 1: This game is identical to Game 0 except that the components of the challenge ciphertext CT^* is computed as follows:

$$c = g_T^\zeta M_b^*, \quad b \xleftarrow{\$} \{0,1\}, \tag{10}$$

$$\boldsymbol{c}_0 = (\omega, \boxed{\tau}, \zeta, 0, \varphi_0)_{\mathbb{B}_0}, \tag{11}$$

for $(t, A_t) \in \Gamma^*$,

$$\boldsymbol{c}_t = (\sigma_t(1,t), \omega(1, A_t), \boxed{\tau(1, A_t)}, 0^2, \boxed{\tau(1, A_t) \cdot \boldsymbol{Z}_t}, 0^2, \varphi_{t,1}, \varphi_{t,2}, 0^2)_{\mathbb{B}}, \tag{12}$$

where $\tau \xleftarrow{\$} \mathbb{F}_q^\times$, $\boldsymbol{Z}_t \xleftarrow{\$} \mathsf{GL}(2, \mathbb{F}_q)$. For all $S_{j,k} \in \mathsf{CV}_{\mathsf{RI}}$,

$$\boldsymbol{c}_{j,k,1} = (\sigma_{j,k,1}(1, d+1), \omega(1, (L_j \| D_k)), \boxed{\tau(1, (L_j \| D_k))}, 0^2,$$
$$\boxed{\tau(1, (L_j \| D_k)) \cdot \boldsymbol{Z}_{d+1,j,k}}, 0^2, \varphi_{j,k,1,1}, \varphi_{j,k,1,2}, 0^2)_{\mathbb{B}}, \tag{13}$$

$$\boldsymbol{c}_{j,k,2} = (\sigma_{j,k,2}(1, d+2), \omega(1, (L_j \| L_k)), \boxed{\tau(1, (L_j \| L_k))}, 0^2,$$
$$\boxed{\tau(1, (L_j \| L_k)) \cdot \boldsymbol{Z}_{d+2,j,k}}, 0^2, \varphi_{j,k,2,1}, \varphi_{j,k,2,2}, 0^2)_{\mathbb{B}}, \tag{14}$$

where $\boldsymbol{Z}_{d+1,j,k}, \boldsymbol{Z}_{d+2,j,k} \xleftarrow{\$} \mathsf{GL}(2, \mathbb{F}_q)$. All other variables are generated as in Game 0.

Game 2-h-1 $(h = 1, \ldots, \widehat{q})$: We denote Game 1 as Game 2-0-3. Game 2-h-1 is the same as Game 2-$(h-1)$-3 other than the components of the h-th queried key for access structure-user identity pair $(\mathbb{S}_h = (\boldsymbol{M}_h, \rho_h), ID_h)$ are constructed as follows:

$$\boldsymbol{k}_0^* = (-s_0, \boxed{-a_0}, 1, \eta_0, 0)_{\mathbb{B}_0^*} \tag{15}$$

for $i = 1, \ldots, \ell$ such that $\rho_h(i) = (t, A_t)$,

$$\boldsymbol{k}_i^* = (\mu_i(t, -1), s_i + \theta_i A_t, -\theta_i, 0^4, \boxed{(a_i + \pi_i A_t, -\pi_i) \cdot \boldsymbol{U}_t}, \eta_{i,1}, \eta_{i,2}, 0^2, 0^2)_{\mathbb{B}^*}, \tag{16}$$

where $\vec{g} \xleftarrow{\$} \mathbb{F}_q^r$, $a_0 = a_0' + a_0''$, $a_0' = \vec{1} \cdot \vec{g}^\mathsf{T}$, $(a_1, \ldots, a_\ell)^\mathsf{T} = \boldsymbol{M}_h \cdot \vec{g}^\mathsf{T}$, $a_0'' \xleftarrow{\$} \mathbb{F}_q$, $\boldsymbol{U}_t = (\boldsymbol{Z}_t^{-1})^\mathsf{T}$ for $\boldsymbol{Z}_t \xleftarrow{\$} \mathsf{GL}(2, \mathbb{F}_q)$, $\pi_i \xleftarrow{\$} \mathbb{F}_q$ for $i = 1, \ldots, \ell$, $\boldsymbol{M}_h$ being an $\ell \times r$ matrix. For all $S_{j,k} \in \mathsf{PV}_u$,

$$\boldsymbol{k}_{j,k,1}^* = (\mu_{j,k,1}(d+1, -1), s_{j,k,1} + \theta_{j,k}(L_j \| D_k), -\theta_{j,k}, 0^4,$$
$$\boxed{(a_{j,k,1} + \pi_{j,k}(L_j \| D_k), -\pi_{j,k}) \cdot \boldsymbol{U}_{d+1,j,k}}, \eta_{j,k,1,1}, \eta_{j,k,1,2}, 0^2, 0^2)_{\mathbb{B}^*}, \tag{17}$$

$$\boldsymbol{k}_{j,k,2}^* = (\mu_{j,k,2}(d+2, -1), s_{j,k,2}((L_j \| L_k), -1), 0^4,$$
$$\boxed{(a_{j,k,2}((L_j \| L_k), -1) \cdot \boldsymbol{U}_{d+2,j,k}}, \eta_{j,k,2,1}, \eta_{j,k,2,2}, 0^2, 0^2)_{\mathbb{B}^*}, \tag{18}$$

where $a_{j,k,1}, a_{j,k,2}, \pi_{j,k} \overset{\$}{\leftarrow} \mathbb{F}_q$ such that $a_{j,k,1} + a_{j,k,2} = a_0''$, $U_{d+1,j,k} = (Z_{d+1,j,k}^{-1})^\intercal$, $U_{d+2,j,k} = (Z_{d+2,j,k}^{-1})^\intercal$ for $Z_{d+1,j,k}, Z_{d+2,j,k} \overset{\$}{\leftarrow} \mathsf{GL}(2, \mathbb{F}_q)$. All the other variables are constructed as in Game 2-$(h-1)$-3.

Game 2-h-2 $(h = 1, \ldots, \widehat{q})$: This game is similar to Game 2-h-1 with the exception that the component k_0^* of the h-th queried key for an access structure-user identity pair $(\mathbb{S}_h = (M_h, \rho_h), ID_h)$ is computed as follows:

$$k_0^* = (-s_0, \boxed{r_0}, 1, \eta_0, 0)_{\mathbb{B}_0^*}, \tag{19}$$

where $r_0 \overset{\$}{\leftarrow} \mathbb{F}_q$, and all the other variables are as in Game 2-h-1.

Game 2-h-3 $(h = 1, \ldots, \widehat{q})$: This game is almost identical to Game 2-h-2 except that the components of the h-th queried key for an access structure-user identity pair $(\mathbb{S}_h = (M_h, \rho_h), ID_h)$ is constructed as follows:

$$k_0^* = (-s_0, r_0, 1, \eta_0, 0)_{\mathbb{B}_0^*} \tag{20}$$

for $i = 1, \ldots, \ell$,

$$k_i^* = (\mu_i(t, -1), s_i + \theta_i A_t, -\theta_i, 0^4, \boxed{0^2}, \eta_{i,1}, \eta_{i,2}, 0^2, 0^2)_{\mathbb{B}^*}, \tag{21}$$

and for all $S_{j,k} \in \mathsf{PV}_u$,

$$k_{j,k,1}^* = (\mu_{j,k,1}(d+1, -1), s_{j,k,1} + \theta_{j,k}(L_j \| D_k), -\theta_{j,k}, 0^4, \boxed{0^2},$$
$$\eta_{j,k,1,1}, \eta_{j,k,1,2}, 0^2, 0^2)_{\mathbb{B}^*}, \tag{22}$$

$$k_{j,k,2}^* = (\mu_{j,k,2}(d+2, -1), s_{j,k,2}((L_j \| L_k), -1), 0^4, \boxed{0^2},$$
$$\eta_{j,k,2,1}, \eta_{j,k,2,2}, 0^2, 0^2)_{\mathbb{B}^*}, \tag{23}$$

where $r_0 \overset{\$}{\leftarrow} \mathbb{F}_q$, and all the other variables are generated as in Game 2-h-2.

Game 3: This game is similar to Game 2-$\widehat{q}$-3 with the only exception that the components c_0 and c of the challenge ciphertext CT^* are computed as follows:

$$c_0 = (\omega, \tau, \boxed{\zeta'}, 0, \varphi_0)_{\mathbb{B}_0}, \tag{24}$$

$$c = g_T^\zeta M_b^*, \tag{25}$$

where $\zeta' \overset{\$}{\leftarrow} \mathbb{F}_q$ (i.e., independent from $\zeta \overset{\$}{\leftarrow} \mathbb{F}_q$), and all other variables are generated as in Game 2-$\widehat{q}$-3.

Let $\mathsf{Adv}_{\mathcal{A}}^{(0)}(\lambda)$, $\mathsf{Adv}_{\mathcal{A}}^{(1)}(\lambda)$, $\mathsf{Adv}_{\mathcal{A}}^{(2\text{-}h\text{-}\jmath)}(\lambda)$ $(h = 1, \ldots, \widehat{q}; \jmath = 1, 2, 3)$ and $\mathsf{Adv}_{\mathcal{A}}^{(3)}(\lambda)$ be the advantage of $\mathcal{A}$ in Game 0, 1, 2-h-$\jmath$ and 3 respectively. Clearly, $\mathsf{Adv}_{\mathcal{A}}^{(0)}(\lambda)$ is equivalent to $\mathsf{Adv}_{\mathcal{A}}^{\mathsf{RABE,IND\text{-}CPA}}(\lambda)$ and $\mathsf{Adv}_{\mathcal{A}}^{(3)}(\lambda) = 0$.

We have evaluated the differences between the pairs of $\mathsf{Adv}_{\mathcal{A}}^{(0)}(\lambda), \mathsf{Adv}_{\mathcal{A}}^{(1)}(\lambda)$, $\{\mathsf{Adv}_{\mathcal{A}}^{(2\text{-}h\text{-}1)}(\lambda), \ldots, \mathsf{Adv}_{\mathcal{A}}^{(2\text{-}h\text{-}3)}(\lambda)\}_{h=1,\ldots,\widehat{q}}$, and $\mathsf{Adv}_{\mathcal{A}}^{(3)}(\lambda)$ in a sequence of lemmas which are presented in the full version. From those lemmas we obtain,

Table 1. Communication and storage comparison

RABE	♯PP	♯SK$_{\mathbb{S},ID}$	♯CT$_{\Gamma,\mathsf{RL}}$	♯param$_{\mathbb{G}}$	Security	Complexity assumptions
[1]	$d+\log N_{\max}+1$ in $\mathbb{G}$, 1 in $\mathbb{G}_T$	$2(\ell+1)\log N_{\max}$ in $\mathbb{G}$	$\sharp\Gamma + \sharp\mathsf{RI}\log\frac{N_{\max}}{\sharp\mathsf{RI}}$ in $\mathbb{G}$, 1 in $\mathbb{G}_T$	q (prime)	Selective	DBDH
[13]	$d+\log N_{\max}+1$ in $\mathbb{G}$, 1 in $\mathbb{G}_T$	$2\ell + 2\log N_{\max}$ in $\mathbb{G}$	$1 + \sharp\Gamma + \sharp\mathsf{RI}\log\frac{N_{\max}}{\sharp\mathsf{RI}}$ in $\mathbb{G}$, 1 in $\mathbb{G}_T$	n (comp.)	Full	SD, GSD, Comp. DH
Ours	111 in $\mathbb{G}$, 1 in $\mathbb{G}_T$	$5 + 16\ell + 16[\log^2 N_{\max} + \log N_{\max}]$ in $\mathbb{G}$	$16\sharp\Gamma + 64\sharp\mathsf{RI} - 27$ in $\mathbb{G}$, 1 in $\mathbb{G}_T$	q (prime)	Full	DLIN

Here, DBDH, SD, GSD and Comp. DH respectively stand for the Decisional Bilinear Diffie-Hellman, Subgroup Decision, Generalized Subgroup Decision and Composite Diffie-Hellman assumptions.

Table 2. Computation comparison

RABE	RABE.Setup	RABE.GenKey	RABE.Encrypt	RABE.Decrypt
[1]	1 in $\mathbb{G}_T$; 1	$(\log N_{\max} + 1)[\ell(d+4)+ \log N_{\max} + 4]$ in $\mathbb{G}$	$1 + (d+4)\sharp\Gamma + \sharp\mathsf{RI}\log\frac{N_{\max}}{\sharp\mathsf{RI}}(\log N_{\max} + 2)$ in $\mathbb{G}$, 1 in $\mathbb{G}_T$	$2\ell + 2$ in $\mathbb{G}$, 1 in $\mathbb{G}_T$; $2\ell + 2$
[13]	$d + \log N_{\max}$ in $\mathbb{G}$, 1 in $\mathbb{G}_T$; 1	$3\ell + \log^2 N_{\max} + 4\log N_{\max} + 4$ in $\mathbb{G}$	$1 + \sharp\Gamma + \sharp\mathsf{RI}\log\frac{N_{\max}}{\sharp\mathsf{RI}}(\log N_{\max} + 2)$ in $\mathbb{G}$, 1 in $\mathbb{G}_T$	2ℓ in $\mathbb{G}$; $2\ell + 2$
Ours	111 in $\mathbb{G}$, 1 in $\mathbb{G}_T$; 1	$10 + 96\ell + 96(\log^2 N_{\max} + \log N_{\max})$ in $\mathbb{G}$	$80\sharp\Gamma + 160\sharp\mathsf{RI} - 49$ in $\mathbb{G}$, 1 in $\mathbb{G}_T$	$16\ell + 16$ in $\mathbb{G}$; $16\ell + 37$

Here, 'x; y' denotes 'x many exponentiations and y many pairings'.

$$\mathsf{Adv}_{\mathcal{A}}^{\mathsf{RABE,IND\text{-}CPA}}(\lambda) = \mathsf{Adv}_{\mathcal{A}}^{(0)}(\lambda)$$

$$\leq \left|\mathsf{Adv}_{\mathcal{A}}^{(0)}(\lambda) - \mathsf{Adv}_{\mathcal{A}}^{(1)}(\lambda)\right| + \sum_{h=1}^{\widehat{q}} \left[\left|\mathsf{Adv}_{\mathcal{A}}^{(2\text{-}(h-1)\text{-}3)}(\lambda) - \mathsf{Adv}_{\mathcal{A}}^{(2\text{-}h\text{-}1)}(\lambda)\right| + \right.$$

$$\left. \sum_{j=1}^{2}\left|\mathsf{Adv}_{\mathcal{A}}^{(2\text{-}h\text{-}j)}(\lambda) - \mathsf{Adv}_{\mathcal{A}}^{(2\text{-}h\text{-}(j+1))}(\lambda)\right|\right] + \left|\mathsf{Adv}_{\mathcal{A}}^{(2\text{-}\widehat{q}\text{-}3)}(\lambda) - \mathsf{Adv}_{\mathcal{A}}^{(3)}(\lambda)\right| + \mathsf{Adv}_{\mathcal{A}}^{(3)}(\lambda)$$

$$\leq \mathsf{Adv}_{\mathcal{B}_1}^{\mathsf{P1}}(\lambda) + \sum_{h=1}^{\widehat{q}}\left[\mathsf{Adv}_{\mathcal{B}_{2\text{-}h\text{-}1}}^{\mathsf{P2}}(\lambda) + \mathsf{Adv}_{\mathcal{B}_{2\text{-}h\text{-}2}}^{\mathsf{P2}}(\lambda)\right] + (4\widehat{q} + 1)/q.$$

Therefore, from Lemmas 1 and 2, of Sect. 2.3 we obtain the upper bound of $\mathsf{Adv}_{\mathcal{A}}^{\mathsf{RABE,IND\text{-}CPA}}(\lambda)$. This completes the proof of Theorem 1. □

5   Efficiency

Tables 1 and 2 compare our RABE scheme with the fully secure RABE scheme [13] and the selectively secure construction [1] which are currently the best known results for RABE (in the key-policy category). We note the following facts:

- Our RABE protocol is the first to achieve constant public parameter size and thus, unlike [1,13], can accommodate large attribute universe and an unbounded number of users.
- Our scheme provides adaptive security instead of selective security achieved in [1] at the expense of some amount of efficiency loss. Note that it is usual to compromise in efficiency in order to achieve better security [4,14].
- Our scheme uses prime order bilinear group as opposed to composite order bilinear group used in the fully secure RABE construction of [13]. As noted by Freeman [6] and several other researchers, the only known instantiation of composite order bilinear group uses elliptic curves (or more generally, abelian varieties) over finite fields. Since the elliptic curve group order must be infeasible to factor, it must be at least 1024 bits. On the other hand, the size of a prime order elliptic curve group that provides an equivalent level of security is only 160 bits which is almost 7 times smaller. This difference in the group order results in great reduction of the ciphertext size when compared in terms of bit length.
- On a more positive note, we employ the subset difference (SD) method, which always provides a smaller covering set compared to the complete subtree (CS) scheme [9], while all previous RABE constructions use CS scheme. Consequently, we could achieve a much smaller ciphertext size, particularly when dealing with large number of users in the system, at the expense of a relatively large private key size, which is primarily due to the large size of private sets of the SD method as opposed to the CS method.
- Regarding computational efficiency of our RABE scheme, note that due to the excessive bit length of the group order, group-operations and pairing computations are prohibitively slow on composite order elliptic curves [6]. In particular, an exponentiation is nearly 25 times slower and a pairing computation is roughly 50 times slower on a 1024 bit composite order elliptic curve than the corresponding operations on a comparable prime order curve [6]. In this light, we can readily see from Table 2 that the computational cost of the encryption algorithms of our RABE scheme is close to that of [13], the only existing RABE scheme with full security to the best of our knowledge, and the decryption algorithm is much faster than that of [13]. However, our key generation algorithm is slower compared to [13].

6   Conclusion

In this paper, we have developed the *first adaptively secure unrestricted* (key-policy) RABE scheme in *prime order bilinear groups* that realizes user revocation through the *subset difference* (SD) scheme – a more efficient variant of the subset cover (SC) framework of Naor et al. [11] as compared to the complete subtree (CS) scheme used in all previous RABE constructions. Due to the application of prime order bilinear groups and the SD scheme, our RABE scheme is highly *broadcast efficient*. It would be interesting to investigate the use of the dynamic version of SD [5] in order to overcome the problem of maintaining a large static binary tree within the state. The possibility to apply SD in designing improved constructions of more advanced primitives such as revocable storage attribute-based encryption (RSABE) or revocable storage predicate encryption (RSPE) [8] is another interesting direction of research.

References

1. Attrapadung, N., Imai, H.: Attribute-based encryption supporting direct/indirect revocation modes. In: Parker, M.G. (ed.) Cryptography and Coding 2009. LNCS, vol. 5921, pp. 278–300. Springer, Heidelberg (2009)
2. Attrapadung, N., Imai, H.: Conjunctive broadcast and attribute-based encryption. In: Shacham, H., Waters, B. (eds.) Pairing 2009. LNCS, vol. 5671, pp. 248–265. Springer, Heidelberg (2009)
3. Boldyreva, A., Goyal, V., Kumar, V.: Identity-based encryption with efficient revocation. In: Proceedings of the 15th ACM Conference on Computer and Communications Security, pp. 417–426. ACM (2008)
4. Boneh, D., Franklin, M.: Identity-based encryption from the Weil pairing. In: Kilian, J. (ed.) CRYPTO 2001. LNCS, vol. 2139, pp. 213–229. Springer, Heidelberg (2001)
5. Chen, W., Ge, Z., Zhang, C., Kurose, J., Towsley, D.: On dynamic subset difference revocation scheme. In: Mitrou, N.M., Kontovasilis, K., Rouskas, G.N., Iliadis, I., Merakos, L. (eds.) NETWORKING 2004. LNCS, vol. 3042, pp. 743–758. Springer, Heidelberg (2004)
6. Freeman, D.M.: Converting pairing-based cryptosystems from composite-order groups to prime-order groups. In: Gilbert, H. (ed.) EUROCRYPT 2010. LNCS, vol. 6110, pp. 44–61. Springer, Heidelberg (2010)
7. Halevy, D., Shamir, A.: The LSD broadcast encryption scheme. In: Yung, M. (ed.) CRYPTO 2002. LNCS, vol. 2442, pp. 47–60. Springer, Heidelberg (2002)
8. Lee, K., Choi, S.G., Lee, D.H., Park, J.H., Yung, M.: Self-updatable encryption: time constrained access control with hidden attributes and better efficiency. In: Sako, K., Sarkar, P. (eds.) ASIACRYPT 2013, Part I. LNCS, vol. 8269, pp. 235–254. Springer, Heidelberg (2013)
9. Lee, K., Lee, D.H., Park, J.H.: Efficient revocable identity-based encryption via subset difference methods. IACR Cryptology ePrint Archive 2014, p. 132 (2014)
10. Liang, X., Lu, R., Lin, X., Shen, X.S.: Ciphertext policy attribute based encryption with efficient revocation. University of Waterloo, Technical report (2010)
11. Naor, D., Naor, M., Lotspiech, J.: Revocation and tracing schemes for stateless receivers. In: Kilian, J. (ed.) CRYPTO 2001. LNCS, vol. 2139, pp. 41–62. Springer, Heidelberg (2001)
12. Okamoto, T., Takashima, K.: Fully secure unbounded inner-product and attribute-based encryption. In: Wang, X., Sako, K. (eds.) ASIACRYPT 2012. LNCS, vol. 7658, pp. 349–366. Springer, Heidelberg (2012)
13. Qian, J., Dong, X.: Fully secure revocable attribute-based encryption. J. Shanghai Jiaotong Univ. (Science) 16, 490–496 (2011)
14. Waters, B.: Dual system encryption: realizing fully secure IBE and HIBE under simple assumptions. In: Halevi, S. (ed.) CRYPTO 2009. LNCS, vol. 5677, pp. 619–636. Springer, Heidelberg (2009)
15. Yu, S., Wang, C., Ren, K., Lou, W.: Attribute based data sharing with attribute revocation. In: Proceedings of the 5th ACM Symposium on Information, Computer and Communications Security, pp. 261–270. ACM (2010)

Weak Keys for the Quasi-Cyclic MDPC Public Key Encryption Scheme

Magali Bardet[✉], Vlad Dragoi, Jean-Gabriel Luque, and Ayoub Otmani

Normandie Univ, France; UR, LITIS, 76821 Mont-saint-aignan, France
{magali.bardet,vlad.dragoi1,jean-gabriel.luque,
ayoub.otmani}@univ-rouen.fr

Abstract. We analyze a new key recovery attack against the Quasi-Cyclic MDPC McEliece scheme. Retrieving the secret key from the public data is usually tackled down using exponential time algorithms aiming to recover minimum weight codewords and thus constructing an equivalent code. We use here a different approach and give under certain hypothesis an algorithm that is able to solve a key equation relating the public key to the private key. We relate this equation to a well known problem the *Rational Reconstruction Problem* and therefore propose a natural solution based on the extended Euclidean algorithm. All private keys satisfying the hypothesis are declared weak keys. In the same time we give a precise number of weak keys and extend our analysis by considering all possible cyclic shifts on the private keys. This task is accomplished using combinatorial objects like Lyndon words. We improve our approach by using a generalization of the Frobenius action which enables to increase the proportion of weak keys. Lastly, we implement the attack and give the probability to draw a weak key for all the security parameters proposed by the designers of the scheme.

Keywords: Quasi-cyclic MDPC codes · McEliece scheme · Rational reconstruction problem · Extended euclidean algorithm

1 Introduction

Moderate Density Parity Check (MDPC) codes were introduced in [MTSB12] in order to propose a public-key encryption scheme following McEliece's general approach [McE78]. These codes can be viewed as Low Density Parity Check (LDPC) codes where the parity-check matrices defining them have higher density. LDPC codes are classically constructed from matrices with constant row weights whereas the codes chosen in [MTSB12] have row weights $O(\sqrt{n \log n})$ assuming n is the length. They can be decoded likewise with Gallager's bit-flipping decoding algorithm. Even if using MDPC codes comes at the cost of a degraded error-correction compared to standard LDPC codes, it is still possible to obtain a probability of decoding failure below an acceptable threshold. Furthermore, because of the presence of low-weight codewords, LDPC codes are vulnerable to key recovery attacks based on Information Set Decoding algorithms

© Springer International Publishing Switzerland 2016
D. Pointcheval et al. (Eds.): AFRICACRYPT 2016, LNCS 9646, pp. 346–367, 2016.
DOI: 10.1007/978-3-319-31517-1_18

(see for instance [HS13], while MDPC codes are purposely designed to resist to such attacks. MDPC codes tend to become a serious choice in cryptography because they display the interesting feature of being less structured than codes that are traditionally encountered in code-based cryptography.

In this work we consider the quasi-cyclic variant of MDPC (QC-MDPC) codes, and instead of searching for relatively small weight codewords, we try to solve an equation relating the public polynomial (public data) to secret polynomials (private key). This equation is related to a well-known problem called the *rational reconstruction problem*, which can be solved for instance by the extended Euclidean algorithm (EEA). Solving this equation, which would give a trapdoor to the corresponding scheme, is expected to be hard in general. Nevertheless, in some cases, the solutions are rather easy to compute. We will call this type of (secret) configurations *weak keys* because they can be recovered efficiently from public data.

The main advantage of our technique is the low complexity of the algorithms that are able to check whether a private key is weak. If the original extended Euclidean algorithm is used then the time complexity is quadratic $O(p^2)$ if p is the length of the input. The first optimizations were proposed by Lehmer [Leh38] in 1938 where the constant factor was improved but the complexity was still quadratic. The first sub-quadratic algorithm was proposed in 1970 by Knuth [Knu71] with complexity $O(p(\log p)^5 \log \log p)$ and shortly after revisited by Schönhage in 1971 [Sch71] who obtained a better complexity $O(p(\log p)^2 \log \log p)$. The Least-Significant-Bit version of the Knuth-Schönhage algorithm is due to Stehlé and Zimmermann in 2004 [SZ04]. Even though the time complexity of this algorithm is not improved the description and the proof of their algorithm is significantly simpler in this case. The average behaviour was studied in [LV06, LV08, CCD+09]. Throughout the paper we call weak keys all pairs of private keys that can be recovered using the EEA algorithm from public data. We extend the collection of weak keys thanks to a group action that preserves the key equation. This permits to consider rather a *weak orbit* whenever the orbit under the action of the group contains at least one weak key.

The main contribution of this paper is to provide a fine analysis of the probability of weak keys and weak orbits for the QC-MDPC scheme, under two different actions. Let p be a prime number and consider a random $(2p, p, \omega)$-QC-MDPC code over $\mathbb{F}_2$ (a precise definition will be given in Sect. 2). Such a code is given by two vectors from $\mathbb{F}_2^p$ with a total Hamming weight ω. A rough estimate shows that, when used in a McEliece scheme, the resulting key is vulnerable to the EEA if the non-zero coefficients are all located at the same block. The probability of getting this configuration is $\frac{\binom{p}{\omega}}{\binom{2p}{\omega}}$. In the article we compute the asymptotic equivalence for the suggested range of parameters in [MTSB12],

$$\omega = \sqrt{2cp \log p}(1 + O(1)) \text{ and } p \to \infty. \tag{1}$$

In this case the probability is equivalent to $p^{-c/2} 2^{-\omega}$.

In Sect. 4 we compute the exact proportion of weak keys and show that it is asymptotically ω times the previous estimate for the conditions in (1):

$$\frac{\omega\binom{p+1}{\omega}}{\binom{2p}{\omega} - (-1)^{\omega/2}\binom{p}{\omega/2}} = \frac{\omega}{p^{c/2}2^{\omega}}\left(1 + O\left(\sqrt{\log^3 p/p}\right)\right). \tag{2}$$

We remark that the cyclic structure of the code defines a natural group action of $(\mathbb{Z}_p, +)$ over the set of public keys. If the coset of a private key contains a weak key, then it is possible to recover the private key by applying EEA to the shifted public key.

To count explicitly the number of weak orbits, we link orbits to Lyndon words and show that counting weak orbits is equivalent to counting Lyndon words with a fixed longuest run value (see Sect. 5 for precise definitions). In [GR61], Gilbert and Riordan count Lyndon words of length p and weight ω. We extend their results and give in Theorem 1 a formula for the number of Lyndon words of length p, weight ω and longuest run less than or equal to k.

This technique permits to increase the quantity of weak keys by a multiplicative factor equal to ω^3 for the conditions in (1), that is to say

$$\omega p^2 \frac{\binom{p-1}{\omega-2}}{\binom{2p}{\omega} + (-1)^{\omega/2+1}\binom{p}{\omega/2}} = \frac{\omega^3}{p^{c/2}2^{\omega}}\left(1 + O\left(\sqrt{\log^3 p/p}\right)\right). \tag{3}$$

In Sect. 6 we define another action of $(\mathbb{Z}_p^*, \times)$ over the set of public keys, that is compatible with the action of $(\mathbb{Z}_p, +)$. We explain how to apply EEA to every element of an orbit under both actions, and show that the attack will succeed if there exists at least one weak key in the orbit of a public key.

We prove that the quantity of keys our algorithm is able to attack is increased using this technique, by a multiplicative factor that is linear in the block length for the conditions in (1), that is to say

$$\frac{\omega p^3 \binom{p-1}{\omega-2}}{\binom{2p}{\omega} + (-1)^{\omega/2+1}\binom{p}{\omega/2}} = \frac{\omega^3 p}{p^{c/2}2^{\omega}}\left(1 + O\left(\sqrt{\log^3 p/p}\right)\right). \tag{4}$$

In Sect. 7 we give numerical values for the proportion of weak keys for all the security parameters suggested by the designers of the cryptosystem. Finaly in Sect. 8 we give experimental timings for our attack.

Due to space constraints, many proofs are just sketched. The full version of this paper is available on the arXiv.org preprint server.

2   QC-MDPC Encryption Scheme

We present here the most relevant material for describing the public-key encryption scheme [MTSB12]. We focus on the quasi-cyclic variant of [MTSB12] which is defined through circulant matrices. Throughout the paper, the weight of a vector or a polynomial refers to the Hamming weight and is denoted by $\| \ \|$.

2.1   QC-MDPC Codes and the Algebra of Circulant Matrices

Definition 1 (Moderate Density Parity Check codes). *A (n, r, w)-code is a linear code defined by a $r \times n$ parity-check matrix ($r < n$) where each row has weight w. A* Moderate Density Parity-Check (MDPC) *code is a (n, r, w)-code with $w = O\left(\sqrt{n \log n}\right)$, when $n \to \infty$.*

Definition 2. *A circulant matrix $\boldsymbol{M}$ of order p is a $p \times p$ matrix obtained by cyclically right shifting its first row $\boldsymbol{m} = (m_0, m_1, \ldots, m_{p-1})$.*

Any circulant matrix is thus completely described by its first row. A circulant matrix is also obtained by cyclically down shifting its first column. It is well-known that the matrix operations of addition and multiplication preserve the circulant structure of matrices.

Proposition 1. *[Dav79]   The algebra of $p \times p$ circulant matrices with entries in a field $\mathbb{K}$ denoted by $\left(\mathfrak{C}_p(\mathbb{K}), +, \times\right)$ is isomorphic to the polynomial algebra $\left(\mathbb{K}[x]/(x^p - 1), +, \cdot\right)$ through the mapping*

$$\mathfrak{C}_p(\mathbb{K}) \longrightarrow \mathbb{K}[x]/(x^p - 1)$$
$$\boldsymbol{M} \longmapsto m(x) = \sum_{i=0}^{p-1} m_i x^i \pmod{x^p - 1}.$$

Corollary 1. *A $p \times p$ circulant matrix $\boldsymbol{M}$ defined by m is invertible if and only if $m(x)$ is coprime to $x^p - 1$. In particular, the weight of m is necessarily odd.*

The algebra of circulant matrices enables to define the algebra of block matrices where the blocks are circulant. Any such matrix can be viewed as a matrix with entries in $\mathbb{K}[x]/(x^p - 1)$. This will define quasi-cyclic codes which represent the unique focus of this article.

Definition 3. *A* Quasi-Cyclic MDPC (QC-MDPC) *code is a MDPC code defined by a block parity-check matrix where each block is a circulant matrix.*

We now have defined all objects that permit to fully describe the scheme [MTSB12]. We will focus however, exclusively on the key generation algorithm since it is the only compound of the scheme that is of interest in this paper.

2.2   QC-MDPC Public-Key Encryption Scheme

The private key is a parity check matrix $\boldsymbol{H}$ of an (n, r, w) QC-MDPC code where $n = n_0 p$ and $r = p$ for some non-negative integer n_0. There exist therefore $p \times p$ circulant matrices $\boldsymbol{H}_1, \ldots, \boldsymbol{H}_{n_0}$ such that

$$\boldsymbol{H} = \left(\boldsymbol{H}_1 \ \boldsymbol{H}_2 \ \cdots \ \boldsymbol{H}_{n_0} \right). \tag{5}$$

This private key is obtained by taking at random the first row of $\boldsymbol{H}$ until $\boldsymbol{H}_{n_0}$ is invertible. The public key is the block parity-check matrix $\boldsymbol{F} \overset{\text{def}}{=} \boldsymbol{H}_{n_0}^{-1}\boldsymbol{H}$, or

$$\boldsymbol{F} = \left(\boldsymbol{H}_{n_0}^{-1}\boldsymbol{H}_1 \ \cdots \ \boldsymbol{H}_{n_0}^{-1}\boldsymbol{H}_{n_0-1} \ \boldsymbol{I}_p \right) \overset{\text{def}}{=} \left(\boldsymbol{F}_1 \ \cdots \ \boldsymbol{F}_{n_0-1} \ \boldsymbol{I}_p \right). \tag{6}$$

Using the isomorphism defined in Proposition 1, the private and public keys are fully described by the sequences $h_1, \ldots, h_{n_0}$ and $f_1, \ldots, f_{n_0-1}$ of polynomials in $\mathbb{K}[x]/(x^p - 1)$ such that for all $i \in \{1, \ldots, n_0 - 1\}$,

$$f_i = \frac{h_i}{h_{n_0}} \quad (\text{mod } x^p - 1). \tag{7}$$

The secret polynomials are taken so that $\sum_{i=1}^{n_0} \|h_i\| = w$.

Hence, the key generation of QC-MDPC scheme can be summarised as follows

– **Private key.** Pick at random $h_1, \ldots, h_{n_0}$ from $\mathbb{K}[x]/(x^p - 1)$ such that $\sum_{i=1}^{n_0} \|h_i\| = w$ and h_{n_0} is prime with $x^p - 1$.
– **Public key.** $f_1, \ldots, f_{n_0-1}$ where $f_i = \dfrac{h_i}{h_{n_0}}$ (mod $x^p - 1$).

2.3   Discussion on the Choice of the Parameters

Since [MTSB12] considers solely *binary* matrices, we assume from now $\mathbb{K} = \mathbb{F}_2$. Furthermore, the weights $\|h_i\|$ are "smoothly" distributed and p is always a prime number for security reasons. During the Key Generation step one must randomly choose the polynomials h_i until at least one of them is invertible. So we might expect, for security reasons, that the designers selected those parameters for which the set of invertible polynomials in the polynomial algebra $\mathbb{K}[x]/(x^p-1)$ is the largest possible. Using a ring isomorphism we give the number of invertible polynomials and thus show which are the proper parameters to be selected.

Proposition 2. *Let p be a prime number and assume $(x - 1) \prod_{i=1}^{d} g_i(x)$ is the decomposition of $x^p - 1$ into irreducible polynomials over $\mathbb{F}_2[x]$ for some $d \geqslant 1$ then $\deg g_i = \frac{p-1}{d}$ for all $i \in \{1, \ldots, d\}$. In particular, the number of invertible polynomials in $\mathbb{F}_2[x]/(x^p - 1)$ equals $\left(2^{(p-1)/d} - 1\right)^d$.*

For the choice of secure parameters, it is recommended to choose p so that the Folding attack [Gen01,Loi01,FOP+14] is inefficient. The most favorable situation is when d is as small as possible, for instance $d = O(1)$ when p tends to infinity. The designers of the scheme considered this option since all the parameters respect this condition. Hence, the number of invertible polynomials in $\mathbb{F}_2[x]/(x^p - 1)$ tends to be 2^{p-1} which is exactly the number of polynomials with an odd Hamming weight. So the probability of choosing an invertible polynomial from the set of polynomials with an odd Hamming weight is $\left(1 - 2^{-(p-1)/d}\right)^d$ which tends to 1 when $d = O(1)$. One very interesting case is when $d = 1$, since it seems to be the most secure choice for the cryptosystem. In Sect. 4 we investigate this particular case.

3   Rational Reconstruction Problem

We are interested in a *key-recovery under a chosen plaintext attack*. When applied on a (pn_0, p, w) QC-MDPC scheme whose public key is the sequence of polynomials $(f_1, \ldots, f_{n_0-1})$, the attack can be reformulated as the problem of finding $(h_1, \ldots, h_{n_0})$ satisfying

$$f_i = \frac{h_i}{h_{n_0}} \quad (\text{mod } x^p - 1) \text{ and } \sum_{i=1}^{n_0} \|h_i\| \leqslant w. \tag{8}$$

This problem can be tackled by applying classical techniques based on exponential algorithms seeking low-weight codewords. It can also be recast as the problem of solving the *rational reconstruction problem* that is described in full details in Sect. 3. The extended Euclidean algorithm solves (8) when there exists an integer $t > 0$ such that $\deg h_i < t \leqslant p$ and $\deg h_{n_0} \leqslant p - t$. Actually, (8) is a special case of a well-known problem called the *Rational Reconstruction problem*. It will be used in Sect. 4 as a general framework within which it is possible to perform a polynomial time key recovery attack.

Remark 1. Because of the bit-flipping decoding algorithm for MDPC codes, an attacker does not necessarily have to find the exact same secret polynomials for decrypting any ciphertext. Indeed, *any* sequence of polynomials satisfying the conditions (8) will lead to an efficient decoding of any ciphertext. It also means that there might exist several equivalent secret keys for a single QC-MDPC scheme.

Definition 4 (Rational reconstruction). *Let g and f be polynomials in $\mathbb{K}[x]$ where $\mathbb{K}$ is a field such that $0 < \deg f < \deg g$. For a given integer r satisfying $1 \leqslant r \leqslant \deg g$, the* rational reconstruction *of f modulo g consists in finding φ and ψ in $\mathbb{K}[x]$ such that $\gcd(\varphi, g) = 1$, $\deg \psi < r$ and $\deg \varphi \leqslant \deg g - r$ and satisfying*

$$\frac{\psi}{\varphi} = f \quad (\text{mod } g). \tag{RR}$$

Remark 2. When $g = x^p$ then we rather speak of Padé approximation.

Note that if (RR) has a solution (φ, ψ) then the quotient ψ/φ is unique. Furthermore if $(\varphi, \psi) \in \mathbb{K}[x]^2$ is a solution of the problem (RR), then it is also a solution to the following problem.

Definition 5. *Let $\mathbb{K}$ be a field, g be a polynomial in $\mathbb{K}[x]$ of degree $p > 0$ and f be in $\mathbb{K}[x]$ of degree $< p$. For a given r with $1 \leqslant r \leqslant p$, the (SRR) problem consists in finding ψ and φ in $\mathbb{K}[x]$ such that $(\varphi, \psi) \neq (0,0)$ and*

$$\varphi f = \psi \quad (\text{mod } g) \quad \text{with } \deg \psi < r \text{ and } \deg \varphi \leqslant p - r. \tag{SRR}$$

Clearly, any solution to (SRR) is solution to (RR) if and only if $\gcd(\varphi, g) = 1$. Moreover, (SRR) always has a non-trivial solution since recovering φ and ψ can be done by solving a linear system of p equations with $r + (p - r + 1) = p + 1$ unknowns representing the coefficients of φ and ψ.

A very efficient way to solve (RR) is to apply the Extended Euclidean Algorithm (EEA) to (f, g). Recall that if we denote by $(\varphi_i, \delta_i, \psi_i)$, with $i \geqslant 0$, the polynomials obtained at the i-th step of $\mathsf{EEA}(f, g)$ then we have $\psi_0 \stackrel{\text{def}}{=} g$, $\psi_1 \stackrel{\text{def}}{=} f$ and for all $i \geqslant 0$:

$$\begin{cases} \psi_i = Q_{i+1}\psi_{i+1} + \psi_{i+2} & \text{with } 0 \leqslant \deg \psi_{i+2} < \deg \psi_{i+1}, \\ \psi_i = \varphi_i f + \delta_i g & \text{with } (\varphi_0, \varphi_1) \stackrel{\text{def}}{=} (0, 1) \text{ and } (\delta_0, \delta_1) \stackrel{\text{def}}{=} (1, 0). \end{cases}$$

We also have the relations $\varphi_{i+2} = -Q_{i+1}\varphi_{i+1} + \varphi_i$ and $\delta_{i+2} = -Q_{i+1}\delta_{i+1} + \delta_i$. We are now able to prove that this approach provides a non-trivial solution. We require the following proposition.

Proposition 3. *At each step $i \geqslant 0$ of $\mathsf{EEA}(f, g)$ it holds that*

$$\deg \varphi_{i+1} = p - \deg \psi_i. \tag{9}$$

The following proposition characterises a solution to (RR) when it exists.

Proposition 4. *Let j be the smallest integer such that $\deg(\psi_j) < r$ then (φ_j, ψ_j) is a non-trivial solution to (SRR). Furthermore, if (φ, ψ) is a solution to (RR) then there exists λ in $\mathbb{K} \backslash \{0\}$ such that $\varphi = \lambda \varphi_j$ and $\psi = \lambda \psi_j$.*

4   Weak Keys

This section is devoted to the identification of private keys $h_1, \ldots, h_{n_0}$ that can be recovered from public key $f_1, \ldots, f_{n_0-1}$ by means of the extended Euclidean algorithm. Since $f_i = \dfrac{h_i}{h_{n_0}} \pmod{x^p - 1}$, the idea of our attack is to start by finding a rational reconstruction of f_1 modulo $x^p - 1$. At each step t of $\mathsf{EEA}(f_1, x^p - 1)$, the attacker checks if the ongoing computed polynomials denoted by $(\psi_t^{(1)}, \varphi_t^{(1)})$ where $\psi_t^{(1)} = f_1 \varphi_t^{(1)}$ satisfy the inequality

$$\|\varphi_t^{(1)}\| + \sum_{i=1}^{n_0-1} \|f_i \varphi_t^{(1)}\| \leqslant w. \tag{10}$$

If such a solution is found then by Proposition 4 we have found (equivalent) secret polynomials. Otherwise, the attacker performs the same attack to f_2 instead of f_1. If this fails again the attack goes on with the other polynomials $f_3, \ldots, f_{n_0-1}$. The main problem is to estimate precisely the number of keys that can be recovered with this technique.

We restrict the study to the case of two blocks $(2p, p, \omega)$ QC-MDPC scheme that is to say $n_0 = 2$. Nevertheless all our results can be extended to $n_0 > 2$. Let p be a prime number and ω an even integer with $1 < \omega < p$. Let $(\omega_1, \omega_2) \in \mathbb{N}^2$ be odd integers such that $\omega_1 + \omega_2 = \omega$. We define the set of *private pairs with fixed weights* by

$$\mathcal{P}_{\omega_1, \omega_2} = \left\{ (h_1, h_2) \in (\mathbb{K}[x]/(x^p - 1))^2 \mid \|h_i\| = \omega_i \text{ and } \omega_i \text{ odd} \right\},$$

and the set of all *private pairs* of a $(2p, p, \omega)$ QC-MDPC scheme by $\mathcal{P}_\omega = \bigcup_{\omega_1+\omega_2=\omega} \mathcal{P}_{\omega_1,\omega_2}$.

Private pairs that can be recovered using the extended Euclidiean algorithm are declared weak pairs.

Definition 6. *A pair $(h_1, h_2) \in \mathcal{P}_\omega$ is called a* weak pair *if*

$$\deg h_1 + \deg h_2 < p. \tag{11}$$

The set of weak pairs is denoted by $\mathcal{W}_\omega = \{(h_1, h_2) \in \mathcal{P}_\omega \mid \deg h_1 + \deg h_2 < p\}$. Similarly, $\mathcal{W}_{\omega_1,\omega_2}$ is defined as $\mathcal{W}_\omega \cap \mathcal{P}_{\omega_1,\omega_2}$.

Remark 3. It is important to notice that *true* collection of private keys of a general $(2p, p, \omega)$ QC-MDPC scheme is actually the set $\mathcal{P}_\omega^* = \bigcup_{\omega_1+\omega_2=\omega} \mathcal{P}_{\omega_1,\omega_2}^*$

where

$$\mathcal{P}_{\omega_1,\omega_2}^* = \left\{ (h_1, h_2) \in \mathcal{P}_{\omega_1,\omega_2} \mid \gcd(h_2, x^p - 1) = 1 \right\}.$$

But in order to simplify our analysis, we will only count weak pairs (h_1, h_2) and not weak keys for a $(2p, p, \omega)$ QC-MDPC scheme. This approximation is also justified by the fact we know from Sect. 2.3 that

$$\lim_{p \to \infty} \left(\sum_{\omega=2}^{2p} |\mathcal{P}_\omega^*| \right) \bigg/ \left(\sum_{\omega=2}^{2p} |\mathcal{P}_\omega| \right) = 1.$$

Remark also that there is one case where the two sets are equal. Indeed if $x^p - 1 = (x-1) \prod_{i=1}^{d} g_i(x)$ is the factorization of $x^p - 1$ into irreducible factors (see Sect. 2.3 for more details) then when $d = 1$ we have $\mathcal{P}_{\omega_1,\omega_2} = \mathcal{P}_{\omega_1,\omega_2}^*$ and $\mathcal{P}_\omega = \mathcal{P}_\omega^*$. For several reasons we consider this case in the article. The first one is that this is the strongest possible case for the QC-MDPC scheme since it avoids folding-type attacks. The second reason is that the number of private keys reaches its maximum since all todd weight polynomials are invertible.

Proposition 5.

$$|\mathcal{W}_{\omega_1,\omega_2}| = \binom{p+1}{\omega_1+\omega_2} \quad and \quad |\mathcal{W}_\omega| = \frac{\omega}{2}\binom{p+1}{\omega}. \tag{12}$$

$$|\mathcal{P}_{\omega_1,\omega_2}| = \binom{p}{\omega_1}\binom{p}{\omega_2} \quad and \quad |\mathcal{P}_\omega| = \frac{1}{2}\left(\binom{2p}{\omega} - (-1)^{\frac{\omega}{2}}\binom{p}{\frac{\omega}{2}}\right). \tag{13}$$

The asymptotic expansion when $\frac{\omega_i^2}{2p} = c_i + O(\frac{1}{\sqrt{p}})$ is

$$\frac{|\mathcal{W}_{\omega_1,\omega_2}|}{|\mathcal{P}_{\omega_1,\omega_2}|} = \sqrt{2\pi\alpha(1-\alpha)}e^{-2\sqrt{c_1 c_2}}\omega^{\frac{1}{2}}2^{-\omega H(\alpha)}(1 + O(1/\sqrt{p}))$$

where $\alpha = 1/(1+\sqrt{c_2/c_1})$ and $H(\alpha) = -\alpha \log_2 \alpha - (1-\alpha)\log_2(1-\alpha)$ is the entropy function. The asymptotic expansion for $\frac{\omega_i^2}{2p} = c_i \log p + O(\sqrt{\log p/p})$ is

$$\frac{|\mathcal{W}_{\omega_1,\omega_2}|}{|\mathcal{P}_{\omega_1,\omega_2}|} = \sqrt{2\pi\alpha(1-\alpha)}p^{-2\sqrt{c_1c_2}}\omega^{\frac{1}{2}}2^{-\omega H(\alpha)}\left(1+O(\sqrt{\log^3 p/p})\right).$$

$$\frac{|\mathcal{W}_\omega|}{|\mathcal{P}_\omega|} = \omega 2^{-\omega} \times \begin{cases} e^{-\frac{c}{2}}\left(1+O(\frac{1}{\sqrt{p}})\right) & if \frac{\omega^2}{2p} = c + O(\frac{1}{\sqrt{p}}), \\ p^{-\frac{c}{2}}\left(1+O(\sqrt{\frac{\log^3 p}{p}})\right) & if \frac{\omega^2}{2p} = c\log p + O(\sqrt{\frac{\log p}{p}}). \end{cases}$$

Proof. Let $(h_1, h_2) \in \mathcal{P}_{\omega_1,\omega_2}$. Then h_i has w_i non-zero coefficients, and a degree less than p, hence $|\mathcal{P}_{\omega_1,\omega_2}| = \binom{p}{\omega_1}\binom{p}{\omega_2}$. For $(h_1, h_2) \in \mathcal{W}_{\omega_1,\omega_2}$ we have $\deg(h_1) + \deg(h_2) < p$. If $k = \deg(h_1)$, then h_1 has a leading coefficient x^k and $\omega_1 - 1$ non-zero coefficients between x^0 and x^{k-1}. The number of such polynomials is $\binom{k}{\omega_1-1}$. Furthermore the number of polynomials h_2 with w_2 non-zero coefficients and $\deg(h_2) < p - k$ equals $\binom{p-k}{\omega_2}$. Using the Gould's formulae [Gou72], we get

$$|\mathcal{W}_{\omega_1,\omega_2}| = \sum_{k=0}^{p-1}\binom{k}{\omega_1-1}\binom{p-k}{\omega-\omega_1} = \binom{p+1}{\omega},$$

$$|\mathcal{P}_\omega| = \sum_{\substack{\omega_1+\omega_2=\omega \\ \omega_i \text{ odd}}}\binom{p}{\omega_1}\binom{p}{\omega_2} = \frac{1}{2}\left[\binom{2p}{\omega} - (-1)^{\frac{\omega}{2}}\binom{p}{\frac{\omega}{2}}\right].$$

As for $\mathcal{W}_\omega$ we obtain:

$$|\mathcal{W}_\omega| = \sum_{\substack{\omega_1+\omega_2=\omega \\ \omega_i \text{ odd}}}\binom{p+1}{\omega} = \binom{p+1}{\omega}\sum_{\substack{\omega_1+\omega_2=\omega \\ \omega_i \text{ odd}}}1 = \frac{\omega}{2}\binom{p+1}{\omega}.$$

For the asymptotic expansion use the Stirling formula and obtain the results.

Corollary 2. *In particular*

$$\frac{|\mathcal{W}_{\omega/2,\omega/2}|}{|\mathcal{P}_{\omega/2,\omega/2}|} = \frac{\binom{p+1}{\omega}}{\binom{p}{\omega/2}^2},$$

with asymptotic equivalence

$$\frac{|\mathcal{W}_{\omega/2,\omega/2}|}{|\mathcal{P}_{\omega/2,\omega/2}|} \sim \begin{cases} \sqrt{\pi}p^{\frac{1}{4}}e^{-2}2^{\frac{1}{4}-2\sqrt{2p}} & if \ \omega = 2\sqrt{2p}, \\ \sqrt{\pi}p^{\frac{1}{4}-2}\log^{\frac{1}{4}}p\,2^{\frac{1}{4}-2\sqrt{2p\log p}} & if \ \omega = 2\sqrt{2p\log p}. \end{cases}$$

The number of weak pairs can be easily increased by considering all possible cyclic shifts on the polynomials (h_1, h_2). We formally define the cyclic shift of a polynomial in terms of group action and explain how we extend the weak pairs to weak orbits.

5   Weak Pairs Derived from the Action of $(\mathbb{Z}_p, +)$

Let $f \in \mathbb{F}_2[x]/(x^p - 1)$ be a public key, and $(h_1, h_2) \in \mathbb{F}_2[x]/(x^p - 1) \times \mathbb{F}_2[x]/(x^p - 1)$ the corresponding private key. We have $f = \frac{h_1}{h_2} \mod (x^p - 1)$. Now assume that there exists $\alpha_1, \alpha_2 \in \mathbb{Z}_p^2$ such that $(x^{\alpha_1} h_1, x^{\alpha_2} h_2)$ is a weak key, then the public key $x^{\alpha_1 - \alpha_2} f = \frac{x^{\alpha_1} h_1}{x^{\alpha_2} h_2}$ can be attacked by EEA, which is equivalent to say that

$$\exists \alpha_1, \alpha_2 \in \mathbb{Z}_p^2 \text{ such that } \deg(x^{\alpha_1} h_1) + \deg(x^{\alpha_2} h_2) < p. \tag{14}$$

Using this idea if our attack does not work on f we repeat it on all p cyclic shifts of f, namely $xf, x^2 f, \ldots, x^{p-1} f$. If there is a shift such that the outgoing polynomials satisfy the weight conditions in (10) then we have succesfully recovered (equivalent) secret polynomials by Proposition 4. As in the previous section we want to estimate precisely the number of keys that can be recovered with this technique.

Definition 7. *The additive group $(\mathbb{Z}_p, +)$ acts on the set of polynomials as:*

$$\mathbb{Z}_p \times \mathbb{F}_2[x]/(x^p - 1) \longrightarrow \mathbb{F}_2[x]/(x^p - 1)$$
$$(\alpha, h) \longmapsto x^\alpha h.$$

The orbit of $h \in \mathbb{F}_2[x]/(x^p - 1)$ under the action of $(\mathbb{Z}_p, +)$ is denoted by $\mathcal{O}_h$.

Definition 8 (Weak orbit). *The set $\mathcal{O}_{h_1} \times \mathcal{O}_{h_2}$ defined by a private key (h_1, h_2) in $\mathbb{F}_2[x]/(x^p - 1)^2$ is called a* weak orbit *if it contains at least one weak key, i.e. satisfies (14).*

Potentially, we would get $p^2 |\mathcal{W}_\omega|$ such keys. But this statement overestimates the real number of weak pairs since it counts several times the same private keys. Nevertheless it gives a first intuition on the quantity of weak pairs that can be recovered using the rational reconstruction.

Lemma 1. *Let $\overline{h_i} = \min \mathcal{O}_{h_i}$ be the minimum polynomial for the lexicographical order of $h_i \in \mathbb{F}_2[x]/(x^p - 1)$. Then the set $\mathcal{O}_{h_1} \times \mathcal{O}_{h_2}$ is a weak orbit if and only if $\deg \overline{h_1} + \deg \overline{h_2} < p$.*

We define the *longest run of zeros of a polynomial* in $\mathbb{F}_2[x]/(x^p - 1)$ by the longest sequence of consecutive zero coefficients. We remark that there is a relation connecting the degree of the minimum polynomial and the longest run of zeros. If k_i denotes the longest run of zeros of $h_i \in \mathbb{F}_2[x]/(x^p - 1)$ we have that $\deg \overline{h_i} = p - k_i - 1$. Since we have the relation between the degree and the longest run of zeros for the minimal polynomial in the equivalence class we can redefine a weak orbit in terms of longest run:

Proposition 6 (Weak orbit). *The set $\mathcal{O}_{h_1} \times \mathcal{O}_{h_2}$ defined by a private key $(h_1, h_2) \in \mathbb{F}_2[x]/(x^p - 1)^2$ is a* weak orbit *if and only if it satisfies the equation:*

$$k_1 + k_2 \geqslant p - 1. \tag{15}$$

At this point we have reduced our key recovery attack to a well-known problem. To count all pairs (h_1, h_2) with the restriction mentioned above, we have to solve another problem: *What is the distribution of the longest run of zeros for the equivalence class of all cyclic shifts of a $\mathbb{K}^p$ vector with fixed Hamming weight?*

Definition 9. *[Lot02] A Lyndon word l is a word satisfying the conditions:*

- *l is a primitive word (i.e. it cannot be written $l = uv$, where u and v commute and $u, v \neq 1$)*
- *l is the smallest element in its conjugacy class for the lexicographical order*

Example 1.

1. Let $\mathcal{O}_{00011} = \{00011, 00110, 01100, 11000, 10001\}$. The Lyndon word here is 00011 since it is the strictly smallest than all the cyclic shifts.
2. Let $\mathcal{O}_{0101} = \{0101, 1010, 0101, 1010\}$. There is no Lyndon word here, since there is no strictly smallest element in the orbit.

An important property is that when p is prime there is a one-to-one mapping between the Lyndon words and the orbits if the weight is different from zero or p. So each equivalence class has p different shifts and the strictly smallest (since it exists) is the Lyndon word.

Theorem 1. *Let p, k, ω be integers, such that $1 \leqslant \omega \leqslant p$ and $k \leqslant p - \omega$. The number of binary Lyndon words with length p, longest run less than or equal to k and weight equal to ω is:*

$$\left| L^{\leqslant k}(p, \omega) \right| = \frac{1}{\omega} \sum_{j \in \mathbb{N}^*, \, j | gcd(p,\omega)} \mu(j) \binom{\frac{\omega}{j}}{\frac{p}{j} - \frac{\omega}{j}}_k, \tag{16}$$

where μ is the Möbius function, defined by $\mu(j) = 0$ if j has a squared prime factor, $\mu(j) = 1$ if j is square-free with an even number of prime factors and $\mu(j) = -1$ otherwise. The standard multinomial coefficient $\binom{j}{i}_k$ is defined as the coefficient of x^i in $\left(1 + x + \cdots + x^k\right)^j$.

The full proof of Theorem 1 is given in Appendix A and it uses a bijection between the Lyndon words with some specific properties on two alphabets: the binary alphabet and an $(k+1)$-ary alphabet. Straightforward we obtain:

Corollary 3. *The number of Lyndon words of length p and Hamming weight equal to ω over the binary alphabet (result already found in [GR61] by Gilbert and Riordan) is:*

$$|L(p, \omega)| = \frac{1}{p} \sum_{j | gcd(p,\omega)} \mu(j) \binom{\frac{p}{j}}{\frac{\omega}{j}}. \tag{17}$$

Corollary 4. *When p is prime we have*

$$\left| L^{\leqslant k}(p, \omega) \right| = \frac{1}{\omega} \binom{\omega}{p - \omega}_k \quad \text{and} \quad |L(p, \omega)| = \frac{1}{p} \binom{p}{\omega}. \tag{18}$$

As we already stated we will consider only the case p prime. Since all the orbits have the same length (p) and each orbit is defined by the corresponding Lyndon word, there is a uniform distribution over the set of Lyndon words when p is prime. So we consider a discrete probability model where the probability space is the set of Lyndon words with length p and weight ω with cardinal $\frac{1}{p}\binom{p}{\omega}$ and the probability of choosing a Lyndon word equals $p/\binom{p}{\omega}$. Furthermore we put a condition on the longest run of each Lyndon word and obtain a different distribution over the same set. In other words we write $L(p,\omega) = \bigcup_{k=\lfloor\frac{p-1}{\omega}\rfloor}^{p-\omega} L^k(p,\omega)$ and denote by $X_{p,\omega}$ a discrete random variable that represents the longest run of zeros of Lyndon words with length p and weight ω. Using Corollary 4 we define:

Definition 10. *The cumulative distribution and mass function for $X_{p,\omega}$ are:*

$$F_{X_{p,\omega}}(k) = \frac{\left|L^{\leqslant k}(p,\omega)\right|}{\left|L(p,\omega)\right|} \text{ and } f_{X_{p,\omega}}(k) = \frac{\left|L^k(p,\omega)\right|}{\left|L(p,\omega)\right|}.$$

Let $Y_{p,\omega_1,\omega_2} = X_{p,\omega_1} + X_{p,\omega_2}$ a discrete random variable that represents the sum of two independent random variables X_{p,ω_1} and X_{p,ω_2}. So the probability of a weak orbit is:

$$P\left(Y_{p,\omega_1,\omega_2} \geqslant p-1\right) = \sum_{k_1+k_2\geqslant p-1} f_{X_{p,\omega_1}}(k_1)f_{X_{p,\omega_2}}(k_2)$$

As p is prime, using Corollary 4 and Definition 10 we get the exact value:

$$P\left(Y_{p,\omega_1,\omega_2} \geqslant p-1\right) = \sum_{k_1+k_2\geqslant p-1} \frac{\binom{\omega_1}{p-\omega_1}_{k_1} - \binom{\omega_1}{p-\omega_1}_{k_1-1}}{\binom{p-1}{\omega_1-1}} \frac{\binom{\omega_2}{p-\omega_2}_{k_2} - \binom{\omega_2}{p-\omega_2}_{k_2-1}}{\binom{p-1}{\omega_2-1}} \quad (19)$$

The first case that seems interesting is when each variable has a longest run greater than or equal to half of the wanted quantity $\frac{p-1}{2}$.

Proposition 7. *Let ω_1 and $\omega_2 \geqslant 2$, then we have:*

$$P\left(X_{p,\omega_1} \geqslant \frac{p-1}{2}\right) P\left(X_{p,\omega_2} \geqslant \frac{p-1}{2}\right) = \omega_1\omega_2 \times \frac{\binom{\frac{p-1}{2}}{\omega_1-1}\binom{\frac{p-1}{2}}{\omega_2-1}}{\binom{p-1}{\omega_1-1}\binom{p-1}{\omega_2-1}}, \quad (20)$$

with asymptotic equivalence

$$\omega_1\omega_2 2^{-\omega} \times \begin{cases} e^{-\frac{c_1+c_2}{2}} & \text{if } \omega_i^2 = c_i p + O(\sqrt{p}), \\ p^{-\frac{c_1+c_2}{2}} & \text{if } \omega_i^2 = c_i p \log p + O(\sqrt{p \log p}). \end{cases}$$

Proof. We apply the formula for the generalized Pascal-DeMoivre coefficient from [Lot02,BBK08]:

$$\binom{\omega}{p-\omega}_k = \sum_{j=0}^{\lfloor\frac{p-\omega}{k+1}\rfloor} (-1)^j \binom{\omega}{j}\binom{p-j(k+1)-1}{\omega-1}.$$

For asymptotic expansion as before use the Stirling approximation for factorials.

Remark 4. We observe that using the shifts increased the probability of a weak key with a multiplicative factor equal to $\omega^{\frac{3}{2}}$. From Sect. 4 when $\omega_1 = \omega_2 = \frac{\omega}{2}$ we have that

$$P\left(X_{p,\omega_1} \geqslant \frac{p-1}{2}\right)^2 \sim \omega^{\frac{3}{2}} \frac{|\mathcal{W}_{\omega_1,\omega_2}|}{|\mathcal{P}_{\omega_1,\omega_2}|}.$$

We step forward and analyze the probability for a weak orbit in the general case. We remark that if either ω_1 or ω_2 equals 1 then the probability of a weak orbit equals 1. But the interesting analysis is when ω_1 and ω_2 are relatively close and $\omega = O\left(\sqrt{p \log p}\right)$.

Proposition 8. *If $\omega_1 \geqslant \omega_2$ and $\omega_i^2 = 2c_i p \log p + O(\sqrt{p \log p})$ then we have*

$$P(Y_{p,\omega_1,\omega_2} \geqslant p-1) \sim \omega_1 \omega_2 \frac{\binom{p-1}{\omega-2}}{\binom{p-1}{\omega_1-1}\binom{p-1}{\omega_2-1}} \quad \text{when } p \to \infty, \tag{21}$$

with asymptotic equivalence

$$P(Y_{p,\omega_1,\omega_2} \geqslant p-1) \sim \omega^2 \sqrt{2\pi\alpha(1-\alpha)} p^{-2\sqrt{c_1 c_2}} \omega^{\frac{1}{2}} 2^{-\omega H(\alpha)}.$$

where $\alpha = 1/(1 + \sqrt{c_2/c_1})$ and $H(\alpha) = -\alpha \log_2 \alpha - (1-\alpha) \log_2(1-\alpha)$

Proof. See Appendix A page 21.

We can easily check that for $\omega_i = \sqrt{c_i p \log p}$ and $c_1 > c_2$ the condition in Proposition 21 is satisfied. Experiments show that if we release the conditions on ω_i the approximation is still sharp. So a deeper investigation of the generalized Pascal-DeMoivre triangles might be used to prove this statement but this is no longer our purpose here.

Corollary 5. *We have the asymptotic equivalences*

$$P(Y_{p,\omega/2,\omega/2} \geqslant p-1) \sim \left(\frac{\frac{\omega}{2}}{\binom{p-1}{\frac{\omega}{2}-1}}\right)^2 \binom{p-1}{\omega-2} \quad \text{when } p \to \infty \text{ and } \omega = o(p),$$

$$P(Y_{p,\omega/2,\omega/2} \geqslant p-1) \sim \sqrt{\pi/2} p^{-\frac{1}{4}} \omega^{\frac{5}{2}} 2^{-\omega} \quad \text{if} \quad \omega_i^2 = \frac{p \log p}{4} + O(\sqrt{\log p/p}).$$

Remark 5. If we recall the results obtained with the first method in Proposition 5 and Corollary 2 we conclude that we gain a multiplicative factor equal to ω^2 using the shifts:

$$P(Y_{p,\omega_1,\omega_2} \geqslant p-1) \sim \omega^2 \times \frac{|\mathcal{W}_{\omega_1,\omega_2}|}{|\mathcal{P}_{\omega_1,\omega_2}|}.$$

Even though only "smooth" repartition is considered in the original article [MTSB12], we continue our analysis in the general case for all possible values $\omega_1 + \omega_2 = \omega$:

Proposition 9. *Let* $Y_{p,\omega} = \sum\limits_{\omega_1+\omega_2=\omega} Y_{p,\omega_1,\omega_2}$ *and* $\omega^2 = p\log p + O(1/\sqrt{p})$. *Then*

$$P(Y_{p,\omega/2,\omega/2} \geqslant p-1) \leqslant P(Y_{p,\omega} \geqslant p-1) \leqslant \omega p^2 \frac{\binom{p-1}{\omega-2}}{\binom{2p}{\omega} + (-1)^{\frac{\omega}{2}+1}\binom{p}{\frac{\omega}{2}}}. \tag{22}$$

The upper bound is asymptotically equivalent to $p^{-\frac{1}{4}}\omega^3 2^{-\omega}$.

Proof. For the upper bound we use Eq. (35) from Appendix A and the formula

$$P(Y_{p,\omega} \geqslant p-1) = \sum\limits_{\omega_1+\omega_2=\omega} P(Y_{p,\omega_1,\omega_2} \geqslant p-1)P(\omega_1,\omega_2)$$

Remark 6. If we recall the result in Sect. 4 we obtain a gain factor that is close to ω^2.

6   Improvements Under the Group Action of $(\mathbb{Z}_p^*, \times)$

In this section we define another group action that leaves the code invariant.

Definition 11. *We denote by " $\cong$ " the equivalence relation corresponding to the cyclic shifts equivalence class. The action of $\mathbb{Z}_p^*$ over $\mathbb{F}_2[x]/(x^p-1)/\cong$ can be defined as follow:*

$$\begin{array}{ccc} \mathbb{Z}_p^* \times (\mathbb{F}_2[x]/(x^p-1)/\cong) & \longrightarrow & (\mathbb{F}_2[x]/(x^p-1)/\cong) \\ (\alpha \quad , \qquad \mathcal{O}_h) & \longmapsto & \alpha \cdot \mathcal{O}_h, \end{array}$$

where $\alpha \cdot \left(\sum\limits_{i=0}^{p-1} a_i x^i\right) = \sum\limits_{i=0}^{p-1} a_i x^{\alpha i}$ *with* $\sum\limits_{i=0}^{p-1} a_i x^i \in \mathcal{O}_h$.

So we start our attack by fixing $\alpha \in \mathbb{Z}_p^*$ and try to find a rational reconstruction of $\alpha \cdot f$ modulo $x^p - 1$. If the algorithm finds a solution (ψ_t, φ_t) where $\psi_t = \alpha \cdot f\varphi_t$ satisfy the inequality

$$\|\varphi_t\| + \|\psi_t\| \leqslant w. \tag{23}$$

then we have found as before (equivalent) secret polynomials.

Otherwise, the attacker performs the same attack to all shifts of f, namely $\alpha \cdot x^j f$. If the attack fails, another α is chosen and the procedure is repeated until the good combination of α and shifts are founded. As before, we want to estimate precisely the number of keys that can be recovered with this technique.

Lemma 2. *The group action previously defined is a ring morphism.*

Proof. We can easily check that $\alpha \cdot (x^a + x^b) = \alpha \cdot x^a + \alpha \cdot x^b$ and $\alpha \cdot (x^{a+b}) = \alpha \cdot x^a \times \alpha \cdot x^b$.

We give now the most relevant properties related to the group action defined above.

Proposition 10. *Let $\alpha \in \mathbb{Z}_p^*$ and $\mathcal{O}_h \in \mathbb{F}_2[x]/(x^p - 1)/ \cong$. The following equivalence holds:*

$$\alpha \cdot \mathcal{O}_h = \mathcal{O}_h \Leftrightarrow \exists\, h^* \in \mathcal{O}_h, \ \alpha \cdot h^* = h^*. \tag{24}$$

Proof. The ($\Leftarrow$) implication comes from the definition of the orbits. For the other implication, let h be an element of the $\mathcal{O}_h$ class so that $\alpha \cdot h \in \mathcal{O}_h$. This means that there exits $j < p$ so that $\alpha \cdot h = x^j h$. Then by setting $k = -j\alpha^{-1}(1 - \alpha^{-1})^{-1}$ we have $\alpha \cdot (x^k h) = x^k h$.

Corollary 6. *Let $h \in \mathbb{F}_2[x]/(x^p - 1)$ and $\overline{\mathcal{O}_h}$ be the orbit of $\mathcal{O}_h$ under the action of $(\mathbb{Z}_p^*, \times)$. Let Γ_h be the subgroup of $(\mathbb{Z}_p^*, \times)$ which stabilizes $\mathcal{O}_h$. Then the cardinality of the orbit $\overline{\mathcal{O}_h}$ is*

$$|\overline{\mathcal{O}_h}| = \frac{p - 1}{|\Gamma_h|}. \tag{25}$$

Proposition 11. *Let $\alpha \in (\mathbb{Z}_p^*, \times)$ and $h \in \mathbb{F}_2[x]/(x^p - 1)$ so that $\|h\| = \omega_1 < p$ and $\alpha \cdot \mathcal{O}_h = \mathcal{O}_h$. Then the order of α divides either ω_1 or $\omega_1 - 1$.*

So only group elements that respect the order property given above can fix elements in the set of polynomials with weight restrictions. Thus a natural consequence is that we can use the Burnside lemma for counting the number of orbits in this case, but this is no longer the purpose here.

As before we say that the set $\overline{\mathcal{O}_{h_1}} \times \overline{\mathcal{O}_{h_2}}$ is a weak orbit if and only if it contains at least one weak pair and denote by $P([Y_{p,\omega}] \geqslant p - 1)$ the probability of a extended weak orbit. We also denote by Γ_{h_1,h_2} the subgroup that stabilize $\mathcal{O}_{h_1} \times \mathcal{O}_{h_2}$. We remark from Proposition 11 that for any pair of polynomials h_i with weight ω_i we have that any $\alpha \in (\mathbb{Z}_p^*, \times)$ that stabilizes the orbit $\mathcal{O}_{h_1} \times \mathcal{O}_{h_2}$ has to satisfy the condition

$$(\mathrm{ord}(\alpha)|\omega_1 \text{ or } \mathrm{ord}(\alpha)|\omega_1 - 1) \quad \text{and} \quad (\mathrm{ord}(\alpha)|\omega - \omega_1 \text{ or } \mathrm{ord}(\alpha)|\omega - \omega_1 - 1).$$

In order to estimate the probability of such weak configurations, two main factors must be taken into consideration: the length of an orbit $\overline{\mathcal{O}_{h_1}} \times \overline{\mathcal{O}_{h_2}}$ and the intersection of two weak orbits.

Proposition 12. *If the intersection of any two weak orbits $\overline{\mathcal{O}_{h_1}} \times \overline{\mathcal{O}_{h_2}} \cap \overline{\mathcal{O}_{h_1^*}} \times \overline{\mathcal{O}_{h_2^*}} = \emptyset$ and $\Gamma_{h_1,h_2} = \{1, -1\}$ for any orbit then we have:*

$$\frac{p-1}{2}\left(\frac{\frac{\omega}{2}}{\binom{p-1}{\frac{\omega}{2}-1}}\right)^2 \binom{p-1}{\omega - 2} \leqslant P([Y_{p,\omega}] \geqslant p - 1) \leqslant \frac{\omega p^3}{2} \frac{\binom{p-1}{\omega-2}}{\binom{2p}{\omega} + (-1)^{\frac{\omega}{2}+1}\binom{p}{\frac{\omega}{2}}}. \tag{26}$$

The asymptotic values for the upper and the lower bound can be computed as in Propositions 8 and 9.

Remark 7. We observe that with this extra group action we improved our probability by a multiplicative factor equal to $p - 1$ in the best case. In the worst case the factor is still linear in the block length (see Proposition 11 and Corrolary 6).

7   Numerical Results

The parameters chosen for the experimental part are those suggested by the designers of the scheme [MTSB12]. The security levels correspond to the best known attacks given in [MTSB12]. The probabilities displayed in Figs. 1 and 2 are computed directly from the formulas given in Corollary 2, Proposition 7, Corollary 5 and Proposition 5.

In Fig. 1 we compute the exact values directly from Corollary 2 and Proposition 7 for the first and the second probability. In the last column we give the asymptotic value of the probability of a weak orbit from Corollary 5. The asymptotic value approaches very precisely the exact value, at least when the exact computation is possible. We used the following procedure to obtain our results:

– We generate the list $L := \left[\binom{\frac{\omega}{2}}{p-\frac{\omega}{2}}_k - \binom{\frac{\omega}{2}}{p-\frac{\omega}{2}}_{k-1} \right]_{k \in \{(p-1)/\frac{\omega}{2}, \ldots, p-\frac{\omega}{2}\}}$.
– We compute the convolution from Eq. 19

$$P\left(Y_{p,\omega_1,\omega_2} \geqslant p-1\right) = \sum_{\substack{k_1+k_2 \geqslant p-1 \\ k_1,k_2 \in \{(p-1)/\frac{\omega}{2}, \ldots, p-\frac{\omega}{2}\}}} L[k_1]L[k_2].$$

The results are amazingly faithful to the asymptotic value in the sense that for all the parameters the exponential factor is the same for the two probabilities up to the last digit. This result is quite amazing since the inequalities used in Appendix A page 21. for the asymptotic expansion are not very sharp. But one of the reasons why the two values are so close might come from the compensation phenomenon when computing the convolution in Eq. 19.

In Fig. 2, we display the probability values for all $\omega_1 + \omega_2 = \omega$. In the first column we compute the exact value of the probability from Proposition 5. Whereas in the next column we compute the asymptotic value of lower bound and the upper bound. In the last column we give only the asymptotic value for the upper

Security level	p	$\frac{\omega}{2}$	$\frac{\lvert \mathcal{W}_{\omega/2,\omega/2} \rvert}{\lvert \mathcal{P}_{\omega/2,\omega/2} \rvert}$ Corollary 2 exact value	$P(X_{p,\frac{\omega}{2}} \geqslant \frac{p-1}{2})^2$ Proposition 7 exact value	$P(Y_{p,\frac{\omega}{2},\frac{\omega}{2}} \geqslant p-1)$ Equation 19 exact value	Corollary 5 asympt. value
80	4801	45	2^{-87}	2^{-78}	$2^{-74.04}$	$2^{-74.04}$
	3593	51	2^{-99}	2^{-90}	$2^{-86.02}$	$2^{-86.02}$
	3079	55	2^{-108}	2^{-98}	$2^{-94.12}$	$2^{-94.12}$
128	9857	71	2^{-139}	2^{-128}	$2^{-124.52}$	$2^{-124.52}$
	7433	81	2^{-159}	2^{-149}	$2^{-145.58}$	$2^{-144.58}$
	6803	85	2^{-167}	2^{-157}	$2^{-153.67}$	$2^{-152.67}$
256	32771	132	2^{-260}	2^{-249}		$2^{-244.3}$
	22531	155	2^{-307}	2^{-295}		$2^{-290.5}$
	20483	161	2^{-319}	2^{-307}		$2^{-302.7}$

Fig. 1. Probability of a weak key (orbit) for the QC-MDPC when $\omega_1 = \omega_2 = \frac{\omega}{2}$.

Security level	p	$\frac{\omega}{2}$	$\frac{|\mathcal{W}_\omega|}{|\mathcal{P}_\omega|}$ Proposition 5 exact value	$P(Y_{p,\omega} \geqslant p-1)$ Proposition 9 bounds Eq. (22)	$P([Y_{p,\omega}] \geqslant p-1)$ Proposition 12 upper bound
80	4801	45	2^{-84}	$[2^{-74}, 2^{-71}]$	2^{-60}
	3593	51	2^{-96}	$[2^{-86}, 2^{-83}]$	2^{-72}
	3079	55	2^{-105}	$[2^{-94}, 2^{-91}]$	2^{-80}
128	9857	71	2^{-136}	$[2^{-125}, 2^{-121}]$	2^{-109}
	7433	81	2^{-156}	$[2^{-145}, 2^{-141}]$	2^{-129}
	6803	85	2^{-164}	$[2^{-153}, 2^{-149}]$	2^{-137}
256	32771	132	2^{-257}	$[2^{-244}, 2^{-241}]$	2^{-227}
	22531	155	2^{-303}	$[2^{-291}, 2^{-287}]$	2^{-273}
	20483	161	2^{-315}	$[2^{-303}, 2^{-299}]$	2^{-285}

Fig. 2. Probability of a weak key, extended weak pairs and improvements on extended weak pairs for the QC-MDPC for all $\omega_1 + \omega_2 = \omega$.

bound. One might think that the upper bound is not very tight and that the exact value of the probability is way lower than the value of the upper bound. Even though we share this concern we want to insist on the following fact. In order to obtain real sharp bounds many unanswered questions concerning the generalized Pascal-DeMoivre triangles are to deal with and this is clearly not the purpose here. Nevertheless the experiments show that the probability is quite close to the upper bound. As p goes to infinity and $\omega = O\left(\sqrt{p \log p}\right)$ the difference between the two values tends to zero. We compute the probabilities for the first cryptographic parameters $p = 4801$ and $\omega = 90$. The exact value for the probability equals $2^{-71.26}$ whereas the upper bound equals $2^{-71.12}$.

8   Complexity and Experimental Timings

The cost of the attack on public key using the two group actions previously defined, is in theory $p - 1$ action of $(\mathbb{Z}_p^*, \times)$ times p action of $(\mathbb{Z}_p, +)$ times the cost of the EEA. This is the worst case scenario and also the case where our attack in applied on a random key (potentially which is not weak).

The first set of parameters that we used were not in the scale of the cryptographic values. More precisely we considered $p = 101$ and $\omega_1 = \omega_2 = 9$. The purpose was to confront the theoretical values for the probabilities of a weak keys and the experimental results. In this sense using MAGMA's random generator we computed 10^5 pair of polynomials for the QC-MDPC scheme and executed the attack on the shifted keys. In theory the probability of finding a weak orbit equals 0.0032. Meanwhile in practice we obtained 317 weak orbits and the time needed to test all the orbits was approximately 6000 s.

In the second part we used the first parameters for the 2^{80} security level which are $p = 4801$ and $\omega = 90$ and consider the most frequent case $\max_{i \in \{1,2\}} \omega_i = 47$. In the first case we applied the EEA on a weak key. In the second part we generated a weak key that we shifted. Therefore we randomly choose an integer $i \in (\mathbb{Z}_p, +)$ and applied the EEA on the ith shift. We repeated the procedure until a weak key was found. In the worst case we had to compute all the p shifts, whereas in average we only needed a small number of trials until the weak key was discovered. The last column corresponds to the following experience. We generated a weak key, then we applied the action of $(\mathbb{Z}_p^*, \times)$ and the we shifted. In this case the procedure is the same: we randomly pick an element of the group $(\mathbb{Z}_p^*, \times)$ and consider the key under the action of this element. Then we apply the Shifted(EEA) until the proper pair of shift and extension in founded. In the worst case we compute all the possible combinations of shifts and extensions.

On a 4-core Intel(R) Xeon(R) CPU ES-2690 @ 2.90 GHz, using MAGMA V2.19-9 we applied two variants of the EEA : the recursive original variant with complexity $O(p^2)$ and the MAGMA implementation using the Knuth–Schönhage version with complexity $O(p \log p^2 \log \log p)$.

	EEA	Shifted(EEA)		Extended(Shifted(EEA))	
	Best	Average	Worst	Average	Worst
Recursive version	0.12 s	4.5 min	9.5 min	5.3 days	1 month
MAGMA version	0.86 ms	2 s	4.1 s	1 h	5 h 30 min

9 Conclusion

The rational reconstruction attack turns out to be a very efficient solution for the key recovery attack on the QC-MDPC scheme. The main advantages of the algorithm is its low complexity, that is sub-quadratic in the code length, and the fact that it can be computed in parallel for several instances of the public key.

We proposed a first technique to estimate the number of private keys that can be recovered with the extended Euclidean algorithm. Furthermore in order to increase the success probability, equivalence classes of the public key have been considered. Formally this operation was defined in terms of two group actions $((\mathbb{Z}_p, +)$ and $(\mathbb{Z}_p^*, \times))$ over the set of polynomials in $\mathbb{F}_2[x]/(x^p - 1)$. Counting equivalence classes turned out to be a combinatorial problem based the theory of Lyndon words. This technique increased the quantity of weak keys by a multiplicative factor equal to ω^2. The second group action $(\mathbb{Z}_p^*, \times)$ increased the number by a multiplicative factor p.

In order to avoid such type of attacks one can easily check if the longest run of the private keys satisfy the conditions given in (15). The designer has to check if the group action previously defined increase or not the longest run in order to insure the security of the key.

We stress out the importance of our counting technique since it can be applied to other cryptographic schemes, for instance the NTRU cryptosystem.

Acknowledgement. We would like to thank the anonymous referees for their careful reading and helpful comments.

A   Appendix

Proof of Theorem 1 First of all we define the variables involved in the theorem. Let p, ω, k be integers, such that $1 \leqslant \omega \leqslant p$ and $k \leqslant p - \omega$. A finite word w is a Lyndon word if w is strictly smaller for the lexicographical order than all of its cyclic shifts. We denote by $\mathcal{L}(\mathcal{A})$ the set of Lyndon words over an alphabet $\mathcal{A}$. Let $\mathcal{B}$ be a binary alphabet, and $\mathcal{L}^{\leqslant k}(\mathcal{B}, p, \omega)$ the set of all Lyndon words with length p, number of ones equal to ω and the longest run of zeros less or equal to k over $\mathcal{B}$. Let $\mathcal{A}_k = \{a_0, a_1, \ldots, a_k\}$ be an alphabet. Monoids $\mathcal{A}_k^*$ and $\mathcal{B}^*$ are endowed with the lexicographic orders satisfying $0 < 1$ and $a_k < \cdots < a_0$. The morphism

$$\varphi : \mathcal{A}_k^* \to (0^*1)^* \subset \mathcal{B}^*$$
$$a_i \to 0^i 1$$

is clearly an order preserving isomorphism. We deduce that $w \in \mathcal{A}_k^*$ is a Lyndon word if and only if $\varphi(w)$ is a Lyndon word (see [Ric03] for details). Setting $\psi(a_{l_0} \ldots a_{l_{j-1}}) = j + \sum_{m=0}^{j-1} l_m$ we obtain $\psi(w) = |\varphi(w)|$.

If we set $\mathcal{L}_\psi(\mathcal{A}_k, \omega, p) = \left\{ l \in \mathcal{L}(\mathcal{A}_k) \ \middle| \ |l| = \omega \text{ and } \psi(l) = p \right\}$ then

$$\varphi\left(\mathcal{L}_\psi(\mathcal{A}_k, \omega, p)\right) = \mathcal{L}^{\leqslant k}(\mathcal{B}, p, \omega).$$

Hence, it suffices to compute $\left| \mathcal{L}_\psi(\mathcal{A}_k, \omega, p) \right|$. We use the fact that the alphabet $\mathcal{A}_k$ is the generating basis for all words in the free monoid $\mathcal{A}_k^*$. In terms of formal series this means

$$\sum_{w \in \mathcal{A}_k^*} w = \frac{1}{1 - \sum\limits_{i=0}^k a_i}. \tag{27}$$

Then we use the Chen-Fox-Lyndon theorem that states that each word can be uniquely expressed as a decreasing product of Lyndon words [KTC58, Lot02]

$$\sum_{w \in \mathcal{A}_k^*} w = \prod_{l \in \mathcal{L}(\mathcal{A}_k)}^{\nearrow} \frac{1}{1 - l}. \tag{28}$$

Sending each letter a_{l_m} to $zx^{l_m + 1}$ one obtains

$$\frac{1}{1 - z \sum\limits_{i=1}^{k+1} x^i} = \prod_{1 \leqslant j \leqslant i}^{\infty} \left(\frac{1}{1 - x^i z^j} \right)^{|\mathcal{L}_\psi(\mathcal{A}_k, j, i)|}. \tag{29}$$

We apply the logarithm in each side of the equality above and develop using the Taylor expansion. In the resulting formula we compare the coefficient of $z^\omega x^p$ in the left hand side and the right hand side and obtain

$$\sum_{\substack{j\mid\omega \\ \frac{\omega}{j}\mid p}} j\left|\mathcal{L}_\psi(A_k, j, \frac{p}{\omega}j)\right| = \binom{\omega}{p-\omega}_k, \tag{30}$$

where $\binom{\omega}{p}_k$ denotes the coefficient of x^p in $(1 + x + x^2 + \cdots + x^k)^\omega$.

We rewrite the last equation as

$$\sum_{j\mid \gcd(\omega,p)} \frac{\omega}{j}\left|\mathcal{L}_\psi(A_k, \frac{\omega}{j}, \frac{p}{j})\right| = \binom{\omega}{p-\omega}_k, \tag{31}$$

and apply the Möbius Inversion [Mob32, Lan09] to find the wanted result.

Proof of Proposition 8 By definition we have:

$$P\left(Y_{p,\omega_1,\omega_2} \geqslant p - 1\right) = \sum_{\omega_2 - 1 \leqslant k \leqslant p - \omega_1} f_{X_{p,\omega_1}}(k)\left(1 - F_{X_{p,\omega_2}}(p - k - 1 - 1)\right).$$

Lemma 3. *Let $\omega \geqslant 2$ and p prime. Then for $k > \lfloor\frac{p-\omega}{2}\rfloor$ we have*

$$f_{X_{p,\omega}}(k) = \frac{\omega\binom{p-k-2}{\omega-2}}{\binom{p-1}{\omega-1}}, \quad F_{X_{p,\omega}}(k-1) = 1 - \frac{\omega\binom{p-k-1}{\omega-1}}{\binom{p-1}{\omega-1}}. \tag{32}$$

For $k \leqslant \lfloor\frac{p-\omega}{2}\rfloor$ the bounds are

$$\frac{\omega\binom{p-k-2}{\omega-2} - \binom{\omega}{2}\left[\binom{p-2k-1}{\omega-1} - \binom{p-2k-3}{\omega-1}\right]}{\binom{p-1}{\omega-1}} \leqslant f_{X_{p,\omega}}(k) \leqslant \frac{\omega\binom{p-k-2}{\omega-2}}{\binom{p-1}{\omega-1}}, \tag{33}$$

$$\frac{\omega\binom{p-k-1}{\omega-1} - \binom{\omega}{2}\binom{p-2k-1}{\omega-1}}{\binom{p-1}{\omega-1}} \leqslant 1 - F_{X_{p,\omega}}(k-1) \leqslant \frac{\omega\binom{p-k-1}{\omega-1}}{\binom{p-1}{\omega-1}}. \tag{34}$$

For the upper bound, this gives

$$P\left(Y_{p,\omega_1,\omega_2} \geqslant p - 1\right) \leqslant \sum_{k=\omega_2-1}^{p-\omega_1} \omega_1 \frac{\binom{p-k-2}{\omega_1-2}}{\binom{p-1}{\omega_1-1}} \omega_2 \frac{\binom{k}{\omega_2-1}}{\binom{p-1}{\omega_2-1}} = \frac{\omega_1\omega_2\binom{p-1}{\omega_1+\omega_2-2}}{\binom{p-1}{\omega_1-1}\binom{p-1}{\omega_2-1}}. \tag{35}$$

For the lower bound, we separate our sum into three different sums, for $k \leqslant \lfloor\frac{p-\omega_1}{2}\rfloor$, $\lfloor\frac{p-\omega_1}{2}\rfloor < k < p - 1 - \lfloor\frac{p-\omega_2}{2}\rfloor = \lceil\frac{p+\omega_2}{2}\rceil - 1$ and $\lceil\frac{p+\omega_2}{2}\rceil - 1 \leqslant k \leqslant p - \omega_1$

and use relations (32), (33) and (34):

$$P\left(Y_{p,\omega_1,\omega_2} \geqslant p-1\right) \geqslant \sum_{k=\omega_2-1}^{p-\omega_1} \omega_1 \frac{\binom{p-k-2}{\omega_1-2}}{\binom{p-1}{\omega_1-1}} \omega_2 \frac{\binom{k}{\omega_2-1}}{\binom{p-1}{\omega_2-1}}$$

$$- \sum_{k=\omega_2-1}^{\lfloor \frac{p-\omega_1}{2} \rfloor} \binom{\omega_1}{2} \frac{\binom{p-2k-1}{\omega_1-1} - \binom{p-2k-3}{\omega_1-1}}{\binom{p-1}{\omega_1-1}} \omega_2 \frac{\binom{k}{\omega_2-1}}{\binom{p-1}{\omega_2-1}}$$

$$- \sum_{k=\lceil \frac{p+\omega_2}{2} \rceil-1}^{p-\omega_1} \binom{\omega_2}{2} \frac{\binom{p-k-2}{\omega_1-2}}{\binom{p-1}{\omega_1-1}} \omega_1 \frac{\binom{2k-p+1}{\omega_2-1}}{\binom{p-1}{\omega_2-1}}$$

We use the relations $\binom{p-2k-1}{\omega_1-1} - \binom{p-2k-3}{\omega_1-1} = \binom{p-2k-2}{\omega_1-2} + \binom{p-2k-3}{\omega_1-2} \leqslant 2\binom{p-2k-2}{\omega_1-2}$

(as $\omega_1 \geqslant 2$), $\dfrac{\omega_1\omega_2}{\binom{p-1}{\omega_1-1}\binom{p-1}{\omega_2-1}} = \dfrac{p^2}{\binom{p}{\omega_1}\binom{p}{\omega_2}}$ and a change of variable $k \to p-k-2$

in the last sum to get

$$\frac{\binom{p}{\omega_1}\binom{p}{\omega_2}}{p^2} P\left(Y_{p,\omega_1,\omega_2} \geqslant p-1\right) \geqslant \binom{p-1}{\omega-2} - \omega_1 \sum_{k=\omega_2-1}^{\lfloor \frac{p-\omega_1}{2} \rfloor} \binom{p-2k-2}{\omega_1-2}\binom{k}{\omega_2-1}$$

$$- \frac{1}{2}\omega_2 \sum_{k=\omega_1-2}^{\lfloor \frac{p-\omega_2}{2} \rfloor-1} \binom{p-2k-3}{\omega_2-1}\binom{k}{\omega_1-2}$$

Now we use the bound $\binom{p-2k-2}{\omega_1-2}\binom{k}{\omega_2-1} \leqslant \binom{p-k-2}{\omega-3}$ and the relation from [Gou72]
$\sum_{k=r}^{s} \binom{a-k}{b} = \binom{a-r+1}{b+1} - \binom{a-s}{b+1} \leqslant \binom{a-r+1}{b+1}$ to get

$$\frac{\binom{p}{\omega_1}\binom{p}{\omega_2}}{p^2} P\left(Y_{p,\omega_1,\omega_2} \geqslant p-1\right) \geqslant \binom{p-1}{\omega-2} - \omega_1 \binom{p-\omega_2}{\omega-2} - \frac{1}{2}\omega_2 \binom{p-\omega_1}{\omega-2}$$

$$\geqslant \binom{p-1}{\omega-2} - \frac{3}{2}\omega_1 \binom{p-\omega_2}{\omega-2}.$$

if $\omega_1 = \max(\omega_1, \omega_2)$. We finally get the bounds

$$1 - \frac{3\omega_1}{2} \frac{\binom{p-\omega_2}{\omega-2}}{\binom{p-1}{\omega-2}} \leqslant \frac{P\left(Y_{p,\omega_1,\omega_2} \geqslant p-1\right)}{p^2 \frac{\binom{p-1}{\omega-2}}{\binom{p}{\omega_1}\binom{p}{\omega_2}}} \leqslant 1. \tag{36}$$

We check that the lower bound tends to 1 when $w_i = O(\sqrt{p \log p})$.

References

[BBK08] Belbachir, H., Bouroubi, S., Khelladi, A.: Connection between ordinary multinomials, fibonacci numbers, bell polynomials and discrete uniform distribution. Ann. Math. Inform. **35**, 21–30 (2008)

[CCD+09] Cesaratto, E., Clément, J., Daireaux, B., Lhote, L., Maume-Deschamps, V., Vallée, B.: Regularity of the euclid algorithm, application to the analysis of fast GCD algorithms. J. Symbolic Comput. **44**(7), 726 (2009)

[Dav79] Davis, P.J.: Circulant Matrices. Pure and applied mathematics. Wiley, New York (1979)

[FOP+14] Faugère, J.C., Otmani, A., Perret, L., de Portzamparc, F., Tillich, J.P.: Folding alternant and goppa codes with non-trivial automorphism groups, submitted, [cs.IT] (2014). arxiv:1405.5101

[Gen01] Gentry, C.: Key recovery and message attacks on NTRU-composite. In: Pfitzmann, B. (ed.) EUROCRYPT 2001. LNCS, vol. 2045, pp. 182–194. Springer, Heidelberg (2001)

[Gou72] Gould, H.W.: Combinatorial identities: a standardized set of tables listing 500 binomial coefficient summations. Morgantown, W Va (1972)

[GR61] Gilbert, E.N., Riordan, J.: Symmetry types of periodic sequences. Illinois J. Math. **5**, 657–665 (1961)

[HS13] Hamdaoui, Y., Sendrier, N.: A non asymptotic analysis of information set decoding. In: Cryptology ePrint Archive, Report /162 (2013)

[Knu71] Knuth, D.E.: The analysis of algorithms. Actes Congr. Internat. Math. **3**, 269–274 (1971). http://cr.yp.to/bib/entries.html#1971/knuth-gcd

[KTC58] Lyndon, R.C., Chen, K.T., Fox, R.H.: Free differential calculus, iv. the quotient groups of the lower central series. Ann. Math. **68**(1), 81–95 (1958)

[Lan09] Landau, E.: Handbuch der Lehre von der Verteilung der Primzahlen. Teubner(1909)

[Leh38] Lehmer, D.H.: Euclid's algorithm for large numbers. Am. Math. Monthly **45**(4), 227–233 (1938)

[Loi01] Loidreau, P.: Codes derived from binary goppa codes. Probl. Inf. Transm. **37**(2), 91–99 (2001)

[Lot02] Lothaire, M.: Algebraic Combinatorics on Words. Encyclopedia of mathematics and its applications. Cambridge University Press, New York (2002)

[LV06] Lhote, L., Vallée, B.: Sharp estimates for the main parameters of the euclid algorithm. In: Correa, J.R., Hevia, A., Kiwi, M. (eds.) LATIN 2006. LNCS, vol. 3887, pp. 689–702. Springer, Heidelberg (2006)

[LV08] Lhote, L., Vallée, B.: Gaussian laws for the main parameters of the euclid algorithms. Algorithmica **50**(4), 497–554 (2008)

[McE78] McEliece, R.J.: A Public-Key System Based on Algebraic Coding Theory, pp. 114–116. Jet Propulsion Lab, DSN Progress Report, 44 (1978)

[Mob32] Möbius, A.F.: Über eine besondere art von umkehrung der reihen. Journal für die reine und angewandte Mathematik **9**, 105–123 (1832)

[MTSB12] Misoczki, R., Tillich, J.-P., Sendrier, N., Barreto, P.S.L.M.: MDPC-McEliece: New McEliece variants from moderate density parity-check codes. IACR Cryptology ePrint Archive, 409 (2012)

[Ric03] Richomme, G.: Lyndon morphisms. Bull. Belg. Math. Soc. Simon Stevin **10**(5), 761–785 (2003)

[Sch71] Schönhage, A.: Schnelle Berechnung von Kettenbruchentwicklungen. (German) [Fast calculation of expansions of continued fractions]. ACTA-INFO, 1, 139–144 (1971)

[SZ04] Stehlé, D., Zimmermann, P.: A binary recursive GCD algorithm. In: Buell, D.A. (ed.) ANTS 2004. LNCS, vol. 3076, pp. 411–425. Springer, Heidelberg (2004)

Author Index

Printed in the United States
By Bookmasters